Homological Methods in Commutative Algebra

GRADUATE STUDIES
IN MATHEMATICS **234**

Homological Methods in Commutative Algebra

Andrea Ferretti

AMERICAN
MATHEMATICAL
SOCIETY
Providence, Rhode Island

2020 *Mathematics Subject Classification*. Primary 13-01, 13D02, 13D07, 13D45, 13H10.

For additional information and updates on this book, visit
www.ams.org/bookpages/gsm-234

Library of Congress Cataloging-in-Publication Data
Names: Ferretti, Andrea, 1981– author.
Title: Homological methods in commutative algebra / Andrea Ferretti.
Description: Providence, Rhode Island : American Mathematical Society, [2023] | Series: Graduate studies in mathematics, 1065-7339 ; volume 234 | Includes bibliographical references and index.
Identifiers: LCCN 2023012887 | ISBN 9781470471286 (hardback) | ISBN 9781470474362 (paperback) | ISBN 9781470474355 (ebook)
Subjects: LCSH: Commutative algebra–Textbooks. | Algebra, Homological–Textbooks. | AMS: Commutative algebra – Instructional exposition (textbooks, tutorial papers, etc.). | Commutative algebra – Homological methods – Syzygies, resolutions, complexes. | Commutative algebra – Homological methods – Homological functors on modules (Tor, Ext, etc.). | Commutative algebra – Homological methods – Local cohomology. | Commutative algebra – Local rings and semilocal rings – Special types (Cohen-Macaulay, Gorenstein, Buchsbaum, etc.).
Classification: LCC QA251.3 .F473 2023 | DDC 512/.44–dc23/eng20230711
LC record available at https://lccn.loc.gov/2023012887

To Matteo, who arrived in our life gracefully
and now is taking all the space in the bed.

Contents

Preface

Commutative algebra is a discipline that draws inspiration from many different fields of mathematics. When I wrote the book *Commutative algebra* [**Fer20**], I tried to convey this by showing the links to various other topics. Sometimes a concept is clarified by its geometric counterpart; in other situations, a computational point of view is more apt; the structure of some rings appearing in number theory is revealed by looking at them as lattices inside $\mathbb{R}^n$; yet other times, topology is the right lens through which a problem can be attacked.

Since the pioneering work of Serre [**Ser56b**] (and many other mathematicians) on regular rings, it has become clear that homological techniques are a powerful tool for the study of rings and modules. Yet, I felt I could not do justice to homological algebra by just adding a chapter on homological methods to an already long book. The present volume is intended to remedy this absence.

The aim of this book is to teach homological algebra first, and then gradually focus on the study of rings and modules. Instead of just constructing derived functors on categories of modules, I decided to dedicate the first four chapters to developing homological algebra on its own. It is my hope that this will help the student who is not familiar with these techniques to learn this beautiful—but sometimes dry—subject, and bring them to a level of proficiency that will allow them to pursue further topics in topology, algebraic geometry, and so on. While the presentation is self contained and reasonably complete, it is quite old style: I resisted the temptation to add at least a chapter on derived categories, which, while fundamental in any modern application of the subject, is tangential to the text.

The second part of the text is dedicated to the application of these techniques in commutative algebra, through the study of projective, injective, and flat modules; the construction of explicit resolutions via the Koszul complex; the properties of regular sequences and their application to the study of regular rings and various other classes of rings that are most easily characterized by their homological properties.

While of course [**Fer20**] and the present volume are naturally complementary, it is hoped that each of them makes sense on its own. The reader of this book is assumed to have a general background in elementary commutative algebra: of course, [**Fer20**] is suitable for the purpose, but familiarity with other books, such as [**AM69**] or [**AK13**] is also helpful.

Since there are many books on commutative algebra, it may be worthwhile to highlight a few of the author's favorite results, which are not always found in comparable texts, such as

(1) the Freyd–Mitchell theorem on embedding small Abelian categories into categories of modules;

(2) the Ischebeck spectral sequences and their application to a common proof of the Auslander–Buchsbaum formula and the equality between depth and injective dimension (these appear as exercises in Chapter 9);

(3) the proof of the Quillen–Suslin theorem on the freeness of projective modules in polynomial rings;

(4) the construction of the Macaulay resultant and its computation using Koszul complexes;

(5) Kunz's characterization of regular local rings in characteristic p, a prime;

(6) the Northcott–Rees theory of irreducible decomposition of ideals and its link with the Cohen–Macaulay and Gorenstein properties; and

(7) the Grothendieck duality for local cohomology.

Of course, this is just a selection of topics according to the taste of the author. We next review the contents of each chapter.

Category theory was originally developed—at least in part—to give a foundation to homological algebra. This is our starting point as well. Chapter 1 introduces the language and tools of category theory. Categories are introduced both as an organizing gadget for other structures, and as algebraic objects themselves. One of the puzzling points when learning about categories is the set theoretic distinction between sets and classes. For this reason, Section 1.2 is dedicated to some background on set theory, both with the ZFC and the NBG formalizations, in order to explain how classes can be introduced as a sound mathematical object. We then move to some standard construction with categories: natural transformation, limits, universal objects, and adjoint pairs, elucidating the various connections between them.

In Chapter 2, we begin the study of Abelian categories. These have formal properties that resemble those of the module categories, and are a suitable place to develop the machinery of derived functors. After setting up the basic definitions, we develop some sheaf theory, both on topological spaces and on categories endowed with a Grothendieck topology. In fact, one of the motivations to set up homological algebra in the context of Abelian categories was to unify the treatment of derived functors for modules and of sheaf cohomology. We study sheaves to give some motivation for the abstract approach, and to provide an interesting example of an Abelian category other than the category of modules on a ring. We then come back to the general settings, where we develop some results on subobjects which allow us to prove standard results such as the snake lemma. The chapter culminates in a proof of the celebrated Freyd–Mitchell theorem, which ensures that (small) Abelian categories can be embedded as full subcategories of the category of left modules over a possibly noncommutative ring.

This result simplifies a lot working with Abelian categories, by enabling proofs by diagram chasing. It is quite common to cite the result in order to simplify the presentation of general Abelian categories, seldom with proof. I decided to include it for various reasons. First, I try to avoid relying on black box results, to a reasonable extent. More importantly, it provides a beautiful example of a fairly nontrivial result on Abelian categories, which provides some substance to a chapter that could otherwise feel rather dull. Finally, it crucially relies on the construction on sheaves on a nontopological site, giving a very nice application of the general theory of sheaves developed before.

Chapter 3 begins the study of derived functors of an additive functor between Abelian categories, assuming that the source has enough injectives or enough projectives. After checking that the notion is well defined, the bulk of the chapter is taken by studying some derived functors in more detail: sheaf cohomology, the Ext and Tor functors and the $\lim^n$ functors. In particular, we show that Ext and Tor can be computed by resolving either of the arguments, and we also develop the construction of the Ext functors as Yoneda extensions.

We end the presentation of homological algebra with Chapter 4 on spectral sequences. These are a fundamental computational tool, with plenty of applications that we try to showcase in the rest of the book. We start from the general to the particular, in the hope that this makes the theory less tricky. After the reader has accepted the definition of an exact pair, which is admittedly not obvious, the construction of the derived exact pair immediately leads to spectral sequences. All examples of spectral sequences in the text are special cases of this construction. By increasing specialization, we review the spectral sequences associated to fibered complexes, double complexes, and finally the Grothendieck spectral sequence for the composition of two derived functors. We end the chapter by including some spectral sequences that link the Ext and Tor functors. Of these, the base change spectral sequences can be obtained from the Grothendieck formalism, while the Ischebeck spectral sequences are derived from scratch.

With Chapter 5, we begin to enter the world of commutative algebra proper. This chapter is dedicated to the Ext functor, and in particular to the study modules that are acyclic for Ext. These are projective or injective modules, according to the side, and have a very different flavor. Projective modules, which are the topic of Section 5.1, are not too far from being locally free, and resemble modules of sections of vector bundles on topological spaces. In constrast, injective modules are divisible, and we have a rich theory of them due to Matlis, based on the notion of injective hull, which we present in Section 5.3. Both can be used to develop a theory of dimension for modules, called the projective (resp., injective) dimension. We study how this quantity changes under various constructions, in particular in the case of quotients, where we provide some change of rings theorems. These are crucial for the study of global dimension of a ring, which is the supremum of either of these quantities. One of the main results of the chapter is Hilbert's syzygy theorem, which in its modern form states that $\operatorname{gl.dim} A[x] = \operatorname{gl.dim} A + 1$ for a Noetherian ring A. We end the chapter by connecting this abstract form to the more classical formulation by Hilbert, in particular proving the Quillen and Suslin theorem stating that projective modules over polynomial rings are in fact free.

In Chapter 6, we focus on the Tor functor, again by characterizing modules that are acyclic for Tor—namely, flat modules. A case of particular interest in geometry is the flatness of B as an A-module when $f\colon A \to B$ is a map of rings. In this case, we say that f is a flat morphism. After showing that localizations and completions are flat, we give various criteria for flatness, and we show various connections between the notions of flat, projective and locally free modules. We end the chapter by showing that flat modules are in fact the direct limit of free modules.

In Chapter 7, we start making things more concrete by borrowing an idea from differential geometry. Namely, by mimicking the construction of the De Rham complex, we construct a complex associated to a sequence of elements in a ring A. We dub this the Koszul complex. By studying the question of when this complex is actually acyclic, one is naturally led to consider regular sequences. These are fundamental to connect the homological and ring-theoretic properties of A, and give rise to yet another invariant of ideals and modules, called the depth. We end the chapter by giving two applications of these ideas: the first one is a formula for multiplicities due to Auslander and Buchsbaum, and the second one is a construction of the Macaulay resultant, a key tool in computational algebra.

Regular rings were introduced in Chapter 10 of [**Fer20**], where we studied their most elementary properties. We tackle them again in Chapter 8, but this time we are better equipped. The key result of the chapter is Serre's theorem, which characterizes regular local rings as local Noetherian rings of finite global dimension. Using this, we are able to analyze the behavior of regularity under many operations, such as localizations, quotient and polynomial extensions. We then turn to the celebrated theorem of Auslander and Buchsbaum stating that local regular rings have unique factorization, and we end the chapter by characterizing regular rings in characteristic p.

Chapter 9 introduces many other classes of rings, that are best understood via their homological properties. Under the assumption of locality, these form a well-known hierarchy:

$$\{\text{regular}\} \subset \{\text{complete intersections}\} \subset \{\text{Gorenstein}\} \subset \{\text{CM}\}.$$

All of the above classes are introduced in the chapter, starting from CM (Cohen–Macaulay) rings (and modules). These are characterized by having depth equal to their dimension. This simple definition is actually equivalent to a myriad of other characterizations—for instance, CM rings are those where the notions of (maximal) regular sequences and systems of parameters are the same. This fact makes CM rings appear essentially everywhere in commutative algebra. Among other things, we prove that CM rings have a particularly simple structure of the lattice of prime ideals. Section 9.4 characterizes them by the number of factors in irreducible decompositions of primary ideals. When this number is 1, we obtain an even more special class, that of Gorenstein rings, which we study in Section 9.7. These are also characterized by having finite injective dimension.

Chapter 10 introduces various types of dualities for modules over a local Noetherian ring. The most elementary, Matlis duality, has a more algebraic flavor, and generalizes the duality for finitely generated vector spaces over a field. A more sophisticated duality, Grothendieck local duality, is the topic of Section 10.7. This is the local analogue of the celebrated Serre duality for sheaves on projective varieties. In order to state the result, we introduce the notion of local cohomology in Sections 10.5 and 10.6, and the canonical module in Section 10.4. By using these notions, we give another characterization of Cohen–Macaulay and Gorenstein rings by their duality properties.

Apart from a general mathematical maturity, the prerequisites for the book are an introduction to commutative algebra, such as [**Fer20**], [**AM69**] or [**AK13**]. Also some familiarity with algebraic geometry, at the level of algebraic varieties, is assumed. The main results that are used in the text are collected in Appendix A for ease of reference. Notice, though, that the appendix is no substitute for a general familiarity with the treated arguments, and should only be used as a reference. Each result stated in the appendix appears in [**Fer20**], and is mentioned with an appropriate reference. We also include a reference to a different source—where possible [**AM69**]—to avoid a strict dependency between the volumes. A few results from [**Fer20**] are not listed in the appendix, but are actually proved again in the book. This is either because a new method allows for a different proof, or because the result logically fits the scope of the text and it does not make sense to omit it.

In the text, we also make some references to some basic algebraic topology or differential geometry—for instance, we mention simplicial and De Rham cohomology. In particular, the very first section draws examples of categories from many branches of mathematics. None of these is a strict requirement: generally these topics are mentioned only as examples or as motivation for a particular construction. The reader that is not familiar with them should not lose much by skipping them, but they can be useful aids for the knowledgeable reader.

Other than cover to cover, there are various ways to read the text, of which we suggest a few. First, Chapter 1 contains a short, self-contained introduction to the most basic ideas in category theory. There are many such introductions, but this is mine. Chapters 1 and 2 together provide a proof of the Freyd–Mitchell theorem, developed from scratch. The short book [**Fre64**] says in its introduction that it provides a geodesic way to the proof of Freyd–Mitchell. It is probably no coincidence that the page count of that book and the first two chapters are more or less the same.

The reader interested in a general introduction to homological algebra can read Chapters 1 to 4. This should give a reasonable introduction to the classical approach, after which one may want to learn about derived and triangulated categories, for instance, from [**GM03**]. Chapter 4 alone makes sense for someone who wants to learn about spectral sequences, and already has some familiarity with derived functors, even if only on categories of modules.

The reader who already has some familiarity with homological algebra and is interested in learning about commutative algebra can start from

Chapter 5, refering back to earlier chapters as needed. The dependency graph among Chapters 5 to 10 is rather linear, but on a first read one can safely skip Sections 5.8, 6.5, 7.5, 7.6, 7.7, 8.6, 9.3, 9.8, and 10.3. This should leave a book about half the size that introduces all relevant ideas, such as depth, regularity, Koszul complex, local duality, Cohen–Macaulay rings, and Gorenstein rings.

Examples in the text usually require only trivial verifications. They are part of the core of the text and should not be skipped. Occasionally, an example requires more work, and it should be considered as an exercise. I frequently punctuate the book with remarks such as "(Prove it!)" to underline the fact that I am skipping some necessary verification, left to the reader.

Exercises vary from simple to hard, and there is (intentionally) no indication to distiguish the level. I suggest, as always, that you try at as many exercises as you can. If some of them look too hard, they can be postponed and tried again after some rumination.

In general, I have tried to avoid depending on exercises for the main body of the text. The cases where I have done so should be easy verifications. On the other hand, many important and subtle counterexamples are presented as series of exercises.

No contribution in this book is original, except of course the usual amount of errors, that should be attributed only to the author. If you spot some of them, you can send an email to ferrettiandrea@gmail.com.

I hope that you will enjoy reading this book as much as I enjoyed writing it!

If I was able to write this book, it is because Massimo Gobbino and Paolo Tilli, when I was young and did not know better, believed in me and persuaded me to undertake the study of mathematics. This turned out to be one of the best choices I made, and I have to really thank them for this. Thanks to Roberto Dvornicich, who instilled in me a lasting love for algebra. I take the opportunity to thank the AMS for their editorial support, especially Ina Mette, who believed in the project and followed it with great patience through many years. Most of all, I want to thank my wife Sbambi, who with her love shows me every day what is really important, and with her patience and understanding has given me the time and peace of mind to do mathematics and finish this book. ⌣

Conventions

A ring is associative and with identity; moreover, rings are *commutative* unless we specify otherwise. It is often useful to do homological algebra for modules over a possibly noncommutative ring, but we will always note when we do so. A ring is *reduced* if it does not have nilpotent elements.

When we speak of a *graded ring* A, we mean that we have a decomposition $A = \bigoplus_{i=0}^{\infty} A_i$, where the A_i are additive subgroups such that $A_i \cdot A_j \subset A_{i+j}$. Over such a ring, a *graded module* M has a decomposition $M = \bigoplus_{i=0}^{\infty} M_i$, where the M_i are additive subgroups such that $A_i \cdot M_j \subset M_{i+j}$. We occasionally consider rings and modules graded over more general monoids, but we always note it explicitly.

For all other conventions, see Appendix A.

Chapter 1

Categories

Ever since the famous Tôhoku paper [**Gro57**] by Grothendieck, it is customary to put the study of homological algebra in the context of Abelian categories. Before studying them, we use this chapter to introduce some categorical language.

Categories are used in all branches of mathematics, and as such our introduction focuses not so much on proving deep results, but rather on providing many examples from various different fields. Even if the reader is not familiar with all topics mentioned in the examples, they should be able to follow enough of them to get a sense of the generality of the concepts presented here.

After the first section defines categories and functors, which are the homomorphisms of these structures, we take a little detour on set theory to put our definitions on a firm ground and steer clear of possible set theoretic paradoxes. Going forward, we introduce in Section 1.3 the concept of natural transformation, which allows to consider functors themselves as objects of appropriate categories. We use this construction to prove the famous Yoneda lemma, which is the category theoretic analogue of Cayley's theorem in the theory of groups.

Section 1.4 introduces the concept of limit and colimit. These are the natural generalization, from the point of view of category theory, of the familiar construction of inverse and direct limits. It turns out that generalizing this concept removes much of the accidental complexity, and limits in general are even simpler to define and work with than direct and inverse limits.

Finally, Section 1.5 introduces the concept of a pair of adjoint functors. These are the generalization of the concept of free object (as in free groups, vector spaces or polynomial algebras). Adjoint functors respect limits and colimits—this is a general fact that comes useful over and over in all branches of mathematics. This section also contains a deeply important result by Freyd on the existence of adjoints to a given functor.

1.1. Categories and functors

To lay the ground for the development of homological algebra, we are going to start by defining categories.

Definition 1.1.1. A *category* $\mathcal{C}$ is given by a *collection* of objects $\mathrm{Obj}(\mathcal{C})$ and the assignment, for every pair of such objects A, B of a *set* $\mathrm{Hom}(A, B)$, called the *morphisms* between A and B.

Moreover, we require the existence of the operation

$$\circ\colon \mathrm{Hom}(B, C) \times \mathrm{Hom}(A, B) \to \mathrm{Hom}(A, C)$$

for every triple of objects A, B, C, called *composition*. We will write, as customary, $f \circ g$ for the composition of $f \in \mathrm{Hom}(B, C)$ with $g \in \mathrm{Hom}(A, B)$. Finally, for every object A there should exist a distinguished morphism $\mathrm{id}_A \in \mathrm{Hom}(A, A)$, called the *identity* of A.

All of such data constitutes a category, provided

(1) the composition law is *associative*, that is, $(f \circ g) \circ h = f \circ (g \circ h)$ whenever such compositions are defined

(2) the identity objects act as left and right identities for the composition, that is, for every pair of objects A, B and every morphism $f \in \mathrm{Hom}(A, B)$ we have $f \circ \mathrm{id}_A = f = \mathrm{id}_B \circ f$

Remark 1.1.2. We have been deliberately vague in saying that $\mathrm{Obj}(\mathcal{C})$ is a collection of objects. In fact, $\mathrm{Obj}(\mathcal{C})$ is not necessarily a set, but is allowed to be a proper class. We will discuss in the next section what this means—we will need some extension of the usual Zermelo–Fraenkel set theory. Nevertheless, $\mathrm{Hom}(A, B)$ is required to be an actual set for each pair of objects A, B. Moreover, we will simply write $A \in \mathcal{C}$ to mean that A is an object of $\mathcal{C}$. A category $\mathcal{C}$ such that $\mathrm{Obj}(\mathcal{C})$ is in fact a set is called *small*.

Notice that some authors allow $\mathrm{Hom}(A, B)$ to be a class itself—in that terminology, our categories would be called *locally small*. All the categories considered in this book will be locally small, without further notice.

As a piece of notation, whenever we want to be specific about the category we will write $\mathrm{Hom}_{\mathcal{C}}(A, B)$ in place of $\mathrm{Hom}(A, B)$. We will also talk of morphisms (or maps, or arrows) $A \to B$ to mean elements of $\mathrm{Hom}(A, B)$ and will even write $f \colon A \to B$ for $f \in \mathrm{Hom}(A, B)$. As we will see, the fact that this notation calls to mind functions is not coincidental.

Definition 1.1.3. Let A, B be objects in a category $\mathcal{C}$, and $f \in \mathrm{Hom}_{\mathcal{C}}(A, B)$. We say that f is an *isomorphism* if it admits an inverse morphism $g \in \mathrm{Hom}_{\mathcal{C}}(B, A)$, that is, $g \circ f = \mathrm{id}_A$ and $f \circ g = \mathrm{id}_B$.

We are now going to give some examples which show the wide applicability of this notion. The reader is not required to be familiar with all of them, but it helps to have seen at least a few.

Example 1.1.4.

(a) The most familiar category is Set, the category having sets as objects and functions between them as morphisms. Composition is defined simply as the usual composition of functions. This is the prototypical example of a category, and the definition of category is meant to capture the basic properties of such structure. By the way, we already see a reason why we did not want to require $\mathrm{Obj}(\mathcal{C})$ to be a set, as the set of all sets is well known to be a paradoxical object in set theory.

(b) Many other categories are modeled on this example, taking sets with additional structure, and maps which preserve this structure. Categories of this kind are called *concrete*, although this is not a precise definition. For instance, we have the category Top, whose object are topological spaces and whose morphisms are *continuous* maps between them. Similarly, we have the category Diff of differentiable (say C^{∞}) manifolds and differentiable maps between them.

(c) Given a field k, the category Vect_k has vector spaces over k as objects, and *linear* maps between vector spaces as morphisms. There are also subcategories $\mathrm{Vect}_k(n)$ consisting of vector spaces of finite dimension n.

(d) Similarly, we have the categories Ring of rings and their homomorphisms, and for a ring A we have the category Mod_A of modules over A and A-linear maps. When A is a noncommutative ring, we will denote by Mod_A the category of *left* A-modules. When A is graded, we will denote by Mod_A the category of graded A-modules, where morphisms have degree 0, while Mod_A^u will be used to denote the category of ungraded A-modules, that is, modules over A regarded as a ring without grading.

(e) The category of groups and their homomorphisms is denoted Grp, while the category of *Abelian* groups is denoted by Ab.

(f) Not all categories arise as sets with extra structure. Every group G can be regarded as a category with a single object $\bullet$, and with morphisms $\mathrm{Hom}(\bullet, \bullet) = G$, where we use the product of G as composition.

(g) More generally, a *groupoid* is a category where every morphism $f \in \mathrm{Hom}(A, B)$ has an inverse $g \in \mathrm{Hom}(B, A)$. This means that $f \circ g = \mathrm{id}_B$ and $g \circ f = \mathrm{id}_A$. With this language, a group is just a groupoid with a single object.

(h) Let S be a topological space. There is a *path category* $\mathrm{Path}(S)$ having for objects the points of S. A morphism between $a, b \in S$ is a path joining a to b, that is a continuous map $\gamma \colon [0, 1] \to S$ such that $\gamma(0) = a$ and $\gamma(1) = b$. Composition is defined by connecting paths. Notice that $\mathrm{Path}(S)$ is a small category.

(i) A small variant of the previous example is the category $\Pi_1(S)$. Objects are again points of S, but morphisms between $a, b \in S$ are *homotopy classes* of paths joining a to b. This is called the *fundamental groupoid* of S, and is in fact a groupoid (Can you see why?).

   This is a generalization of the fundamental group $\pi_1(S, a)$, which takes into account all basepoints at once. The theory of the fundamental groupoid is in some sense simpler than the fundamental group, since many constructions that are akward for $\pi_1(S, a)$ because of the dependency on a choice of a basepoint become canonical in the context of $\Pi_1(S)$.

(j) Let S be a topological space. The open sets of S form the objects of a category $\mathrm{Top}(S)$, with a single morphism $U \to V$ whenever U is a subset of V.

(k) More generally, let $X, <$ be any partially ordered set. We can form an *order category* having X as objects and a single morphism $x \to y$ for a pair $x, y \in X$ such that $x \leq y$. We usually denote this category simply by X, or $\mathrm{Ord}(X)$ in case we want to avoid confusion. We recover the previous item by considering a topology as a set of open sets, partially ordered by inclusion.

(l) The *homotopy category* hTop has topological spaces as objects, but its morphisms are not continuous maps. Instead, a morphism in hTop is an equivalence class of continuous maps with respect to homotopy.

(m) A *Boolean algebra* A is a set endowed with two associative and commutative binary operations $\vee$ and $\wedge$, a unary operation $\neg$ and two elements $\bot$ and $\top$ satisfying the following properties. First, $\bot$

is an identity for $\vee$, $\top$ is an identity for $\wedge$. The two operations are required to distribute over each other, that is, for all $a, b, c \in A$, $a \vee (b \wedge c) = (a \vee b) \wedge (a \vee c)$ and vice versa. Finally, for all elements $a \in A$, one must have $a \vee \neg a = \bot$ and $a \wedge \neg a = \top$. A typical example is the power set of a set X, with the operations union, intersection, and complement. Boolean algebras form a category Bool, where maps are functions that preserve the operations.

(n) Let M, N be two n-dimensional manifolds. A *cobordism* between M and N is a $n + 1$-dimensional manifold with boundary W such that $\delta W = M \sqcup N$ is the disjoint union of M and N.

   The (unoriented) *bordism category* Bord(n) has n-dimensional manifolds as objects, and cobordisms as morphisms. Assume we have three n-dimensional manifolds M, N and P, with cobordisms W between M and N, and Z between N and P. The composition $Z \circ W$ is obtained by gluing W and Z along N in a smooth way, to obtain a cobordism between M and P (see [**Hir97**, Chapter 7] for a precise definition).

(o) If $\mathcal{C}$ is any category, we can construct its *opposite* category $\mathcal{C}^{op}$ by choosing Obj$(\mathcal{C}^{op}) := $ Obj$(\mathcal{C})$, and Hom$_{\mathcal{C}^{op}}(A, B) := $ Hom$_{\mathcal{C}}(B, A)$. The composition in $\mathcal{C}^{op}$ is just the same as the composition in $\mathcal{C}$, but with its arguments reversed.

(p) Let $\mathcal{C}, \mathcal{D}$ be two categories. We have a *product category* $\mathcal{C} \times \mathcal{D}$. Objects of $\mathcal{C} \times \mathcal{D}$ are pairs (C, D), with $C \in \mathcal{C}$ and $D \in \mathcal{D}$. For $C, C' \in \mathcal{C}$ and $D, D' \in \mathcal{D}$, we set Hom$_{\mathcal{C} \times \mathcal{D}}((C, D), (C', D')) := $ Hom$_{\mathcal{C}}(C, C') \times $ Hom$_{\mathcal{D}}(D, D')$, where the composition is componentwise.

(q) Let $\mathcal{C}$ be a category, $A \in \mathcal{C}$ any object. There is a category $\mathcal{C}(A, -)$, called a Hom-category, defined as follows. Objects of $\mathcal{C}(A, -)$ are morphisms $A \to B$ in $\mathcal{C}$, for each object $B \in \mathcal{C}$. A morphism in $\mathcal{C}(A, -)$ between objects $f \colon A \to B$ and $g \colon A \to C$ is just a morphism $h \colon B \to C$ of $\mathcal{C}$ that makes the following diagram commutative

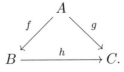

Similarly, there is a category $\mathcal{C}(-, A)$ whose objects are the morphisms $B \to A$ in $\mathcal{C}$, for each object $B \in \mathcal{C}$. Can you define the morphisms of $\mathcal{C}(-, A)$?

(r) The *category of simplexes* Δ has $\mathbb{N}$ as objects. For $n \in \mathbb{N}$, we denote $[n] := \{0, \ldots, n\}$. A morphism between m and n in Δ is a nondecreasing map $[m] \to [n]$.

Notice that categories have a dual nature, and the theory always reflects that. On the one hand, some categories seem to encode properties of entire classes of mathematical structures: such is the case for Set, Top, and Bord(n) for instance. On the other hand, categories are algebraic structures themselves, and can be investigated in the same way we do for groups or rings.

When investigating some new algebraic structure, the first step is almost always to define a suitable class of morphisms. Categories are no exception.

Definition 1.1.5. Let $\mathcal{C}$, $\mathcal{D}$ be two categories. A *covariant functor* F between $\mathcal{C}$ and $\mathcal{D}$ is a way to assign to each object $A \in \text{Obj}(\mathcal{C})$ an object $F(A) \in \text{Obj}(\mathcal{D})$, and to each morphism $f \in \text{Hom}_\mathcal{C}(A, B)$ a morphism $F(f) \in \text{Hom}_\mathcal{D}(F(A), F(B))$, in such a way that

(1) for all $A \in \text{Obj}(\mathcal{C})$ we have $F(\text{id}_A) = \text{id}_{F(A)}$, and

(2) $F(f \circ g) = F(f) \circ F(g)$ whenever the composition is defined.

Remark 1.1.6. We have gone out of our way to avoid saying that a functor is a map—again, this is because the objects of a category can fail to be a set. After we discuss the set theoretic foundation for such matters, we will start being a little less precise with language. In any case will write $F \colon \mathcal{C} \to \mathcal{D}$ to denote a functor F between $\mathcal{C}$ and $\mathcal{D}$.

In many important cases, there is a natural way to transfer objects from one category to another, but the natural action on morphisms reverses the arrows.

Definition 1.1.7. Let $\mathcal{C}$, $\mathcal{D}$ be two categories. A *contravariant functor* F between $\mathcal{C}$ and $\mathcal{D}$ is a way to assign to each object $A \in \text{Obj}(\mathcal{C})$ an object $F(A) \in \text{Obj}(\mathcal{D})$, and to each morphism $f \in \text{Hom}_\mathcal{C}(A, B)$ a morphism $F(f) \in \text{Hom}_\mathcal{D}(F(B), F(A))$, in such a way that

(1) for all $A \in \text{Obj}(\mathcal{C})$ we have $F(\text{id}_A) = \text{id}_{F(A)}$, and

(2) $F(f \circ g) = F(g) \circ F(f)$ whenever the composition is defined.

Remark 1.1.8. The datum of a contravariant functor $\mathcal{C} \to \mathcal{D}$ is exactly equivalent to the datum of a *covariant* functor $\mathcal{C}^{op} \to \mathcal{D}$. For this reason, we will often state many results only in terms of covariant functors, and when we mention a functor without further qualification, we will always mean a covariant one.

For the same reasons that there is no set of all sets, there is also no class of all classes. This precludes us from defining a category of categories, since its objects would not even be a class. But we can restrict to *small* categories. If $\mathcal{C}$ and $\mathcal{D}$ are small categories, the class of all functors between them is a set, and this allows us to give the following definition.

Definition 1.1.9. The category of small categories, denoted Cat, is the category whose objects are small categories and whose morphisms are covariant functors between them.

We now turn to some examples of functors.

Example 1.1.10.

(a) Each group can be seen as a set if we just forget the group structure. This gives a natural covariant functor Grp $\to$ Set, which is called a *forgetful* functor. Similarly we have forgetful functors Ring $\to$ Set, Top $\to$ Set, and so on for the various concrete categories.

(b) Not all forgetful functors have to target Set. For instance, every ring is also an Abelian group for $+$, and this gives a forgetful functor Ring $\to$ Ab.

(c) Let A be a graded ring. We have a functor $U\colon \mathrm{Mod}_A \to \mathrm{Mod}_A^u$ that forgets the grading.

(d) Associating to each continuous map its homotopy class gives a homotopy functor Top $\to$ hTop.

(e) Many interesting functors on hTop are defined in algebraic topology. For instance, for every Abelian group G we have the n-th homology functor $h_n(G)\colon$ hTop $\to$ Ab. This associates to each space S the singular homology group $H_n(S,G)$, and to each continuous function $f\colon S \to T$ the induced map in homology $H_n(S,G) \to H_n(T,G)$.

(f) Similarly, homotopy groups define functors. Since computing homotopy groups requires the choice of a basepoint, these are not defined on hTop. Rather, they are defined on a category of *pointed spaces*, whose objects are pairs (S,s) of a topological space S and a point $s \in S$ (what are the morphisms?).

(g) Cohomology defines *contravariant* functors. Namely, for each Abelian group we have the n-th cohomology functor $h^n(G)$ which associates to each space S the group $H^n(S,G)$. The maps induced on cohomology reverse arrows, which is the reason why $h^n(G)$ is contravariant.

When the coefficients are a ring A, the cup product defines a ring structure on the direct sum $H^*(S, A) := \bigoplus_n H^n(S, A)$, and this induces a contravariant functor $h^*(A)\colon \mathrm{hTop} \to \mathrm{Ring}$.

(h) Every category $\mathcal{C}$ has a contravariant functor $\mathcal{C} \to \mathcal{C}^{op}$ which can be considered the universal contravariant functor (see Definition 1.4.1).

(i) Let A be a ring, M an A-module. There is a covariant functor $M \otimes -$ from Mod_A to itself, which sends N to $M \otimes N$. The action on the morphism sends the homomorphism $f\colon N_1 \to N_2$ to the map $\mathrm{id}_M \otimes f$. Symmetrically, there is another functor $- \otimes M\colon \mathrm{Mod}_A \to \mathrm{Mod}_A$.

(j) Let $\mathcal{C}$ be any category, A an object of $\mathcal{C}$. There is a covariant functor $h^A := \mathrm{Hom}(A, -)\colon \mathcal{C} \to \mathrm{Set}$, which sends $B \in \mathcal{C}$ to $\mathrm{Hom}(A, B)$. Given a morphism $f\colon B \to C$ in $\mathcal{C}$, we define

$$h^A(f)\colon \ \mathrm{Hom}(A, B) \longrightarrow \mathrm{Hom}(A, C),$$

$$g \longmapsto f \circ g.$$

In other words, h^A acts on morphisms associating to f the composition with f.

Symmetrically, there is a functor $h_A := \mathrm{Hom}(-, A)\colon \mathcal{C} \to \mathrm{Set}$, but this time it is contravariant. This is because given $f\colon B \to C$ in $\mathcal{C}$ we have the natural map

$$h_A(f)\colon \ \mathrm{Hom}(C, A) \longrightarrow \mathrm{Hom}(B, A),$$

$$g \longmapsto g \circ f.$$

In this case, composition on the right inverts the direction of the arrows. A functor of the form h_A is called *representable*. Dually, h^A is called a *corepresentable* functor.

(k) Let k be a field. There is a functor $\mathrm{Set} \to \mathrm{Vect}_k$ which associates to every set S the vector space formally generated by S, that is the set of formal linear combinations

$$a_1 s_1 + \cdots + a_r s_r,$$

where $a_1, \ldots, a_r \in k$ and $s_1, \ldots, s_r \in S$. This can also be identified with the vector space of function $S \to k$ with finite support. This defines a functor because a linear map between vector spaces is determined uniquely once its values are known on a basis.

(l) Let V, W be vector spaces over the field k. Every linear map $L\colon V \to W$ induces a map between the dual vector spaces

$$L^{\vee}\colon\ W^{\vee} \longrightarrow V^{\vee},$$

$$f \longmapsto f \circ L.$$

This remark says that there is a contravariant functor

$$\vee\colon \mathrm{Vect}_k \to \mathrm{Vect}_k$$

that takes V to $V^{\vee}$.

By composition, there is also a *covariant* functor $\mathrm{Vect}_k \to \mathrm{Vect}_k$ that amounts to taking the *double dual* of a vector space.

(m) Let V be a vector space over the field k, $n \in N$. We can form the vector space $\Lambda^n V$ of antisymmetric tensors, and a linear map $V \to W$ induces a map on tensors $\Lambda^n V \to \Lambda^n W$. This defines a functor $\Lambda^n\colon \mathrm{Vect}_k \to \mathrm{Vect}_k$.

(n) Let Diff be the category of differentiable manifolds and VDiff the category of differentiable vector bundles. Objects of VDiff are maps $E \to B$ in Diff, where E is a vector bundle over B, and the map from $E \to B$ to $E' \to B'$ is a commutative square

$$\begin{array}{ccc} E & \longrightarrow & E' \\ \downarrow & & \downarrow \\ B & \longrightarrow & B'. \end{array}$$

The tangent bundle construction defines a functor $T\colon \mathrm{Diff} \to \mathrm{VDiff}$. Namely, for every differentiable manifold M, the tangent bundle $TM \to M$ is an object of VDiff. Given a morphism $f\colon M \to N$, the differential of f defines a morphism $df\colon TM \to TN$, which is compatible with projections and linear on the fibers, and we define Tf to be the associated commutative square.

As is customary with algebraic structures, one can also define substructures.

Definition 1.1.11. Let $\mathcal{C}$ be a category. A *subcategory* $\mathcal{C}'$ of $\mathcal{C}$ is a category such that each object of $\mathcal{C}'$ is also an object of $\mathcal{C}$, and for each pair of objects A, B of $\mathcal{C}'$ there is an inclusion $\mathrm{Hom}_{\mathcal{C}'}(A, B) \subset \mathrm{Hom}_{\mathcal{C}}(A, B)$.

If, moreover, there is an equality $\mathrm{Hom}_{\mathcal{C}'}(A, B) = \mathrm{Hom}_{\mathcal{C}}(A, B)$ for each pair of objects A, B of $\mathcal{C}'$, we say that $\mathcal{C}'$ is a *full* subcategory of $\mathcal{C}$—in this case, the two categories are only distinguished by their objects.

More generally, let $F\colon \mathcal{C} \to \mathcal{D}$ be a (covariant) functor. We say that F is *full* if for each pair of objects $A, B \in \mathcal{C}$, the induced map $\operatorname{Hom}(A, B) \to \operatorname{Hom}(F(A), F(B))$ is surjective. We say that F is *faithful* if for each such pair of objects the induced map between morphisms is injective. A functor that is both full and faithful is called *fully faithful*.

A functor $F\colon \mathcal{C} \to \mathcal{D}$ which is injective on objects and faithful is called an *embedding*. If it is moreover a full functor, it is called a *full embedding*. Thus, a full embedding allows us to see $\mathcal{C}$ as a full subcategory of $\mathcal{D}$.

1.2. Sets and classes

Many treatments over categories gloss over the actual details of what it means that the objects of a categories can be a proper class. While most of the work with categories can be done disregarding foundational issues, this might leave one with lingering questions. Putting these matters on stable ground requires a small detour on set theory. This section is less formal than the rest of the book and offers few proofs, but hopefully it can shed some light on the assumptions that need to be made in order to make working with categories more rigorous.

Most importantly, a distinction must be made between two kinds of collections: sets and classes. All sets will be classes, but the converse will not be true—classes that are not sets will be called *proper* classes. We want to make sense of statements like $x \in S$ for two entities x and S, where x is a *set*. In other words, we will allow sets to be members of other sets or classes, but classes will not necessarily be members of other classes. In particular, we will define the class of all sets, but not the set of all sets or the class of all classes.

There are two approaches to define a suitable theory: one is to change the axioms of the usual Zermelo–Fraenkel set theory with choice (ZFC) and obtain a different axiomatic theory that distinguishes between sets and classes. The second approach is to require the existence of a sufficiently large cardinal κ, in such a way that operations on sets of cardinality less than κ produce other sets of cardinality less than κ. In this way sets of cardinality less than κ will form a model of ZFC, and we will redefine sets as small sets, and classes as arbitrary sets.

Both approaches have merit, but before discussing them it is worthwhile to review the usual axioms of ZFC in an informal way. First, the *axiom of extensionality*, states that two sets having the same elements are actually the same set.

(ZFC 1) If for all x, we have $x \in y \iff x \in z$, then $y = z$.

Given this axiom, if an empty set exists, it is unique and we can give it the name $\emptyset$. The second axiom, the *axiom of foundation*, requires every nonempty set to have an element disjoint from itself.

(ZFC 2) For all sets $x \neq \emptyset$, there is $y \in x$ such that there is no z for which $z \in x$ and $z \in y$.

The third axiom, the *axiom of restricted comprehension* is actually an axiom schema. It allows us to define subsets of a set.

(ZFC 3) If ϕ is any first-order formula with a free variable and z is a set, there exists a set y such that $x \in y \iff x \in z$ and $\phi(x)$.

By the axiom of extensionality, this set is in fact unique, and we denote it by $\{x \in z \mid \phi(x)\}$. If a set z exists at all, there exists an empty set, defined for instance as $\{x \in z \mid \texttt{false}\}$. The *axiom of union* allows us to define the union of a family of sets.

(ZFC 4) If z is a set, there exists a set y such that $x \in y$ if and only if there exists $z' \in z$ such that $x \in z'$. Again, this is unique by extensionality and we denote it by $y = \cup z$.

So far there are not many ways to obtain sets other than $\emptyset$. The *axiom of pairing* allows for the creation of sets that consist of at least two elements.

(ZFC 5) Given x, y there is a set z such that $z' \in z \iff z' = x$ or $z' = y$. As usual, we denote such set by $\{x, y\}$.

Notice that this also defines the one element set $\{x\}$ by taking $y = x$. We can now define an ordered pair (x, y) as the set $\{\{x\}, \{x, y\}\}$ and a triple $(x, y, z) := (x, (y, z))$. Next is the *axiom of power set*. As usual, $x \subset y$ means that for all sets u, $u \in x \implies u \in y$.

(ZFC 6) For a set x there is a set y such that $z \in y \iff z \subset x$. As usual, we denote such set by $\mathcal{P}(x)$.

Given pairs and the power set, we can now define the product $x \times y$ as a subset of $\mathcal{P}(x \cup y)$, where of course $x \cup y := \cup\{x, y\}$. The product is the subset of $\mathcal{P}(x \cup y)$ consisting of ordered pairs—it exists by the *axiom of restricted comprehension* and is unique by extensionality. At this point we can define a function $f \colon x \to y$ as a triple (x, y, Γ) such that $\Gamma \subset x \times y$ and for all $a \in x$ there is exactly one $b \in y$ such that $(a, b) \in \Gamma$. The next *axiom of replacement* is again a schema of axioms and it looks a little technical.

(ZFC 7) Assume given a binary relation p such that for all x there is a unique y such that $p(x, y)$ holds. If z is a set, there is a set w such that $y \in w \iff \exists x \in z : p(x, y)$.

Replacement is a way to build sets out of other sets, meaning that ouf of z, we can define a w such that it only contains elements that are related to those in z via a given relation p.

The cautious reader has surely noticed that our axioms so far only allow us to define finite sets. We will say that a set x is infinite if there exists a function $f\colon x \to x$ which is injective but not surjective (with the usual definitions of injective and surjective). A set that is not infinite will be called finite. The *axiom of infinity* allows us to define such sets.

(ZFC 8) There exists a set x that is infinite.

Finally, ZFC would not be complete without the *axiom of choice*.

(ZFC 9) Assume x is a set and for all $y \in x$, $y \neq \emptyset$. Then there is a *choice function* $f\colon x \to \cup x$, that is, a function f such that $f(y) \in y$ for all $y \in x$.

We will not belabor the consequences of those axioms, which should be familiar to the reader. One can prove from them that sets are well founded, develop the arithmetic of cardinal and ordinal numbers, prove Zorn's lemma, the principle of transfinite induction; and so on.

We will now contrast these axioms with a different theory due to von Neumann, Bernays, and Gödel (NBG). Elements in this theory will be called *classes*, with sets being the classes that are members of other classes. In particular, there will be no axiom of pairing for classes—it can happen that a class X is not a member of any other class. By convention, we will denote sets by small letters and classes by capital letters. Extensionality holds unmodified.

(NBG 1) If for all X, we have $X \in Y \iff X \in Z$, then $Y = Z$.

We say that a class x is a set if there exists a class Y such that $x \in Y$. The axiom of pairing and foundation are identical to ZFC, but only holds for sets.

(NBG 2) Given *sets* x, y there is a set z such that $z' \in z \iff z' = y$ or $z' = y$. As usual, we denote such set by $\{x, y\}$.

(NBG 3) For all *sets* $x \neq \emptyset$, there is $y \in x$ such that there is no z for which $z \in x$ and $z \in y$.

As in ZFC, we can now define pairs, triples, and so on. There is no need to have the axiom schema of restricted comprehension, because we are allowed to define quite large classes. The *axiom of membership* defines the class of all pairs with a membership relation. This allows us to treat the predicate $\in$ as an object *inside* our theory.

(NBG 4) There exists a class E such that $(x, y) \in E \iff x \in y$.

Unlike our previous examples, this is not determined by extensionality, since the definition only determines the *pairs* that are members of E. The next axioms allow us to define intersections and complements.

(NBG 5) Given classes X, Y there exists a class Z such that $x \in Z \iff x \in X$ and $x \in Y$. This class is unique by extensionality, and denoted $X \cap Y$.

(NBG 6) Given a class X, there exists a class Y such that a set $x \in Y \iff x \notin X$. Such class is denoted X^c. Notice that the complement here is absolute, not relative to some ambient set.

The two axioms allow us to define $\emptyset := X \cap X^c$ for any class X (at least one class exists, by the axiom of membership). The complement of $\emptyset$ is the class V of all sets. Next we have the *axiom of domain*.

(NBG 7) Given a class X, there exists a class Y such that $x \in Y \iff \exists y : (x, y) \in X$. This is unique, as usual, and denoted $\mathrm{Dom}(X)$.

The next axioms are a little technical.

(NBG 8) Given a class X, there exists a class Y such that $u \in Y$ if and only if there exist x, y with $u = (x, y)$ and $x \in X$. This class is denoted $X \times V$.

(NBG 9) Given a class X, there is a class Y such that $(x, y, z) \in Y \iff (y, z, x) \in X$.

(NBG 10) Given a class X, there is a class Y such that $(x, y, z) \in Y \iff (x, z, y) \in X$.

The previous axioms allow us to prove a result that takes the place of the axiom of restricted comprehension.

Theorem (Class existence, [**Gö08**]). *Let ϕ be a first-order formula, whose quantifiers only involve sets, with free variables $x_1, \ldots, x_m$ and $Y_1, \ldots, Y_n$. For all classes $Y_1, \ldots, Y_n$ there exists a class A such that $(x_1, \ldots, x_m) \in A$ if and only if $\phi(x_1, \ldots, x_m, Y_1, \ldots, Y_n)$ holds.*

Notice that the comprehension here is in fact unrestricted—the class

$$\{(x_1, \ldots, x_m) \mid \phi(x_1, \ldots, x_m, Y_1, \ldots, Y_n)\}$$

is defined regardless of an ambient set. The class existence theorem is due to Gödel—for a simple self-contained proof, see [**Ban20**, Section 7].

Functions in NBG are defined differently, since there is no need to restrict their domain. A function f is simply a subclass $f \subset V \times V$ such that if

$(x, y) \in f$ and (x, z) we have $y = z$. We can then recover their domain as $\mathrm{Dom}(f)$. The image is defined using the class existence theorem as

$$\mathrm{im}(f) := \{y \mid \exists x, (x, y) \in f\}.$$

More generally, for a class X we can define the image

$$f[X] := \{y \mid \exists x \in X, (x, y) \in f\}.$$

We use square brackets to avoid ambiguity when X is a set x—$f(x)$ denotes the only y (if any) with $(x, y) \in f$, while $f[x]$ denotes the image of x via f. The next *axiom of replacement* is slightly different from the version in ZFC, due to the fact that functions are defined globally.

(NBG 11) If f is a function and x is a set, the image $f[x]$ is also a set.

The class existence theorem also allows us to define the power and union of a class. The next axioms deal with these constructions.

(NBG 12) If x is a set, the union $\cup x$ is a set.

(NBG 13) If x is a set, the power class $\mathcal{P}(x)$ is a set.

The axiom of infinity in NBG also takes a slightly different form.

(NBG 14) There is a set $x \neq \emptyset$ such that for all $y \in x$ there is $z \in x$ such that $y \subsetneq z$.

It can be proved that such a set is, in fact, infinite. Finally, we have an equivalent of the axiom of choice, which can be made unrestricted.

(NBG 15) There is a function f such that for all $x \neq \emptyset$ we have $f(x) \in x$.

An important consequence of the axioms presented here is the following result.

Theorem 1.2.1 (Separation). *If x is a set and $Y \subset x$ is a class, then Y is a set.*

Proof. Define the class

$$F := \{(y, z) \mid y \in Y \text{ and } y = z\}.$$

Then F is a function and $F[x] = Y$, which is a set by the axiom of replacement. $\square$

Given this, we see that all traditional constructions of set theory never produce a proper class. In fact, it it not difficult to prove that if $\mathcal{U}$ is a model of NBG, the sets in $\mathcal{U}$ are a model of ZFC. In other words, NBG is an extension of ZFC, in that sets in NBG satisfy all ZFC axioms. In fact, more is known:

Theorem (Cohen, [**Coh08**]). *The theory* NBG *is a conservative extension of* ZFC, *that is, the theorems of* NBG *that only involve sets are theorems of* ZFC.

A consequence of this result is that ZFC proves a contradiction if and only if NBG does (for instance, take the contradiction $\emptyset \neq \emptyset$). By the completeness of first order logic ([**Sri13**, Chapter 4]), it follows that ZFC has a model if and only if NBG does. Hence the two theories are in some sense equally strong.

There is a second approach to constructing classes that does not require changing the set theory we work with. It has the advantage of working in the ZFC framework, but the disadvantage of requiring a strictly stronger assumption.

Definition 1.2.2. We say that a cardinal κ is *strongly inaccessible* if:

(1) κ is uncountable;

(2) for all cardinals $\lambda < \kappa$, $2^\lambda < \kappa$ (recall that 2^λ is the cardinality of $\mathcal{P}(x)$ for a set x of cardinality λ); and

(3) if x is a set of cardinality $< \kappa$, and the same holds for members $y \in x$, then the cardinality of $\cup x$ is again $< \kappa$.

In other words, there is not a way to get to cardinality κ using set theoretic constructions starting from sets of smaller cardinality. It is not difficult to prove the following.

Proposition 1.2.3. *Let κ be a strongly inaccessible cardinal in a model $\mathcal{U}$ of* ZFC. *Let $\mathcal{U}_\kappa$ be the subset of $\mathcal{U}$ consisting of all sets x such that for all chains*

$$x_n \in x_{n-1} \in \cdots \in x_1 \in x_0 = x$$

we have $|x_i| < \kappa$ for $i = 0, \ldots, n$. Then $\mathcal{U}_\kappa$ is itself a model of ZFC.

The model $\mathcal{U}_\kappa$ is called a *Grothendieck universe*. Given an inaccessible cardinal κ, one can redefine members of $\mathcal{U}_\kappa$ as sets, and call a class any element of $\mathcal{U}$. The axioms of ZFC still hold for $\mathcal{U}_\kappa$, but from the outside we are able to see that there are bigger collections.

As we hinted, the existence of a strongly inaccessible cardinal cannot be proved in ZFC, and in fact it is independent of it. The reason is that using an inaccessible cardinal one can, as above, prove the existence of a model of ZFC. Since Proposition 1.2.3 can be formalized in ZFC, if one could prove the existence of a strongly inaccessible cardinal in ZFC, it would follow that ZFC could prove its own consistency, which contradicts Gödel's second incompleteness theorem ([**Sri13**, Section 7.5.]).

The price to pay for working with Grothendieck universes is that one has to make stronger logical assumptions. The advantage is of course that one can keep working in the familiar ZFC setting, where the following axiom is added.

(ZFC 10) There exists a strongly inaccessible cardinal κ.

Both approaches discussed in this section serve well as a foundational setting to develop category theory. The preference of the author goes to Grothendieck universes, but which setting is chosen is inconsequential to the rest of the book. For a much more thorough discussion of the set theoretic foundations for category theory, including many other approaches, see [**Shu08**].

1.3. Natural transformations

A famous quote from the first page of [**Fre64**] reads:

> It is not too misleading, at least historically, to say that categories are what one must define in order to define functors, and that functors are what one must define in order to define natural transformations.

Categories and functors are what we used to define the stage of action, but natural transformations are the leading actors of category theory. The remarkable observation by MacLane [**EM45**] is that not only are there mathematical theories, that define categories, and relations between these theories, that define functors, but in a number of places there are nontrivial relations between functors themeselves.

Definition 1.3.1. Let $\mathcal{C}$, $\mathcal{D}$ be two categories, and $F, G \colon \mathcal{C} \to \mathcal{D}$ be (covariant) functors between them. A *natural transformation* (or *functor morphism*) f between F and G is the datum, for each object C of $\mathcal{C}$, of a morphism

$$f_C \colon F(C) \to G(C)$$

in $\mathcal{D}$ such that the diagram

$$\begin{array}{ccc} F(C_1) & \xrightarrow{f_{C_1}} & G(C_1) \\ {\scriptstyle F(g)}\downarrow & & \downarrow{\scriptstyle G(g)} \\ F(C_2) & \xrightarrow[f_{C_2}]{} & G(C_2) \end{array}$$

commutes for all choices of objects C_1 and C_2 of $\mathcal{C}$ and all morphisms $g \in \mathrm{Hom}(C_1, C_2)$.

The natural transformation f is called a *natural isomorphism* if moreover f_C is an isomorphism for all objects C of $\mathcal{C}$.

As usual, examples abound.

Example 1.3.2.

(a) Let $\vee^2 \colon \mathrm{Vect}_k \to \mathrm{Vect}_k$ be the double dual functor defined in Example 1.1.10(1). Given a vector space V, every element $v \in V$ determines a functional $\alpha(v) \colon V^\vee \to k$ defined by

$$\alpha(v)(f) = f(v).$$

This defines a map $V \to (V^\vee)^\vee$, which is easily shown to be injective. Putting all these maps together we obtain a natural transformation $\mathrm{id} \to \vee^2$ between the identity and double dual functors.

(b) One may be tempted to do the same for the dual functor $\vee$ which sends V to $V^\vee$, but the issue is that $\vee$ is contravariant, while id is covariant, so there is not even clear what this would be. This is what one formally means when saying that there is a natural inclusion of a vector space into its double dual, but there is no natural map between a vector space and its dual (even in the finite-dimensional case, when they are abstractly isomorphic).

(c) For a ring A let $\mathrm{GL}^n(A)$ be the group of invertible $n \times n$ matrices with coefficients in A. Clearly, GL^n can be seen as a functor Ring $\to$ Grp. Another functor $-^* \colon$ Ring $\to$ Grp is the unit functor sending A to A^*.

The fact that the determinant is defined by a universal formula, independent of A, can be expressed by saying that det is a natural transformation between GL^n and $-^*$.

(d) Let hTop$_\bullet$ be the category of *pointed* topological spaces, with homotopy classes of continuos maps as morphisms. On hTop$_\bullet$ we have two sets of functors from algebraic topology: the homotopy groups π_n and the homology groups H_n, say with coefficients in $\mathbb{Z}$. For a space X, an element of $\pi_n(X)$ is defined by the homotopy class of a (pointed) continuous map $f \colon S^n \to X$. Denoting by $[S^n]$ the fundamentcal class, this defines an element $f_*([S^n]) \in H_n(X, \mathbb{Z})$. This construction defines a natural transformation $\pi_n \to H_n(-, \mathbb{Z})$, called the Hurewicz map.

For $n = 1$, the Hurewicz map $\pi_1(X) \to H_1(X, \mathbb{Z})$ is the abelianization map ([**Hat02**, Theorem 2.A.1]), hence this particular natural transformation can be described in purely algebraic terms.

(e) Let M be a real manifold of dimension n. There are (at least) two ways to construct cohomology groups with real coefficients for M:

the singular cohomology groups $H^k(M, \mathbb{R})$, and the De Rham cohomology $H^k_{DR}(M)$. By standard approximation theorems, the group $H^k(M, \mathbb{R})$ can be defined using cohomology classes of cocycles defined on C^∞ singular cycles $\Delta_k \to M$, where Δ_k is the standard k-simplex.

Elements of $H^k_{DR}(M)$ are classes of k-differential forms on M. Let ω be such a differential form, and $f \colon \Delta_k \to M$ a C^∞ function defining a k-cycle on M. We can take its pullback to Δ_k, and the integral

$$f \mapsto \int_{\Delta_k} f^*(\omega)$$

defines a linear functional on C^∞ cycles. This is a k-cocycle on M, and the class of this cocycle depends on ω only up to coboundaries, by Stokes's theorem. It follows that this construction defines a map

$$H^k_{DR}(M) \to H^k(M, \mathbb{R}),$$

and the compatibility of pullbacks ensures that this is a natural transformation $H^k_{DR} \to H^k(\cdot, \mathbb{R})$.

In fact, the De Rham theorem states that this map is an isomorphism ([**Voi02**, Remark 4.48]).

In many places in algebra, we can get some information on an algebraic structure by observing the action of the structure on itself. For instance, for groups, this leads to Cayley's theorem that presents any group G as a subgroup of the group of permutations $\mathfrak{S}(G)$. For rings, the same approach shows that every (possibly noncommutative) ring with unity A can be embedded as a subring of $\mathrm{End}(A, +)$, the ring of endomorphism of the additive group of A.

For categories, this leads to the famous lemma of Yoneda. Let $\mathcal{C}$ be a category, and $X \in \mathcal{C}$. We have seen that X determines a functor $h_X \colon \mathcal{C}^{op} \to$ Set. If Y is another object of $\mathcal{C}$, with a morphism $f \in \mathrm{Hom}(X, Y)$, there is a natural transformation f_* between h_X and h_Y. In fact, for any object C of $\mathcal{C}$, an element of $\mathrm{Hom}(C, X)$ can be composed with f to yield and element of $\mathrm{Hom}(C, Y)$. If D is another object and $g \colon C \to D$, we get the square

$$
\begin{array}{ccc}
\mathrm{Hom}(D, X) & \longrightarrow & \mathrm{Hom}(D, Y) \\
\downarrow & & \downarrow \\
\mathrm{Hom}(C, X) & \longrightarrow & \mathrm{Hom}(C, Y),
\end{array}
$$

and both ways from $\mathrm{Hom}(D, X)$ to $\mathrm{Hom}(C, Y)$ lead to the function $h \mapsto f \circ h \circ g$. Hence the square is commutative and f_* is a natural transformation.

Assume now that $\mathcal{C}$ is small. In this case, functors $\mathcal{C}^{op} \to \mathrm{Set}$ form their own category, which we denote $\mathrm{Fun}(\mathcal{C}^{op}, \mathrm{Set})$. Morphisms in $\mathrm{Fun}(\mathcal{C}^{op}, \mathrm{Set})$ are natural transformations. To every object X of $\mathcal{C}$, we can associate an object h_X of $\mathrm{Fun}(\mathcal{C}^{op}, \mathrm{Set})$. The above construction associates to every morphism $X \to Y$ in $\mathcal{C}$ a morphism $h_X \to h_Y$ in $\mathrm{Fun}(\mathcal{C}^{op}, \mathrm{Set})$. In other words, we have defined a functor

$$h_\bullet \colon \mathcal{C} \to \mathrm{Fun}(\mathcal{C}^{op}, \mathrm{Set}).$$

Theorem 1.3.3 (Yoneda lemma). *The functor $h_\bullet$ is a full embedding.*

Proof. If X and Y are distinct objects of $\mathcal{C}$, the sets $\mathrm{Hom}(X, X)$ and $\mathrm{Hom}(X, Y)$ are distinct by definition of a category, thus $h_\bullet$ is injective on objects.

For any pair of objects X, Y, let $f \in \mathrm{Hom}(X, Y)$, and f_* the induced natural transformation between h_X and h_Y. Then $f_*(\mathrm{id}_X) = f$, thus $h_\bullet$ is faithful, as f can be recovered uniquely from f_*.

Finally, let g be a natural transformation between h_X and h_Y. Then the above line suggests that we take $f := g(\mathrm{id}_X) \in \mathrm{Hom}(X, Y)$. The fact that g is natural shows that in fact $g(h) = f \circ h = f_*(h)$ for all objects C and all $h \in \mathrm{Hom}(C, X)$. It follows that $g = f_*$; hence $h_\bullet$ is full. $\qquad\square$

The Yoneda lemma allows us to reconstruct an object X of $\mathcal{C}$ using the data of the sets $\mathrm{Hom}(A, X)$ for all objects $A \in \mathcal{C}$. It can also happen that this is redundant, and given the data of the sets $\mathrm{Hom}(A, X)$ for *some* objects A is enough. In particular, when a single object A is enough, the functor h_A is enough to reconstruct the whole category.

Definition 1.3.4. Let $\mathcal{C}$ be a category. An object A of $\mathcal{C}$ is called a *generator* if the functor $h^A \colon \mathcal{C} \to \mathrm{Set}$ is an embedding. It is called a *cogenerator* if the functor $h_A \colon \mathcal{C}^{op} \to \mathrm{Set}$ is an embedding.

Generator objects will be a key ingredient in the proof of the Freyd–Mitchell embedding theorem in Section 2.8.

Remark 1.3.5. Let $\mathcal{C}$ be a category, $F \colon \mathcal{C}^{op} \to \mathrm{Set}$ be any functor, not necessarily representable. As in the Yoneda lemma, we have a bijection $F(X) \cong \mathrm{Hom}(h_X, F)$. In fact, given a natural transformation $g \colon h_X \to F$, we obtain the element $f = g(\mathrm{id}_X) \in F(X)$, and we can recover g from f exactly as in the proof.

This statement is also sometimes called the Yoneda lemma—we obtain the earlier version by choosing $F = h_Y$.

The functor $h_\bullet$ used in the Yoneda lemma is sometimes called the *Yoneda* functor. It allows us to see every small category $\mathcal{C}$ as a full subcategory of the functor category $\mathrm{Fun}(\mathcal{C}^{op}, \mathrm{Set})$. In some sense, it suggests a general way to construct "generalized objects" of a category. For every particular category $\mathcal{C}$, one can see an element of $\mathrm{Fun}(\mathcal{C}^{op}, \mathrm{Set})$ as a generalized object of $\mathcal{C}$, and try to translate notions that apply to objects of $\mathcal{C}$ to objects of $\mathrm{Fun}(\mathcal{C}^{op}, \mathrm{Set})$. This philosophy has been extensively used by Grothendieck and his school, especially to define and study moduli problems (see, for instance, the discussion in [**HM98**, Section 1.A]).

Natural transformations can also be used to define the correct notion of two categories being "essentially the same". For most algebraic objects, the right notion is that of an isomorphism. It turns out that for categories, this notion is often too strict.

Definition 1.3.6. Let $\mathcal{C}, \mathcal{D}$ be two categories. A functor $F \colon \mathcal{C} \to \mathcal{D}$ is called an *isomorphism* if there exists a functor $G \colon \mathcal{D} \to \mathcal{C}$ such that the composition $G \circ F$ is the identity functor $\mathrm{id}_{\mathcal{C}}$ of $\mathcal{C}$ and vice versa $F \circ G = \mathrm{id}_{\mathcal{D}}$.

Example 1.3.7. Let Vect_k^* be the category of finite-dimensional vector spaces over the field k, $\vee^2 \colon \mathrm{Vect}_k^* \to \mathrm{Vect}_k^*$ the double dual functor defined in Example 1.1.10(l). One would like to state that $\vee^2$ is an isomorphism, as every finite-dimensional vector space is naturally isomorphic to its double dual. But a vector space is *not* literally its own double dual, hence $\vee^2$ is not an isomorphism. In fact, $\vee^2$ is not even surjective on objects.

To repair this and many other similar examples, we weaken the above notion a bit.

Definition 1.3.8. Let $\mathcal{C}, \mathcal{D}$ be two categories. A functor $F \colon \mathcal{C} \to \mathcal{D}$ is called an *equivalence of categories* if there exists a functor $G \colon \mathcal{D} \to \mathcal{C}$ and natural isomorphisms

$$\epsilon_{\mathcal{C}} \colon G \circ F \to \mathrm{id}_{\mathcal{C}} \ \text{ and } \ \epsilon_{\mathcal{D}} \colon F \circ G \to \mathrm{id}_{\mathcal{D}}.$$

Thus, under this definition, we do not require that for an object C of $\mathcal{C}$ we have $G(F(C)) = C$, but there should be a natural isomorphism between C and $G(F(C))$.

Definition 1.3.9. Let $F \colon \mathcal{C} \to \mathcal{D}$ be a functor. We say that F is *essentially surjective* if for every object D of $\mathcal{D}$ there is an object C of $\mathcal{C}$ such that $D \cong F(C)$.

The following criterion is straightforward, but it is a good exercise to make sure that one is not lost in definitions.

Proposition 1.3.10. *Let $F \colon \mathcal{C} \to \mathcal{D}$ be a functor. Then F is an equivalence if and only if F is fully faithful and essentially surjective.*

Example 1.3.11.

(a) The double dual functor $v^2\colon \mathrm{Vect}_k^* \to \mathrm{Vect}_k^*$ is not an isomorphism, but is an equivalence of categories between Vect_k^* and itself.

(b) Let k-Alg be the category of algebras over the field k, and k-Alg_{rf} the subcategory of finitely generated, reduced k-algebras. Let Aff_k be the category of affine algebraic varieties over k, with regular maps as morphism. Then, as discussed in Theorem A.9.4, there is a correspondence between these two categories. Namely, let $V \subset \mathbb{A}^n(k)$ be an affine variety—to V we can attach the k-algebra

$$R(V) := k[x_1, \dots, x_n]/I(V),$$

where $I(V)$ is the ideal of polynomials vanishing on V. In the other direction, if A is a finitely generated, reduced, k-algebra, we can write

$$A \cong k[x_1, \dots, x_n]/I$$

for some reduced ideal I, and then to A we can attach the zero locus $V(I) \subset \mathbb{A}^n(k)$.

In fact, there are two contravariant morphisms $\mathrm{Aff}_k \to k$-Alg_{rf} and k-$\mathrm{Alg}_{rf} \to \mathrm{Aff}_k$ that realize an equivalence of categories between k-Alg_{rf} and Aff_k^{op}.

(c) Let L/K be a Galois extension of fields. The Galois correspondence [**Fer20**, Section A.5] is an order inverting bijection between subfields of L containing K and subgroups of the Galois group $Gal(L/K)$.

Both sets are partially ordered, and thus can be regarded as categories as in Example 1.1.4(k). In this way, the Galois correspondence becomes an isomorphism of categories between the order category of intermediate extensions of L/K and the opposite of the order category of subgroups of $Gal(L/K)$.

(d) A *Boolean ring* A is a ring such that $a^2 = a$ for all $a \in A$. Given a Boolean ring, we obtain a Boolean algebra (see Example 1.1.4(m)) with the same underlying set, with operations

$$a \wedge b = a \cdot b \text{ and } a \vee b = a + b + a \cdot b,$$

where $\top = 1$ and $\bot = 0$. Conversely, starting from a Boolean algebra A, we get a Boolean ring on the same underlying set by declaring

$$a \cdot b = a \wedge b \text{ and } a + b = (a \vee b) \wedge \neg(a \wedge b)$$

(check this!). This correspondence gives an equivalence of categories between the category of Boolean rings and that of Boolean algebras.

(e) Let S be a compact Hausdorff space, and $C(S)$ be the algebra of $\mathbb{C}$-valued continuous functions. It is easy to prove that $C(S)$, with the sup topology, is a C^*-algebra with unit (see [**Fol94**, Section 1.1]), and each continuous function $f\colon S \to T$ induces a pullback homomorphism $f^*\colon C(T) \to C(S)$.

On the other hand, to each C^*-algebra A with unit we can associate its spectrum $\sigma(A)$, consisting of homomorphism $A \to \mathbb{C}$. The spectrum inherits a topology by its inclusion into dual Banach space A^*, endowed with weak* topology, and it turns out that $\sigma(A)$ is compact and Hausdorff.

If we denote by CHTop the category of compact Hausdorff spaces and C^*-Alg the category of C^*-algebras, these constructions define two contravariant functors $F\colon \text{CHTop} \to C^*$-Alg and $G\colon C^*$-Alg $\to$ CHTop. The Gelfand–Naimark theorem ([**Fol94**, Theorem 1.20]) states that F and G are equivalences between CHTop^{op} and C^*-Alg.

A similar equivalence of categories exists between (the opposite category of) locally compact, Hausdorff spaces and non unital C^*-algebras.

1.4. Limits

In many contexts in mathematics, the notion of a universal property arises. In this section, we want to capture this phenomenon with a general definition, and generalize it through the use of *limits*. As a guiding example, we will use the notion of tensor product.

Let A be a ring, and M, N two A-modules. The tensor product $M \otimes_A N$, or simply $M \otimes N$, comes equipped with an A-bilinear map

$$\otimes\colon M \times N \to M \otimes N.$$

Moreover, $\otimes$ is *universal* among such maps. Namely, if P is any other A-module equipped with a bilinear map

$$b\colon M \times N \to P,$$

there exists a unique morphism $M \otimes N \to P$ of A-modules that makes the diagram

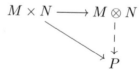

commute.

We can rephrase this in categorical terms. Let $\mathcal{C}$ be the category whose object are A-modules P equipped with a bilinear map $b\colon M \times N \to P$. A morphism between $b\colon M \times N \to P$ and $b'\colon M \times N \to P'$ is given by a map $P \to P'$ such that the diagram

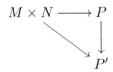

is commutative. Then we know that $M \otimes N$ is a object of $\mathcal{C}$, and moreover it has a unique morphism towards any other object of $\mathcal{C}$. This is a notion that deserves a name.

Definition 1.4.1. Let $\mathcal{C}$ be a category, C an object of $\mathcal{C}$. We say that C is *initial* if for any object D of $\mathcal{C}$, there is a unique map $C \to D$. Symmetrically, C is *final* if for any object D of $\mathcal{C}$, $\mathrm{Hom}(D, C)$ consists of a single morphism. An object that is either initial or final is called *universal*.

Clearly, an initial object of $\mathcal{C}$ corresponds to a final object of $\mathcal{C}^{op}$; and conversely. With this notion, the universality of the tensor product can be rephrased by saying that $M \otimes N$ is initial in the category of A-modules equipped with bilinear maps from $M \times N$. This is not a unique situation by any means, and we can find many other example of universal objects.

Example 1.4.2.

(a) The product $M \times N$ is an A-module equipped with two projection maps $\pi_1\colon M \times N \to M$ and $\pi_2\colon M \times N \to M$. If P is any other A-module equipped with two maps $f\colon P \to M$ and $g\colon P \to N$, the pair (f, g) defines a morphism $P \to M \times N$. This makes $M \times N$ a final object of the category of A-modules equipped with maps towards M and N.

(b) Similarly, the direct sum $M \oplus N$ is initial among A-modules equipped with maps *from* M and N.

(c) The above situation is not unique to modules. For instance, in the category Top of topological spaces, the product $S \times T$ of two spaces is endowed with two maps $S \times T \to S$ and $S \times T \to T$, and is in fact a final object of the corresponding category. The initial object of the category of spaces equipped with maps *from* S and T is instead the disjoint union $S \sqcup T$.

(d) Polynomial rings also satisfy a universal property. Namely, let A and B be rings. Given a morphism $f\colon A \to B$ and an element $b \in B$, there exists a unique morphism $f'\colon A[x] \to B$ such that $f'|_A = f$ and $f'(x) = b$.

Let $\mathcal{C}$ be the category whose object are triples (B, b, g), where B is a ring, $b \in B$, and $g\colon A \to B$. A morphism in $\mathcal{C}$ is a morphism

of rings that preserves the distinguished element and commutes
with the morphism from A. The polynomial ring $A[x]$ has a distinguished element x and a natural map $\iota\colon A \to A[x]$, hence the
triple $(A[x], x, \iota)$ is an object of $\mathcal{C}$. The universal property of rings
amounts to saying that this triple is an initial object in $\mathcal{C}$.

(e) Let A be a ring, $I \subset A$ an ideal. The quotient ring A/I is initial
in the category of A-algebras $f\colon A \to B$ such that $f(I) = 0$. Here
morphisms are homomorphisms of A-algebras.

(f) Similarly, let A be a ring and $S \subset A$ a multiplicative set. Then the
localization $S^{-1}A$ is initial in the category of A-algebras $f\colon A \to B$
such that $f(S) \subset B^*$.

Objects characterized by a universal property are essentially uniquely
determined. We can now make this precise.

Proposition 1.4.3. *Let $\mathcal{C}$ be a category with an initial object C. If C' is
any other initial object, there is a* unique *isomorphism $C \to C'$ in $\mathcal{C}$.*

Dually, the same result holds for final objects.

Proof. The fact that C and C' are initial defines uniquely two maps, $f\colon C \to
C'$ and $g\colon C' \to C$. The composition $g \circ f$ is the unique element of $\mathrm{Hom}(C, C)$,
which is id_C, and similarly $f \circ g = \mathrm{id}_{C'}$.   □

There is a special case of universal objects, that can be seen at the same
time as a generalization. For this, we recall from [**Fer20**, Section 7.4] the
notion of direct and inverse limits. In the setting described in volume 1, we
are given a partially ordered set I, which we think as indices. The set is
directed, which means that for every $i, j \in I$ we can find $k \in I$ such that
$i \le k$ and $j \le k$. We are also given A-modules M_i for each $i \in I$, and for
each pair $i < j$, a map $M_i \to M_j$. These maps are called *compatible* if for
every $i < j < k$ the diagram

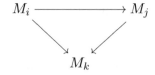

commutes. This datum is called a *direct system*.

Associated to the direct system, there is an A-module $M = \varinjlim M_i$,
called the *direct limit* of the M_i, equipped with maps $M_i \to M$ such that

for every $i < j$ the diagram

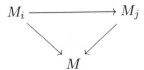

commutes. Moreover, in the language just introduced, M is initial among all modules equipped with compatible maps from the M_i.

Example 1.4.4. Lest this seems excessively abstract, we recall that this is the process one learns when summing fractions. Here the index set is $\mathbb{Z}_+$, with the partial order given by divisibility. For every $n > 0$, the additive group

$$M_n = \left\{ \frac{a}{n} \mid a \in \mathbb{Z} \right\}$$

is isomorphic to $\mathbb{Z}$ itself. Whenever m divides n, there is a natural map $M_m \to M_n$. In order to sum two element a/m with a b/n, we find a common multiple $k = k_m m = k_n n$, and we identify $\frac{a}{m} = \frac{ak_m}{k}$ and $\frac{b}{n} = \frac{bk_n}{k}$, then we perform the usual sum inside M_k. This exhibits the additive group of the rationals $\mathbb{Q}$ as the direct limit of the direct system $\{M_n\}$.

Of course, there is a dual notion of *inverse limit*. For this, we need an inverse system, that is, for all $i < j$ we will have a map $M_j \to M_i$, and these maps will be compatible. Given an inverse system, its inverse limit $\varprojlim M_i$ is an A-module M equipped with maps $M \to M_i$ that are compatible with the inverse system. Moreover, M is final among all A-modules equipped with such maps.

Example 1.4.5. The typical example here is the ring of p-adic integers, which (additively) can be constructed as the inverse limit of the system of groups $\mathbb{Z}/p^n\mathbb{Z}$, equipped with their projection maps

$$\mathbb{Z}/p^n\mathbb{Z} \to \mathbb{Z}/p^m\mathbb{Z}$$

whenever $m < n$.

In the light of our categorical language, a few constraints that we have added look a little artificial. For one thing, there is no reason to limit ourselves to A-modules. The same definitions would work for sets, groups, rings, and in fact any category. Less obviously, the requirement of an ordered set of indices is also artificial. In fact, let $\mathrm{Ord}(I)$ be the order category associated to I. The datum of a direct system in $\mathcal{C}$ is exactly the same as that of a covariant functor $\mathrm{Ord}(I) \to \mathcal{C}$. All complicated requirements about the commutativity of diagrams are simply subsumed by the commutativity of diagrams required by functoriality. Even better, an inverse system can be seen simply as a contravariant functor $\mathrm{Ord}(I) \to \mathcal{C}$. This expresses quite clearly the relation between the two notions.

When we take this point of view, there is nothing special in taking an order category as the source of the functor. We might as well consider any category $\mathcal{I}$ as the starting point. We still think of $\mathcal{I}$ as an indexed family of objects with arrows connecting them. A functor $\mathcal{I} \to \mathcal{C}$ is going to single out some objects of $\mathcal{C}$, together with some morphisms between them. The exact shape of this structure depends on how we choose $\mathcal{I}$.

Definition 1.4.6. Let $\mathcal{I}$ and $\mathcal{C}$ be two categories, $F\colon \mathcal{I} \to \mathcal{C}$ a functor. Consider an object C of $\mathcal{C}$ endowed with morphisms

$$f_i\colon C \to F(i)$$

for any object i of $\mathcal{I}$, in such a way that for all morphisms $m \in \mathrm{Hom}(i,j)$ the triangle

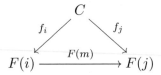

commutes. Such an object (together with the system of maps) is called a *cone* over F. The cone C is called the *limit* of F if C is universal with respect to this property, that is, for all other objects D of $\mathcal{C}$ with morphisms

$$g_i\colon D \to F(i)$$

there is a unique map $u \in \mathrm{Hom}(D,C)$ such that $g_i = f_i \circ u$ for all objects i of $\mathcal{I}$.

Remark 1.4.7. As hinted above, there is a relation between the concept of limit and that of universal object. On the one hand, an initial object of $\mathcal{C}$ is a special case of a limit, in fact the limit of the empty functor from the empty category. Vice versa, the definition of limit makes it clear that a limit is the initial object in a suitable category.

Namely, given a functor $F\colon \mathcal{I} \to \mathcal{C}$, we can consider a related category $\mathcal{C}_F$ of cones over F. Objects of $\mathcal{C}_F$ are objects C of $\mathcal{C}$ endowed with morphisms $f_i\colon C \to F(i)$ for all objects i of $\mathcal{I}$. Morphisms in $\mathcal{C}_F$ are just ordinary morphisms in C that commute with the maps f_i. The limit of F is then a final object of $\mathcal{C}_F$. By Proposition 1.4.3, it follows that the limit of F, if it exists, is unique up to a unique isomorphisms. By a slight a abuse of notation, we can give it a name and denote it $\lim F$.

Clearly there is also a dual notion.

Definition 1.4.8. Let $F\colon \mathcal{I} \to \mathcal{C}$ be a functor. Consider an object C of $\mathcal{C}$ endowed with morphisms

$$f_i\colon F(i) \to C$$

for any object i of $\mathcal{I}$, in such a way that for all morphisms $m \in \mathrm{Hom}(i,j)$ the triangle

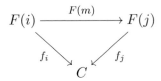

commutes. Such an object (together with the system of maps) is called a *cocone* over F. The cocone C is called the *colimit* of F if C is universal with respect to this property (write this explicitly!).

Again, the colimit of F, if it exists, is unique up to a unique isomorphism, and we can denote it $\mathrm{colim}\, F$.

Example 1.4.9.

(a) The direct limit of a direct system is a colimit in this sense, where the category $\mathcal{I}$ of indices is just an order category. Dually, an inverse limit is a limit.

(b) Consider the category $\mathcal{I}$ with two objects $\bullet_1$ and $\bullet_2$, and only the identity morphisms. A functor $\mathcal{I} \to \mathcal{C}$ is given by the choice of two objects C_1, C_2 of $\mathcal{C}$. The limit of F is an object C with two maps $\pi_1\colon C \to C_1$ and $\pi_2\colon C \to C_2$, universal with respect to these maps. This is called the *product* of C_1 and C_2, and usually denoted $C_1 \times C_2$.

The colimit of F is an object C with two maps $\iota_1\colon C_1 \to C$ and $\iota_2\colon C_2 \to C$, which is again universal. This is called the *coproduct* of C_1 and C_2, and denoted $C_1 \sqcup C_2$.

In the categories Set and Top, the coproduct is just the disjoint union. Instead, in algebraic categories such as Ring, Mod_A, and Vect_k, the coproduct is the direct sum, which in this case agrees with the product.

(c) More generally, let $\mathcal{I}$ be a *discrete* category, that is, a category whose only morphisms are the identities. The limit of a functor $\mathcal{I} \to \mathcal{C}$ is called the *product* of the objects indexed by $\mathcal{I}$. Its colimit is called the *coproduct* of these objects. We will denote the product of F as

$$\prod_{i \in \mathcal{I}} F(i)$$

and the coproduct as

$$\coprod_{i\in\mathcal{I}} F(i).$$

(d) Consider the category $\mathcal{I}$ with three objects and two nonidentity morphisms depicted below:

A functor $F\colon \mathcal{I} \to \mathcal{C}$ amounts to the diagram

$$\begin{array}{c} A \\ \downarrow f \\ B \xrightarrow{\;g\;} X. \end{array}$$

The limit of F, if it exists, is an object sitting in a commutative square

$$\begin{array}{ccc} E & \longrightarrow & A \\ \downarrow & & \downarrow \\ B & \longrightarrow & X \end{array}$$

and universal with respect to such diagrams. In this case, E is called the *fibered product* of f and g, or the pullback of f via g.

For instance, in Set, the fibered product of f and g is

$$\{(a,b) \in A \times B \mid f(a) = g(b)\},$$

that is the preimage of the diagonal $\Delta \subset X \times X$ via $f \times g$. The same construction gives a fibered product in many other categories, such as Top or Mod_A.

(e) Dually, a functor from

amounts to the diagram

$$\begin{array}{c} X \xrightarrow{\;f\;} A \\ \downarrow g \\ B \end{array}$$

in $\mathcal{C}$. The colimit of this functor is called the *fibered coproduct* of f and g, or the *pushout* of f along g.

(f) Consider the category $\mathcal{I}$ with two objects and two nonidentity arrows depicted below

$$\bullet \rightrightarrows \bullet.$$

A functor $\mathcal{I} \to \mathcal{C}$ amounts to a diagram

$$A \underset{g}{\overset{f}{\rightrightarrows}} X$$

in $\mathcal{C}$. The limit of such a functor is called the *equalizer* of f and g, denoted by $\text{eq}(f, g)$. Clearly, equalizers are a special case of fibered products, in the case where A and B coincide. In categories such as Mod_A or Grp, when g is the 0 morphism, the equalizer of f and g is the kernel of f.

The colimit of this functor is called the *coequalizer* of f and g, and is a special case of the fibered coproduct. When g is the 0 morphism, the coequalizer of f and g is the cokernel of f.

Many of these constructions are familiar from various branches of mathematics—for instance, the pullback operation for vector bundles, or kernels and cokernels of homomorphisms—but the language of limits allows us to view them through a common lens.

We already know that limits and colimits enjoy a uniqueness property. Existence is not always guaranteed though. For instance, there is in general no product in the category of fields. In fact, all homomorphisms of fields are injective. This implies that the product of two fields k_1 and k_2 should be a common subfield of both, something that is impossible if for instance k_1 and k_2 have different characteristics. It is then interesting to investigate, for a given category, what kind of limits and colimits exist.

Definition 1.4.10. A category $\mathcal{C}$ is called *complete* if every functor $\mathcal{I} \to \mathcal{C}$, where $\mathcal{I}$ is any small category, has a limit. It is called *cocomplete* if every functor $\mathcal{I} \to \mathcal{C}$, again with $\mathcal{I}$ small, has a colimit.

Remark 1.4.11. The restriction on the size of $\mathcal{I}$ is crucial here. In fact, a category $\mathcal{C}$ that admits all limits (even all products) indexed by categories of the same size as $\mathcal{C}$ is isomorphic to a preorder, that is, $\text{Hom}_\mathcal{C}(A, B)$ has at most one member for all $A, B \in \mathcal{C}$. The argument is simple, although one has to be a little careful about the underlying foundation to formalize it, and it appears as Theorem 2.1 in [**Shu08**] (see also Exercise 37).

The next result greatly simplifies the question of the existence of limits and colimits. We say that $\mathcal{C}$ has arbitrary products if for all small, discrete categories $\mathcal{I}$ and all functors $F\colon \mathcal{I} \to \mathcal{C}$, the limit of F exists. We say that

$\mathcal{C}$ has equalizers if all functors $\mathcal{I} \to \mathcal{C}$, where $\mathcal{I}$ is the category

$$\bullet \rightrightarrows \bullet.$$

have a limit. Dually, we define the notion of $\mathcal{C}$ having arbitrary coproducts and coequalizers.

Theorem 1.4.12. *Let $\mathcal{C}$ be a category, and assume that $\mathcal{C}$ has arbitrary products and equalizers. Then $\mathcal{C}$ is complete. Dually, if $\mathcal{C}$ has arbitrary coproducts and coequalizers, $\mathcal{C}$ is cocomplete.*

Proof. It is enough to prove the results for limits. Let $\mathcal{I}$ be any small category and $F \colon \mathcal{I} \to \mathcal{C}$ a functor. Define products

$$C := \prod_{i \in \mathcal{I}} F(i) \text{ and } D := \prod_{\substack{i,j \in \mathcal{I} \\ f \in \mathrm{Hom}(i,j)}} F(j).$$

Notice that the first one is the product over all object in $\mathcal{I}$, and the second one the product over all morphisms in $\mathcal{I}$. In any case, they are both products of objects of the form $F(i)$ for objects i of $\mathcal{I}$. There are two natural morphisms $C \to D$.

Every object $F(j)$ in the second product appears also in the first product, hence we have the corresponding projection map

$$C = \prod_{i \in \mathcal{I}} F(i) \to F(i).$$

Applying the universal property of products, we get a morphism $t \colon C \to D$.

Similarly, for every morphism $f \in \mathrm{Hom}(i, j)$ we have a projection map from C that we can compose with $F(f)$ to obtain

$$C = \prod_{i \in \mathcal{I}} F(i) \longrightarrow F(i) \xrightarrow{F(f)} F(j).$$

Using again the universal property, we obtain another morphism $s \colon C \to D$.

We now let E be the equalizer of s and t, and check that E is a limit of F. By construction, E is equipped with a morphism to C, hence, composing with projections, with a morphism $e_i \colon E \to F(i)$ for all $i \in \mathcal{I}$. If $f \in \mathrm{Hom}(i, j)$ is any morphism, the triangle

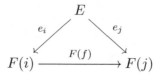

commutes, by virtue of E being an equalizer. Finally, if M is any other object equipped with morphisms $m_i \colon M \to F(i)$, the universal property gives a morphism $m \colon M \to C$. The fact that the morphisms m_i are compatible with F says that the compositions $s \circ m$ and $t \circ m$ are the same, hence we get a map $M \to E$ by the universal property of equalizers. □

Corollary 1.4.13. *The categories* Set, Grp, Ab, Mod$_A$, *and* Vect$_k$ *are complete and cocomplete.*

Proof. Each of these categories has arbitrary products. We have already remarked that Set has fibered products (hence equalizers), and it is easy to check that the set theoretic equalizers of two maps f, g in each of the categories Grp, Ab, Mod$_A$, and Vect$_k$ is a subgroup (respectively submodule, subspace). The proof that these categories are cocomplete is similar, but for categories Grp, Ab, Mod$_A$, and Vect$_k$, the coequalizer of f and g is simply the cokernel of $f - g$ (or $x \mapsto f(x)g(x)^{-1}$ in the case of Grp). □

When defining direct and inverse limits, we required the index set to be directed. We can generalize this requirement for arbitrary limits and colimits.

Definition 1.4.14. Let $\mathcal{I}$ be a category. We say $\mathcal{I}$ is *filtered* if

(1) $\mathcal{I}$ is not empty;
(2) for every pair $i, j \in \mathcal{I}$ there exist an object $k \in \mathcal{I}$ such that $\mathrm{Hom}(i, k)$ and $\mathrm{Hom}(j, k)$ are nonempty; and
(3) given two objects $i, j \in \mathcal{I}$ and two morphisms $f, g \in \mathrm{Hom}(i, j)$ there exists an object k and a morphism $h \in \mathrm{Hom}(j, k)$ such that $h \circ f = h \circ g$.

Dually, we say $\mathcal{I}$ is *cofiltered* if $\mathcal{I}^{op}$ is filtered.

We have corresponding notions for limits

Definition 1.4.15. Let $F \colon \mathcal{I} \to \mathcal{C}$ be a functor. If $\mathcal{I}$ if filtered, we say that colim F is a *filtered colimit*. If $\mathcal{I}$ if cofiltered, we say that lim F is a *cofiltered limit*.

The main reason for introducing filtered colimits is that they commute with finite limits with values in Set. To make this notion precise, let us start with a bifunctor

$$\mathcal{I} \times \mathcal{J} \to \mathcal{C},$$

where we regard both $\mathcal{I}$ and $\mathcal{J}$, which we assume to be small, as categories of indices. For all $i \in \mathcal{I}$, we can define

$$F_i = F(i, -) \colon \mathcal{J} \to \mathcal{C},$$

and similarly for $j \in \mathcal{J}$ we have

$$F^j = F(-, j) \colon \mathcal{I} \to \mathcal{C}.$$

Assume that for all i, $\lim F_i$ exists, and for all j, $\operatorname{colim} F^j$ exists (for instance, $\mathcal{C}$ is complete and cocomplete). Then we have the morphisms

$$\lim F_i \to F(i,j) \to \operatorname{colim} F^j.$$

By the universal property, this induces a morphism

$$(1.4.1) \qquad \lambda \colon \operatorname*{colim}_{i \in \mathcal{I}} \lim_{j \in \mathcal{J}} F \to \lim_{j \in \mathcal{J}} \operatorname*{colim}_{i \in \mathcal{I}} F,$$

again assuming these limits and colimits exist. We can now state the main result on filtered colimits.

Theorem 1.4.16. *Let $F \colon \mathcal{I} \times \mathcal{J} \to \mathrm{Set}$ be a bifunctor, where $\mathcal{I}$ and $\mathcal{J}$ are small. If $\mathcal{I}$ is filtered and $\mathcal{J}$ is finite, the natural morphism λ in (1.4.1) is an isomorphism.*

By a finite category, we just mean a category with finitely many objects and morphisms. Our approach will follow [**Bor94**, Section 2.13]. To start proving the theorem, we will need an explicit description of limits and colimits with values in Set, similarly to the explicit description that one derives for direct and inverse limits (see for instance [**Fer20**, Section 7.4]).

Given a finite category $\mathcal{J}$ and a functor $F \colon \mathcal{J} \to \mathrm{Set}$, we can explicitly describe the set $\lim F$ as

$$(1.4.2) \qquad \lim F = \{(x_j \in F(j))_{j \in \mathcal{J}} \mid F(f)(x_i) = x_j \text{ for all } f \colon i \to j\}.$$

The set $\lim F$ thus defined is clearly a cone over F via the projections $(x_j)_{j \in \mathcal{J}} \to x_j$. Moreover, for every other cone C with functions $g_j \colon C \to F(j)$, there is a natural function $C \to \lim F$ which is defined by g_j on the j-th coordinate.

The description of a colimit is in general more complicated, but it simplifies in the case of a filtered colimit. Namely, given $F \colon \mathcal{I} \to \mathrm{Set}$, where $\mathcal{I}$ is small and filtered,

$$(1.4.3) \qquad \operatorname*{colim} F = \bigsqcup_{i \in \mathcal{I}} F(i)/\sim,$$

where $\bigsqcup$ denotes disjoint union and $\sim$ is the equivalence relation given by $x_i \in F(i) \sim x_j \in F(j)$ if and only if there are $f \colon i \to k$ and $g \colon j \to k$ such that $F(f)(i) = F(g)(j)$. That $\sim$ is indeed an equivalence relation follows from the hypothesis that $\mathcal{I}$ is filtered (check this!). Knowing this, the quotient by $\sim$ makes (1.4.3) into a cocone, which is clearly universal.

Before turning to the proof, we need one more fact.

Lemma 1.4.17. *Let $\mathcal{J}$ be a finite category and $\mathcal{I}$ a filtered category. Then every functor $F \colon \mathcal{J} \to \mathcal{I}$ has a cocone.*

Proof. By induction from Definition 1.4.14(2), we find an object $i \in \mathcal{I}$ endowed with maps $f_j \colon F(j) \to i$ for all $j \in \mathcal{J}$. The object i may not be a cocone: if $F(j) = F(j')$ for some $j \neq j' \in \mathcal{J}$, it may happen that f_j and $f_{j'}$ are not equal, but we can remedy this by finitely many applications of Definition 1.4.14(3). □

Proof of Theorem 1.4.16. Using (1.4.2) and (1.4.3), we can describe λ in (1.4.1) in set theoretic terms. An element of $\mathrm{colim}_{i \in \mathcal{I}} \lim_{j \in \mathcal{J}} F$ is an equivalence class $\overline{(x_j)}$, where the tuple (x_j) has components $x_j \in F(i,j)$ for some fixed i. An element of $\lim_{j \in \mathcal{J}} \mathrm{colim}_{i \in \mathcal{I}} F$ is a tuple $(\overline{x_j})$, which has a component $\overline{x_j}$ for each $j \in \mathcal{J}$, which is itself an equivalence class of elements $x_j \in F(i,j)$ for some fixed i. Moreover, the components of a tuple (x_j) belong to the limit if and only if they satisfy the consistency requirement that $F(f)(x_i) = x_j$ for all $f \colon i \to j$. The map λ is then given by

$$\lambda\left(\overline{(x_j)}\right) = (\overline{x_j}).$$

To check that λ is injective, assume that $\lambda\left(\overline{(x_j)}\right) = \lambda\left(\overline{(y_j)}\right)$. We can take for representatives two tuples (x_j) and y_j where $x_j \in F(i_1, j)$ and $y_j \in F(i_2, j)$. Because of Definition 1.4.14(2), up to replacing (x_j) and (y_j) with a different representative of the same class, we can assume that $i_1 = i_2 = i$. Since $\lambda\left(\overline{(x_j)}\right) = \lambda\left(\overline{(y_j)}\right)$, we have $\overline{x_j} = \overline{y_j}$ for all j. In other words, for all j, we have a map $f_j \colon i \to k_j$ such that $F^j(f_j)(x_j) = F^j(f_j)(y_j)$. We can apply Lemma 1.4.17 to the finite diagram consisting of all maps $i \to k_j$. Then we find a common $f \colon i \to k$ such that $F^j(f)(x_j) = F^j(f)(y_j)$ for all j simultaneously, i.e., $\overline{(x_j)} = \overline{(y_j)}$.

To check that λ is surjective, let us take a tuple $(\overline{x_j})$. Again, we choose a representative $x_j \in F(i_j, j)$ for all j. Lemma 1.4.17 gives us a morphism $i_j \to i$ such that, if y_j is the image of x_j in $F(i,j)$, the tuple (y_j) is consistent, so $(y_j) \in \lim_{j \in \mathcal{J}} F_i$. Then the tuple (y_j) satisfies $\lambda\left(\overline{(y_j)}\right) = (\overline{x_j})$. □

Remark 1.4.18. Theorem 1.4.16 applies to the category Ab and other algebraic categories as well. To see this, assume that $F \colon \mathcal{I} \times \mathcal{J} \to$ Ab is a bifunctor. In (1.4.2), the subset defining $\lim F$ is a subgroup of the product of all $F(j)$. In (1.4.3), the group operation is well defined on the quotient, using that $\mathcal{I}$ is filtered. The map λ in (1.4.1) is then a group homomorphism, and by the result above it is bijective, hence it is an isomorphism.

1.5. Adjoint pairs

Let $\mathcal{C}, \mathcal{D}$ be two categories, with functors $F \colon \mathcal{C} \to \mathcal{D}$ and $G \colon \mathcal{D} \to \mathcal{C}$. In this section, we investigate a particular relation that can hold between Hom sets involving objects in the image of F and objects in the image of G. To express this, notice that F defines a functor

$$\mathrm{Hom}_{\mathcal{D}}(F-,-)\colon \mathcal{C} \times \mathcal{D}^{op} \to \mathrm{Set}\,.$$

Here $\mathcal{C} \times \mathcal{D}^{op}$ is the product category of $\mathcal{C}$ and $\mathcal{D}^{op}$. Similarly, G defines a functor

$$\mathrm{Hom}_{\mathcal{C}}(-,G-)\colon \mathcal{C} \times \mathcal{D}^{op} \to \mathrm{Set}\,.$$

To unravel the definition, $\mathrm{Hom}_{\mathcal{D}}(F-,-)$ is defined on objects as

$$\mathrm{Hom}_{\mathcal{D}}(F-,-)(C,D) := \mathrm{Hom}_{\mathcal{D}}(F(C),D),$$

while

$$\mathrm{Hom}_{\mathcal{C}}(-,G-)(C,D) := \mathrm{Hom}_{\mathcal{D}}(C,G(D)).$$

We leave it to the reader to find the action of these two functors on morphisms.

Definition 1.5.1. A natural isomorphism Ψ between $\mathrm{Hom}_{\mathcal{D}}(F-,-)$ and $\mathrm{Hom}_{\mathcal{C}}(-,G-)$ is called an *adjunction* between F and G. In this case, we say that (F,G) form an *adjoint pair*, or that F is left-adjoint to G and G is right-adjoint to F. We also write $F \dashv G$ (or $G \vdash F$) to indicate that F is left adjoint of G. We sometimes denote the adjoint pair by the notation

$$F\colon \mathcal{C} \rightleftarrows \mathcal{D}\colon G.$$

Before giving examples, we are going to expand the definition a little bit. The notion of adjunction requires a bijective correspondence

$$\Psi_{C,D}\colon \ \mathrm{Hom}_{\mathcal{D}}(F(C),D) \cong \mathrm{Hom}_{\mathcal{C}}(C,G(D))$$

for all objects $C \in \mathcal{C}$ and $D \in \mathcal{D}$. Moreover, this correspondence has to be natural both in C and D. This means that a morphism $f \in \mathrm{Hom}_{\mathcal{C}}(C_1,C_2)$ induces a commutative square

$$\begin{array}{ccc}
\mathrm{Hom}_{\mathcal{D}}(F(C_2),D) & \xrightarrow{\ \Psi_{C_2,D}\ } & \mathrm{Hom}_{\mathcal{C}}(C_2,G(D)) \\
{\scriptstyle F(c_f)}\Big\downarrow & & \Big\downarrow{\scriptstyle c_f} \\
\mathrm{Hom}_{\mathcal{D}}(F(C_1),D) & \xrightarrow{\ \Psi_{C_1,D}\ } & \mathrm{Hom}_{\mathcal{C}}(C_1,G(D)),
\end{array}$$

where $c_f(g) = g \circ f$ for any morphism g with source C_2. A similar commutative square exists for any morphism $g \in \mathrm{Hom}_{\mathcal{D}}(D_1,D_2)$.

For a fixed pair F,G we can employ a more straightforward notation. Given objects $C \in \mathcal{C}$ and $D \in \mathcal{D}$, and a morphism $f \in \mathrm{Hom}_{\mathcal{D}}(F(C),D)$, we will denote

$$f^{\flat} = \Psi_{C,D}(f) \in \mathrm{Hom}_{\mathcal{C}}(C,G(D))$$

the corresponding morphism in $\operatorname{Hom}_{\mathcal{C}}(C, G(D))$. Similarly, a morphism $g \in \operatorname{Hom}_{\mathcal{C}}(C, G(D))$ defines

$$g^{\#} = \Psi_{C,D}^{-1}(g) \in \operatorname{Hom}_{\mathcal{D}}(F(C), D).$$

In particular, the identity $\operatorname{id}_{F(C)}$ defines

$$\eta_C = \operatorname{id}_{F(C)}^{\flat} \in \operatorname{Hom}_{\mathcal{C}}(C, G(F(C))),$$

which is called the *unit* of the pair F, G. Symmetrically,

$$\epsilon_D = \operatorname{id}_{G(D)}^{\#} \in \operatorname{Hom}_{\mathcal{D}}(F(G(D)), D)$$

is called the *counit* of the pair F, G.

Putting together all the units gives a natural transformation $\eta \colon \operatorname{id}_{\mathcal{C}} \to G \circ F$. Dually, we also have a natural transformation $\epsilon \colon F \circ G \to \operatorname{id}_{\mathcal{D}}$. We still call these natural transformations the unit and counit of the pair F, G.

Given a morphism $f \in \operatorname{Hom}_{\mathcal{D}}(F(C), D)$, the identity $f = f \circ \operatorname{id}_{F(C)}$ implies that $f^{\flat} = G(f) \circ \eta_C$ by naturality. This shows that the unit η is enough to recover the bijection $\Psi_{C,D}$ for all objects C, D, hence the adjunction. Again, we can see this from the other side: the identity $g^{\#} = \epsilon_D \circ F(g)$ shows that the counit ϵ is enough to determine the adjunction.

Example 1.5.2.

(a) Let V, W be vector spaces over the field k, $\mathcal{B}$ a basis of V. A linear map $V \to W$ is uniquely determined by a set theoretic function $\mathcal{B} \to W$. We can rephrase this by saying that we have an adjoint pair

$$B \colon \operatorname{Set} \rightleftarrows \operatorname{Vect}_k \colon F,$$

where F is the forgetful functor and B is the functor which associates to a set S the vector space formally generated over k by S.

(b) In the same way, there is an adjoint pair

$$G \colon \operatorname{Set} \rightleftarrows \operatorname{Grp} \colon F,$$

where F is the forgetful functor and G is the functor defined by letting $G(S)$ be the free group over S. More generally, for each category $\mathcal{C}$ with a forgetful functor $F \colon \mathcal{C} \to \operatorname{Set}$, we can see the left adjoint of F (if it exists) as the functor constructing free objects in $\mathcal{C}$.

(c) For three sets S, T, U, we have a natural bijective correspondence between $\operatorname{Hom}_{\operatorname{Set}}(S \times T, U)$ and $\operatorname{Hom}_{\operatorname{Set}}(S, \operatorname{Hom}_{\operatorname{Set}}(T, U))$ (can you see why?). This can be rephrased by saying that the two functors of sets $- \times T$ and $\operatorname{Hom}_{\operatorname{Set}}(T, -)$ are an adjoint pair.

(d) Similarly, for modules M, N, P over the ring A we have a natural isomorphism

$$\operatorname{Hom}(M \otimes_A N, P) \cong \operatorname{Hom}(M, \operatorname{Hom}(N, P)),$$

which makes the functors $- \otimes N$ and $\operatorname{Hom}(N, -)$ an adjoint pair over Mod_A.

(e) For a group G, let $\mathbb{Z}[G]$ be the associated group ring. Elements of $\mathbb{Z}[G]$ are formal finite linear combinations $\sum_i a_i g_i$, where $a_i \in \mathbb{Z}$ and $g_i \in G$. The multiplication is inherited from that of G, making $\mathbb{Z}[G]$ into a noncommutative ring with identity. If A is any other ring, a morphism $f \colon \mathbb{Z}[G] \to A$ is uniquely defined by assigning the elements $f(g)$ for $g \in G$, with the only restriction that these have to be invertible elements. In other words, we get a natural bijection

$$\operatorname{Hom}(G, A^*) \cong \operatorname{Hom}(\mathbb{Z}[G], A),$$

making the group ring functor $\mathbb{Z}[-] \colon \operatorname{Grp} \to \operatorname{Ring}$ and the unit functor $-^* \colon \operatorname{Ring} \to \operatorname{Grp}$ into an adjoint pair (here we have used Ring to the denote the category of noncommutative rings).

(f) Let $f \colon A \to B$ be a morphism of rings. Via f, each B-module can be seen as an A-module, so we have a natural forgetful functor $F \colon \operatorname{Mod}_B \to \operatorname{Mod}_A$.

   In the other direction, to every A-module M we can associate the B-module $B \otimes_A M$. If N is a B-module, every morphism of A-modules $g \colon M \to N$ can be extended to a morphism of B-modules $B \otimes_A M \to N$ by sending $b \otimes m$ to $b \cdot g(m)$. This can be rephrased by saying that the functor $B \otimes_A -$ is left adjoint to F. The functor $B \otimes_A -$ is called the *extension of scalars* from A to B.

(g) In the same setting as the previous point, let M be an A-module. If we regard B as an A-module via f, the A-module $\operatorname{Hom}_A(B, M)$ has a natural structure of B-module. Namely, for $b \in B$ and $g \colon B \to M$, define $(b \cdot g)(b') = g(bb')$.

   If N is a B-module and $g \colon N \to M$ a morphism of A-modules, there is a unique associated morphism of B-modules $N \to \operatorname{Hom}_A(B, M)$. This sends n to the morphism $b \to g(bn)$. This operation can be inverted: a morphism of B-modules $h \colon N \to \operatorname{Hom}_A(B, M)$ has an associated morphism of A-modules $g \colon N \to M$ defined by $g(n) = h(n)(1)$. This can be rephrased by saying that the functor $\operatorname{Hom}_A(B, -)$ is *right* adjoint to F. The functor $\operatorname{Hom}_A(B, -)$ is called *coextension of scalars* from A to B.

(h) Let $\mathcal{I}$ be a small category, and assume that $\mathcal{C}$ admits all limits of functors $\mathcal{I} \to \mathcal{C}$, so there is a functor (see Exercise 15)

$$\lim \colon \operatorname{Fun}(\mathcal{I}, \mathcal{C}) \to \mathcal{C}.$$

In the other direction, there is the diagonal functor

$$\Delta \colon \mathcal{C} \to \operatorname{Fun}(\mathcal{I}, \mathcal{C}),$$

which assigns to $C \in \mathcal{C}$ the constant functor $\Delta(C)$. This is defined on $\mathcal{I}$ by sending all objects to C and all morphisms to id_C.

We claim that the limit functor $\lim$ is a right adjoint to Δ. To make this clear, take $C \in \mathcal{C}$ and $F \colon \mathcal{I} \to \mathcal{C}$. An element of $\operatorname{Hom}(\Delta(C), F)$ is a natural transformation. This assigns to every $i \in \mathcal{I}$ a morphism $\Delta(C)(i) = C \to F(i)$, and to every morphism $f \colon i \to j$, a morphism $F(i) \to F(j)$ (commutativity follows naturally). Thus, an element of $\operatorname{Hom}(\Delta(C), F)$ is just a way to see C as a cone over F. For each such diagram, the universal property gives a unique morphism $C \to \lim F$. This defines a bijection between $\operatorname{Hom}(\Delta(C), F)$ and $\operatorname{Hom}(C, \lim F)$, which is easily checked to be natural in both F and C.

In the examples, we have silently implied that the adjoint functor is unique. As usual, this is because it is universal in an appropriate category.

Proposition 1.5.3. *Let $F \colon \mathcal{C} \to \mathcal{D}$ be a functor and $G, G' \vdash F$ two right adjoints of F. Then there is a natural isomorphism $G \cong G'$. Similarly, any two left adjoint of a fixed functor are naturally isomorphic.*

Proof. If G is right adjoint to F, there is a counit ϵ, which is a natural transformation $\epsilon \colon F \circ G \to \operatorname{id}_\mathcal{D}$. Let $\mathcal{F}$ be the category whose objects are pairs (H, ϕ), where $H \colon \mathcal{D} \to \mathcal{C}$ is a functor and $\phi \colon F \circ H \to \operatorname{id}_\mathcal{D}$ is a natural transformation. A morphism between (H, ϕ) and (H', ϕ') is a natural transformation $\sigma \colon H \to H'$ such that the triangle

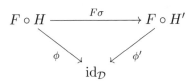

commutes. The definition of right adjoint implies that (G, ϵ) is a terminal object of $\mathcal{F}$. Namely, given a pair (H, ϕ), for each $D \in \mathcal{D}$ we have morphism $\phi_D \colon F(H(D)) \to D$. By adjunction, we get a map $\phi_D^\flat \colon H(D) \to G(D)$, and these maps fit together into a natural transformation $H \to G$ which is the required morphism in $\mathcal{F}$. Being terminal, the pair (G, ϵ) is unique up to isomorphism. $\qquad\square$

Having a kind of uniqueness, it is natural to ask if adjoint functors always exist. There is a natural obstruction to this, namely the following result, which says that left adjoints preserve colimits and right adjoints preserve limits.

Proposition 1.5.4. *Let* $F\colon \mathcal{C} \rightleftarrows \mathcal{D}\colon G$ *be an adjoint pair and* $T\colon \mathcal{I} \to \mathcal{C}$ *be a functor with a colimit* $C = \operatorname{colim} T$. *Then* $F(C)$ *is a colimit of* $F \circ T$. *Dually, if* $S\colon \mathcal{I} \to \mathcal{D}$ *has a limit* $= \lim S$, *then* $G(D)$ *is a limit of* $G \circ S$.

Proof. As usual, we only prove the first statement. Being a colimit, C is endowed with maps $c_i\colon T(i) \to C$ for all $i \in \mathcal{I}$. This defines maps $F(c_i)\colon F(T(i)) \to F(C)$, showing that $F(C)$ is a cocone.

If D is another cocone, with maps $d_i\colon F(T(i)) \to D$, by adjunction we have $d_i^\flat\colon T(i) \to G(D)$. Since C is a colimit, we get the universal map $f\colon C \to G(D)$, and adjunction again gives $f^{\#}\colon F(C) \to D$. It is a simple check that $f^{\#}$ makes all the relevant diagrams commute, thus $F(C)$ is a colimit of $F \circ T$. $\qquad\square$

The situation is common enough to warrant a definition.

Definition 1.5.5. A functor $G\colon \mathcal{D} \to \mathcal{C}$ is called *continuous* if it preserves small limits, that is, for all functors $S\colon \mathcal{I} \to \mathcal{D}$, where $\mathcal{I}$ is small, having a limit D, $G(D)$ is (with induced cone structure) a limit of $G \circ S$. Dually, a functor $F\colon \mathcal{C} \to \mathcal{D}$ that preserves small colimits is called *cocontinuous*.

Using this terminology, the proposition above states that right adjoints are continuous, while left adjoints are cocontinuous.

Remark 1.5.6. In the previous definition we remarked that $G(D)$ must be a limit with induced cone structure. In fact, limits are not just objects, but they are endowed with maps from $S(i)$ for all $i \in \mathcal{I}$. The maps $S(i) \to D$ induce maps $G(S(i)) \to G(D)$, and what we require is that these maps make $G(D)$ into a limit. A similar remark holds for cocontinuous functors.

Example 1.5.7. Let $F\colon \mathcal{I} \to \mathcal{C}$ be a functor having a limit C with maps $c_i\colon C \to F(i)$. For any object $D \in \mathcal{C}$, giving a morphism $D \to C$ is equivalent to giving a collection of compatible morphisms $D \to F(i)$. We can restate this as a bijection

$$\operatorname{Hom}_{\mathcal{C}}(D, C) \cong \lim \operatorname{Hom}_{\mathcal{C}}(D, F(i)),$$

where the right-hand limit is in Set. This says exactly that h^D preserves limits. In other words, corepresentable functors are continuous. Similarly, representable functors are continuous as functors $\mathcal{C}^{op} \to$ Set.

We can now state the main result on the existence of adjoints, known as the *adjoint functor theorem*.

Theorem 1.5.8 (Freyd). *Let $G\colon \mathcal{D} \to \mathcal{C}$ a functor that preserves limits. Assume that $\mathcal{D}$ is complete and that $\mathcal{C}$ satisfies the following condition. For every object $C \in \mathcal{C}$ there exists a set of objects $\{D_i\}_{i \in I}$ of $\mathcal{D}$ and morphisms $f_i\colon C \to G(D_i)$ such that every morphism $f\colon C \to G(D)$ for some object $D \in \mathcal{D}$ can be factored as*

$$C \xrightarrow{\ f_i\ } G(D_i) \xrightarrow{\ G(d)\ } G(D)$$

for some index i and some morphism $d \in \mathrm{Hom}(D_i, D)$. Then G has a left adjoint.

Of course, the theorem has a dual version for the existence of right adjoints, which we do not take the trouble to spell out explicitly. The condition in the theorem is a smallness requirement, in that we require the family $\{D_i\}$ to be a set. It is named the *solution set condition*.

Proof. Fix an object $C \in \mathcal{C}$ and consider the category $\mathcal{J}_C$ whose objects are pairs (D, ϕ), where D is an object of $\mathcal{D}$ and $\phi \in \mathrm{Hom}_{\mathcal{C}}(C, G(D))$. Morphisms in $\mathcal{J}_C$ between (D, ϕ) and (D', ϕ') are just morphisms $d\colon D \to D'$ that make the triangle

$$
\begin{array}{ccc}
 & C & \\
\phi \swarrow & & \searrow \phi' \\
G(D) & \xrightarrow{\ G(d)\ } & G(D')
\end{array}
$$

commute. This gives us a functor $\pi_C\colon \mathcal{J}_C \to \mathcal{D}$ that just forgets the map ϕ.

Assume first that $\mathcal{D}$ is small, so that $\mathcal{J}_C$ is small as well. Then, since $\mathcal{D}$ is complete, π_C has a limit, and we can define $F(C) = \lim \pi_C$. It is easy to check that a morphism $C \to C'$ induces a natural transformation $\pi_C \to \pi_{C'}$, hence a map $F(C) \to F(C')$ guaranteed by the universal property of limits. This makes F into a functor. In fact, F is a left adjoint to G.

To see this, let (D, ϕ) be any pair in $\mathcal{J}_C$, so $\phi\colon C \to G(D)$. Since $F(C) = \lim \pi_C$ and G preserves limits, $G(F(C))$ is the limit of $G \circ \pi_C$. Notice that $G \circ \pi_C(D, \phi) = G(D)$, so this gives a map $G(F(C)) \to G(D)$. The universal property of the limit gives a map $C \to G(F(C))$, because C is a cone over $G \circ \pi_C$ by definition. Hence a map $h\colon F(C) \to D$ induces

$$C \longrightarrow G(F(C)) \xrightarrow{\ G(h)\ } G(D),$$

so we have a function $\mathrm{Hom}(F(C), D) \to \mathrm{Hom}(C, G(D))$.

Moreover, an object $D \in \mathcal{D}$ with a map $\phi\colon C \to G(D)$ is just an element of $\mathcal{J}_C$, and we get a map $F(C) \to D$ just because $F(C)$ is a cone. This gives the reverse map $\mathrm{Hom}(C, G(D)) \to \mathrm{Hom}(F(C), D)$, proving that F is indeed left adjoint to G.

The issue is that small categories are rarely complete. If $\mathcal{D}$ is not small, then π_C is a functor from a large category, and as such needs not have a limit. This is where the solution set condition enters the picture.

We modify the above proof as follows. Instead of considering the functor $\pi_C \colon \mathcal{J}_C \to \mathcal{D}$, we consider the small category $\mathcal{J}'_C$ whose objects are pairs (D_i, ϕ), with $\phi \colon C \to D_i$. As above, we have a functor π'_C, and we define $F(C) := \lim \pi'_C$. The above argument goes mostly unchanged, save for the last paragraph, where we need to transform a map $C \to G(D)$ into a map $F(C) \to D$.

An object $D \in \mathcal{D}$ with a map $\phi \colon C \to G(D)$ is not an element of $\mathcal{J}'_C$, but can be factored as a map $C \to G(D_i)$ for some i, followed by $G(d)$ for some $d \colon D_i \to D$. This gives a map $F(C) \to D_i$, and composing with d we get the desired $F(C) \to D$. We leave the details of this construction to the reader. $\square$

1.6. Exercises

1. Which ones of the functors in Example 1.1.10 are full? Which ones are faithful?

2. Let M be a model of NBG. Prove that sets in M form a model for ZFC.

3. Let G be a finite group, $G - \mathrm{Mod}_k$ the category of G-modules over the field k, that is, vector spaces over the field k endowed with a linear action of G. Morphisms in $G - \mathrm{Mod}_k$ are linear maps that are G-equivariant. Let $k[G]$ be the noncommutative ring consisting of formal linear combinations $\sum_{g \in G} a_g g$, where $a_g \in k$, with multiplication extended by linearity from the multiplication of G. Let $\mathrm{Mod}_{k[G]}$ be the category of left $k[G]$-modules. Show that the categories $G - \mathrm{Mod}_k$ and $\mathrm{Mod}_{k[G]}$ are *isomorphic*.

4. Prove that Top is both complete and cocomplete.

The following two exercises are inspired by [**Lei06**].

5. Let $\mathcal{B}$ be the category whose objects are Banach spaces, and whose morphisms are linear maps $f \colon X \to Y$ such that $\|f(x)\|_Y \le \|x\|_X$ for all $x \in X$. For $X, Y \in \mathcal{B}$, we have $X \oplus Y \in \mathcal{B}$, endowed with the norm

$$\|(x, y)\|_{X \oplus Y} := \frac{\|x\|_X + \|y\|_Y}{2}.$$

Let $\mathcal{C}$ be the category whose objects are triples (X, ξ, u), where X is a Banach space, $\xi \colon X \oplus X \to X$ is a morphism in $\mathcal{B}$ and $u \in X$ is an element with $\|u\|_X \le 1$. A morphism from (X, ξ, u) to (Y, ζ, v) in $\mathcal{C}$ is a morphism

$f \colon X \to Y$ in $\mathcal{B}$ such that $f(u) = v$ and the diagram

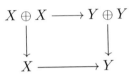

commutes.

Prove that $\mathcal{C}$ has an initial object, namely $(L^1([0,1]), \gamma, 1)$, where γ is the concatenation function

$$\gamma(f,g)(t) = \begin{cases} f(2t) \text{ for } t \in [0, 1/2) \\ g(2t-1) \text{ for } t \in (1/2, 1]. \end{cases}$$

6. Let $\mathcal{C}$ be the category of Exercise 5. Notice that $(\mathbb{R}, m, 1) \in \mathcal{C}$, where m is the mean function $m(x,y) := (x+y)/2$. By the above exercise, there is a unique map $(L^1([0,1]), \gamma, 1) \to (\mathbb{R}, m, 1)$, which arises from a map $s \colon L^1([0,1]) \to \mathbb{R}$. Prove that

$$s(f) = \int_0^1 f.$$

This allows us to define $L^1([0,1])$ and the integral purely from categorical properties of Banach spaces, without developing measure theory first.

7. Let X, Y be two partially ordered sets. A *monotone Galois connection*, or simply Galois connection, between X and Y is a pair of monotone (i.e., order preserving) functions $f \colon X \to Y$ and $g \colon Y \to X$ such that

$$f(x) \le y \text{ if and only if } x \le g(y)$$

for all $x \in X$ and $y \in Y$. Prove that a monotone Galois connection between x and Y is the same as an adjunction between $\mathrm{Ord}(X)$ and $\mathrm{Ord}(Y)$.

8. An *antitone* Galois connection is just a monotone Galois connection between X and Y with the inverted order. If L/K is a normal extension of fields, prove that the usual Galois correspondence [**Fer20**, Section A.5] is an antitone Galois connection between the set of intermediate extensions L'/K (so that $L' \subset L$) and the set of subgroups of the Galois group $Gal(L/K)$.

9. Let f, g be a Galois connection between X and Y (see Exercise 7). Prove that f determines g uniquely; and conversely.

10. Prove the *Lambek–Moser theorem*: there is a bijective correspondence between the set of Galois connections f, g between $\mathbb{N}$ and $\mathbb{N}$ and the set of pairs of complementary infinite subsets $F, G \subset \mathbb{N}$ such that $0 \in F$. Such correspondence is given by associating to the Galois connection f, g the sets

$$F := \{f(m) + m \mid m \in \mathbb{N}\} \text{ and}$$
$$G := \{g(n) + n + 1 \mid n \in \mathbb{N}\}.$$

Derive from this a formula for the n-th non square number in $\mathbb{N}$. The Lambek–Moser theorem appeared in [**LM54**] and later in [**Lam94**]. See also the exposition in [**Bak19**].

11. Let A and B be sets, and consider the power sets $\mathcal{P}(A)$ and $\mathcal{P}(B)$ order categories. Given a function $f\colon A \to B$, there is a functor $f^{-1}\colon \mathcal{P}(B) \to \mathcal{P}(A)$ taking inverse images. Prove that f^{-1} has both a left and a right adjoint and compute them.

12. Check that the pairs defined in Example 1.5.2 are indeed adjoint—in particular, that the bijections of morphisms are natural. What are the unit and counit maps?

13. Let $F\colon \mathrm{Top} \to \mathrm{Set}$ be the forgetful functor. Check that F has both a left adjoint and a right adjoint.

14. Let G be a group, which we regard as a category $\mathcal{G}$ with a single object. Thus, a functor $T\colon \mathcal{G} \to \mathrm{Set}$ is the same as a set X with an action of G. Prove that the limit of T is X^G, the set of fixed points of the action, while the colimit is X/G, the set of orbits.

15. Let $\mathcal{I}$ be a small category, and assume that $\mathcal{C}$ admits all limits of functors $\mathcal{I} \to \mathcal{C}$. Let $F, G\colon \mathcal{I} \to \mathcal{C}$ be two functors. Prove that a natural transformation from F to G induces a morphism $\lim F \to \lim G$ in $\mathcal{C}$. Thus, it is well defined, a *limit functor*
$$\mathrm{Fun}(\mathcal{I}, \mathcal{C}) \to \mathcal{C}.$$

Exercises 16–26 define group objects and monoid objects in categories, and study their first properties. It turns out that monoid objects in a suitable category are related to adjunctions.

16. Let $\mathcal{C}$ be a category having binary products and a terminal object $\bullet$. We can mimic the definition of group by replacing sets and functions with objects and morphisms of $\mathcal{C}$. Namely, a group object $G \in \mathcal{C}$ is an object together with

    (a) a morphism $e\colon \bullet \to G$,

    (b) a multiplication map $m\colon G \times G \to G$, and

    (c) an inverse map $-^{-1}\colon G \to G$.

Notice that we cannot speak of elements of G, hence the identity is replaced by a morphism $\bullet \to G$. These data are required to satisfy certain axioms, expressing the fact that e is a left and right identity for m, m is associative, and $-^{-1}$ provides inverses.

Write these requirements explicitly as the commutativity of suitable diagrams. Check that your definition makes sense by proving that group objects in Top are topological groups and group objects in Diff are Lie groups.

17. Let G be a group object in Grp (see the previous exercise). Prove that G is an Abelian group (this is an abstract form of the Eckmann–Hilton argument).

We also can define monoid objects in categories, but these can be treated in a slightly more general context. For the following exercises, we need a definition.

Definition 1.6.1. A *monoidal category* is a category $\mathcal{C}$ endowed with

    (1) a functor $\otimes\colon \mathcal{C} \times \mathcal{C} \to \mathcal{C}$, called the tensor product;

    (2) an object $I \in \mathcal{C}$, called the unit;

    (3) a natural isomorphism
$$\alpha\colon (-\otimes -)\otimes - \to -\otimes(-\otimes -),$$
    called the *associator*;

    (4) a natural isomorphism $\lambda\colon I\otimes - \to -$, called the *left unitor*;

    (5) a natural isomorphism $\rho\colon -\otimes I \to -$, called the *right unitor*.

These need to satisfy some coherence conditions:

    (1) for all $C, D \in \mathcal{C}$ the diagram

$$
\begin{array}{ccc}
(C\otimes I)\otimes D & \xrightarrow{\ \alpha_{C,I,D}\ } & C\otimes(I\otimes D) \\
& \searrow{\scriptstyle\rho_C\otimes\mathrm{id}_D} \qquad \swarrow{\scriptstyle\mathrm{id}_C\otimes\lambda_D} & \\
& C\otimes D &
\end{array}
$$

    commutes;

    (2) for all $A, B, C, D \in \mathcal{C}$, the diagram

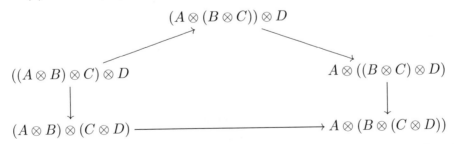

    commutes (we have omitted arrow names, but you can guess them).

These axioms look complicated, but in reality tensor product is usually not associative literally, but only up to isomorphism. The triangle and pentagon identities can be used to generate many other natural identities, much like in a monoid the associativity for three elements can be used to omit parenthesis in every finite product without ambiguity.

18. Let $\mathcal{C}$ be a category with finite products. Let $\bullet$ be the terminal object (which is the empty product). Prove that $\mathcal{C}$ is a monoidal category, where $C \otimes D := C \times D$ and $I = \bullet$. This is called a Cartesian monoidal category.

19. Prove that Vect_k and Mod_A (k a field, A a commutative ring) are monoidal categories under the usual tensor product.

20. Let $\mathcal{C}$ be a category and $\text{End}(\mathcal{C}) = \text{Fun}(\mathcal{C}, \mathcal{C})$ be the category of endofunctors of $\mathcal{C}$. Prove that $\text{End}(\mathcal{C})$ has a monoidal structure where $\otimes$ is the operation of composition of functors.

21. Let $\mathcal{C}$ be a monoidal category. A monoid object is an object $M \in \mathcal{C}$ equipped with two maps $m\colon M \otimes M \to M$ and $e\colon I \to M$. These must satisfy certain axioms, analogous to those of a monoid, expressed as the commutativity of suitable diagrams. Write out such axioms and check that your definition makes sense by proving that a monoid object in the Cartesian monoidal category Set (see Exercise 18) is in fact a regular monoid.

22. Let $\mathcal{C}$ be a category and $\text{End}(\mathcal{C}) = \text{Fun}(\mathcal{C}, \mathcal{C})$ be the category of endofunctors of $\mathcal{C}$, with the monoidal structure defined in Exercise 20. A monoid in $\text{End}(\mathcal{C})$ is thus a functor $T\colon \mathcal{C} \to \mathcal{C}$ with two natural transformations $e\colon \text{id} \to T$ and $m\colon T^2 \to T$, satisfying appropriate compatibility conditions. Write explicitly these conditions. Such a functor is called a *monad*.

23. Let F, G be a pair of adjoint functors between categories $\mathcal{C}$ and $\mathcal{D}$. Prove that the composition $T := G \circ F$ is a monad.

24. As a converse to the previous exercise, let $T\colon \mathcal{C} \to \mathcal{C}$ be any monad. Prove that there is a category $\mathcal{D}$ and a pair of adjoint functors F, G between $\mathcal{C}$ and $\mathcal{D}$ such that $T = G \circ F$. (Define $\text{Obj}(\mathcal{D}) = \text{Obj}(\mathcal{C})$ and $\text{Hom}_\mathcal{D}(A, B) = \text{Hom}_\mathcal{C}(A, T(B))$. Prove that $\mathcal{D}$ is in fact a category, called the *Kleisli category* of T.)

25. Let $L\colon \text{Set} \to \text{Set}$ be the functor of finite lists, which on objects is given by

$$L(S) := \{\emptyset\} \cup S \cup S^2 \cup \cdots .$$

Check that L is indeed a functor by writing its action on morphism, and prove that L is a monad, called the *list monad*.

In the same fashion, prove that the functor $\mathcal{P}\colon \text{Set} \to \text{Set}$ defined by $\mathcal{P}(X) = \{\text{subsets of } X\}$ is a monad, called the *power set monad*.

26. For a set S, define a finite probability distribution on S as a function $p\colon S \to [0, 1]$ with finite support such that $\sum_{s \in S} p(s) = 1$. Define $P(S) := \{\text{finite probability distributions on } S\}$. Define P on morphisms so that P becomes a *covariant* functor Set $\to$ Set. Prove that P is a monad, called the *probability monad*.

27. Give a direct proof of Proposition 1.5.3 using the fact that for each pair of objects $C \in \mathcal{C}$ and $D \in \mathcal{D}$, the bijection

$$\Psi_{C,D} \colon \operatorname{Hom}(F(C), D) \to \operatorname{Hom}(C, G(D))$$

determines uniquely the functor $\operatorname{Hom}(-, G(D))$, and then applying the Yoneda lemma.

28. Let $\mathcal{C}, \mathcal{D}$ be two categories with morphisms $F \colon \mathcal{C} \to \mathcal{D}$ and $G \colon \mathcal{D} \to \mathcal{C}$. Let η be a natural transformation between $\operatorname{id}_{\mathcal{C}}$ and $G \circ F$. Find necessary and sufficient conditions for η to be the unit of an adjunction.

29. Let $F \colon \mathcal{C} \rightleftarrows \mathcal{D} \colon G$ be an adjoint pair and $C \in \operatorname{Obj}(\mathcal{C})$. Let η be the unit of the pair and ϵ its counit. Prove that $\operatorname{id}_{F(C)} = \epsilon_{F(C)} \circ F(\eta_C)$. State and prove a similar result for the counit.

Exercises 30–35 introduce the category of simplicial sets and its basic properties. For a leisurely introduction to simplicial sets and their use in topology, see [**Fri08**].

30. Let Δ the category of simplexes defined in Example 1.1.4(r). Let $d_i \colon [n] \to [n+1]$ be the morphisms given by

$$d_i(0, \dots, n) = (0, \dots, i-1, i+1, \dots, n),$$

which we call a *face*. Conversely, let $s_i \colon [n+1] \to [n]$ be the morphisms given by

$$s_i(0, \dots, n+1) = (0, \dots, i-1, i, i, i+1, \dots, n+1),$$

which we call a degeneracy. Show that every morphism in Δ is a finite composition of face and degeneracy maps, and that these satisfy the following identities:

$$d_i d_j = d_{j-1} d_i \text{ if } i < j,$$
$$d_i s_j = s_{j-1} d_i \text{ if } i < j,$$
$$d_j d_j = d_{j+1} s_i = \operatorname{id},$$
$$d_i s_j = s_j d_{i-1} \text{ if } i > j - 1, \text{ and}$$
$$s_i s_j = s_{j+1} s_i \text{ if } i \leq j.$$

31. A contravariant functor $X \colon \Delta \to \operatorname{Set}$ is called a *simplicial set*. By the previous exercise, this is the same as giving the datum of a collection of sets X_n and face and degeneracy functions $d_i \colon X_{n+1} \to X_n$ and $s_i \colon X_n \to X_{n+1}$ satisfying some identities.

Recall that a *simplicial complex* is a subset $S \subset \mathcal{P}([n])$ such that if $F \in S$ and $F' \subset F$, then $F' \in S$ (F' is a face of F). Show that every simplicial complex gives rise to a simplicial set.

32. Using natural transformations as morphisms, simplicial sets form a category of their own, which we denote SSet. Show that there is a functor

$$\mathcal{S}\colon \mathrm{Top} \to \mathrm{SSet}$$

which assigns to each topological space X the singular set $\mathcal{S}(X)$ defined by $\mathcal{S}(X)([n]) = \mathrm{Hom}_{\mathrm{Top}}(\Delta_n, X)$, where Δ_n is the standard n-simplex. This is the same set that is used to generate n-chains in the theory of singular homology.

33. As a counterpart to the previous exercise, we construct a *realization functor*

$$|{-}|\colon \mathrm{SSet} \to \mathrm{Top},$$

which assigns to each simplicial set S a topological space $|S|$, obtained by gluing together all the simplexes described by S.

To describe $|S|$, let X_n be the set $S([n])$, considered as a discrete topological space, and define

$$|S| := \bigsqcup_{i=0}^{\infty} X_n \times \Delta_n / \sim\ .$$

Here $\sim$ is the following equivalence relation. Using convex combinations, extend d_i and s_i to continuous maps $D_i\colon \Delta_n \to \Delta_{n+1}$ and $S_i\colon \Delta_{n+1} \to \Delta_n$, which can be seen as inclusion of the i-th face and projection to the i-th face respectively. Then $\sim$ is generated by the identifications

$$(x, D_i(p)) \sim (d_i(x), p) \text{ and } (x, S_i(p)) \sim (s_i(x), p)$$

for $x \in X_n$ and $p \in \Delta_{n\pm1}$.

Show that $|{-}|$ is well defined and is in fact a functor.

34. Show that the singular simplex and geometric realization functors defined in the previous two exercises are adjoint to each other. Namely, there is a natural bijection

$$\Psi_{S,X}\colon \mathrm{Hom}_{\mathrm{Top}}(|S|, X) \cong \mathrm{Hom}_{\mathrm{SSet}}(S, \mathcal{S}(X))$$

for a simplicial set S and a topological space X.

35. Let $\mathcal{C}$ be a small category. Show that there is a simplicial set $N(\mathcal{C})$ such that

$$N(\mathcal{C})_k = \{k\text{-tuples of composable morphisms}\}.$$

Here by a k-tuple of composable morphisms we mean a collection of maps $f_i \in \mathrm{Hom}(C_{i-1}, C_i)$ for $i = 1, \dots, k$, for some objects $C_0, \dots, C_k \in \mathcal{C}$. Define suitable face and degeneracy maps that make $N(\mathcal{C})$ into a simplicial set. The simplicial set $N(\mathcal{C})$ is called the *nerve* of $\mathcal{C}$.

36. Prove that the morphism of (1.4.1) need not be an isomorphism when $\mathcal{I}$ is not filtered, even assuming that $\mathcal{C} = \mathrm{Set}$ and $\mathcal{I}, \mathcal{J}$ are finite.

37. Let $\mathcal{C}$ be a complete small category. Prove that for all $A, B \in \mathcal{C}$, $\mathrm{Hom}(A, B)$ has at most one element. (Consider morphisms $X \to \prod_f Y$, where f ranges over all morphisms in $\mathcal{C}$).

Chapter 2

Abelian Categories

This chapter develops the theory of Abelian categories. Abelian categories are characterized by axioms that mimic the properties of the category Mod_A. They were introduced in [**Buc55**] under the name of *exact categories*, and since the Tôhoku paper by Grothendieck ([**Gro57**]) they have been considered the natural setting in which one can develop homological algebra. If $\mathcal{C}$ is Abelian, one can define complexes and exact sequences of objects of $\mathcal{C}$, and develop the theory of homology and derived functors – which we will do in the next chapter. As it turns out, Abelian categories are not only formally similar to categories of modules: a celebrated result of Freyd and Mitchell, first exposed in [**Fre64**], states that a small Abelian category can be embedded as a subcategory of Mod_A—for a possibly noncommutative ring A—in a way that preserves exact sequences.

We start our exposition by introducing additive and Abelian categories. Additive categories resemble Abelian categories, but their axioms are weaker, and it is easier to treat them first. Abelian categories and their elementary properties are introduced in Section 2.2, where we also define injective and projective objects.

The central part of the chapter is devoted to sheaves, first in the context of a topological space. Sheaves are an important topological construction, and they deserve a treatise of their own, such as [**Ive86**]. Here, we limit ourselves to their construction and basic properties, in order to provide a nontrivial example of Abelian category. In Section 2.4, we extend the theory to study sheaves on a site, which is a categorical abstraction of a topological space, sufficient to carry out the theory of sheaves. This will be used to prove that the category of Ab-valued left-exact functors on an Abelian category is also Abelian, a crucial fact that we will need in the proof of the Freyd–Mitchell theorem.

In Section 2.5, we come back to the general setting of Abelian categories and develop some standard lemmas on exact sequences. These are a suite of simple facts—known as the five lemma, the nine lemma and the snake lemma—that are usually proved on a category of modules by the technique of diagram chasing. In a general Abelian category, objects do not have elements, and one must find an alternative approach: for this, we develop a theory of subobjects, and study how they behave with respect to direct and inverse images. This is not the only possible strategy: another common approach is to embed an Abelian category via the Freyd–Mitchell theorem and then prove the results for modules. In the exercises, we describe two other possibilities: the first one, used in [**ML71**], uses *members*, which are certain morphisms up to an equivalence relation, while the second one recognizes a certain kind of subdiagrams, called *salamanders*, and relies on the exactness of a suitable sequence extracted from such diagrams.

Sections 2.6 and 2.7 are devoted to the study of projective and injective objects. These are objects of an Abelian category such that the associated representable (resp., corepresentable) functor is exact, and they turn out to be central in many constructions in homological algebra. We prove various existence results for injective objects, and in particular existence results for injective hulls, which are a sort of minimal injective extension of objects in an Abelian category. This is a prelude to Chapter 5, which specializes to the study of projective and injective *modules*.

In the last section, we put everything together to give a proof of the Freyd–Mitchell embedding theorem. This is an important cornerstone of homological algebra, which shows the prominence of the categories of modules, which are the central object of study of this book.

2.1. Additive categories

So far, we have been considering categories of varied nature. In this chapter, we are going to specialize to the type of categories that are most similar to the categories of modules over a ring.

Definition 2.1.1. Let $\mathcal{A}$ be a category. We say that $\mathcal{A}$ is an *additive category* if it satisfies the following axioms.

(1) For all objects $A, B \in \mathcal{A}$, the set $\mathrm{Hom}(A, B)$ has the structure of an Abelian group (which we will write in additive notation), in such a way that the composition

$$\mathrm{Hom}(B, C) \times \mathrm{Hom}(A, B) \to \mathrm{Hom}(A, C)$$

is additive on both arguments for $A, B, C \in \mathcal{A}$.

Notice that in particular the set $\mathrm{Hom}(A, B)$ is never empty, as it contains the 0 element.

(2) There is an object which is both initial and terminal for $\mathcal{A}$. Since universal objects are unique up to a unique isomorphism, we will choose one such object and denote it by 0.

A consequence of this is that the identity id_0 is the only element of $\mathrm{Hom}_\mathcal{A}(0, 0)$. For any other object $A \in \mathcal{A}$, we will denote by 0 the unique morphisms $0 \to A$ and $A \to 0$.

(3) In $\mathcal{A}$, there exist binary products and coproducts.

Remark 2.1.2. The symmetry in the axioms implies that $\mathcal{A}^{op}$ is an additive category whenever $\mathcal{A}$ is.

Remark 2.1.3. If A, B are objects of an additive category, the set $\mathrm{Hom}(A, B)$ has the structure of an Abelian group. In particular, we can take all (co)representable functors h_A and h^A to have values in Ab instead of Set.

With only these simple axioms, we can start defining the analogue of some standard constructions in module theory. First, some simple consequences of the definition.

Let us fix two objects $A, B \in \mathcal{A}$ and consider the projection maps $p_A \colon A \times B \to A$, $p_B \colon A \times B \to B$, and inclusion maps $i_A \colon A \to A \sqcup B$, $i_B \colon B \to A \sqcup B$. Applying the universal property of the product to id_A and 0 yields a map

$$(\mathrm{id}_A, 0) \colon A \to A \times B,$$

and similarly for B. We can then apply the universal property of the coproduct to the two maps $(\mathrm{id}_A, 0)$ and $(0, \mathrm{id} B)$ to obtain a morphism $\sigma_{A,B} \colon A \sqcup B \to A \times B$. By construction, this has the property that

(2.1.1) $p_A \circ \sigma_{A,B} \circ i_A = \mathrm{id}_A,$

and similarly for B. The other composition is

(2.1.2) $p_B \circ \sigma_{A,B} \circ i_A = 0,$

and conversely.

In the other direction, we have the morphism

$$\tau_{A,B} := i_A \circ p_A + i_B \circ p_B \colon A \times B \to A \sqcup B.$$

Applying (2.1.1) and (2.1.2) we get

$$\tau_{A,B} \circ \sigma_{A,B} \circ i_A = i_A,$$

and similarly for B. This means that $A \sqcup B$, endowed with the two maps $\tau_{A,B} \circ \sigma_{A,B} \circ i_A$ and $\tau_{A,B} \circ \sigma_{A,B} \circ i_B$, is a coproduct of A and B, and by uniqueness we get $\tau_{A,B} \circ \sigma_{A,B} = \mathrm{id}_{A \sqcup B}$. A dual reasoning (check it!) shows that $\sigma_{A,B} \circ \tau_{A,B} = \mathrm{id}_{A \times B}$.

In conclusion, the two maps are inverse to each other, and $A \times B$ and $A \sqcup B$ are canonically isomorphic. In other words, the same object is both a product and a coproduct.

Definition 2.1.4. Let $\mathcal{A}$ be an additive category. Given objects $A, B \in \mathcal{A}$, we denote by $A \oplus B$ an object that is both a product and coproduct of A and B. This is called the *direct sum* of A and B.

The direct sum is endowed with the maps in the following square

$$
\begin{array}{ccc}
A \oplus B & \xrightarrow{p_B} & B \\
{\scriptstyle p_A} \downarrow & & \downarrow {\scriptstyle i_B} \\
A & \xrightarrow{i_A} & A \oplus B.
\end{array}
$$

The previous discussion can be summarized by the following.

Proposition 2.1.5. *Let $\mathcal{A}$ be an additive category, $A, B \in \mathcal{A}$. Then*

$$
\begin{aligned}
p_A \circ i_A &= \mathrm{id}_A, \\
p_B \circ i_B &= \mathrm{id}_B, \\
i_A \circ p_A + i_B \circ p_B &= \mathrm{id}_{A \oplus B}, \\
p_B \circ i_A &= 0, \ and \\
p_A \circ i_B &= 0.
\end{aligned}
$$

Example 2.1.6.

   (a) For a possibly noncommutative ring A, the category Mod_A of left A-modules is additive.

   (b) The category of topological Abelian groups is additive.

   (c) If $\mathcal{C}$ is a small category, the functor category $\mathrm{Fun}(\mathcal{C}, \mathrm{Ab})$ is additive.

Of course, we have a corresponding notion for morphisms.

Definition 2.1.7. Let $\mathcal{A}, \mathcal{B}$ be two additive categories. A functor $F \colon \mathcal{A} \to \mathcal{B}$ is called *additive* if for every pair of objects $A, B \in \mathcal{A}$, the function $\mathrm{Hom}_{\mathcal{A}}(A, B) \to \mathrm{Hom}_{\mathcal{B}}(F(A), F(B))$ is a group homomorphism.

Remark 2.1.8. It is easy to check that the identities in Proposition 2.1.5 are enough to characterize the direct sum, that is, every object endowed with maps i_A, i_B, p_A, p_B satisfying these identities is a product and coproduct of A and B. Since additive functors preserve these identities, it follows that $F(A \oplus B) \cong F(A) \oplus F(B)$ for any additive functor F. More explicitly, the map $F(A \oplus B) \to F(A) \oplus F(B)$ arising from the maps obtained applying F to p_A and p_B is an isomorphism.

Conversely, a functor F such that $F(A \oplus B) \cong F(A) \oplus F(B)$ in this natural way is necessarily additive. To see this, take $f, g \in \mathrm{Hom}(A, B)$ and express $f + g$ as the composition

$$A \xrightarrow{(id_A, id_A)} A \oplus A \xrightarrow{(f, g)} B \oplus B \xrightarrow{p_1 + p_2} B.$$

A useful consequence of the above remark is

Proposition 2.1.9. *Let $F \colon A \rightleftarrows B \colon G$ be a pair of adjoint functors between additive categories. Then F and G are both additive.*

Proof. By Theorem 1.5.4, F preserves colimits, in particular the coproduct. By the remark, this implies that F is additive. The statement for G follows by duality or using the fact that the direct sum is also a product. $\square$

The first constructions that we can carry out in additive categories are kernels and cokernels.

Definition 2.1.10. Let $f \colon A \to B$ be a morphism in an additive category. The *kernel* of f, denoted $\ker f$, is the equalizer of f and 0. Dually, the *cokernel* of f, denoted $\mathrm{coker}\, f$, is the coequalizer of f and 0.

Notice that $\ker f$ is not just an object, but is the datum of an object K and a morphism $K \to A$ such that the composition

$$K \longrightarrow A \xrightarrow{f} B$$

is 0, and universal with respect to this property.

Definition 2.1.11. Let $f \colon A \to B$ be a morphism in an additive category. We say that f is a *monomorphism* (or that f is *monic*) if $\ker f = 0$ and an *epimorphism* (or that f is *epic*) if $\mathrm{coker}\, f = 0$.

Kernels need not exist in an additive category, but when they do exist, they are representable.

Proposition 2.1.12. *Let A be an additive category, $f \colon A \to B$ a morphism. Let $F \colon A \to \mathrm{Ab}$ be the functor defined by*

$$F(X) = \ker(\mathrm{Hom}(X, A) \to \mathrm{Hom}(X, B)).$$

Then f has a kernel if and only if F is representable, in which case $F \cong h_{\ker f}$.

Proof. Let $K = \ker f$, fitting in the universal diagram

$$K \xrightarrow{\ k\ } A \xrightarrow{\ f\ } B.$$

Then for $X \in \mathcal{A}$ we get the functions

$$\mathrm{Hom}(X, K) \xrightarrow{\ k \circ -\ } \mathrm{Hom}(X, A) \xrightarrow{\ f \circ -\ } \mathrm{Hom}(X, B),$$

and clearly the composition of the two arrows is 0. Moreover, a function $g\colon X \to A$ lies in the kernel of $\mathrm{Hom}(X, A) \to \mathrm{Hom}(X, B)$ exactly when $f \circ g = 0$. By the universal property of the kernel, this means that $g = k \circ g'$, for some $g'\colon X \to K$. This shows that $\mathrm{Hom}(X, K)$ is exactly the kernel of $\mathrm{Hom}(X, A) \to \mathrm{Hom}(X, B)$, so $F \cong h_K$.

Vice versa, assume that $F \cong h_K$ for some object $K \in \mathcal{A}$. In particular, there is a natural transformation $h_K \to h_A$. By the Yoneda lemma, this comes from a morphism $k\colon K \to A$. Then it is a matter of inverting the implications above to prove that K is the kernel of f. □

Remark 2.1.13. By duality, it follows immediately that $\mathrm{coker}\, f$, when it exist, is corepresentable. Namely, in $\mathcal{A}^{op}$, $\mathrm{coker}\, f$ becomes a kernel, and thus it represents the functor $F\colon \mathcal{A}^{op} \to \mathrm{Ab}$ defined by

$$F(X) = \ker(\mathrm{Hom}_{\mathcal{A}^{op}}(X, B) \to \mathrm{Hom}_{\mathcal{A}^{op}}(X, A)).$$

In other words, $\mathrm{coker}\, f$ corepresents the functor $G\colon \mathcal{A} \to \mathrm{Ab}$ defined by

$$G(X) = \ker(\mathrm{Hom}_{\mathcal{A}}(B, X) \to \mathrm{Hom}_{\mathcal{A}}(A, X)).$$

It is *false*, though, that $\mathrm{coker}\, f$ represents the functor

$$F(X) = \mathrm{coker}(\mathrm{Hom}(X, A) \to \mathrm{Hom}(X, B)),$$

as the following example shows (see also Exercise 1).

Example 2.1.14. In the category $\mathcal{A} = \mathrm{Ab}$, take $A = B = \mathbb{Z}$, and consider the map f given by multiplication by n. Then $\mathrm{coker}\, f = \mathbb{Z}/n\mathbb{Z}$. But there are no nontrivial morphisms $\mathbb{Z}/n\mathbb{Z} \to \mathbb{Z}$, so if we take $X = \mathbb{Z}/n\mathbb{Z}$, we get $\mathrm{Hom}(X, A) = \mathrm{Hom}(X, B) = 0$; hence

$$\mathrm{coker}(\mathrm{Hom}(X, A) \to \mathrm{Hom}(X, B)) = 0.$$

Lemma 2.1.15. *Let $\mathcal{A}$ be an additive category that admits kernels, $f\colon A \to B$ a morphism in $\mathcal{A}$, and $k\colon K \to A$ the kernel of f. Then $\ker k = 0$. Dually, the cokernel of a cokernel also vanishes.*

Proof. Let $k' \colon K' \to K$ be the kernel of K. In the diagram

$$
\begin{array}{ccc}
K' \xrightarrow{\ k'\ } K \longrightarrow 0 \\
\downarrow \qquad k\downarrow \qquad \downarrow \\
0 \longrightarrow A \xrightarrow{\ f\ } B,
\end{array}
$$

the two squares are pullbacks, by definition of kernel. It is a simple verification that then the outer rectangle is also a pullback, hence $K' = 0$. $\square$

Assume that kernels and cokernels exist in $\mathcal{A}$, and let $f \colon A \to B$ be a morphism. We can take $K = \ker f$, with its associated morphism $k \colon K \to A$, and let $X = \operatorname{coker} k$. Or we can go the other direction, taking $L = \operatorname{coker} f$, with associated morphism $l \colon B \to L$, and let $Y = \ker l$. This is summarized in the following diagram (the dotted lines are explained below).

(2.1.3)
$$
\begin{array}{ccc}
K \xrightarrow{\ k\ } A \xrightarrow{\ x\ } X \\
\qquad h\downarrow \quad f \\
g \quad Y \xrightarrow{\ y\ } B \xrightarrow{\ l\ } L
\end{array}
$$

In this situation, there is a canonical morphism $X \to Y$, defined as follows. Since $l \circ f = 0$, we get a morphism $h \colon A \to Y$ such that $f = y \circ h$. By construction, $y \circ h \circ k = f \circ k = 0$, so the composition $h \circ k$ factors though $\ker y$, which vanishes by Lemma 2.1.15. Since $h \circ k = 0$, we get the required arrow $X \to Y$ from the universal property of the cokernel.

In fact, one can prove that $\ker g = 0$ and $\operatorname{coker} g = 0$ (Exercise 2). Even granting this, it is not necessarily the case that g is an isomorphism, as it will become clear in the next section.

2.2. Abelian categories

The setting of additive categories provides us with the language to talk about kernels and cokernels, but in order to do something meaningful with complexes, exact sequences, and in general do homological algebra, we need something more.

Definition 2.2.1. Let $\mathcal{A}$ be an additive category. We say that $\mathcal{A}$ is an *Abelian category* if $\mathcal{A}$ admits kernels and cokernels, and moreover for every morphism $f \colon A \to B$ the canonical map g of (2.1.3) is an isomorphism.

Expanding a little, this means that every morphism $f \colon A \to B$ of $\mathcal{A}$ gives rise to a sequence

(2.2.1)
$$
K \xrightarrow{\ k\ } A \xrightarrow{\ f_1\ } I \xrightarrow{\ f_2\ } B \xrightarrow{\ c\ } C,
$$

where (K, k) is the kernel of f, (C, c) its cokernel; I is both the cokernel of k and the kernel of c, and $f = f_2 \circ f_1$. This is called the *canonical decomposition* associated to f. The object I (together with the maps f_1 and f_2) is called the *image* of f, denoted im f.

We start by exploring the simplest consequences of this property.

Proposition 2.2.2. *Let $\mathcal{A}$ be an Abelian category, $f \colon A \to B$ a morphism of $\mathcal{A}$ such that* $\ker f = \operatorname{coker} f = 0$. *Then f is an isomorphism.*

Proof. The cokernel of $0 \to A$ is A, while the kernel of $B \to 0$ is B.   □

Corollary 2.2.3. *If $f \colon A \to B$ is a monomorphism, then A, f is the kernel of its cokernel. Dually, every epimorphism is the cokernel of its kernel.*

Proof. In the canonical decomposition (2.2.1), the image I is both the kernel of c and the cokernel of k. The map $A \to I$ is an epimorphism, and is a monomorphism exactly when f is. In this case, it is an isomorphism by Proposition 2.2.2, so A is a kernel of c. Similarly, if f is an epimorphism, $I \to B$ is an isomorphism, so B is a cokernel of k.   □

Example 2.2.4.

(a) Let A be a possibly noncommutative ring. The category Mod_A of left A-modules is an Abelian category.

(b) The category of topological Abelian groups is not Abelian. In fact, the concepts of kernel and cokernel in this category are purely algebraic and do not see the topological structure. For instance, we can see $\mathbb{R}$ as a topological group both with the Euclidean and with the discrete topology. The identity, seen as a map $\mathbb{R}_{disc} \to \mathbb{R}_{euc}$, is a continuous homomorphism with 0 kernel and cokernel, but is not an isomorphism.

(c) An important example of Abelian category is given by sheaves, which we will treat in detail in the next section.

(d) Let $\mathcal{J}$ be a small category and $\mathcal{A}$ an Abelian category. The functor category $\operatorname{Fun}(\mathcal{J}, \mathcal{A})$ inherits from $\mathcal{A}$ the structure of Abelian category, where all operations are defined pointwise. For instance, let $F, G \colon \mathcal{J} \to \mathcal{A}$ be two functors. Then $\operatorname{Hom}(F, G)$ consists of natural transformations between F and G. Two natural transformations s, t define morphism $s_A, t_A \colon F(A) \to G(A)$ for an object $A \in \mathcal{J}$. Since $\operatorname{Hom}(F(A), G(A))$ has the structure of an Abelian group, we can define the sum $s + t$ pointwise by $(s + t)_A := s_A + t_A$. Similarly, kernels and cokernels are defined pointwise. We invite the reader to actually check that all operations are well defined.

(e) Let $\mathcal{A}, \mathcal{B}$ be two Abelian categories, with $\mathcal{A}$ small. The category $\mathrm{Add}(\mathcal{A}, \mathcal{B})$ consisting of additive functors is an Abelian subcategory of $\mathrm{Fun}(\mathcal{A}, \mathcal{B})$.

Proposition 2.2.5. *Let $f\colon A \to B$ be a morphism in an Abelian category, $I = \operatorname{im} f$. Then the natural morphism $I \to B$ is a monomorphism and the natural morphism $A \to I$ is an epimorphism.*

Proof. This is an immediate consequence of Lemma 2.1.15. $\qquad\square$

Remark 2.2.6. Let $f, g\colon A \to B$ be two morphisms in an Abelian category. Then the equalizer of f and g exists, and equals $\ker f - g$. Similarly, the coequalizer is $\operatorname{coker} f - g$.

An additive category $\mathcal{A}$ admits binary products and has a terminal object, hence it has all finite products by induction. Dually, it has all finite coproducts. A slight modification of the proof of Theorem 1.4.12, together with the above remark, implies the following result.

Proposition 2.2.7. *Let $\mathcal{A}$ be an Abelian category. Then $\mathcal{A}$ admits all finite limits and colimits.*

Example 2.2.8. Let $f\colon A \to C$ and $g\colon B \to C$ be morphisms in the Abelian category $\mathcal{A}$. Then the fibered product can be identified with

$$\ker\left(A \oplus B \xrightarrow{(f,-g)} C \right).$$

A similar description is available for the fibered coproduct.

At this point we are able to extend the usual notions of complexes and exact sequences of modules to the context of Abelian categories.

Definition 2.2.9. Let $\mathcal{A}$ be an Abelian category,

$$(A_\bullet, f_\bullet) = \cdots \xrightarrow{f_{k-1}} A_k \xrightarrow{f_k} A_{k+1} \xrightarrow{f_{k+1}} \cdots$$

a possibly infinite family of morphisms between objects of $\mathcal{A}$. We say that $(A_\bullet, f_\bullet)$ is a *complex* if $f_{k+1} \circ f_k = 0$ for all k.

A morphism between two complexes $(A_\bullet, f_\bullet)$ and $(B_\bullet, g_\bullet)$ consists of morphisms $h_k\colon A_k \to B_k$ such that all squares

$$\begin{array}{ccc} A_k & \xrightarrow{f_k} & A_{k+1} \\ {\scriptstyle h_k}\downarrow & & \downarrow{\scriptstyle h_{k+1}} \\ B_k & \xrightarrow{g_k} & B_{k+1} \end{array}$$

commute. With this definition, complexes form a category of their own, so we can speak of isomorphic complexes and so on.

Let $(A_\bullet, f_\bullet)$ be a complex, and consider the canonical decomposition of $f_{k-1}\colon A_{k-1} \to A_k$. This looks like

$$\ker f_{k-1} \longrightarrow A_{k-1} \longrightarrow \operatorname{im} f_{k-1} \longrightarrow A_k \longrightarrow \operatorname{coker} f_{k-1}.$$

By construction the composition $\operatorname{im} f_{k-1} \to A_k \to A_{k+1}$ is 0, and this induces a natural map $\operatorname{im} f_{k-1} \to \ker f_k$.

Definition 2.2.10. The complex $(A_\bullet, f_\bullet)$ is said to be *exact* at A_k if the natural map $\operatorname{im} f_{k-1} \to \ker f_k$ is an isomorphism. The complex $(A_\bullet, f_\bullet)$ is said to be exact if it is exact everywhere.

An exact complex of the form

$$0 \longrightarrow A \longrightarrow B \longrightarrow C \longrightarrow 0$$

(implied to be 0 on the right and on the left of these terms) is called a *short exact sequence*. In this situation, we will say that A is a *subobject* of B and C is a *quotient object* of B. Abusing notation, we may even write $C \cong B/A$.

Notice that in the above situation, we can recover A as the kernel of $B \to C$, and conversely C as the cokernel of $A \to B$.

Remark 2.2.11. Any (long) exact sequence $(A_\bullet, f_\bullet)$ gives rise to various short exact sequences of the form

$$0 \longrightarrow \ker f_k \longrightarrow A_k \longrightarrow \operatorname{im} f_k \longrightarrow 0$$

which are joined together by the isomorphism $\operatorname{im} f_k \cong \ker f_{k+1}$.

Given any two objects A, B of $\mathcal{A}$, we can always form the exact sequence

$$(2.2.2) \qquad 0 \longrightarrow A \xrightarrow{i_A} A \oplus B \xrightarrow{p_B} B \longrightarrow 0.$$

Definition 2.2.12. A short exact sequence that is isomorphic to (2.2.2) is called *split*.

The following result should be familiar from the case of modules, and is a good exercise to become familiar with Abelian categories.

Proposition 2.2.13. *Let*

$$0 \longrightarrow A \xrightarrow{f} B \xrightarrow{g} C \longrightarrow 0$$

be a short exact sequence. The following are equivalent:

(i) *the sequence is split;*

(ii) *there is a map $h\colon B \to A$ such that $h \circ f = \operatorname{id}_A$; and*

(iii) *there is a map $s\colon C \to B$ such that $g \circ s = \operatorname{id}_C$.*

We record another simple fact on exact sequences that we will need in the proof of Freyd–Mitchell theorem.

Lemma 2.2.14. *Let $s\colon B \to C$ be an epimorphism. Then there is an exact sequence*

$$0 \longrightarrow B \longrightarrow B \times_C B \xrightarrow{\ p_1 - p_2\ } B \xrightarrow{\ s\ } C \longrightarrow 0,$$

where p_1 and p_2 are the projections on the two components.

Proof. The two identities induce a map $\Delta\colon B \to B \times_C B$. Clearly both compositions $(p_1 - p_2) \circ \Delta$ and $s \circ (p_1 - p_2)$ are 0. Since $p_1 \circ \Delta = \mathrm{id}_B$, Δ is a monomorphism. Exactness at $B \times_C B$ is immediate.

To check exactness at B, let $f\colon T \to B$ be any morphism such that $s \circ f = 0$. Then we can consider the two maps f and 0 from T to B. Their composition with s is the same, so we get a map $t\colon T \to B \times_C B$ from the universal property such that $f = p_1 \circ t = (p_1 - p_2) \circ t$. $\square$

Since functors respect composition, it is immediate to check that any additive functor sends complexes to complexes. It is not always the case that they respect exact sequences, though. The study of how additive functors fail to preserve exactness, and how to repair that, lies at the heart of homological algebra.

Definition 2.2.15. Let $F\colon \mathcal{A} \to \mathcal{B}$ be an additive functor between Abelian categories. We say that F is *exact* if for every short exact sequence in $\mathcal{A}$

$$0 \longrightarrow A \longrightarrow B \longrightarrow C \longrightarrow 0$$

the induced complex

$$0 \longrightarrow F(A) \longrightarrow F(B) \longrightarrow F(C) \longrightarrow 0$$

is also exact. If it is exact, save perhaps at $F(C)$, we say that F is *left exact*. If it is exact, save perhaps at $F(A)$, we say that F is *right exact*.

By splitting a long exact sequence into short ones, it follows immediately that if $A_\bullet$ is *any* exact sequence in $\mathcal{A}$ and F is exact, then $F(A_\bullet)$ is also an exact sequence in $\mathcal{B}$.

Remark 2.2.16. If $F\colon \mathcal{A} \to \mathcal{B}$ is a *contravariant* additive functor, we say that F is left (resp., right) exact when the associated functor $\mathcal{A}^{op} \to \mathcal{B}$ is left (resp., right) exact. In other words, we invert arrows in $\mathcal{A}$, so left and right refer to the objects in $\mathcal{B}$. So if F is a contravariant left-exact functor, then

$$0 \longrightarrow F(C) \longrightarrow F(B) \longrightarrow F(A)$$

is exact, while if F is contravariant and right exact,

$$F(C) \longrightarrow F(B) \longrightarrow F(A) \longrightarrow 0$$

is exact.

The prototypical example of left-exact functors is given by representable and corepresentable ones.

Proposition 2.2.17. *Let $\mathcal{A}$ be an Abelian category, $X \in \mathcal{A}$. The functors h_X and h^X (with target Ab) are both left exact.*

Proof. It is enough to prove that h^X is left exact. So let

$$0 \longrightarrow A \xrightarrow{f} B \xrightarrow{g} C \longrightarrow 0$$

be any short exact sequence. This means that (A, f) is a kernel of g, so evey morphism $h \colon X \to B$ such that $g \circ h = 0$ factors through A. This proves that the sequence

$$0 \longrightarrow \mathrm{Hom}(X, A) \longrightarrow \mathrm{Hom}(X, B) \longrightarrow \mathrm{Hom}(X, C)$$

is exact at $\mathrm{Hom}(X, B)$. Moreover, if $h \colon X \to A$ satisfies $f \circ h = 0$, h factors through $\ker f = 0$, which proves that the sequence is exact at $\mathrm{Hom}(X, A)$.
$\square$

Example 2.2.18. Even in Ab, the Hom functors are in general not right exact. For instance, in Ab we have the exact sequence

$$0 \longrightarrow \mathbb{Z} \xrightarrow{\cdot n} \mathbb{Z} \longrightarrow \mathbb{Z}/n\mathbb{Z} \longrightarrow 0,$$

but $\mathrm{Hom}(\mathbb{Z}/n\mathbb{Z}, \mathbb{Z}) = 0$, showing that $h^{\mathbb{Z}/n\mathbb{Z}}$ is not right exact. Similarly, $h_{\mathbb{Z}/n\mathbb{Z}}$ is not right exact, because the map $\mathrm{Hom}(\mathbb{Z}, \mathbb{Z}/n\mathbb{Z}) \to \mathrm{Hom}(\mathbb{Z}, \mathbb{Z}/n\mathbb{Z})$ induced by multiplication by n is 0.

In fact, a converse of Proposition 2.2.17 holds.

Proposition 2.2.19. *Let $\mathcal{A}$ be an Abelian category,*

(2.2.3) $$0 \longrightarrow A \longrightarrow B \longrightarrow C \longrightarrow 0$$

a complex in $\mathcal{A}$. If

$$0 \longrightarrow \mathrm{Hom}(X, A) \longrightarrow \mathrm{Hom}(X, B) \longrightarrow \mathrm{Hom}(X, C)$$

is exact for all $X \in \mathcal{A}$, then (2.2.3) is left exact. Dually, if

$$0 \longrightarrow \mathrm{Hom}(C, X) \longrightarrow \mathrm{Hom}(B, X) \longrightarrow \mathrm{Hom}(A, X)$$

is exact for all $X \in \mathcal{A}$, then (2.2.3) is right exact.

Proof. By duality, we only need to prove the first statement. Taking $X = \ker(A \to B)$, we see that $\ker(A \to B) = 0$, hence $A \to B$ is a monomorphism. Taking $X = \ker(B \to C)$, we see that the natural morphism $X \to B$ is induced from a morphism $X \to A$—in other words, that $\ker(B \to C) = \operatorname{im}(A \to B)$. $\qquad\square$

This can be used to derive a simple but important result, that is proved as Proposition 1.4.11 in [**Fer20**].

Corollary 2.2.20. *Let A be a ring, N an A-module. The functors $N \otimes -$ and $- \otimes N$ are both right exact on Mod_A.*

Proof. By symmetry, we prove the result for $N \otimes -$. The crucial point is the adjunction

$$\operatorname{Hom}(M \otimes_A N, P) \cong \operatorname{Hom}(M, \operatorname{Hom}(N, P)),$$

described in Example 1.5.2(d). Let

$$0 \longrightarrow M_1 \longrightarrow M_2 \longrightarrow M_3 \longrightarrow 0$$

be a short exact sequence of A-modules. Using the adjunction and the fact $\operatorname{Hom}(-, \operatorname{Hom}(N, P))$ is left exact, we find that

$$0 \longrightarrow \operatorname{Hom}(M_3 \otimes N, P) \longrightarrow \operatorname{Hom}(M_2 \otimes N, P) \longrightarrow \operatorname{Hom}(M_1 \otimes N, P)$$

is exact for all A-modules P. The conclusion follows from Proposition 2.2.19. $\qquad\square$

Another important example of exact functor is localization. For a ring A and a multiplicative set $S \subset A$, it is proved in [**Fer20**, Proposition 1.6.17] that the localization functor $\mathrm{Mod}_A \to \mathrm{Mod}_{S^{-1}A}$ is exact. If you have not seen the proof, this is a simple but useful exercise. A useful consequence is the following fact, which is [**Fer20**, Corollary 1.6.20], but we prove again here for reference.

Proposition 2.2.21. *Let $f \colon M \to N$ be a map of A-modules. Then f is injective (resp., surjective) if and only if the induced map $f_P \colon M_P \to N_P$ is injective (resp., surjective) for all prime (or even maximal) ideals $P \subset A$.*

Proof. An A-module M is 0 if and only if $M_P = 0$ for all prime (or maximal) ideals $P \subset A$—this follows at once by considering the annihilator ideal $\operatorname{Ann}(M)$. The conclusion follows by applying this fact and the exactness of localization to $\ker f$ and $\operatorname{coker} f$ respectively. $\qquad\square$

2.3. Sheaves

So far, the only examples of Abelian categories that we have seen are categories of modules, or small variations on them. If these were the only examples, there would be no compelling reason to work in the generality of Abelian categories. The main reason for doing so is to provide a foundation that also work for sheaves.

Sheaves are ubiquitous in all branches of geometry: they provide a natural language to talk about additional structures that can exist on a topological space. While they are not strictly necessary for our treatment of commutative algebra, it would be a shame to omit them completely. In this section, we give a quick introduction to this language.

Let X be a topological space. Intuitively, a sheaf S on X is some kind of construction that associates to every open set $U \subset X$ an object $S(U)$, and such that two operations are allowed:

(i) a restriction map $S(V) \to S(U)$ whenever there is an inclusion of open sets $U \subset V$, and

(ii) a gluing process that makes it possible to reconstruct $S(U)$ from the datum of $S(U_i)$ for an open covering $\{U_i\}$ of U.

To start being more precise, we will consider the first item.

Definition 2.3.1. Let X be a topological space, $\mathcal{C}$ a category. A *presheaf* S on X with values in $\mathcal{C}$ is the datum of

(1) an object $S(U) \in \mathcal{C}$ for every open set $U \subset X$, and

(2) a restriction map $r_{VU} \colon S(V) \to S(U)$ for every inclusion of open sets $U \subset V$.

These should satisfy the compatibility condition that $r_{WU} = r_{VU} \circ r_{WV}$ for open sets $U \subset V \subset W$.

A moment's thought shows that the datum a presheaf on X with values in $\mathcal{C}$ is the same as a contravariant functor $\mathrm{Top}(X) \to \mathcal{C}$, where $\mathrm{Top}(X)$ is the order category of the open sets of X; see Example 1.1.4(j) and (k). In particular, presheaves on X with values in $\mathcal{C}$ form a category, with natural transformations as functors, which we will denote $\mathrm{Psh}_{\mathcal{C}}(X)$, or simply $\mathrm{Psh}(X)$ when $\mathcal{C}$ is implicit.

The gluing requirement is more subtle. To understand it, it is worth keeping in mind the prototypical example of sheaf, which is the sheaf C of continuous functions $X \to \mathbb{R}$. This is defined by

$$C(U) := \{f \colon U \to \mathbb{R} \mid f \text{ is continuous}\},$$

with the obvious restriction maps. Then C is a presheaf of rings, with an important additional property. Given an open set U and an open covering $\{U_i\}_{i \in I}$ of U, a function $f \in C(U)$ is determined by the datum of all its restrictions $f_i = r_{UU_i}(f)$. Moreover, a collection of functions $\{f_i \in C(U_i)\}$ can be glued together to get a function on U, *provided* these functions agree on the intersections $U_{ij} = U_i \cap U_j$.

To simplify notation, let us denote $r_i = r_{UU_i}$ and $r_{ij} = r_{U_iU_{ij}}$. Then what we are saying is that for each collection $\{f_i \in C(U_i) \mid i \in I\}$ such that $r_{ij}(f_i) = r_{ji}(f_j)$, there is a unique $f \in C(U)$ such that $f_i = r_i(f)$.

We can rephrase that by introducing a category $\mathcal{I}$ with objects $I \sqcup I \times I$ and maps $i \to (i, j)$ for all $i, j \in I$. Then we have a functor $F \colon \mathcal{I} \to C$ defined by

$$F(i) = C(U_i), F(i, j) = C(U_{ij}) \text{ and } F(i \to (i, j)) = r_{ij},$$

and our observation is that $C(U)$, together with the maps $\{r_i\}$, is a limit of this functor.

If C admits arbitrary products, this is made more transparent by the remark that

$$C(U) \xrightarrow{\; r_i \;} \prod_i C(U_i) \underset{r_{ji}}{\overset{r_{ij}}{\rightrightarrows}} \prod_{i,j} C(U_i \cap U_j)$$

exhibits $C(U)$ as the equalizer of the upper and lower maps.

This property of functions in $C(U)$ being determined locally is something which also happens for differentiable functions, holomorphic functions and so on, and it is the concept that we are trying to capture.

Definition 2.3.2. Let I be a set. The pair category over I, denoted $\mathrm{Pair}(I)$, has objects $i \in I$ and $(i, j) \in I \times I$. The only nonidentity morphisms of $\mathrm{Pair}(I)$ are maps $i \to (i, j)$ for all $i \neq j \in I$.

Definition 2.3.3. Let $S \colon \mathrm{Top}(X)^{op} \to C$ be a presheaf on X, with restriction maps $r_{UV} \colon S(U) \to S(V)$. We say that S is a *sheaf* if it has the following property. For all open sets $U \subset X$ and open coverings $\{U_i\}_{i \in I}$ of U, consider the functor $F \colon \mathrm{Pair}(I) \to C$ defined by

$$F(i) = S(U_i), F((i, j)) = S(U_{ij}) \text{ and } F(i \to (i, j)) = r_{U_iU_{ij}}.$$

Then $S(U) = \lim F$ (with its restriction maps). Elements of $S(U)$ are called *sections* of S on U.

To simplify notation when no ambiguity can arise, given open sets $V \subset U$, a presheaf S and $s \in S(U)$, we will denote $r_{UV}(s) = s|_V$.

Example 2.3.4.

(a) For every pair of topological spaces X, Y, we have the sheaf C_Y defined by $C_Y(U) = \{f : U \to Y \mid f \text{ is continuous}\}$ for an open set $U \subset X$. We simply write C for the sheaf $C_\mathbb{R}$.

(b) For every pair of C^k manifolds X, Y (k may be ∞), we have the sheaf C_Y^k defined by $C_Y^k(U) = \{f : U \to Y \mid f \text{ is of class } C^k\}$ for $U \subset X$.

(c) For every pair of complex manifolds X, Y, we have the sheaf $\mathcal{O}_Y$ defined by $\mathcal{O}_Y(U) = \{f : U \to Y \mid f \text{ is holomorphic } \}$ for $U \subset X$. We simply write $\mathcal{O}$ for the sheaf $\mathcal{O}_\mathbb{C}$.

(d) Corresponding to the previous items, we have sheaves of nonzero continuous functions C^* on any topological space X and nonzero holomorphic functions $\mathcal{O}^*$ on any complex manifold X.

(e) Let $\pi : E \to X$ be a surjective map of topological spaces. With an abuse of notation that hopefully will cause no harm, we write C_E for the sheaf of continuous *sections* of π, namely

$$C_E(U) = \{f : X \to E \mid f \text{ is continuous and } \pi \circ f = id_X\}.$$

This is especially common when $E \to X$ is a vector bundle over X, and is the reason why we call elements of a sheaf "sections".

(f) In the same way, we have the sheaves of C^k (resp., holomorphic) sections of maps of C^k (resp., complex) manifolds.

(g) In particular, for each smooth manifold M, we have the sheaf Ω^p of smooth p-forms on M. These are sections of $\Lambda^p T^* M$, where $TM \to M$ is the tangent bundle of M.

(h) For a pair X, Y of algebraic varieties, endowed with the Zariski topology (Section A.9), we can define similar sheaves of regular functions $U \to Y$, or regular sections of a map $Y \to X$.

(i) To get an example of a presheaf that is not a sheaf, we have to involve a condition that is not local. A typical example is the following. Let X be a topological space, and define

$$C_b(U) = \{f : U \to \mathbb{R} \mid f \text{ is continuous and bounded } \}.$$

Then, C_b is not a sheaf in general. For instance, we can cover $\mathbb{R}$ with intervals and take—say, the identity function $\mathbb{R} \to \mathbb{R}$. This is bounded on each interval, but not bounded globally.

(j) Let X be a topological space, μ a measure on the σ-algebra of Borel subsets of X (the σ-algebra generated by open sets). For a similar reason as the previous point, the presheaf $\mathcal{L}^p$ defined by

$$\mathcal{L}^p(U) = \left\{ f \colon U \to \mathbb{R} \mid \int_U |f|^p d\mu < \infty \right\} / \sim,$$

where $\sim$ is almost everywhere equality, is not a sheaf. The condition that $|f|^p$ has finite integral is global, and can hold on a covering of U without being true on U itself.

Since every sheaf is a presheaf, we can form a category of sheaves $\mathrm{Sh}_{\mathcal{C}}(X)$ on the topological space, which is just the full subcategory of $\mathrm{Psh}_{\mathcal{C}}(X)$ where the objects are sheaves. Sometimes we will just write $\mathrm{Sh}(X)$ when $\mathcal{C}$ is implicit from the context. More explicitly, a morphism $f \colon S \to T$ (with values in $\mathcal{C}$) is a collection of morphisms $f_U \colon S(U) \to T(U)$ for each open set $U \subset X$ such that

$$\begin{array}{ccc} S(V) & \xrightarrow{f_V} & T(V) \\ \downarrow & & \downarrow \\ S(U) & \xrightarrow{f_U} & T(U) \end{array}$$

commutes for every inclusion $U \subset V$ of open sets.

Let $x \in X$ be a point, S a presheaf on X with values in $\mathcal{C}$. Assuming that $\mathcal{C}$ has small colimits, we can define the *stalk* of S at x as

$$S_x := \varinjlim_{x \in U} S(U).$$

That is, the family of open sets $U \ni x$ forms a direct system, and its direct limit is S_x. Hence, an element of S_x can be identified with an element of $S(U)$ for some $U \ni x$, modulo the equivalence relation that identifies $s \in S(U)$ with $r_{UV}(s) \in S(V)$ whenever $U \supset V \ni x$.

This is a common construction when considering the sheaf of continuous or differentiable functions. In this case, an element of S_x is called a *germ* of a continuous (or differentiable) function at x. For instance, when X is a complex manifold and $S = \mathcal{O}_X$, the sheaf of holomorphic functions, then the stalk $\mathcal{O}_{X,x}$ can be identified with the set of convergent power series in x (this is just a way of rephrasing the fact that every holomorphic function has a unique expansion in power series which converges on a small radius).

Let X be a topological space, $\mathcal{C}$ a category. Since every sheaf is in particular a presheaf, there is a natural inclusion $p \colon \mathrm{Sh}_{\mathcal{C}}(X) \to \mathrm{Psh}_{\mathcal{C}}(X)$. When $\mathcal{C} = \mathrm{Set}$, we can describe an adjoint of this functor, by means of stalks.

Let S be a presheaf of sets on X, so that in particular for every point $x \in X$ we have the stalk S_x. The disjoint union

$$\widehat{X}_S := \bigsqcup_{x \in X} S_x$$

is endowed with a projection $\pi \colon \widehat{X}_S \to X$—just define $\pi(f) = x$ if $f \in S_x$. We give $\widehat{X}_S$ the coarsest topology that makes π continuous. Namely, for an open set $U \in X$ and an element $f \in S(U)$, we have corresponding germs $f_x \in S_x$ for all $x \in U$. The topology on $\widehat{X}_S$ is generated by sets of the form

$$U_f := \{ f_x \mid x \in U \}$$

for all such pairs U, $f \in S(U)$.

Definition 2.3.5. The space $\widehat{X}_S$ is called the *étalé space* associated to the presheaf S on X.

The étalé space defines a sheaf $C_{\widehat{X}_S}$ on X of continuous sections, explicitly given by

$$C_{\widehat{X}_S}(U) = \{ f \colon U \to \widehat{X}_S \mid f \text{ is continuous and } \pi \circ f = \mathrm{id}_U \}.$$

Fix a point $x \in X$ and a germ $f_x \in S_x$. This is the class of some element $f \in S(U)$ for an open set $U \ni x$. The restriction $\pi|_{U_f} \colon U_f \to U$ is by construction a homeomorphism. Moreover, there is a map of presheaves $\iota \colon S \to C_{\widehat{X}_S}$, defined by $(\iota f)(x) = f_x$ for $U \subset X$ open and $f \in S(U)$.

Proposition 2.3.6. *If S is a sheaf, the map $\iota \colon S \to C_{\widehat{X}_S}$ is an isomorphism.*

Proof. Fix an open set $U \subset X$, and consider the map $\iota_U \colon S(U) \to C_{\widehat{X}_S}(U)$. Assume that $\iota(f) = \iota(g)$ for some pair $f, g \in S(U)$. This means that for all $x \in U$ we have $f_x = g_x$, so there is an open set U_x with $x \in U_x \subset U$ such that $r_{UU_x}(f) = r_{UU_x}(g)$. Since this holds for all $x \in U$, we can find a covering of U by open sets V such that $r_{UV}(f) = r_{UV}(g)$, and this implies that $f = g$ since S is a sheaf.

Similarly, take any element $F \in C_{\widehat{X}_S}(U)$. This determines, for every point $x \in U$, a germ f_x of section of π. This in turn, is defined on an open set $U_x \ni x$. If we take two such germs f_x, f_y defined open sets U_x, U_y, their restrictions agree on $U_x \cap U_y$ because π is a local homeomorphism. These elements then glue to a unique $f \in S(U)$ satisfying $\iota(f) = F$, showing that ι is surjective. $\qquad\square$

The above construction shows that for any presheaf of sets we have a canonical way to construct a sheaf in such a way that it is an isomorphism on sheaves.

Definition 2.3.7. Let X be a topological space, $S \in \mathrm{Psh}_{\mathrm{Set}}(X)$. The sheaf of continuous sections of the étalé space $\pi \colon \widehat{X}_S \to X$ is called the sheaf *associated* to S, and sometimes denoted aS.

The above construction is called *sheafification*. We can also extend it to morphisms: given a morphism of presheaves $F \colon S \to T$, we get a morphism $aF \colon aS \to aT$ by
$$(aF)(x \to f_x) := (x \to (Ff)_x),$$
which is a continuous section of $\widehat{X}_T \to X$. This makes a into a functor $\mathrm{Psh}(X) \to \mathrm{Sh}(X)$.

Proposition 2.3.8. *The associated sheaf functor a is left adjoint to the forgetful functor* $\mathrm{Sh}(X) \to \mathrm{Psh}(X)$.

Proof. We have to prove that for a presheaf P and a sheaf S, we have a natural bijection
$$\Psi_{P,S} \colon \mathrm{Hom}_{\mathrm{Sh}(X)}(aP, S) \cong \mathrm{Hom}_{\mathrm{Psh}(X)}(P, S).$$
Now, certainly there is a well-defined natural map
$$\mathrm{Hom}_{\mathrm{Psh}(X)}(P, S) \xrightarrow{\ a\ } \mathrm{Hom}_{\mathrm{Sh}(X)}(aP, aS) \xrightarrow{\ \cong\ } \mathrm{Hom}_{\mathrm{Sh}(X)}(aP, S)$$
given by the natural isomorphism between aS and S. In the other direction, the morphism of presheaves $\iota \colon P \to aP$ induces a map
$$\mathrm{Hom}_{\mathrm{Sh}(X)}(aP, S) = \mathrm{Hom}_{\mathrm{Psh}(X)}(aP, S) \xrightarrow{\ \circ \iota\ } \mathrm{Hom}_{\mathrm{Psh}(X)}(P, S).$$
It is immediate to check that these two maps are inverse of each other. $\square$

Remark 2.3.9. Our construction of sheafification works for the category Set. It is easy to extend it to work on many categories which consist of sets with additional structure. For concreteness, let $\mathcal{C}$ be any of the categories Ring, Mod_A for a ring A, Vect_k for a field k, Ab or Grp.

As we have seen in Example 1.5.2(b), the forgetful functor $F \colon \mathcal{C} \to \mathrm{Set}$ has a left adjoint which is the free functor $\mathrm{Set} \to \mathcal{C}$. It follows that F preserves limits. Since the sheaf condition is expressed in terms of limits, it follows that a presheaf $S \in \mathrm{Psh}_{\mathcal{C}}(X)$ which, when regarded as a presheaf of sets, is a sheaf of sets, is indeed a sheaf with value in $\mathcal{C}$.

Given a presheaf $P \in \mathrm{Psh}_{\mathcal{C}}(X)$, the stalks of P at a point $x \in X$—as sets—are defined as a direct limit of $P(U)$ for $U \ni x$. In all the above cases, it follows that the stalk P_x inherits in an obvious way the operations, so that P_x is an element of $\mathcal{C}$. We can then carry out the above procedure of sheafification using the étalé space and obtain an associated sheaf aP. Since aP is a sheaf of sets and a presheaf with values in $\mathcal{C}$, it is a sheaf with values in $\mathcal{C}$.

Example 2.3.10. Let X be a topological space, S a set. The constant presheaf C_S defined by $C_S(U) = S$ is not a sheaf, as soon as S has at least two elements and X has two disjoint open sets. In fact, if U, V are disjoint open sets of X and $u, v \in S$ are distinct, the two sections $u \in C_S(U)$ and $v \in C_S(V)$ fail to satisfy the gluing condition. The compatibility is trivially satisfied, since $U \cap V = \emptyset$, but there is no element in $C_S(U \cup V) = S$ that restricts to u on U and to v on V.

The sheaf associated to C_S is the sheaf of *locally constant* functions with values in S (this is immediate from the definition of étalé space). In particular, $aC_S(U)$ is the disjoint union of one copy of S per each connected component of U. This sheaf is called the *locally constant* sheaf with values in S, denoted by $\underline{S}$, or sometimes simply by S, when no confusion can arise.

After this detour on sheaves, we come back to the main topic of our chapter, and show that in many cases we can construct Abelian categories out of sheaves. Let $\mathcal{O}$ be a sheaf of rings over the topological space X. A (pre)sheaf $\mathcal{M}$ of $\mathcal{O}$-modules is a (pre)sheaf of Abelian groups such that $\mathcal{M}(U)$ is a $\mathcal{O}(U)$-module. We will denote the categories of presheaves and sheaves of $\mathcal{O}$-modules by $\mathrm{Psh}_{\mathcal{O}}(X)$ and $\mathrm{Sh}_{\mathcal{O}}(X)$. As in our earlier example, there is a sheafification functor

$$a\colon \mathrm{Psh}_{\mathcal{O}}(X) \to \mathrm{Sh}_{\mathcal{O}}(X).$$

Example 2.3.11.

(a) Let $\pi\colon E \to X$ be a topological real vector bundle. Sections of E can be multiplied by continuous functions $X \to \mathbb{R}$. This gives $C_E(X)$ the structure of a module over $C(X)$.

(b) In a similar fashion, if $\pi\colon E \to X$ is a vector bundle of smooth manifolds, there is a structure of $C^\infty(X)$-module for the sheaf of sections $C_E^\infty(X)$.

(c) If $\pi\colon E \to X$ is a vector bundle of complex manifolds, there is a structure of $\mathcal{O}(X)$-module for the sheaf of sections $\mathcal{O}_E(X)$.

(d) The same holds in the algebraic category: the sheaf of sections of a vector bundle of algebraic varieties is a sheaf of modules over the regular functions.

The aim of this section is to show that the category of sheaves of $\mathcal{O}$-modules $\mathrm{Sh}_{\mathcal{O}}(X)$ is an Abelian category, in a fairly nontrivial way. To do this, we need to define the kernel and cokernel of a morphism $f\colon S \to T$ in $\mathrm{Sh}_{\mathcal{O}}(X)$. The naive approach would be to define

$$(\ker f)(U) := \ker(f_U\colon S(U) \to T(U)),$$
$$(\mathrm{coker}\, f)(U) := \mathrm{coker}(f_U\colon S(U) \to T(U)).$$

This works only in part, because coker f, thus defined, can fail to be a sheaf.

Example 2.3.12. Take $X = \mathbb{C}$ and consider the sheaf $\mathcal{O}$ of holomorphic functions and $\mathcal{O}^*$ of nonzero holomorphic functions. The exponential gives a map

$$\exp\colon \ \mathcal{O} \longrightarrow \mathcal{O}^*.$$

$$f \longmapsto e^f$$

We denote by

$$C(U) := \operatorname{coker}(\exp_U \colon \mathcal{O}(U) \to \mathcal{O}^*(U))$$

the presheaf defined as the naive cokernel.

Let $U \subset \mathbb{C}$ be a disk not containing 0. Then the complex logarithm is defined on U, hence locally on U we can write $z = e^{L(z)}$, where L is a branch of the logarithm. This means that the class of z in $C(U)$ is 0, since z is in the image of $\exp_U$.

On the other hand, this is not true on an annulus going around 0, say

$$A := \{z \mid 1 < |z| < 2\}.$$

This entails that the class of z in $C(A)$ is not zero. Since A can be covered by disks not meeting 0, this implies that C cannot be a sheaf.

On the other hand, we have

Proposition 2.3.13. *Let X be a topological space, $f\colon S \to T$ a map of sheaves of Abelian groups. Then the naive kernel K defined as the presheaf*

$$K(U) := \ker(f_U \colon S(U) \to T(U))$$

is a sheaf.

Notice that this applies to sheaves of A-modules for a ring A, or to sheaves of $\mathcal{O}$-modules for a sheaf of rings $\mathcal{O}$ over X.

Proof. Let $U \subset X$ be open, $\{U_i\}$ a covering of U, and $\{s_i \in K(U_i)\}$ a family of sections of K over U_i, compatible over the intersections $U_i \cap U_j$. Then in particular the s_i are sections of S, hence they glue uniquely to give a section $s \in S(U)$. Moreover $f(s) = 0$, since $f(s_i) = 0$ and T is a sheaf. It follows that $s \in K(U)$. □

Definition 2.3.14. Let X be a topological space, $f\colon S \to T$ a map of sheaves of Abelian groups (or modules over a ring, or over a sheaf of rings). The kernel $\ker f$ is the sheaf

$$(\ker f)(U) := \ker(f_U \colon S(U) \to T(U)),$$

while the cokernel $\operatorname{coker} f$ is the sheaf *associated* to the presheaf

$$C(U) := \operatorname{coker}(f_U \colon S(U) \to T(U)).$$

This allows us to finish the construction of our example.

Proposition 2.3.15. *Let X be a topological space. The category of sheaves* $\mathrm{Sh}_{\mathrm{Ab}}(X)$ *is Abelian.*

Proof. First, notice that the category of presheaves $\mathrm{Psh}_{\mathrm{Ab}}(X)$, with the naive kernels and cokernels, is trivially Abelian. The category $\mathrm{Sh}_{\mathrm{Ab}}(X)$ is clearly additive, so we need to prove the existence of the canonical decomposition for morphisms of sheaves.

As a first step, we prove that kernels and cokernels, as defined in Definition 2.3.14, satisfy the respective universal properties (hence they are actual kernels and cokernels in $\mathrm{Sh}_{\mathrm{Ab}}(X)$). For kernels this is immediate: they satisfy the universal property in the larger category of presheaves. For cokernels, this follows because cokernels in $\mathrm{Psh}_{\mathrm{Ab}}(X)$ satisfy the universal property, and the associated sheaf functor a, being a left adjoint, preserves colimits.

Finally, let $f\colon S \to T$ be a morphism of sheaves. Looking at S and T as presheaves, we get the canonical decomposition

$$0 \longrightarrow K \longrightarrow S \longrightarrow I \longrightarrow T \longrightarrow C \longrightarrow 0$$

in $\mathrm{Psh}_{\mathrm{Ab}}(X)$. Applying the associated sheaf functor a and taking into account Proposition 2.3.13, we obtain

$$0 \longrightarrow K \overset{k}{\longrightarrow} S \longrightarrow aI \longrightarrow T \overset{c}{\longrightarrow} aC \longrightarrow 0.$$

Here we use the fact that a is exact (prove it!). By definition, K and aC are respectively the kernel and cokernel of f. It remains to check that aI is the kernel of c and the cokernel of k, and this follows again from the equivalent property of I and the fact that a is exact. $\square$

Corollary 2.3.16. *Let X be a topological space. For every ring A, the category* $\mathrm{Sh}_{\mathrm{Mod}_A}(X)$ *is Abelian, and for every sheaf of rings $\mathcal{O}$, the category* $\mathrm{Sh}_{\mathrm{Mod}_{\mathcal{O}}}(X)$ *is Abelian.*

Unlike presheaves, the category of sheaves of Abelian groups on X actually encodes nontrivial information about the topology of X, due to the fact that cokernels are computed via sheafification. In fact, this phenomenon is what makes the functor of sections not exact, and gives rise—as we will see—to the theory of sheaf cohomology.

Proposition 2.3.17. *Let X be a topological space, $\Gamma\colon \mathrm{Sh}_{\mathrm{Ab}}(X) \to \mathrm{Ab}$ the functor of global sections, defined by $\Gamma(S) = S(X)$. Then Γ is left exact.*

Proof. Let

$$0 \longrightarrow S \overset{f}{\longrightarrow} T \overset{g}{\longrightarrow} U$$

be an exact sequence of sheaves on X. Then (S, f) is the kernel of g, so by our definition of kernel $S(X)$ is the kernel of $g_X \colon T(X) \to U(X)$, and

$$0 \longrightarrow S(X) \xrightarrow{f_X} T(X) \xrightarrow{g_X} U(X)$$

is exact as well. □

2.4. Sites

An important realization of Grothendieck was that, in order to define sheaves, one does not need a topological space after all. For presheaves this is clear as they are just contravariant functor from the category of open sets. For the gluing condition to make sense, one only needs a suitable notion of covering, and that can be easily axiomatized. In this section, we are going to show that, given a suitable axiomatization of coverings, one can carry out essentially all the theory of sheaves. The only thing that we will need to rework is the operation of sheafification, since our construction in the topological setting made use of the étalé space, which does not generalize well to this broader context.

Definition 2.4.1. Let $\mathcal{C}$ be a category, and assume that $\mathcal{C}$ has fibered products. A *Grothendieck topology* τ on $\mathcal{C}$ is the datum, for each object $U \in \mathcal{C}$, of a class $\tau(U)$ of sets of morphisms $\{f_i \colon U_i \to U\}_{i \in I}$, called *coverings* of U. Such families of coverings should satisfy the following properties:

(1) if $\{f_i \colon U_i \to U\}$ is a covering of U and $g \colon V \to U$ is a morphism in $\mathcal{C}$, the pullback family $\{U_i \times_U V \to V\}$ is a covering of V;

(2) if $\{f_i \colon U_i \to U\}$ is a covering of U and, for every i, $\{g_{ij} \colon U_{ij} \to U_i\}$ is a covering of U_i, the compositions $\{f_i \circ g_{ij} \colon U_{ij} \to U\}$ form a covering of U; and

(3) the identity map $\{\mathrm{id}_U \colon U \to U\}$ is a covering of U.

A category endowed with a Grothendieck topology is called a *site*.

The category of open sets of a topological space is clearly a site with the usual notion of covering. There are some other useful examples.

Example 2.4.2.

(a) The *big site* associated to a topological space S is the category Top $/S$ of topological spaces endowed with a map to S. If $U \to S$ is an object, a covering is a family of morphisms $f_i \colon U_i \to U$ such that each f_i is an embedding, $f_i(U_i)$ is open in U, and the open sets $\{f_i(U_i)\}$ cover U.

(b) The most important example of a Grothendieck topology is the *étale topology* on schemes. We do not have the background to treat it here, but the interested reader is advised to look at [**Mil13**].

It is easy to copy almost verbatim our definition of sheaf and presheaf for a site.

Definition 2.4.3. Let $X = (\mathcal{C}, \tau)$ be a site, $\mathcal{D}$ a category. A *presheaf* on X with values in $\mathcal{D}$ is just a functor $S \colon \mathcal{C}^{op} \to \mathcal{D}$. The presheaf S is called a *sheaf* if it has the following property. For every covering family $\{f_i \colon U_i \to U\}_{i \in I}$ of X, consider the functor $F \colon \mathrm{Pair}(I) \to \mathcal{D}$ defined by

$$F(i) = S(U_i), F(i,j) = S(U_{ij}) \text{ and } F(i \to (i,j)) = U_i \to U_i \times_U U_j.$$

Then $S(U) = \lim F$ (with its restriction maps). Elements of $S(U)$ are called *sections* of S on U.

If the category $\mathcal{D}$ has arbitrary products, we can rephrase the sheaf condition by saying that for all coverings $\{f_i \colon U_i \to U\}_{i \in I}$, $S(U)$ is the equalizer of the map of products

$$(2.4.1) \qquad \prod_i S(U_i) \rightrightarrows \prod_{i,j} S(U_i \times_U U_j),$$

where the upper map is induced by the pullback of f_i and the lower map by the pullback of f_j. We will describe this succintly with the following diagram:

$$S(U) \xrightarrow{f_i} \prod_i S(U_i) \rightrightarrows \prod_{i,j} S(U_i \times_U U_j).$$

Just as in the case of topological spaces, if $X = (\mathcal{C}, \tau)$ is a small site (meaning that $\mathcal{C}$ is small) and $\mathcal{D}$ is a category, we can define the category $\mathrm{Psh}_{\mathcal{D}}(X)$ of $\mathcal{D}$-valued presheaves on X, and its subcategory $\mathrm{Sh}_{\mathcal{D}}(X)$ of sheaves.

For presheaves on sites, we do not have the notion of point (at least, not in an obvious way), therefore our construction of sheafification using stalks will not work. To describe an alternative approach, we introduce some terminology.

Definition 2.4.4. Let $X = (\mathcal{C}, \tau)$ be a site, P a presheaf on X, with values in Set. We say that S is *separated* if for every object $U \in \mathcal{C}$ and covering $\{f_i \colon U_i \to U\}_{i \in I}$, the natural map

$$P(U) \xrightarrow{f_i} \prod_i P(U_i)$$

is injective.

By definition, every sheaf is separated, so separated presheaves are an intermediate notion between presheaves and sheaves. They are especially useful because of the *plus construction*. Given a presheaf of sets P on X, define the presheaf

$$(2.4.2) \qquad P^+(U) := \varinjlim \mathrm{eq}\left(\ \prod P(U_i) \rightrightarrows \prod P(U_i \times_U U_j)\ \right),$$

where eq denotes the equalizer and the direct limit is over all coverings $\{\ U_i \to U\}$.

Remark 2.4.5. Since the family of coverings of U is a priori a proper class, it may very well happen that the direct limit does not exist. A presheaf P such that the limit in (2.4.2) exists is called *basically bounded* [**Wat75**]. We will carry out the sheafification procedure only for basically bounded presheaves. This does not create any problems when $\mathcal{C}$ is small, or even equivalent to a small category (in which case we say that $\mathcal{C}$ is *essentially small*).

Theorem 2.4.6. *Let $X = (\mathcal{C}, \tau)$ be a site, $P \in \mathrm{Psh}(X)$ a presheaf, and assume that P^+ is defined.*

(i) *The presheaf P^+ is separated.*

(ii) *If P is separated, the presheaf P^+ is a sheaf.*

In particular, for every presheaf P, P^{++} (when defined) is a sheaf.

Proof. First, let us spell out the definition of $P^+(U)$ for $U \in \mathcal{C}$. An element of $P^+(U)$ is the datum of a covering $\{f_i \colon U_i \to U\}$ and an element

$$s \in \mathrm{eq}\left(\ \prod P(U_i) \rightrightarrows \prod P(U_i \times_U U_j)\ \right),$$

modulo a certain equivalence relation. In turn, s is given by section $\{s_i \in P(U_i)\}$ such that $P(p_i)(s_i) = P(p_j)(s_j)$, where p_i and p_j are the two projections from $U_i \times_U U_j$ to its factors. Two such data $\{U_i \to U\}, \{s_i\}$ and $\{U_i' \to U\}, \{s_i'\}$ represent the same element of the direct limit if they agree on a common refinement—explicitly, there is a covering $\{V_{ij}\}$ with maps $v_{ij} \colon V_{ij} \to U_i$ and $v_{ij}' \colon V_{ij} \to U_j'$ such that $P(v_{ij})(s_i) = P(v_{ij}')(s_j')$.

For (i), let $s, t \in P^+(U)$ two sections. Assume that $\{g_i \colon V_i \to U\}$ is a covering such that $P^+(g_i)(s) = P^+(g_i)(t)$ for all i. Both s and t can be represented on the same covering $\{U_i\}$. By hypothesis, these datum become the same on the common refinement $\{U_i \times_U V_j\}$, hence $s = t$.

For (ii), assume that P is separated, and let $\{f_i \colon U_i \to U\}$ be a covering, and for each i, $s_i \in P^+(U_i)$. Assume that the sections s_i agree on the common restrictions, so that the pullback of s_i and s_j on $U_{ij} := U_i \times_U U_j$ are the same. Up to a refinement, we can assume that s_i is represented by a datum $\{s_{ij}\}$ on the covering $\{U_{ij}\}$. Putting together all maps $\{U_{ij} \to U\}$

gives a covering of U, with an associated datum $s_{ij} \in P(U_{ij})$. This datum is compatible, that is, the pullback of s_{ij} and s_{kl} on $U_{ij} \times_U U_{kl}$ is the same. (Why?) Hence, it represents some element $s \in P^+(U)$. By construction, the pullback of s and of s_i to each U_{ij} are the same element of $P^+(U_{ij})$, hence—since P is separated—s restricts to s_i on U_i.   □

We can now use the plus construction to define sheafification.

Definition 2.4.7. Let X be a site, P a presheaf of sets on X. The *sheaf associated* to P (when defined) is $aP := P^{++}$.

By the above result, $aP = P^{++}$ is always a sheaf. Moreover, there is a map of presheaves $\iota \colon P \to aP$. This is because for an object U, $\mathrm{id}_U \colon U \to U$ is a covering, hence $P(U)$ is one of the terms appearing in the direct limit defining $P^+(U)$.

Remark 2.4.8. Assume that X is a small site, so that aP is defined for *every* presheaf of sets P. By the universal property of direct limit, a morphism of presheaves $P_1 \to P_2$ induces a morphism $aP_1 \to aP_2$, so that a becomes a functor $\mathrm{Psh}(X) \to \mathrm{Sh}(X)$.

As in the case of topological spaces, we can prove

Proposition 2.4.9. *If S is a sheaf, the map $\iota \colon S \to aS$ is an isomorphism.*

Proof. Clear, since for every covering $\{U_i \to U\}$, $S(U)$ is the equalizer of (2.4.1), hence all terms appearing in the direct limit defining $S^+(U)$ are isomorphic to each other.   □

We can also generalize Proposition 2.3.8, with exactly the same proof.

Proposition 2.4.10. *Let X be a small site. The associated sheaf functor a is left adjoint to the forgetful functor $\mathrm{Sh}(X) \to \mathrm{Psh}(X)$.*

As in the case of topological spaces, we can define the sheafification functor also for presheaves with values in Ab, in Mod_A for a ring A, or in Vect_k for a field k. In such situations, we can also define kernels and cokernels.

Definition 2.4.11. Let X be a small site, $f \colon S \to T$ a map of sheaves of Abelian groups (or modules over a ring). The kernel $\ker f$ is the sheaf

$$(\ker f)(U) := \ker(f_U \colon S(U) \to T(U)),$$

while the cokernel $\mathrm{coker}\, f$ is the sheaf *associated* to the presheaf

$$C(U) := \mathrm{coker}(f_U \colon S(U) \to T(U)).$$

Using Definition 2.4.11 we can generalize Proposition 2.3.15, again with the same proof.

Proposition 2.4.12. *Let X be a small site. The category $\mathrm{Sh}_{\mathrm{Ab}}(X)$ is Abelian, and so is the category $\mathrm{Sh}_{\mathrm{Mod}_A}(X)$ for a ring A.*

One easy but useful fact is

Proposition 2.4.13. *Let X be a small site. The functor $a\colon \mathrm{Psh}_{\mathrm{Ab}}(X) \to \mathrm{Sh}_{\mathrm{Ab}}(X)$ is exact.*

Proof. The functor a is clearly additive. Being a left adjoint, it preserves colimits, in particular cokernels. To see that it preserves kernels, notice that the construction of the associated sheaf requires taking a filtered colimit, and that commutes with finite limits by Theorem 1.4.16 and Remark 1.4.18. $\square$

2.5. Standard lemmas

Having at our disposal complexes and exact sequences, we can start doing the basics of homological algebra in Abelian categories. There are a few standard results that are used over and over, and they are a good opportunity to learn the basic tricks of the trade. We will revisit these results in the exercises, using different techniques.

Lemma 2.5.1 (Five lemma). *Let*

$$
\begin{array}{ccccccccc}
A & \longrightarrow & B & \longrightarrow & C & \longrightarrow & D & \longrightarrow & E \\
\downarrow{\scriptstyle\alpha} & & \downarrow{\scriptstyle\beta} & & \downarrow{\scriptstyle\gamma} & & \downarrow{\scriptstyle\delta} & & \downarrow{\scriptstyle\epsilon} \\
A' & \longrightarrow & B' & \longrightarrow & C' & \longrightarrow & D' & \longrightarrow & E'
\end{array}
$$

be a commutative diagram in an Abelian category, where the rows are exact sequences. Assume that β, δ are isomorphisms, α is an epimorphism, and ϵ is a monomorphism. Then γ is an isomorphism as well.

The five lemma is an immediate consequence of the following *four lemma*, together with its dual statement.

Lemma 2.5.2 (Four lemma). *Let*

$$
\begin{array}{ccccccc}
A & \xrightarrow{a} & B & \xrightarrow{b} & C & \xrightarrow{c} & D \\
\downarrow{\scriptstyle\alpha} & & \downarrow{\scriptstyle\beta} & & \downarrow{\scriptstyle\gamma} & & \downarrow{\scriptstyle\delta} \\
A' & \xrightarrow{a'} & B' & \xrightarrow{b'} & C' & \xrightarrow{c'} & D'
\end{array}
$$

be a commutative diagram in an Abelian category, where the rows are exact sequences. Assume that β, δ are monomorphisms and α is an epimorphism. Then γ is a monomorphism as well.

There are various ways one can prove these standard lemmas. Our approach will follow [**Ann**] and use the language of subobjects. Let $\mathcal{A}$ be an Abelian category and A an object of $\mathcal{A}$. Recall that a subobject of A is a monomorphism $K \to A$ and a quotient of A is an epimorphism $A \to C$. From Corollary 2.2.3, we immediately obtain

Proposition 2.5.3. *Let A be an object of an Abelian category. There is a bijective correspondence between subobjects and quotients of A, given by $K \mapsto \operatorname{coker}(K \to A)$, and vice versa $C \mapsto \ker(A \to C)$.*

Given two subobjects $u \colon U \to A$ and $v \colon V \to A$ of A, if there is a morphism $f \colon U \to V$ such that $u = v \circ f$, then f is necessarily a monomorphism (why?), so U is a subobject of V. In this case, we will sometimes abuse notation and write $U \subset V$.

Lemma 2.5.4. *Let A be an object of an Abelian category, and $u \colon U \to A$, $v \colon V \to A$ two subobjects. If $f \colon U \to V$ and $g \colon V \to U$ are such that $u = v \circ f$ and $v = u \circ g$, then f and g are isomorphisms.*

Proof. We have a chain of subobjects

$$ U \xrightarrow{\ f\ } V \xrightarrow{\ g\ } U \xrightarrow{\ u\ } A, $$

and it is enough to show that $g \circ f \colon U \to U$ is an isomorphism. Since all the maps are monomorphisms, it is enough to show that $g \circ f$ is an epimorphism. But by constructions, $u \circ g \circ f = u$, hence u and $u \circ g \circ f$ have the same image inside A. $\qquad\square$

Lemma 2.5.4 allows us to put a partial order on subobjects of A modulo isomorphism. We will hence be able to talk about the smallest or largest subobject satisfying certain conditions.

Since Abelian categories have finite limits, if U, V are subobjects of A, we can take the fibered product of U and V over A.

Definition 2.5.5. Let A be an object of an Abelian category, $U \to A$ and $V \to A$ two subobjects. The intersection $U \cap V$ is the fibered product $U \times_A V$, with its structure morphism $U \cap V \to A$. The sum $U + V$ is the image of the map $U \oplus V \to A$.

Of course, these are made to mimic the analogous definitions for modules, so before using them it is a good idea to check that they behave in the intuitive way.

Proposition 2.5.6. *Let U, V be subobjects of A. Then $U \cap V$ and $U + V$ are subobjects of A.*

Proof. The claim is clear for $U + V$, which is the image of a morphism to A. Let $K \to U \cap V$ be a map such that the composition $K \to A$ is 0. Then the maps $K \to U$ and $K \to V$ are both 0, since $U \to A$ and $V \to A$ are monomorphisms, hence $K \to U \cap V$ is 0 as well, by the universal property of the fibered product. $\qquad\square$

Definition 2.5.7. Let $f \colon A \to B$ be a morphism in an Abelian category. If $U \to A$ is a subobject of A, the *image* of U, denoted $f(U)$, is defined to be the image of the morphism $U \to A \to B$. If $V \to B$ is a subobject of B, the *preimage* of V, denoted $f^{-1}(V)$, is the fibered product $A \times_B V$.

In the above setting, $f(U)$ is a subobject of B by definition, while $f^{-1}(V)$ is a subobject of A by the universal property of fibered products.

Lemma 2.5.8. *Let $f \colon A \to B$ be a morphism in an Abelian category, $V \to B$ a subobject of B. Then $f^{-1}(V)$ is the largest subobject $U \to A$ such that $f(U) \subset V$.*

Proof. A subobject $U \to A$ satisfies $f(U) \subset V$ if and only if it fits into a commutative diagram

$$
\begin{array}{ccc}
U & \longrightarrow & A \\
\downarrow & & \downarrow f \\
V & \longrightarrow & B.
\end{array}
$$

Every such object has a unique map to $f^{-1}(V)$ by the universal property of fibered product. $\qquad\square$

This fact has some immediate corollaries.

Corollary 2.5.9. *Let $f \colon A \to B$ be a morphism in an Abelian category. Then $\ker f = f^{-1}(0)$ (here we see 0 as a subobject of B via the unique map $0 \to B$).*

Corollary 2.5.10. *Let $f \colon A \to B$ be a morphism in an Abelian category. Then $U \subset f^{-1}(f(U))$.*

For the following, notice that to each object A we can attach a category of subobjects $\mathrm{Sub}(A)$. A morphism $f \colon A \to B$ determines functors $f_* \colon \mathrm{Sub}(A) \to \mathrm{Sub}(B)$ and $f^* \colon \mathrm{Sub}(B) \to \mathrm{Sub}(A)$ given by direct and inverse image respectively. We say that either of these functors F is injective on objects, up to isomorphism, if $F(U) \cong F(V)$ implies $U \cong V$ (as subobjects of A or B respectively).

Proposition 2.5.11. *Let $f \colon A \to B$ be a morphism in an Abelian category. Then f is a monomorphism if and only if f_* is injective on objects, up to isomorphism.*

Proof. Let $K = \ker f$. Clearly $f(K) = f(0) = 0$, so if f is not a monomorphism, f_* is not injective. Conversely, if f is a monomorphism, for each subobject $U \to A$, the composition $U \to A \to B$ is itself a subobject, hence it equals the image of U, and f_* is injective. □

Corollary 2.5.12. *If $f\colon A \to B$ is a monomorphism, for every subobject $U \to A$, we have $f^{-1}(f(U)) \cong U$.*

Proof. By Lemma 2.5.8, U and $f^{-1}(f(U))$ have the same image, and by Proposition 2.5.11, f_* is injective on objects. □

We have similar results connecting epimorphisms with inverse images.

Proposition 2.5.13. *Let $f\colon A \to B$ be an epimorphism in an Abelian category. Then f^* is injective on objects, up to isomorphism.*

Proof. Let $V_1 \to B$ and $V_2 \to B$ be two subobjects. We have corresponding cokernels, which we denote by B/V_i, and by Corollary 2.5.9, $f^{-1}(V_i)$ is the kernel of the composition $A \to B/V_i$. If $f^{-1}(V_1) = f^{-1}(V_2)$, the maps $A \to B/V_1$ and $A \to B/V_2$ have the same kernel. Since f is an epimorphism, these are both quotients of A. By the bijection between subobjects and quotients, we get an isomorphism $B/V_1 \cong B/V_2$ commuting with the map from A. Since f is an epimorphism, they also commute with the map from B, hence B/V_1 and B/V_2 are isomorphic quotient objects of B. Dualizing again, we get that V_1 and V_2 are isomorphic subobjects of B. □

Corollary 2.5.14. *Let $f\colon A \to B$ be an epimorphism in an Abelian category, $V \to B$ a subobject. Then $f(f^{-1}(V)) \cong V$.*

Proof. By Proposition 2.5.13, it is enough to show that the two objects have the same preimage. By Lemma 2.5.8, $f^{-1}(f(f^{-1}(V)))$ is the largest subobject $U \to A$ such that $f(U) \subset f(f^{-1}(V))$. It is immediate to check that the same property holds for $f^{-1}(V)$, from which we conclude that $f^{-1}(f(f^{-1}(V))) \cong f^{-1}(V)$. □

Corollary 2.5.15. *Let $f\colon A \to B$ be a morphism in an Abelian category, $V \to B$ a subobject. Then $f(f^{-1}(V)) \cong V \cap \operatorname{im} f$.*

Proof. Apply the above Corollary to $A \to \operatorname{im} f$, which is an epimorphism. □

Corollary 2.5.16. *Let $f\colon A \to B$ be a morphism in an Abelian category, $U \to A$ and $V \to B$ two subobjects. Then $f(U \cap f^{-1}(V)) \cong f(U) \cap V$.*

Proof. Apply Corollary 2.5.16 to the composition $U \to A \to B$. □

With these preliminaries, we are now able to prove the four lemma, Lemma 2.5.2. The following is similar to a standard diagram chasing proof, but instead of elements, we chase subobjects, using the properties that we have proved so far.

Proof of Lemma 2.5.2. Let $u_c \colon U_c \to C$ be a subobject. It is enough to show that if $\gamma \circ u_c = 0$, then $U_c = 0$. For such a subobject, the composition $U_c \to D$ is 0, since δ is a monomorphism, hence $\operatorname{im} u_c \subset \operatorname{im} b$ by exactness.

Consider the subobject $U_b := b^{-1}(U_c) \to B$. Then the composition $U_b \to C'$ is 0. Moreover, $U_b \to B'$ is a subobject of B', since β is a monomorphism. This subobject is contained in $\operatorname{im} a'$ by exactness. Let $U_{a'} := (a')^{-1}(U_b) \to A'$ be the preimage in A', and let $U_a = \alpha^{-1}(U_{a'}) \to A$.

Since α is an epimorphism, by Corollary 2.5.14 we have $\alpha(U_a) \cong U_{a'}$. Since $U_b \to B'$ is contained in $\operatorname{im} a'$, by Corollary 2.5.15, we find that $a'(U_a') = U_b$. By commutativity, the image of $U_a \to A$ inside B' is $U_b \to B'$ as well. By Proposition 2.5.11, it follows that the image of $U_a \to A$ inside B is $U_b \to B$. We conclude that the image of $U_b \to B$ inside C is 0, and applying Corollary 2.5.15 again, we find that $U_c \to C$ is the 0 object. $\qquad\square$

Another result in the same lines is the *nine lemma*.

Lemma 2.5.17 (Nine lemma). *Let*

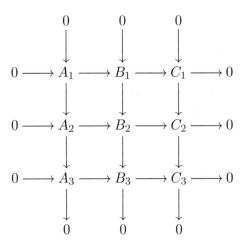

be a commutative diagram in an Abelian category. Assume that the columns and the middle row are exact. Then the top row is exact if and only if the bottom row is.

Like the five lemma, it follows easily by applying a slightly weaker result together with its dual version.

Lemma 2.5.18 (Half nine lemma). *Let*

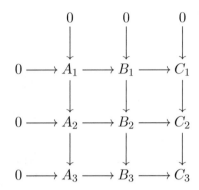

be a commutative diagram in an Abelian category. Assume that the columns and the two bottom rows are exact. Then the top row is exact as well.

Proof. First we prove that $A_1 \to B_1$ is a monomorphism. Let $u\colon U \to A_1$ be a subobject, and assume that the composition $U \to B_1$ is 0. Then $U \to B_2 = 0$ as well, and since $A_2 \to B_2$ and $A_1 \to A_2$ are monomorphisms, u is 0.

The composition $A_1 \to B_1 \to C_1$ is 0, because $C_1 \to C_2$ is a monomorphism and $A_2 \to B_2 \to C_2$ is 0.

Finally, let $u\colon U \to B_1$ be a subobject such that the image in C_1 is 0. Then the image in C_2 is 0 as well, so the subobject $U \to B_2$ is contained in the image of $A_2 \to B_2$. Let $v\colon V \to A_2$ be the pullback of $U \to B_2$, and $w\colon W \to A_1$ its pullback to A_1. Then W and U have the same image in B_2, and since $B_1 \to B_2$ is a monomorphism, they have the same image in B_1. $\qquad\square$

Corollary 2.5.19. *Let $s\colon B \to C$ be an epimorphism, and $f\colon Z \to C$ be any morphism in an Abelian category. Then the pullback $s'\colon Z \times_C B \to Z$ is an epimorphism as well. Moreover, $\ker s \cong \ker s'$.*

Proof. We can complete s to an exact sequence

$$0 \longrightarrow A \longrightarrow B \longrightarrow C \longrightarrow 0$$

and notice that there is a morphism $(0, s)\colon A \to Z \times_C B$. There is a map $Z \oplus B \to C$ given by $f \circ \pi_Z - s \circ \pi_B$, and its kernel is $Z \times_C B$ by definition.

Putting all these maps together gives the diagram

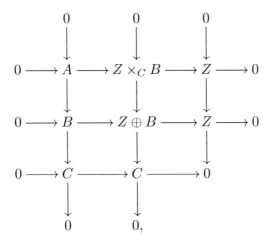

where the columns and the two lower rows are exact. It follows that the upper row is exact as well. □

With similar techniques, we can prove some other standard results.

Lemma 2.5.20 (Snake lemma). *Let*

$$A \xrightarrow{\;f\;} B \xrightarrow{\;g\;} C \longrightarrow 0$$
$$\downarrow{\alpha} \qquad \downarrow{\beta} \qquad \downarrow{\gamma}$$
$$0 \longrightarrow A' \xrightarrow[f']{} B' \xrightarrow[g']{} C'$$

be a commutative diagram with exact rows in an Abelian category. Then there is an exact sequence

$$\ker \alpha \longrightarrow \ker \beta \longrightarrow \ker \gamma \xrightarrow{\;\delta\;} \operatorname{coker} \alpha \longrightarrow \operatorname{coker} \beta \longrightarrow \operatorname{coker} \gamma.$$

The result was famously proved in the movie [**Wei80**], albeit only for the case of modules over a ring.

Remark 2.5.21. If moreover $A \to B$ is a monomorphism, so is $\ker \alpha \to \ker \beta$. Similarly if $B' \to C'$ is an epimorphism, so is $\operatorname{coker} \beta \to \operatorname{coker} \gamma$. Hence if we start from a morphism between short exact sequences, the sequence in the snake lemma can be extended with 0 on both edges.

We will not prove the snake lemma right now, although this certainly can be done with the methods developed so far. Rather, we will give a different proof in Section 2.8, by reducing to the case of modules. We develop even more approaches to the proof of these kinds of results in the exercises.

2.6. Projectives and injectives

In Proposition 2.2.17 we have seen that representable and corepresentable functors are left exact. It is worthwhile to give a name to objects X such that h_X or h^X are exact even on the right.

Definition 2.6.1. Let $\mathcal{A}$ be an Abelian category. An object $P \in \mathcal{A}$ is called *projective* if h^P is exact. An object $I \in \mathcal{A}$ is called *injective* if h_I is exact.

Expanding the definition, we see that P is projective if for any commutative diagram

(2.6.1)

$$
\begin{array}{ccc}
 & & P \\
 & \swarrow & \downarrow \\
A & \longrightarrow B & \longrightarrow 0
\end{array}
$$

with exact row, the dashed line can always be filled in. In other words, if $A \to B$ is an epimorphism, any morphism $P \to B$ can be lifted to A.

Symmetrically, I is injective if for any commutative diagram

(2.6.2)

$$
\begin{array}{ccc}
0 \longrightarrow A & \longrightarrow & B \\
\downarrow & \swarrow & \\
I & &
\end{array}
$$

with exact row, the dashed line can always be filled in. In other words, if $A \to B$ is a monomorphism, any morphism $A \to I$ can be extended to B.

Remark 2.6.2. Let $\{I_\alpha\}$ be a family of injective objects. If the product $\prod I_\alpha$ exists, it is also injective by its universal property. Dually, the coproduct of projective objects, when it exists, is projective. The coproduct of injective objects is not necessarily injective, and the product of projective objects is not necessarily projective; see the Bass–Papp (5.3.1) and Chase theorems in Section 5.3 for an explicit characterization of this phenomenon in the category of A-modules.

Definition 2.6.3. Let $\mathcal{A}$ be an Abelian category. We say that $\mathcal{A}$ has *enough projectives* if any object $A \in \mathcal{A}$ admits an epimorphism $P \to A$ from a projective object P. $\mathcal{A}$ has *enough injectives* if any object $A \in \mathcal{A}$ admits a monomorphism $A \to I$ into an injective object I.

Our aim in this section is to give some characterization of projective and injective objects in familiar categories, and prove some existence results that guarantees that the categories we will be working with have enough injectives (and sometime enough projectives). In Chapter 5, we will study projective and injective modules in much more detail.

Projective modules are especially easy to characterize.

Proposition 2.6.4. *Let A be a (possibly noncommutative) ring. A left A-module M is projective if and only if it is a direct summand of a free A-module.*

Proof. If P is projective, choose a free A-module F with an epimorphism $F \to P$. Then the associated exact sequence splits by Proposition 2.2.13.

Vice versa, assume that $P \oplus Q = F$, where F is free. If we have a diagram like (2.6.1), we get a map $F \to B$ by the map $P \to B$ and the 0 morphisms $Q \to B$. This lifts to a map $F \to A$ by lifting it on a basis of F and extending it by A-linearity. The restriction is the desired map $P \to B$.   □

Since every A-module is a quotient of a free A-module, we get

Corollary 2.6.5. *Let A be a (possibly noncommutative) ring. The category Mod_A has enough projectives.*

The same proof works in the graded case. Recall from Definition 1.1.4(d) that when A is a graded ring, we consider the category Mod_A of graded A-modules and the category Mod_A^u of ungraded ones. There is a functor $U \colon \mathrm{Mod}_A \to \mathrm{Mod}_A^u$ that forgets the grading.

Definition 2.6.6. Let $A = \bigoplus_{i=0}^{\infty} A_i$ be a graded ring. A graded module $M = \bigoplus_{i=0}^{\infty} M_i$ is called *free* if it has a basis consisting of homogeneous elements, that is, $M \cong \bigoplus_{\alpha} A m_{\alpha}$, where each m_{α} is homogeneous.

With the same proof as above, we obtain:

Proposition 2.6.7. *Let A be a graded ring. A graded A-module M is projective if and only if it is a direct summand of a free A-module.*

Corollary 2.6.8. *Let $A = \bigoplus_{i=0}^{\infty} A_i$ be a graded ring, and $P = \bigoplus_{i=0}^{\infty} P_i$ a graded A-module. Then P is projective in Mod_A if and only if it is projective in Mod_A^u.*

Proof. Every epimorphism $s \colon M \to N \to 0$ of graded A-modules is also an epimorphism of ungraded A-modules. If P is graded and projective in Mod_A^u, every graded morphism $f \colon P \to N$ can be lifted as a morphism $g \colon P \to M$, which a priori is ungraded.

Let $b = \deg f$. For a homogeneous $p \in P_k$, let $n = f(p) \in N_{k+b}$, and write $g(p) = \sum_i m_i$, where $m_i \in M_i$. Then $n = s(m) = \sum_i s(m_i)$, which implies that $s(m_i) = 0$ unless $i = k$. We then define $g' \colon P \to M$ by letting $g'(p) = m_k$ for a homogeneous p as above, and extending by linearity. It is easy to check that g' is well defined, and clearly it is a homogeneous lift of f, hence P is projective as a graded A-module.

Vice versa, assume that P is projective in Mod_A. Then P is a direct summand of a free graded A-module by Proposition 2.6.7. But every graded free A-module is also free as an ungraded module, hence P is projective in Mod_A^u. □

Another immediate consequence is:

Corollary 2.6.9. *For a graded ring A, the category Mod_A of graded A-modules has enough projectives.*

Remark 2.6.10. Notice that all of the above results work as well for rings and modules, that are graded over $\mathbb{Z}$ instead of $\mathbb{N}$.

Injective objects are generally more complicated. The following result from [**Bae40**] was the starting point for the study of injective Abelian groups. To state it, we need to recall a definition.

Definition 2.6.11. Let M be an A-module. We say that M is *divisible* if for all $m \in M$ and nonzero $a \in A$, there is $m' \in M$ such that $am' = m$.

Theorem 2.6.12 (Baer's criterion). *Let A be a ring, M an A-module. Then M is injective if and only if every homomorphism $I \to M$, where $I \subset A$ is an ideal, can be extended to a homomorphism $A \to M$.*

Proof. One implication is trivial. Conversely, assume that M has the stated property. Given a diagram

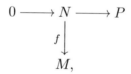

we can view N as a submodule of P. Consider the pairs (N', f'), where $N \subset N' \subset P$ and $f' \colon N' \to M$ extends f. By Zorn's lemma, we find a maximal extension (N_0, f_0), and we only need to prove that $N_0 = P$.

If this is not the case, take any $p \in P \setminus N_0$ and define $g \colon A \to P$ by $g(1) = p$. The ideal $I := g^{-1}(N_0) \subset A$ is proper, so the map $I \to N_0 \to M$ extends to a map $h \colon A \to M$. Setting $N_1 = N_0 + A \cdot p$, we can extend f_0 to $f_1 \colon N_1 \to M$ by defining $f_1(p) = h(1)$, contradicting the maximality of N_0. □

Corollary 2.6.13. *Let A be a principal ideal domain. An A-module is injective if and only if it is divisible.*

Proof. By hypothesis, every ideal $I \subset A$ is generated by a single element $a \in I$. The condition that a morphism $f \colon I \to M$ extends to A amounts to saying that $m = f(a)$ is divisible by a in M. □

Remark 2.6.14. In fact, the same proof shows that injective modules are always divisible, without any hypothesis on A.

Corollary 2.6.15. *The category* Ab *has enough injectives.*

Proof. Let G be an Abelian group, and choose an epimorphism $\bigoplus_{i \in I} \mathbb{Z} \to G$ for some index set I. This gives an exact sequence

$$0 \longrightarrow K \longrightarrow \bigoplus_{i \in I} \mathbb{Z} \longrightarrow G \longrightarrow 0$$

which we can embed into

$$0 \longrightarrow K \longrightarrow \bigoplus_{i \in I} \mathbb{Q} \longrightarrow G' \longrightarrow 0$$

by embedding each copy of $\mathbb{Z}$ into a copy of $\mathbb{Q}$ and letting G' be the quotient. The map $G \to G'$ is a monomorphism by the four lemma (Lemma 2.5.2), and G' is divisible because it is a quotient of a divisible group. $\square$

This result is the stepping stone to proving that many other categories have enough injectives, a fact that will be critical in the next chapter.

Lemma 2.6.16. *Let* $F \colon \mathcal{A} \rightleftarrows \mathcal{B} \colon G$ *be an adjunction between Abelian categories, where F and G are additive. If F is left exact, then G preserves injective objects.*

Proof. Let $I \in \mathcal{A}$ be injective. We have an isomorphism of (contravariant) functors

$$\operatorname{Hom}(-, G(I)) \cong \operatorname{Hom}(F(-), I),$$

and the latter is right exact. $\square$

Proposition 2.6.17. *Let* $F \colon \mathcal{A} \rightleftarrows \mathcal{B} \colon G$ *be an adjunction between Abelian categories, where F and G are additive. Assume that F is exact and faithful, and that $\mathcal{B}$ has enough injectives. Then $\mathcal{A}$ has enough injectives.*

Proof. Given $A \in \mathcal{A}$, take a monomorphism $F(A) \to I$, where $I \in \mathcal{B}$ is injective. By adjunction, we get a morphism $f \colon A \to G(I)$, and $G(I)$ is injective by Lemma 2.6.16. Let $K = \ker f$, so that we have the exact sequence

$$0 \longrightarrow K \longrightarrow A \stackrel{f}{\longrightarrow} G(I).$$

Applying F, which is exact, we get the exact sequence

$$0 \longrightarrow F(K) \longrightarrow F(A) \stackrel{F(f)}{\longrightarrow} F(G(I)).$$

The map $F(A) \to I$ factors as $F(A) \to F(G(I)) \to I$, using the unit of the adjunction, hence $F(f)$ is a monomorphism. This means that $F(K) = 0$, and since F is faithful, $K = 0$ as well. $\square$

We can apply this result to the forgetful functor $\mathrm{Mod}_A \to \mathrm{Ab}$. This is clearly exact and faithful, and by Example 1.5.2(g), it has a right adjoint, namely coextension of scalars. Since Ab has enough injectives by Corollary 2.6.15, we conclude

Corollary 2.6.18. *Let A be a commutative ring. The category Mod_A has enough injectives.*

Let $\mathcal{I}$ be a small category, and $\mathcal{A}$ a complete Abelian category. Then $\mathrm{Fun}(\mathcal{I}, \mathcal{A})$ is an Abelian category (Example 2.2.4(d). We want to prove that if $\mathcal{A}$ has enough injectives, then the same holds for $\mathrm{Fun}(\mathcal{I}, \mathcal{A})$. To this aim, we introduce another adjunction.

Fix objects $i \in \mathcal{I}$ and $A \in \mathcal{A}$. We define a functor $i_*(A) \colon \mathcal{I} \to \mathcal{A}$ by

$$i_*(A)(j) := \prod_{f \colon j \to i} A,$$

where we take a copy of A for each morphism $f \in \mathrm{Hom}(j, i)$. For a morphism $g \in \mathrm{Hom}(j, k)$, we obtain a map $g \circ \colon \mathrm{Hom}(k, i) \to \mathrm{Hom}(j, i)$, and this induces a morphism $i_*(A)(j) \to i_*(A)(k)$. In fact, by the universal property of the product, this is the same as a morphism $i_*(A)(j) \to A$ for every $f \colon k \to i$, and we take for this the projection on the factor associated to $g \circ f \colon j \to i$.

Lemma 2.6.19. *For every functor $F \colon \mathcal{I} \to \mathcal{A}$, there is a natural bijection $\mathrm{Hom}(F, i_*(A)) \cong \mathrm{Hom}(F(i), A)$.*

Proof. To give a natural transformation $F \to i_*(A)$ amounts to giving certain maps $F(j) \to i_*(A)(j)$ for all $j \in \mathcal{I}$. By naturality and the definition of $i_*(A)$, these are uniquely determined by the composition $F(i) \to i_*(A)(i) \to A$, where the last map is the projection on the factor corresponding to id_i. $\square$

Let us fix $i \in \mathcal{I}$. This determines the functor $i_* \colon \mathcal{A} \to \mathrm{Fun}(\mathcal{I}, \mathcal{A})$, as well as an evaluation functor $ev_i \colon \mathrm{Fun}(\mathcal{I}, \mathcal{A}) \to \mathcal{A}$ defined by $ev_i(F) = F(i)$. We can rephrase the lemma by saying that ev_i is left adjoint to i_*. We cannot directly use Proposition 2.6.17, since a single functor ev_i is not necessarily faithful. Still, this is enough for the following result.

Proposition 2.6.20. *Let $\mathcal{I}$ be a small category, and $\mathcal{A}$ a complete Abelian category with enough injectives. Then $\mathrm{Fun}(\mathcal{I}, \mathcal{A})$ has enough injectives. In particular, if B is a ring, the category $\mathrm{Fun}(\mathcal{I}, \mathrm{Mod}_B)$ has enough injectives.*

Proof. By Lemmas 2.6.19 and 2.6.16, $i_*(A)$ is injective whenever A is. Given a functor $F \colon \mathcal{I} \to \mathcal{A}$, choose for each $i \in \mathcal{I}$ a monomorphism $F(i) \to A_i$, where $A_i \in \mathcal{A}$ is injective. Then $i_*(A_i)$ is injective, and we have a monomorphism of functors $F \to \prod_{i \in \mathcal{I}} i_*(A_i)$. $\square$

This can be immediately rephrased in terms of presheaves.

Corollary 2.6.21. *Let X be a small site, and $\mathcal{A}$ a complete Abelian category with enough injectives. Then $\mathrm{Psh}_{\mathcal{A}}(X)$ has enough injectives.*

Proving that sheaves have enough injectives, on the other hand, is a much more subtle task. The case of sheaves on topological spaces is simpler, though; see Exercise 9.

Theorem 2.6.22. *Let X be a small site. Then $\mathrm{Sh}_{\mathrm{Ab}}(X)$ has enough injectives.*

The idea behind the theorem is the following. Start from a sheaf S_0. By Corollary 2.6.21, we can embed S_0 into an injective presheaf I_0. If I_0 was a sheaf, since it is injective as a presheaf, it would be a fortiori injective in $\mathrm{Sh}(X)$. Even if I_0 is not a sheaf, we can consider the associated sheaf $S_1 = aI_0$ and the morphism $S_0 \to I_0$ induces a morphism $S_0 \to S_1$. Since the functor a is exact and $S_0 \to I_0$ is monic in $\mathrm{Psh}(X)$, $S_0 \to S_1$ is a monomorphism as well. We can iterate this process a transfinite number of times, definining—for each ordinal α—$S_{\alpha+1} = aI_\alpha$, where $S_\alpha \to I_\alpha$ is monic and I_α is an injective presheaf. For a limit ordinal α, we put $S_\alpha = \mathrm{colim}_{\beta < \alpha} S_\beta$. Here we use the fact that Ab is cocomplete.

This defines an increasing family $\{S_\alpha\}$ of sheaves, indexed by ordinals, in such a way that there is a monomorphism $S_\alpha \to S_\beta$ for every pair of ordinals $\alpha < \beta$. We will check that eventually one of the S_α is injective. To do so, we need an analogue of Baer's criterion 2.6.12. Just as any Abelian group can be seen as a $\mathbb{Z}$-module, every sheaf of Abelian groups can be seen as a sheaf of modules over a suitable sheaf of rings.

Let $U \in X$. We denote by $\underline{\mathbb{Z}}_U$ the sheaf associated to the presheaf
$$\mathbb{Z}_U(V) := \bigoplus_{f\colon V \to U} \mathbb{Z}.$$
For a sheaf $S \in \mathrm{Sh}_{Ab}(X)$ and an object $U \in X$ we have
$$\mathrm{Hom}(\underline{\mathbb{Z}}_U, S) \cong \mathrm{Hom}(\mathbb{Z}, S(U)) \cong S(U),$$
since every such morphism is determined uniquely by the morphism $\mathbb{Z} \to S(U)$ associated to id_U. With these notations, the analogue of Baer's criterion is:

Lemma 2.6.23. *Let $I \in \mathrm{Sh}_{\mathrm{Ab}}(X)$, and assume that for all objects $U \in X$, given a subsheaf $F \subset \underline{\mathbb{Z}}_U$ and a morphism $f\colon F \to I$, there exists an extension of f to $\underline{\mathbb{Z}}_U$. Then I is injective.*

We omit the proof, since it is a simple modification of the arguments in Theorem 2.6.12. The crucial point is that all subsheaves of all sheaves of the form $\underline{\mathbb{Z}}_U$ as $U \in X$ varies form a set. This is enough to guarantee that the sheaf S_α is injective for an ordinal α large enough.

Proof of Theorem 2.6.22. Let $\{T_i\}$ be the set of all subsheaves of some $\underline{\mathbb{Z}}_U$, and $T := \bigoplus T_i$. We claim that the family of sheaves S_α has the following property. There exists an ordinal α such that all maps $T \to S_\alpha$ factor through S_β for some $\beta < \alpha$.

Given this claim, we argue that S_α is injective. By Lemma 2.6.23, it is enough to prove the extension property for the case of a subsheaf $F \subset \underline{\mathbb{Z}}_U$ and a map $f \colon F \to S_\alpha$. By the above, f factors as a composition $F \to S_\beta \to S_\alpha$ for some $\beta < \alpha$. But then we have an extension $\underline{\mathbb{Z}}_U \to S_{\beta+1}$ by construction of $S_{\beta+1}$, and we can compose this with the inclusion $S_{\beta+1} \subset S_\alpha$.

It remains to prove the claim. A map $T \to S_\alpha$ is determined by maps of Abelian groups $T(U) \to S_\alpha(U)$ for all $U \in X$. For each fixed U, the map $T(U) \to S_\alpha(U)$ factors through $S_\beta(U)$ as soon as β has cofinality bigger than the cardinality of $T(U)$. The same happens simultaneously for all $U \in X$ as soon as the cofinality of β is bigger than the sum of the cardinality of $T(U)$ for all $U \in X$. $\qquad\square$

2.7. Essential extensions

In this section, we are going to prove that certain categories admit a sort of minimal injective extension of objects. This will turn out to be a useful tool in the study of modules, and will also play a role in the proof of the Freyd–Mitchell theorem.

Definition 2.7.1. Let $\mathcal{A}$ be an Abelian category. A monomorphism $A \to B$ in $\mathcal{A}$ is called an *essential extension* if for every other nonzero monomorphism $A' \to B$ the intersection $A \cap A' \neq 0$. If, moreover, B is injective, the monomorphism $A \to B$ is called an *injective hull* (or *injective envelope*) of A.

The aim of this section is to show that the injective hull of an object, when it exists, is essentially unique, and to give conditions for its existence. The starting point is the following characterization of essential extensions.

Lemma 2.7.2. *Let* $i \colon A \hookrightarrow B$ *be a monomorphism in* $\mathcal{A}$. *Then the following are equivalent:*

   (i) *i is an essential extension; and*

   (ii) *for all morphisms $g \colon B \to T$ with $T \in \mathcal{A}$, g is monic if and only if $g \circ i$ is monic.*

Proof. If $g\colon B \to T$ is a morphism such that $g \circ i$ is monic, then $K = \ker g$ is a subobject of B such that $A \cap K = 0$. Hence if i is essential, we must have $K = 0$, that is, g is monic.

Conversely, assume that i is not essential, so that there is a nontrivial subobject $K \subset B$ such that $A \cap K = 0$. Then $A \to B/K$ is monic, but $B \to B/K$ is not.   □

Remark 2.7.3. We will use repeatedly the following construction. Let $i\colon A \to B$ be an essential extension, E an injective object. If we are given a monomorphism $A \to E$, this can be extended to B, since E is injective. Moreover, the extension $B \to E$ remains monic by Lemma 2.7.2.

To go further, it is useful to introduce a condition on $\mathcal{A}$, that will be satisfied in the cases we are interested in.

Definition 2.7.4. Let $\mathcal{A}$ be an Abelian category. We say that $\mathcal{A}$ *has complements* if for every subobject $A \to B$ there exists a subobject $C \to B$ such that $A \cap C = 0$ and C is maximal with respect to this condition.

Example 2.7.5.

(a) Every concrete category (for instance Mod_A) has complements, by Zorn's lemma.

(b) More generally, assume that for a given object $A \in \mathcal{A}$, we can take a set S of subobjects of A such that every subobject of A is isomorphic to an object in S. Then $\mathcal{A}$ has complements by Zorn's lemma.

(c) In particular, every small Abelian category has complements. Moreover, if $\mathcal{A}$ is small, $\mathrm{Fun}(\mathcal{A}, \mathrm{Mod}_A)$ has complements. In fact, to give a subfunctor of $F\colon \mathcal{A} \to \mathrm{Mod}_A$ amount to choosing a submodule of $F(A)$ for all $A \in \mathcal{A}$ (morphisms are determined by F), and so the collection of subfunctors is a set.

Lemma 2.7.6. *Let $\mathcal{A}$ be an Abelian category with complements, $A \in \mathcal{A}$. Then A is injective if, and only if, every essential extension $A \subset B$ is an isomorphism.*

Proof. If A is injective and $i\colon A \to B$ is monic, then i splits and we have a decomposition $B = A \oplus A'$, showing that i is not essential unless $A' = 0$.

Conversely, assume that all essential extensions of A are trivial. Let $i\colon A \to B$ be any monomorphism, and choose a maximal subobject $A' \subset B$ such that $A \cap A' = 0$. The map $A \to B/A'$ is injective, and by maximality of A' it is an essential extension. It follows that $A \to B/A'$ is an isomorphism, which implies that $B \cong A \oplus A'$. Since any inclusion of A into a bigger object splits, A is injective.   □

We can now prove existence and uniqueness of the injective hull. The result is due to Eckmann and Schopf [**ES53**] in the context of A-modules. The proof of existence here comes from [**Gab62**].

Theorem 2.7.7 (Eckmann, Schopf). *Let $\mathcal{A}$ be an Abelian category with complements, $A \in \mathcal{A}$. If $A \hookrightarrow E$ and $A \hookrightarrow E'$ are two injective hulls, there is an isomorphism $E \cong E'$ that makes the diagram*

commute. If moreover $\mathcal{A}$ has enough injectives, then there exists an injective hull E of A.

Since the injective hull of A is unique up to an isomorphism that preserves A, we can give it a name and denote it $E(A)$.

Proof. Assume that E and E' are two injective hulls of A. By Remark 2.7.3, we find an injective homomorphism $f\colon E \to E'$. Then f is an essential extension as well, and since E is injective it must be an isomorphism by Lemma 2.7.6.

For existence, let J be any injective object with a monomorphism $j\colon A \to J$. Choose a maximal subobject Q of J such that $Q \cap A = 0$. Then, by applying existence of complements to J/Q, choose a maximal subobject I of J such that $j(A) \subset I$ and $I \cap Q = 0$. If we denote $p\colon J \to J/Q$ the projection, it follows that p induces an isomorphism between I and $p(I)$. By construction $p(I) \to J/Q$ is an essential extension. By Remark 2.7.3, the inverse isomorphism $p(I) \to I$ can be extended to a monomorphism $h\colon J/Q \to J$, so $h(J/Q)$ is an essential extension of I. It follows that $I = h(J/Q)$, hence the projection p splits and $J = I \oplus Q$. In particular, I is injective as well (why?), and $A \to I$ is an essential extension. $\square$

Definition 2.7.8. Let $\mathcal{A}$ be an Abelian category. We say that $\mathcal{A}$ has *injective hulls* if all objects A of $\mathcal{A}$ admit an injective hull.

By Theorem 2.7.7, every category with enough injectives and complements has injectives hulls. There is also a converse.

Proposition 2.7.9. *Let $\mathcal{A}$ be a category with injective hulls. Then $\mathcal{A}$ has enough injectives and has complements. In particular, injective hulls are unique.*

Proof. The first claim is obvious. For the second, let $i\colon A \to B$ be any monomorphism, E an injective hull of A. Since E is injective, we get a morphism $f\colon B \to E$, and then $\ker f$ is a complement of A in B. The last claim is Theorem 2.7.7. □

It is useful to rephrase Remark 2.7.3 in the special case of the injective hull.

Proposition 2.7.10. *Let $\mathcal{A}$ be an Abelian category with injective hulls. Any monomorphism $A \to I$ with I injective can be extended this to a monomorphism $E(A) \to I$. In other words, the injective hull $E(A)$ is the minimal injective extension of A.*

Corollary 2.7.11. *Let A, B be two objects of $\mathcal{A}$. Then there is a natural isomorphism $E(A \oplus B) \cong E(A) \oplus E(B)$.*

Proof. The object $E(A) \oplus E(B)$ contains both A and B, so there is a natural monomorphism $A \oplus B \subset E(A) \oplus E(B)$. By Proposition 2.7.10, we get an inclusion $E(A \oplus B) \subset E(A) \oplus E(B)$. Since $E(A \oplus B)$ is injective and contains A, it contains $E(A)$. Symmetrically, it also contains $E(B)$, hence the inclusion is an equality. □

2.8. The Freyd–Mitchell theorem

So far, we have treated Abelian categories as abstract objects. Even proving fairly simple results such as the five lemma has required quite a lot of work. In concrete categories such as Mod_A, the proof of the five lemma is substantially simpler, due to the fact that one can perform diagram chasing using elements of the modules.

It turns out that, in fact, the category of modules over a ring is in a sense the fundamental example. This is because every other Abelian category admits an embedding into a category of modules (over a possibly non-commutative ring), in a way that preserves exact sequences. This famous result of Freyd and Mitchell, first appeared in [**Fre64**].

Theorem 2.8.1 (Freyd–Mitchell). *Let $\mathcal{A}$ be a small Abelian category. Then there exist a noncommutative ring A and an exact, fully faithful embedding $F\colon \mathcal{A} \to \mathrm{Mod}_A$.*

Before going into the details of the proof, we expose the general strategy. The basic observation, that allows us to produce a ring, is the following.

Remark 2.8.2. Let A be an object of the Abelian category $\mathcal{A}$. With the operation of composition, the set $\mathrm{Hom}(A, A)$ becomes a noncommutative ring. Moreover, $\mathrm{Hom}(A, B)$ is a left module over $\mathrm{Hom}(A, A)$ for any other object B.

Let us fix a small Abelian category $\mathcal{A}$. The remark leads us to consider the corepresentable functor $h^A \colon \mathcal{A} \to \mathrm{Ab}$ for a suitable object A. This will be an embedding exactly when A is a generator. But we cannot assume that $\mathcal{A}$ has a generator, and to start we will need something that will work on an arbitrary Abelian category.

A natural step is to consider the Yoneda embedding. Given an object $A \in \mathcal{A}$, we have the associated representable functor h_A. Since this is an additive functor, we can consider a variation of the Yoneda embedding

$$h_\bullet \colon \quad \mathcal{A} \longrightarrow \mathrm{Add}(\mathcal{A}^{op}, \mathrm{Ab}),$$

$$A \longmapsto h_A,$$

where $\mathrm{Add}(\mathcal{A}^{op}, \mathrm{Ab})$ is the full subcategory of $\mathrm{Fun}(\mathcal{A}^{op}, \mathrm{Ab})$ consisting of *additive functors*.

It is easy to check that $\mathrm{Add}(\mathcal{A}^{op}, \mathrm{Ab})$ is in fact an Abelian category, but we run into the issue that $h_\bullet$ is not exact. In fact, given an exact sequence

$$0 \longrightarrow A \longrightarrow B \longrightarrow C \longrightarrow 0$$

in $\mathcal{A}$, the associated sequence

$$0 \longrightarrow h_A \longrightarrow h_B \longrightarrow h_C \longrightarrow 0$$

is *not* exact. This is completely analogous to Example 2.2.18, where we have seen that the functors h_X are not right exact in general (see also Lemma 2.8.4).

The way out of this conundrum is to observe that the corepresentable functors are not only additive, but also left exact. We can then consider the subcategory $\mathcal{L}(\mathcal{A}^{op}, \mathrm{Ab})$ of $\mathrm{Add}(\mathcal{A}^{op}, \mathrm{Ab})$ consisting of left exact functors. This turns out to be an Abelian category, although not trivially. What's more, the Yoneda embedding with values in $\mathcal{L}(\mathcal{A}^{op}, \mathrm{Ab})$ will turn out to be exact. This is the first step towards the result.

We are now reduced to finding an embedding for the category $\mathcal{L}(\mathcal{A}^{op}, \mathrm{Ab})$. For this, we will prove two basic facts about $\mathcal{L}(\mathcal{A}^{op}, \mathrm{Ab})$: it is complete and it has an injective cogenerator, hence its dual has a projective generator. If P is such a projective generator, the functor h^P is an exact embedding into the category of modules over the ring $\mathrm{Hom}(P, P)$, thereby concluding the proof of the Freyd–Mitchell theorem.

With this strategy in mind, we now turn to the details of the proof.

Definition 2.8.3. The category $\mathcal{L}(\mathcal{A}, \mathcal{B})$ is the full subcategory of $\mathrm{Add}(\mathcal{A}, \mathcal{B})$ where objects are left exact functors.

Let $\mathcal{A}$ be a small Abelian category. Since $\mathrm{Add}(\mathcal{A}, \mathrm{Ab})$ is itself Abelian, we can discuss whether various functors between it, $\mathcal{A}$ and Ab are exact.

Lemma 2.8.4. *Let*

$$0 \longrightarrow A \longrightarrow B \longrightarrow C \longrightarrow 0$$

be a short exact sequence in $\mathcal{A}$. Then

(2.8.1) $$0 \longrightarrow h_A \longrightarrow h_B \longrightarrow h_C$$

is exact. In other words, the Yoneda functor $h_\bullet \colon \mathcal{A} \to \mathrm{Add}(\mathcal{A}^{op}, \mathrm{Ab})$ is left exact.

Proof. Since the structure of Abelian category on $\mathrm{Add}(\mathcal{A}, \mathrm{Ab})$ is defined pointwise, this is the same as saying that

$$0 \longrightarrow \mathrm{Hom}(X, A) \longrightarrow \mathrm{Hom}(X, B) \longrightarrow \mathrm{Hom}(X, C)$$

is exact for all $X \in \mathcal{A}$, which follows since h^X is left exact by Proposition 2.2.17. $\qquad\square$

Lemma 2.8.5. *Let $\mathcal{A}, \mathcal{B}$ be Abelian categories, with $\mathcal{A}$ small, A an object of $\mathcal{A}$. Then the functor $E_A \colon \mathrm{Add}(\mathcal{A}, \mathcal{B}) \to \mathcal{B}$ given by evaluation at A (that is, $E_A(F) := F(A)$) is exact.*

Proof. This follows at once from the fact that the Abelian structure on $\mathrm{Add}(\mathcal{A}, \mathcal{B})$ is defined pointwise. $\qquad\square$

Proposition 2.8.6. *Let $\mathcal{A}$ be a small Abelian category, $F \colon \mathcal{A} \to \mathrm{Ab}$ an additive functor. If F is injective in $\mathrm{Add}(\mathcal{A}, \mathrm{Ab})$, then F is right exact. If moreover F preserves monomorphisms, it is exact.*

Proof. Let

$$0 \longrightarrow A \longrightarrow B \longrightarrow C \longrightarrow 0$$

be a short exact sequence in $\mathcal{A}$. By the dual of Lemma 2.8.4, we obtain that

$$0 \longrightarrow h^C \longrightarrow h^B \longrightarrow h^A$$

is exact. Since F is injective,

$$\mathrm{Hom}(h^A, F) \longrightarrow \mathrm{Hom}(h^B, F) \longrightarrow \mathrm{Hom}(h^C, F) \longrightarrow 0$$

is also exact, and by Remark 1.3.5 this is the same as

$$F(A) \longrightarrow F(B) \longrightarrow F(C) \longrightarrow 0.$$

This shows that F is right exact. If moreover F preserves monomorphisms, it is exact. $\qquad\square$

In view of this result, it makes sense to try to understand under what conditions an additive functor preserves monomorphisms. One such condition is as follows.

Lemma 2.8.7. *Let $F \to G$ be an essential extension in $\mathrm{Add}(\mathcal{A}, \mathrm{Ab})$. If F preserves monomorphisms, then so does G.*

Proof. Assume that G does not preserve monomorphisms, and $m\colon A \to B$ be a monomorphism of $\mathcal{A}$ such that $G(m)\colon G(A) \to G(B)$ is not injective. This means that we can find a nonzero $a \in G(A)$ such that $G(m)(a) = 0$. We will use a to define a subfunctor (i. e. a subobject) F' of G.

Given X in $\mathcal{A}$, define

$$F'(X) := \{x \in G(X) \mid x = G(t)(a) \text{ for some } t\colon A \to X\}.$$

Clearly $F'(X)$ is well defined as a set, and is closed under addition and negation, so $F'(X)$ is a subgroup of $G(X)$. By restriction of G, F' is also defined on morphisms, so it is a subfunctor of G. Moreover, $a \in F'(A)$, so F' is not the 0 functor.

Since $F \to G$ is essential, the intersection $F \cap F'$ is not 0. Unwinding the definition, this means that there exists an object X such that $F(X) \cap F'(X) \neq 0$ (both are subgroups of $G(X)$). But then F cannot preserve monomorphisms as well. □

To go further, we will need to give an Abelian structure to $\mathcal{L}(\mathcal{A}, \mathrm{Ab})$. It is clear, in the same way as it is for $\mathrm{Add}(\mathcal{A}, \mathrm{Ab})$, that $\mathcal{L}(\mathcal{A}, \mathrm{Ab})$ is an additive category. Our next step is to define kernels and cokernels.

Kernels are easy. If $F, G\colon \mathcal{A} \to \mathrm{Ab}$ are left exact functors and $t\colon F \to G$ is a morphism of functors, then we can define a functor $K(A) := \ker t_A$ for an object A of $\mathcal{A}$. Then K is left exact as well, thanks to the half nine lemma, Lemma 2.5.18. It is then immediate to check that K is a kernel for t.

Cokernels are a more delicate matter. To define them, we realize $\mathcal{L}(\mathcal{A}, \mathrm{Ab})$ as the category of sheaves on a suitable site. Given $C \in \mathcal{A}$, define a covering of C to be a one element family $\{B \to C\}$, where $B \to C$ is an epimorphism. This defines a Grothendieck topology thanks to Corollary 2.5.19. We let X be the site given by $\mathcal{A}$ endowed with this topology.

By definition, $\mathrm{Psh}_{\mathrm{Ab}}(X) = \mathrm{Fun}(\mathcal{A}^{op}, \mathrm{Ab})$. A presheaf S is a sheaf if and only if for all $C \in \mathcal{A}$ and all epimorphisms $B \to C$ we have the equalizer diagram

$$S(C) \longrightarrow S(B) \rightrightarrows S(B \times_C B).$$

Thus, if S is additive, S is a sheaf on X if and only if the sequence

$$(2.8.2) \qquad 0 \longrightarrow S(C) \longrightarrow S(B) \xrightarrow{S(p_1 - p_2)} S(B \times_C B)$$

is exact for all epimorphism $B \to C$. If we let $A := \ker(B \to C)$, then by Lemma 2.2.14, we see that A is the image of $p_1 - p_2\colon B \times_C B \to B$. Hence, exactness of (2.8.2) is the same as exactness of

$$0 \longrightarrow S(C) \longrightarrow S(B) \longrightarrow S(A).$$

In other words, an additive functor is a sheaf for this topology if and only if it is left exact. We can use this to prove

Theorem 2.8.8. *Let $\mathcal{A}$ be a small Abelian category. Then the category $\mathcal{L}(\mathcal{A}, \mathrm{Ab})$ is Abelian and has enough injectives.*

Proof. It is not restrictive to prove the first claim for $\mathcal{A}^{op}$. The sheafification functor $a \colon \mathrm{Psh}_{\mathrm{Ab}}(X) \to \mathrm{Sh}_{\mathrm{Ab}}(X)$ restricts to a functor $\mathrm{Add}(\mathcal{A}^{op}, \mathrm{Ab}) \to \mathcal{L}(\mathcal{A}^{op}, \mathrm{Ab})$, which is again left adjoint to the natural inclusion $\mathcal{L}(\mathcal{A}^{op}, \mathrm{Ab}) \to \mathrm{Add}(\mathcal{A}^{op}, \mathrm{Ab})$. Knowing this, the the proof that $\mathcal{L}(\mathcal{A}^{op}, \mathrm{Ab})$ is Abelian proceeds exactly as in Proposition 2.3.15.

For the second claim, let $F \colon \mathcal{A} \to \mathrm{Ab}$ be a left exact functor. Since $\mathrm{Add}(\mathcal{A}, \mathrm{Ab})$ has enough injectives and has complements, it has injective hulls by Theorem 2.7.7. Let $F \to E(F)$ be the injective hull in $\mathrm{Add}(\mathcal{A}, \mathrm{Ab})$. By Lemma 2.8.7, $E(F)$ preserves monomorphisms, hence it is exact by Proposition 2.8.6. Then $E(F) \in \mathcal{L}(\mathcal{A}, \mathrm{Ab})$, and since it is injective in $\mathrm{Add}(\mathcal{A}, \mathrm{Ab})$, it is a fortiori injective in $\mathcal{L}(\mathcal{A}, \mathrm{Ab})$. $\qquad\square$

Remark 2.8.9. Here it is important that the target category is Ab. In fact, $\mathcal{L}(\mathcal{A}, \mathcal{B})$ can fail to be an Abelian category for other choices of $\mathcal{B}$, even simple ones such as Ab^{op}. For examples, see [**Pos17**, Example 2.7] or [**Pos18**].

Now that we know that $\mathcal{L}(\mathcal{A}, \mathrm{Ab})$ is Abelian, our next task is to verify that the Yoneda functor $h_\bullet \colon \mathcal{A} \to \mathcal{L}(\mathcal{A}^{op}, \mathrm{Ab})$ is in fact exact. To do this, we will exploit the fact that $\mathcal{L}(\mathcal{A}, \mathrm{Ab})$ has an injective cogenerator. This will be constructed as a suitable product, so we discuss products first.

Proposition 2.8.10. *Let $\mathcal{A}, \mathcal{B}$ be Abelian categories, with $\mathcal{A}$ small and $\mathcal{B}$ complete. The categories $\mathrm{Fun}(\mathcal{A}, \mathcal{B})$, $\mathrm{Add}(\mathcal{A}, \mathcal{B})$ and $\mathcal{L}(\mathcal{A}, \mathcal{B})$ are complete.*

Proof. By Theorem 1.4.12, it is enough to prove that all these categories have equalizers and arbitrary products. If F, G are functors in any of these categories, the equalizer of F and G is simply $\ker F - G$, which exists and is well defined on all of these categories.

The product of functors can be defined pointwise since $\mathcal{B}$ is complete, and the product of additive (resp., left exact) functors is itself additive (resp., left exact). $\qquad\square$

Corollary 2.8.11. *Let $\mathcal{A}$ be a small Abelian category. The category $\mathcal{L}(\mathcal{A}, \mathrm{Ab})$ has a generator.*

Proof. We prove the equivalent fact that $\mathcal{L}(\mathcal{A}^{op}, \mathrm{Ab})$ has a generator. Let $G\colon \mathcal{A} \to \mathrm{Ab}$ be the product of all functors h_A for $A \in \mathcal{A}$ (this is well-defined, since $\mathcal{A}$ is small). We claim that G is a generator, that is, the functor $h^G\colon \mathcal{L}(\mathcal{A}^{op}, \mathrm{Ab}) \to \mathrm{Ab}$ is an embedding.

For a functor $L \in \mathcal{L}(\mathcal{A}^{op}, \mathrm{Ab})$, giving a morphism $G \to L$ amounts to giving a morphism $h_A \to L$ for all $A \in \mathcal{A}$. By Remark 1.3.5, $\mathrm{Hom}(h_A, L)$ can be identified with $L(A)$. Thus,

$$\mathrm{Hom}(G, L) \cong \prod \mathrm{Hom}(h_A, L) \cong \prod L(A),$$

and so $h^G(L)$ determines $L(A)$ for all A. Moreover, a morphism $L \to L'$ determines a map $\mathrm{Hom}(G, L) \to \mathrm{Hom}(G, L')$, which in coordinates can be identified with the natural transformation $L(A) \to L'(A)$. This tells us that morphisms can also be determined by the functor h^G. $\qquad\square$

Remark 2.8.12. Let $\mathcal{A}$ be an Abelian category with a generator G. Then for every object $A \in \mathcal{A}$, a subobject $U \to A$ is determined by the subset $\mathrm{Hom}(G, U) \subset \mathrm{Hom}(G, A)$ (here we use that h_G is left exact). In particular, the subobjects of A form a set.

The following result is a crucial point in the proof of the Freyd–Mitchell theorem.

Theorem 2.8.13. *Let $\mathcal{A}$ be a small Abelian category. Then $\mathcal{L}(\mathcal{A}, \mathrm{Ab})$ has an injective cogenerator.*

Proof. By Corollary 2.8.11, $\mathcal{L}(\mathcal{A}, \mathrm{Ab})$ has a generator G. By the above remark, G has only a set of subobjects. Hence the product of all quotients of G is well defined, call it P. Finally, let $P \to I$ be a monomorphism with I injective.

Take any nonzero morphism $A \to B$ of $\mathcal{L}(\mathcal{A}, \mathrm{Ab})$. Since G is a generator, we can find a morphism $G \to A$ such that $G \to A \to B$ is nonzero. Let J be the image of $G \to B$. Then J is a quotient of G, so there is a map $J \to P \to I$ which is a monomorphism. Since I is injective, a monomorphism $J \to I$ lifts to a map $B \to I$, so that the diagram

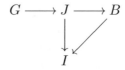

commutes. In particular the composition $G \to B \to I$ is not 0, and a fortiori the same is true for $A \to B \to I$. This implies that h_I is injective on $\mathrm{Hom}(A, B)$, so I is a cogenerator. $\qquad\square$

This fact has two important consequences.

Theorem 2.8.14. *Let $\mathcal{A}$ be a small Abelian category. Then the Yoneda functor $h_\bullet \colon \mathcal{A} \to \mathcal{L}(\mathcal{A}^{op}, \mathrm{Ab})$ is a fully faithful exact embedding.*

Proof. We already know from the Yoneda lemma (Lemma 1.3.3) that $h_\bullet$ is a fully faithful embedding, and by Proposition 2.2.17 it takes values in $\mathcal{L}(\mathcal{A}^{op}, \mathrm{Ab})$. All we need to check is that $h_\bullet$ is exact.

Let I be an injective cogenerator for $\mathcal{L}(\mathcal{A}^{op}, \mathrm{Ab})$, and

$$0 \longrightarrow A \longrightarrow B \longrightarrow C \longrightarrow 0$$

an exact sequence in $\mathcal{A}$. Then we know that at least

$$0 \longrightarrow h_A \longrightarrow h_B \longrightarrow h_C$$

is exact in $\mathcal{L}(\mathcal{A}^{op}, \mathrm{Ab})$, and since I is injective,

$$(2.8.3) \qquad \mathrm{Hom}(h_C, I) \longrightarrow \mathrm{Hom}(h_B, I) \longrightarrow \mathrm{Hom}(h_A, I) \longrightarrow 0$$

is exact as well. By the Yoneda lemma, this is the same as

$$I(C) \longrightarrow I(B) \longrightarrow I(A) \longrightarrow 0.$$

Since I is left exact, this sequence is exact also on the left, which means that the same holds for (2.8.3). Since I is a cogenerator, h_I is an embedding, hence this implies that

$$0 \longrightarrow h_A \longrightarrow h_B \longrightarrow h_C \longrightarrow 0$$

is exact as well. $\square$

The second important consequence of Theorem 2.8.13 is that the opposite category of $\mathcal{L}(\mathcal{A}, \mathrm{Ab})$ has a projective generator. This is enough to embed it into a category of modules.

Theorem 2.8.15. *Let $\mathcal{A}$ be an Abelian category with a projective generator P, and consider the ring $A = \mathrm{Hom}(P, P)$. Then the functor h^P is an exact embedding of $\mathcal{A}^{op}$ into Mod_A. Moreover, if $\mathcal{A}$ is complete and $\mathcal{A}' \subset \mathcal{A}$ is a small Abelian subcategory, we can choose the projective generator P in such a way that h^P is fully faithful on $\mathcal{A}'$.*

Proof. The functor h^P is exact because P is projective, and an embedding because P is a generator. For every object M of $\mathcal{A}$, the Abelian group $\mathrm{Hom}(P, M)$ has the structure of a left A-module by left composition.

It remains to prove that we can choose P so that h^P is full on $\mathcal{A}'$. Fix any projective generator P_0 of $\mathcal{A}$. Since $\mathcal{A}$ is complete, for every object $X \in \mathcal{A}$ we can consider the object $P_X := \bigoplus_{f \in \mathrm{Hom}(P_0, X)} P_0$, a direct sum of copies of P_0, one for each morphism $P_0 \to X$. By construction, we get a morphism $P_X \to X$, which equals f on the P_0 component associated to f. This is an epimorphism, because P_0 is a generator. Moreover P_X, being a direct sum of copies of P_0, is itself a projective generator. Since $\mathcal{A}'$ is small, we can even consider $P := \bigoplus_{X \in \mathcal{A}} P_X$. Then P is a projective generator of $\mathcal{A}$ and admits an epimorphism to all objects $X \in \mathcal{A}'$. We will see that this is sufficient to prove that h^P is full on $\mathcal{A}'$.

Take any morphism $f \colon h^P(X) \to h^P(Y)$ for some objects $X, Y \in \mathcal{A}$, and choose epimorphisms $P \to X$ and $P \to Y$. Let K be the kernel of the first map. By applying h^P, which is exact, we get a diagram

$$
\begin{array}{ccccccccc}
0 & \longrightarrow & h^P(K) & \longrightarrow & A & \longrightarrow & h^P(X) & \longrightarrow & 0 \\
 & & & & \big\downarrow{\scriptstyle g} & & \big\downarrow{\scriptstyle f} & & \\
 & & & & A & \longrightarrow & h^P(Y) & \longrightarrow & 0,
\end{array}
$$

since $h^P(P) = A$. Here with find a lift g of f, since A is free on itself. The map g must be multiplication on the right by some element $a \in A$, namely $g(x) = xa$. But then a is an element of $A = \mathrm{Hom}(P, P)$, so we can consider the diagram

$$
\begin{array}{ccccccccc}
0 & \longrightarrow & K & \longrightarrow & P & \longrightarrow & X & \longrightarrow & 0 \\
 & & & & \big\downarrow{\scriptstyle a} & & & & \\
 & & & & P & \longrightarrow & Y & \longrightarrow & 0
\end{array}
$$

in $\mathcal{A}$. The composition $K \to Y$ is 0, since it is mapped to 0 by h^P, which is faithful, hence we get an induced map $h \colon X \to Y$ making

$$
\begin{array}{ccccccccc}
0 & \longrightarrow & K & \longrightarrow & P & \longrightarrow & X & \longrightarrow & 0 \\
 & & & & \big\downarrow{\scriptstyle a} & & \big\downarrow{\scriptstyle h} & & \\
 & & & & P & \longrightarrow & Y & \longrightarrow & 0
\end{array}
$$

commute. Applying h^P we get an analogous diagram in Mod_A, and we conclude that $h^P(h) = f$ because $A \to h^P(X)$ is epic. $\qquad\square$

We are finally ready to conclude the proof of the Freyd–Mitchell embedding theorem.

Proof of Theorem 2.8.1. Let $\mathcal{A}$ be a small Abelian category. By Theorem 2.8.14, $\mathcal{A}$ has a fully faithful exact embedding into $\mathcal{B} := \mathcal{L}(\mathcal{A}^{op}, \mathrm{Ab})$. In turn, Theorem 2.8.15 gives an exact functor $\mathcal{B} \to \mathrm{Mod}_A$ for a suitable ring A, which can be made a fully faithful embedding on $\mathcal{A}$. $\qquad\square$

Apart from its theoretical interest, the Freyd–Mitchell embedding can be used to prove many results on Abelian categories such as those in Section 2.5. As an application, we are going to prove the snake lemma, Lemma 2.5.20. The idea is that using the embedding theorem, one can prove the result in the category Mod_A for some ring A, and there one can use diagram chasing techniques using elements of modules.

Remark 2.8.16. Theorem 2.8.1 only works for small categories, so one may wonder whether this line of reasoning is sound for large ones. If $\mathcal{A}$ is any Abelian category, a diagram in $\mathcal{A}$ can be specified as a functor $D\colon \mathcal{J} \to \mathcal{A}$ from a small category of indices $\mathcal{J}$. The image of $\mathcal{J}$ is clearly small, and there is a minimal Abelian category containing this image, which is itself small (Exercise 4). Hence, results involving D and objects constructed from it can usually be proved assuming that the ambient category is small.

Proof of Lemma 2.5.20. As we argued above, it is enough to prove the result in Mod_A. We define the connecting homomorphism $\delta\colon \ker\gamma \to \mathrm{coker}\,\alpha$. Let $c \in \ker\gamma$, and lift it to an element $b \in B$ such that $g(b) = c$. By commutativity, $\beta(b) \in \ker g' = \mathrm{im}\,f'$, so we can take $a' \in A'$ such that $f'(a') = b'$. We let $\delta(c) := a' \pmod{\mathrm{im}\,\alpha}$.

To see that this is well defined, notice that we have taken two choices. The first choice involved b—another choice for a lift would differ by $f(a)$ for some $a \in A$, and the resulting class modulo the image of α would not change. The second choice was a', which is unique since f' is injective.

Notice that $a' = 0$ if and only if $b \in \ker\beta$, so that the six-step sequence is exact at $\ker\gamma$. Moreover, the image of δ is the same as $(f')^{-1}\,\mathrm{im}\,\beta$, hence the sequence is also exact at $\mathrm{coker}\,\alpha$. Exactness at $\ker\beta$ and $\mathrm{coker}\,\beta$ follows immediately from the exactness at B and B' of the two rows. $\qquad\square$

2.9. Exercises

1. Let $\mathcal{A}$ be an additive (or even Abelian) category, $f\colon A \to B$ a morphism of $\mathcal{A}$. Show that the functor $F\colon \mathcal{A} \to \mathrm{Ab}$ defined by

$$F(X) = \mathrm{coker}(\mathrm{Hom}(X, A) \to \mathrm{Hom}(X, B))$$

can fail to be representable.

2. Prove that the map g in (2.1.3) satisfies $\ker g = 0$ and $\mathrm{coker}\, g = 0$.

3. Prove Proposition 2.2.13.

4. Let $\mathcal{A}$ be an Abelian category and $\mathcal{A}'$ a small subcategory of $\mathcal{A}$. Prove that there is a small Abelian category containing $\mathcal{A}'$, and find bounds on its cardinality.

5. Let X be a small site. Prove that the plus construction

$$-^+\colon \mathrm{Psh}(X) \to \mathrm{SepPsh}(X)$$

is *not* adjoint to the natural inclusion $\mathrm{SepPsh}(X) \to \mathrm{Psh}(X)$, where we denote by $\mathrm{SepPsh}(X)$ the category of separated presheaves on X.

6. Let

(2.9.1) $$0 \longrightarrow A \longrightarrow B \longrightarrow C \longrightarrow 0$$

be a *complex* of sheaves of Abelian groups on a topological space X. Prove that (2.9.1) is exact if and only if it is exact on stalks, that is, for each point $p \in X$ the complex of Abelian groups

(2.9.2) $$0 \longrightarrow A_p \longrightarrow B_p \longrightarrow C_p \longrightarrow 0$$

is exact.

7. Let X be a topological space, $\mathcal{A}$ an Abelian category. Fix an object $A \in \mathcal{A}$ and a point $x \in X$. Prove that the presheaf A_x defined by

$$A_x(U) = \begin{cases} A \text{ if } x \in U \\ 0 \text{ otherwise} \end{cases}$$

is a sheaf on X, called the *skyscraper* sheaf with stalk A at x. Assuming that X is a T_1 topological space, prove that in fact the stalk of A_x at $y \in X$ is A if $x = y$ and 0 otherwise.

8 (**[She09]**). Let $X = \{a, b, c, d\}$ be a topological space with four points, having the nontrivial open sets $\{b\}$, $\{c\}$, $\{b, c\}$, $\{a, b, c\}$, $\{b, c, d\}$. Define a presheaf F on X by

$$F(\emptyset) = 0$$
$$F(\{b\}) = F(\{c\}) = \mathbb{Z}/2\mathbb{Z}$$
$$F(U) = \mathbb{Z} \text{ if } |U| \geq 2$$

with the obvious restriction maps. Compute F^+ and check that F^+ is separated, but is not a sheaf.

9. Let X be a topological space, $\mathcal{A}$ a complete category with enough injectives. Prove that $\mathrm{Sh}_{\mathcal{A}}(X)$ has enough injectives, by embedding every sheaf S into the product $\prod_{x \in X} S_x$ of skyscraper sheaves at all points $x \in X$.

10. Let X, Y be two topological spaces, $f \colon X \to Y$ a continuous function. Given a presheaf $F \in \mathrm{Psh}_{\mathcal{C}}(X)$, define a presheaf $f_* F \in \mathrm{Psh}_{\mathcal{C}}(Y)$ by the assignment

$$f_* F(U) := F(f^{-1}(U)).$$

Show that f_* is well defined, and that it sends sheaves to sheaves, thereby defining a functor $f_* \colon \mathrm{Sh}_{\mathcal{C}}(X) \to \mathrm{Sh}_{\mathcal{C}}(Y)$.

11. Let X, Y be two topological spaces, $f \colon X \to Y$ a continuous function. Given a sheaf $G \in \mathrm{Sh}_{\mathrm{Set}}(Y)$, define a presheaf $f^{-1} G \in \mathrm{Psh}_{\mathrm{Set}}(X)$ by the assignment

$$f^{-1} G(U) := \varinjlim_{V \supset f(U) \text{ open}} G(V).$$

Show that f^{-1} is well defined and that f^{-1} is, in fact, a sheaf, so that we get a functor $f^{-1}\colon \mathrm{Sh}_{\mathrm{Set}}(Y) \to \mathrm{Sh}_{\mathrm{Set}}(X)$.

12. Let X, Y be two topological spaces, $f\colon X \to Y$ a continuous function, and f^{-1} the functor defined in Exercise 11. Show that f^{-1} preserves stalks, that is, for a sheaf $G \in \mathrm{Sh}_{\mathrm{Set}}(Y)$ and $x \in X$ we have

$$(f^{-1}G)_x \cong G_{f(x)}.$$

Using Exercise 6, conclude that f^{-1} is exact as a functor on Ab-valued sheaves.

13. Let X, Y be two topological spaces, $f\colon X \to Y$ a continuous function. Let f_* be the functor defined in Exercise 10 f^{-1} be the functor defined in Exercise 11. Show that f^{-1} and f_* are adjoint, so that for $F \in \mathrm{Sh}_{\mathrm{Set}}(X)$ and $G \in \mathrm{Sh}_{\mathrm{Set}}(Y)$ there is a natural isomorphism

$$\mathrm{Hom}(f^{-1}G, F) \cong \mathrm{Hom}(G, f_*F).$$

The next two exercises follow [**Gra62**] to construct the sheafification for presheaves with values in categories more general than Set.

14. Let X be a topological space, $\mathcal{C}$ a cocomplete category, $F \in \mathrm{Psh}_{\mathcal{C}}(X)$ a presheaf. Show that for $x \in X$ there is a well-defined stalk F_x. If moreover $\mathcal{C}$ is complete, show that the presheaf defined by

$$\widetilde{F}(U) := \prod_{x \in U} F_x$$

is, in fact, a sheaf and that there is a functorial embedding $F \hookrightarrow \widetilde{F}$.

15. Let X be a topological space, $\mathcal{C}$ a complete and cocomplete category. Given a presheaf $F \in \mathrm{Psh}_{\mathcal{C}}(X)$, let $\widetilde{F}$ be the sheaf defined in Exercise 14. Let $\mathcal{F}$ be the family of all subsheaves of $\widetilde{F}$, up to isomorphism. Show that it makes sense to order $\mathcal{F}$ by the relation of subobject, and that $\mathcal{F}$ has a smallest object aF. Show that one can use this construction to define a functor $a\colon \mathrm{Psh}_{\mathcal{C}}(X) \to \mathrm{Sh}_{\mathcal{C}}(X)$, which is left adjoint of the natural inclusion.

16. Does the converse of Proposition 2.5.13 hold? That is, if $f\colon A \to B$ is a morphism such that f^* is injective on objects, up to isomorphism, does this imply that f is an epimorphism?

The following two exercises generalize the classical *isomorphism theorems* for modules over a ring.

17. Let $\mathcal{A}$ be an Abelian category, $X \in \mathcal{A}$ and U, V two subobjects of X. Prove that $(U + V)/V$ and $U/(U \cap V)$ are isomorphic subobjects of X/V.

18. Let $\mathcal{A}$ be an Abelian category, $X \in \mathcal{A}$ and U, V two subobjects of X with $U \subset V$. Prove that $(X/V)/(U/V)$ and X/U are isomorphic quotients of X.

19. Prove the snake lemma, Lemma 2.5.20, using the techniques of Section 2.5.

Exercises 20–24 give an alternative approach to diagram chasing in Abelian categories via the *salamander lemma* from [**Ber07**] (see also the presentation in [**Ger07**]). For these exercises, we will fix an infinite double complex in an Abelian category, of which we show a small bit

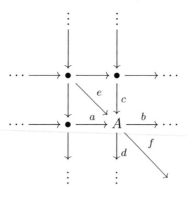

For such a diagram, denote

- the *vertical homology* $A^{\|} := \ker d / \operatorname{im} c$;
- the *horizontal homology* $_{\|}A := \ker b / \operatorname{im} a$;
- the *receptor* $^{\square}A := (\ker b \cap \ker d) / \operatorname{im} e$;
- the *donor* $A_{\square} := \ker f / (\operatorname{im} a + \operatorname{im} c)$.

20. Let A be an object in a double complex as above. Prove that there are well-defined maps $^{\square}A \to A^{\|}$, $A^{\|} \to A_{\square}$, $^{\square}A \to_{\|} A$, $_{\|}A \to A_{\square}$. If $A \to B$ is one of the arrows of the complex, there is also a map $A_{\square} \to^{\square} B$.

21. Prove the *salamander lemma*: if $A \to B$ is a horizontal arrow in the double complex, as pictured here

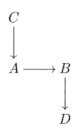

then there is a six-term exact sequence

$$C_\square \longrightarrow {}_\| A \longrightarrow A_\square \longrightarrow {}^\square B \longrightarrow {}_\| B \longrightarrow {}^\square D,$$

called a *salamander* centered at $A \to B$. State and prove a similar result for vertical arrows.

22. Use the salamander lemma to prove the four (hence also the five) lemma.

23. Use the salamander lemma to prove the nine lemma.

24. Use the salamander lemma to prove the snake lemma.

Exercises 25–30 give an alternative approach to diagram chasing in Abelian categories via the use of *members*, as developed in [**ML71**].

25. Let $\mathcal{A}$ be a small Abelian category, $A \in \mathcal{A}$. Given two morphisms $f: X \to A$ and $g: Y \to A$, say that $f \sim g$ if there exists epimorphisms $s: T \to X$ and $t: T \to Y$ such that $f \circ s = g \circ t$. Prove that $\sim$ is an equivalence relation on the set of morphisms with target A. A *member* of A is an equivalence class of morphisms to A. If a is a member of A, we write $a \in_m A$.

26. Prove that the functions are well defined on members (Exercise 25), that is, if $f: A \to B$ and $a \in_m A$, then $f(a) := f \circ a$ is a well-defined member of B. Prove that the following diagram chasing rules apply:

    (a) $f: A \to B$ is a monomorphism if and only if, for all members $a \in_m A$, $f(a) = 0$ implies $a = 0$.

    (b) $f: A \to B$ is a monomorphism if and only if, for all members $a, b \in_m A$, $f(a) = f(b)$ implies $a = b$.

    (c) $f: A \to B$ is an apimorphism if and only if, for all members $b \in_m B$, there exists $a \in_m A$ such that $f(a) = b$.

    (d) $f: A \to B$ is 0 if and only if, for all members $a \in_m A$, $f(a) = 0$.

    (e) The sequence $A \xrightarrow{\ f\ } B \xrightarrow{\ g\ } C$ is exact if and only if $g \circ f = 0$ and for all $b \in_m B$ such that $g(b) = 0$ there exists $a \in A$ such that $f(a) = b$.

27. Let $\mathcal{A}$ be a small Abelian category, $A \in \mathcal{A}$. Prove that the following form of subtraction is available for members of A (notice that members are *not* an Abelian group). Given a morphism $f: A \to B$ and $a, b \in_m A$ such that $f(a) = f(b)$, there is $d \in_m A$ such that

    (a) $f(d) = 0$;

    (b) if $g: A \to C$ has $g(b) = 0$, then $g(a) = g(d)$;

    (c) if $g: A \to C$ has $g(a) = 0$, then $g(b) = -g(d)$.

28. Use members in Abelian categories to prove the four (hence also the five) lemma.

29. Use members in Abelian categories to prove the nine lemma.

30. Use members in Abelian categories to prove the snake lemma.

31. Let $\mathcal{A}$ be any category, and assume that there is an object $0 \in \mathcal{A}$ which is both initial and terminal. Then we can define kernels and cokernels via their universal property. We can also define $f : A \to B$ to be a monomorphism if 0 is a kernel and an epimorphism if 0 is a cokernel. Now assume that

    (a) $\mathcal{A}$ has binary products and coproducts;

    (b) every morphism in $\mathcal{A}$ has a kernel and a cokernel;

    (c) every monomorphism is a kernel;

    (d) every epimorphism is a cokernel.

Prove that $\mathcal{A}$ is an Abelian category. Notice in particular that you will have to show that for each pair of objects $A, B \in \mathcal{A}$, the set $\operatorname{Hom}(A, B)$ has the structure of an Abelian group.

32. Let X be a topological space. Prove that the forgetful functor $\mathrm{Sh}_{\mathrm{Ab}}(X) \to \mathrm{Sh}(X)$ admits a left adjoint $G : \mathrm{Sh}(X) \to \mathrm{Sh}_{\mathrm{Ab}}(X)$. If S is any sheaf of sets, the sheaf $G(S)$ is called the free Abelian sheaf generated by S. What is this operation when X is a point?

33. Let $\mathcal{A}$ be an Abelian category, $f : A \to B$ a morphism of $\mathcal{A}$, and assume that f has two distinct factorizations $f = e_1 m_1 = e_2 m_2$, where e_1, e_2 are epimorphisms, and m_1, m_2 are monomorphisms. Show that there is a commutative diagram

$$
\begin{array}{ccc}
A \xrightarrow{\;e_1\;} T_1 \xrightarrow{\;m_1\;} B \\[2pt]
\Big\downarrow{\scriptstyle id_A} \quad\; \Big\downarrow{\scriptstyle t} \quad\;\; \Big\downarrow{\scriptstyle id_B} \\[2pt]
A \xrightarrow{\;e_2\;} T_2 \xrightarrow{\;m_2\;} B,
\end{array}
$$

where t is an isomorphism.

34. Let $F : \mathcal{A} \to \mathcal{B}$ be an additive functor between Abelian categories. Prove that F is left exact if and only if it preserves all finite limits, and right exact if and only if it preserves all finite colimits.

35. Use Exercise 34 to give a different proof of Corollary 2.2.20.

Chapter 3

Derived Functors

This chapter introduces one of the central concepts in homological algebra, derived functors. From the beginning of the study of homology of topological spaces, it was observed that many homology theories admit some form of a long exact sequence in homology. Trying to capture this in an abstract setting led to the formulation of the zig-zag lemma, Lemma 3.1.9.

From an algebraic point of view, this lemma allows to construct long exact sequences in a lot of other situations. Many natural functors between Abelian categories are not exact, but are left or right exact—often as a consequence of some kind of adjunction. The theory of derived functors allows to extend a left exact sequence

$$0 \longrightarrow F(A) \longrightarrow F(B) \longrightarrow F(C),$$

obtained by applying a left-exact functor F to an exact sequence, and give it a canonical continuation as a long exact sequence, at least as long as the source category has enough injectives. A dual formulation holds for right-exact functors. This leads to a general procedure of concocting useful long exact sequences of homology, that often result in nontrivial information on the original functor.

We begin by exploring some general properties of the categories of complexes, and the homological properties of morphisms between complexes, leading to the aforementioned zig-zag lemma. Section 3.2 uses this machinery to define right- and left-derived functors—under the assumption of existence of enough injectives or projectives. Derived functors can also be used to give another take on the adjunction between $\mathcal{L}(\mathcal{A}, \mathrm{Ab})$ and $\mathrm{Add}(\mathcal{A}, \mathrm{Ab})$ that was crucial in the proof of Theorem 2.8.8, albeit only in a special case.

The next sections study in more detail various examples of derived functors. Section 3.3 presents some easy results that simplify the computation of derived functors, and uses them to obtain some vanishing results on the cohomology of sheaves. These, in turn, allow us to relate cohomology of sheaves to De Rham cohomology.

Section 3.4 studies the functors Ext^i and Tor_i, which are probably the most prototypical of all derived functors. These are the derived functors of $\text{Hom}(-,-)$ and $-\otimes-$ respectively. Since both of them are bifunctors, a slight complication is that one can compute derived functors with respect to either variable. A crucial point in this section is that the end result is the same. The rest of the section explores some consequences of this fact, and finally gives an alternative construction of the groups Ext^i due to Yoneda, that does not require the existence of enough injectives or projectives.

Finally, Section 3.5 studies the derived functors of limits, in a very special case of inverse limits of towers of objects. The aim here is to give a more explicit way to compute these higher limits, and prove some vanishing result.

3.1. Categories of complexes

Let $\mathcal{A}$ be an Abelian category. We will define a few categories of complexes with objects in $\mathcal{A}$.

Definition 3.1.1. The *category of complexes* $\text{Kom}(\mathcal{A})$ has for objects doubly infinite complexes of objects in $\mathcal{A}$

$$\cdots \longrightarrow A_{n-1} \longrightarrow A_n \longrightarrow A_{n+1} \longrightarrow \cdots .$$

A morphism between complexes $A_\bullet = \{A_n\}$ and $B_\bullet = \{B_n\}$ consists of morphisms $A_n \to B_n$ in $\mathcal{A}$ such that the square

$$\begin{array}{ccc} A_n & \longrightarrow & A_{n+1} \\ \downarrow & & \downarrow \\ B_n & \longrightarrow & B_{n+1} \end{array}$$

commutes for all n. A complex $A_\bullet$ is called *bounded below* (resp., *bounded above*) if $A_n = 0$ for all $n \ll 0$ (resp., $n \gg 0$), and *bounded* if it is bounded both below and above. The subcategories of bounded below, bounded above and bounded complexes will be denoted by $\text{Kom}^+(\mathcal{A})$, $\text{Kom}^-(\mathcal{A})$ and $\text{Kom}^b(\mathcal{A})$ respectively.

Given a complex $A_\bullet$, its homology is a measure of the failure to be exact.

Definition 3.1.2. Let $A_\bullet$ be a complex over the Abelian category $\mathcal{A}$. The *n-th homology object $H_n(A_\bullet)$* is the quotient

$$H_n(A_\bullet) := \frac{\ker A_n \to A_{n+1}}{\operatorname{im} A_{n-1} \to A_n},$$

where—as usual—the quotient is the cokernel of the natural monomorphism

$$\operatorname{im}(A_{n-1} \to A_n) \to \ker(A_n \to A_{n+1}).$$

If $H_n(A_\bullet) = 0$ for all n, we say that $A_\bullet$ is *acyclic*.

It is common to denote $B_n = \operatorname{im}(A_{n-1} \to A_n)$ (for boundary) and $Z_n = \ker(A_n \to A_{n+1})$ (for cycle, or Zyclus in German). With this notation, B_n is a subobject of Z_n and $H_n(A_\bullet) = Z_n/B_n$.

It is immediate to check that a morphism of complexes $f \colon A_\bullet \to B_\bullet$ induces a well-defined map $H_n(A_\bullet) \to H_n(B_\bullet)$ for all n, which we still denote by f, or f_n.

Definition 3.1.3. Let $f \colon A_\bullet \to B_\bullet$ be a morphism of complexes. If the induced map $H_n(A_\bullet) \to H_n(B_\bullet)$ is an isomorphism for all n, we say that f is a *quasi-isomorphism*.

Of course, most early examples of complexes come from topology.

Example 3.1.4.

(a) Let S be a topological space, G an Abelian group. As usual, Δ_n denotes the n-simplex and $f_i \colon \Delta_{n-1} \to \Delta_n$ the i-th face map. A map $\Delta_n \to S$ is called an n-dimensional singular simplex. The n-th singular chain group of S with coefficients in G is

$$C_n(S; G) := \left\{ \sum a_s s \right\},$$

where we take formal finite linear combinations of singular simplexes s with coefficients $a_s \in G$.

There is a natural boundary map

$$\partial_n \colon \ C_n(S; G) \longrightarrow C_{n-1}(S; G),$$

$$s \longmapsto \sum_{i=0}^{n} (-1)^i s \circ f_i,$$

where s is a singular simplex, and it is a routine verification that $\partial_{n-1} \circ \partial_n = 0$. Hence $C_\bullet(S; G)$ is a complex, called the singular complex of S.

The homology of this complex is denoted by $H_n(S; G)$, and is in fact the singular homology of S with coefficients in G.

(b) Similarly, if M is a smooth (C^∞) manifold, the exterior derivative operator $d\colon \Lambda^n(M) \to \Lambda^{n+1}(M)$ between differential forms defines the De Rham complex $\Lambda^\bullet(M)$. Its homology is the De Rham cohomology of M.

The example of singular homology inspires an algebraic definition. Let $s_0, s_1\colon \Delta_n \to S$ be two singular simplexes. A homotopy between s_0 and s_1 is a function $s\colon \Delta_n \times I \to S$ defined on the prism $\Delta_n \times I$, which agrees with s_0 on the bottom face and s_1 on the top face. The prism has a structure of simplicial complex by a suitable subdivision. The subdivision is a little tricky to get right, but since we do not need it, we refer the reader to [**Hat02**, Theorem 2.10], and only show an illustration of the case $n = 2$. The boundary of the prism consists of two copies of Δ_n with opposite orientation, plus $\partial \Delta_n \times I$. By this observation, we obtain the relation

$$\partial s = s_1 - s_0 + h,$$

where h is a homotopy between ∂s_0 and ∂s_1. This geometric observation prompts the following purely algebraic definition.

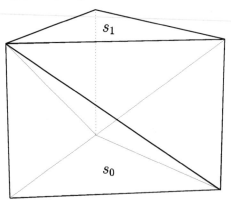

Definition 3.1.5. Let $(A_\bullet, a_\bullet)$ and $(B_\bullet, b_\bullet)$ be two complexes, and let $f, g\colon A_\bullet \to B_\bullet$ two morphisms. A *chain homotopy* h between f and g is a sequence of maps $h_n\colon A_n \to B_{n-1}$ such that

$$h_{n+1} \circ a_n - b_{n-1} \circ h_n = g_n - f_n.$$

If there exists a chain homotopy between f and g, we say that f and g are homotopic.

It is often more convenient to omit indices and denote the boundary maps a by d_A and b by d_B, so that the homotopy condition reads

$$h \circ d_A - d_B \circ h = g - f.$$

Homotopies are useful because of the following algebraic property.

Proposition 3.1.6. *Let $f, g\colon A_\bullet \to B_\bullet$ two homotopic morphisms. Then f and g induce the same maps on homology, that is, the maps $f\colon H_n(A) \to H_n(B)$ and $g\colon H_n(A) \to H_n(B)$ are equal.*

Proof. Let h be a homotopy between f and g. On $\ker A_n \to A_{n+1}$, we have the equality $f - g = d_B \circ h$, hence $\operatorname{im}(f - g) \subset \operatorname{im} d_B$. $\qquad\square$

This fact is often used in a slightly different form.

Definition 3.1.7. Let $f\colon A_\bullet \to B_\bullet$ be a morphism of complexes. We say that f is a *homotopy equivalence* if there exists a morphism $g\colon B_\bullet \to A_\bullet$ such that $f \circ g$ is homotopic to id_B and $g \circ f$ is homotopic to id_A.

With this terminology, an immediate consequence of Proposition 3.1.6 is

Corollary 3.1.8. *A homotopy equivalence $f\colon A_\bullet \to B_\bullet$ is a quasi-isomorphism.*

Let $\mathcal{A}$ be an Abelian category. It is easy to check that $\operatorname{Kom}(\mathcal{A})$ is also an Abelian category, where kernels and cokernels are taken element-wise. That is, for a morphism of complexes $F\colon A_\bullet \to B_\bullet$, the sequence of kernels $K_i := \ker F_i$ is itself a complex, and $K_\bullet$ is a kernel of F. Dually, the sequence of cokernels $C_i := \operatorname{coker} F_i$ is a complex, and $C_\bullet$ is a cokernel of F. All axioms of Abelian category are checked easily. Henceforth, we will speak of short exact sequences of complexes and so on. A useful result is the *zig-zag lemma*.

Lemma 3.1.9 (Zig-zag lemma). *Let*

$$0 \longrightarrow A_\bullet \longrightarrow B_\bullet \longrightarrow C_\bullet \longrightarrow 0$$

be an exact sequence of complexes. Then there are connecting morphisms $\delta_n\colon H_n(C) \to H_{n+1}(A)$ such that the sequence

$$\cdots \longrightarrow H_n(A) \longrightarrow H_n(B) \longrightarrow H_n(C) \overset{\delta_n}{\longrightarrow} H_{n+1}(A) \longrightarrow H_{n+1}(B) \longrightarrow H_{n+1}(C) \longrightarrow \cdots$$

is exact.

Proof. Denote by $a_n\colon A_n \to A_{n+1}$ the complex morphisms in $A_\bullet$, and similarly for b_n, c_n. Since $A_\bullet$ is a complex, we have an induced morphism $\widetilde{a}_n\colon \operatorname{coker} a_n \to \ker a_{n+1}$, and by definition $H_{n+1}(A) \cong \operatorname{coker} \widetilde{a}_n$. But, crucially, we also have $H_n(A) \cong \ker \widetilde{a}_n$.

Two applications of the snake lemma (Lemma 2.5.20) show that the rows of the diagram

$$
\begin{array}{ccccccc}
\operatorname{coker} a_n & \longrightarrow & \operatorname{coker} b_n & \longrightarrow & \operatorname{coker} c_n & \longrightarrow & 0 \\
\downarrow & & \downarrow & & \downarrow & & \\
0 \longrightarrow \ker a_{n+1} & \longrightarrow & \ker b_{n+1} & \longrightarrow & \ker c_{n+1} & &
\end{array}
$$

are exact. Yet another application of the snake lemma then gives the exactness of the six term sequence

$$
\begin{array}{ccc}
H_n(A) \longrightarrow H_n(B) \longrightarrow H_n(C) \\
 \delta_n \\
\hookrightarrow H_{n+1}(A) \longrightarrow H_{n+1}(B) \longrightarrow H_{n+1}(C).
\end{array}
$$

Joining all these six-term sequences yields the conclusion. $\qquad\square$

A useful perspective on quasi-isomorphisms is given by the construction of mapping cones.

Definition 3.1.10. Let $f\colon A_\bullet \to B_\bullet$ be a morphism of complexes. The *mapping cone* of f, denoted $C(f)$, is the complex defined by $C(f)_n = A_{n+1} \oplus B_n$, with differential given by

$$
d_{C(f)} = \begin{pmatrix} -d_A & 0 \\ f & d_B \end{pmatrix}
$$

relative to this decomposition.

The relation $d_{C(f)}^2 = 0$ follows from the fact that f is a morphism of complexes. To simplify notation, it is common to denote the shifted complex $A[k]_n = A_{n+k}$, with differential $(-1)^k d_A$, so that $C_\bullet = A[1]_\bullet \oplus B_\bullet$, with differential

$$
d_{C(f)} = \begin{pmatrix} d_{A[1]} & 0 \\ f & d_B \end{pmatrix}.
$$

By construction, the inclusion $B \to C(f)$ and the projection $C(f) \to A[1]$ are morphisms of complexes (why?), so that we have the exact sequence

$$
0 \longrightarrow B \longrightarrow C(f) \longrightarrow A[1] \longrightarrow 0 \,.
$$

Taking the associated long exact sequence, we obtain:

Proposition 3.1.11. *Let* $f\colon A_\bullet \to B_\bullet$ *be a morphism of complexes. Then there is a long exact sequence*

$$\cdots \longrightarrow H_n(A) \longrightarrow H_n(B) \longrightarrow H_n(C(f))$$
$$\hookrightarrow H_{n+1}(A) \longrightarrow H_{n+1}(B) \longrightarrow H_{n+1}(C(f)) \longrightarrow \cdots.$$

A priori, the map $H_n(A) \to H_n(B)$ is the connecting map from the zig-zag lemma, but it is an immediate verification that it agrees with the map induced by f itself.

Corollary 3.1.12. *Let* $f\colon A_\bullet \to B_\bullet$ *be a morphism of complexes. Then* f *is a quasi-isomorphism if and only if* $C(f)$ *is acyclic.*

3.2. Derived functors

Let $F\colon \mathcal{A} \to \mathcal{B}$ be a left-exact functor between Abelian categories. Given a short exact sequence in $\mathcal{A}$

$$0 \longrightarrow A \longrightarrow B \longrightarrow C \longrightarrow 0$$

we get a left-exact sequence

$$(3.2.1) \qquad\qquad 0 \longrightarrow F(A) \longrightarrow F(B) \longrightarrow F(C)$$

but so far we have no way to measure the extent to which this fails to be right exact. The definition of the right derived functors $R^i F$ will allow us to extend (3.2.1) to a long exact sequence

$$0 \longrightarrow F(A) \longrightarrow F(B) \longrightarrow F(C)$$
$$\hookrightarrow R^1 F(A) \longrightarrow R^1 F(B) \longrightarrow R^1 F(C)$$
$$\hookrightarrow R^2 F(A) \longrightarrow R^2 F(B) \longrightarrow R^2 F(C) \longrightarrow \cdots.$$

Resolutions are the fundamental building block to define derived functors.

Definition 3.2.1. Let $A \in \mathcal{A}$ be an object of an Abelian category. A *(right) resolution* of A is a complex $C_\bullet$ such that $C_i = 0$ for $i < 0$ and the diagram

$$0 \longrightarrow A \longrightarrow C_0 \longrightarrow C_1 \longrightarrow C_2 \longrightarrow \cdots$$

is exact. In other words, $C_\bullet$ is concentrated in nonnegative degree, is exact except at C_0, and $A \cong \ker(C_0 \to C_1)$. If all C_i are injective objects, we say that $C_\bullet$ is an *injective resolution* of A.

Remark 3.2.2. We can look at A as a trivial complex concentrated in degree 0. Then a resolution of A is just a complex concentrated in nonnegative degree which is quasi-isomorphic to A.

There is a dual notion of resolution using complexes concentrated in nonpositive degrees—this will be called a *left resolution*. In particular, a *projective resolution* of A will be a complex $P_\bullet$ concentrated in nonpositive degrees such that

$$\cdots \longrightarrow P_{-2} \longrightarrow P_{-1} \longrightarrow P_0 \longrightarrow A \longrightarrow 0$$

is exact, and each P_i is projective. This way, a left (resp., projective) resolution of A corresponds to a right (resp., injective) resolution of A in the opposite category. For this reason, we will concentrate on right resolutions, leaving to the reader to formulate suitable dual statements.

Clearly, a necessary condition for the existence of injective resolutions is that $\mathcal{C}$ should have enough injectives, since the first step of an injective resolution $A \to I_\bullet$ is a monomorphism $A \to I_0$. It is easy to check that this is also sufficient.

Proposition 3.2.3. *Let $\mathcal{A}$ be an Abelian category with enough injectives. Then every object of $\mathcal{A}$ admits an injective resolution.*

Proof. Start with a monomorphism $A \to I_0$, where I_0 is injective, and let $C_0 = \mathrm{coker}(A \to I_0)$, so that

$$0 \longrightarrow A \longrightarrow I_0 \longrightarrow C_0 \longrightarrow 0$$

is exact. Then, find a monomorphism $C_0 \to I_1$ with I_1 injective, having cokernel C_1, and so on, recursively. By construction, $I_\bullet$ will be an injective resolution of A. $\qquad\square$

Of course, there are many possible injective resolutions of a given object. The question of uniqueness arises. We can deal with this question in slightly more generality.

Proposition 3.2.4. *Let $\mathcal{A}$ be an Abelian category, $s \colon A \to B$ a morphism. Let*

$$0 \longrightarrow A \longrightarrow C_0 \longrightarrow C_1 \longrightarrow C_2 \longrightarrow \cdots$$

be a right resolution of A, and

$$0 \longrightarrow B \longrightarrow I_0 \longrightarrow I_1 \longrightarrow I_2 \longrightarrow \cdots$$

be a complex, where the I_k are injectives. Then there exists a map of complexes $C_\bullet \to I_\bullet$ that extends s, and such a map is unique up to chain homotopy.

Proof. We construct a morphism $f_n \colon C_n \to I_n$ by induction. Since I_0 is injective, the map $A \to B \to I_0$ can be extended to C_0, as in the following diagram

$$
\begin{array}{ccccc}
0 & \longrightarrow & A & \longrightarrow & C_0 & \longrightarrow & \cdots \\
& & \downarrow{\scriptstyle s} & & \downarrow{\scriptstyle f_0} & & \\
0 & \longrightarrow & B & \longrightarrow & I_0 & \longrightarrow & \cdots.
\end{array}
$$

Having constructed f_n, let $W_n = \operatorname{coker}(C_{n-1} \to C_n)$ and $Z_n = \operatorname{coker}(I_{n-1} \to I_n)$. Notice that f_n induces a morphism $W_n \to Z_n$ by the universal property of W_n. Since $C_\bullet$ is exact in positive degree, $W_n \to C_{n+1}$ is a monomorphism. This allows us to extend the map $W_n \to I_{n+1}$ in the following diagram

$$
\begin{array}{ccccc}
0 & \longrightarrow & W_n & \longrightarrow & C_{n+1} & \longrightarrow & \cdots \\
& & \downarrow{\scriptstyle f_n} & & \vdots{\scriptstyle f_{n+1}} & & \\
& & Z_n & \longrightarrow & I_{n+1} & \longrightarrow & \cdots,
\end{array}
$$

thereby defining f_{n+1}.

For uniqueness, let f, g be two morphisms of complexes that extend s. We need to construct morphisms $h_n \colon C_n \to I_{n-1}$ that satisfy

$$(3.2.2) \qquad h_{n+1} d_C - d_I h_n = g_n - f_n$$

for $n \geq 1$. We start by defining $h_0 = 0$. Let W_n be the image of $C_n \to C_{n+1}$. Equation (3.2.2) determines h_{n+1} on W_n, and since $W_n \to C_{n+1}$ is a monomorphism, this can be extended to a map $h_{n+1} \colon C_{n+1} \to I_n$. $\qquad\square$

Corollary 3.2.5. Let $I_\bullet, J_\bullet$ be two injective resolutions of A. Then there is a quasi-isomorphism $f \colon I_\bullet \to J_\bullet$ such that

$$
\begin{array}{ccccccc}
0 & \longrightarrow & A & \longrightarrow & I_0 & \longrightarrow & I_1 & \longrightarrow & \cdots \\
& & \downarrow{\scriptstyle \operatorname{id}_A} & & \downarrow{\scriptstyle f_0} & & \downarrow{\scriptstyle f_1} & & \\
0 & \longrightarrow & A & \longrightarrow & J_0 & \longrightarrow & J_1 & \longrightarrow & \cdots
\end{array}
$$

commutes.

Proof. Proposition 3.2.4 gives two morphisms of complexes $f \colon I_\bullet \to J_\bullet$ and $g \colon J_\bullet \to I_\bullet$. The same result, applied to $g \circ f$, gives a chain homotopy between $\operatorname{id}_{I_\bullet}$ and $g \circ f$, hence $g \circ f$ induces the identity on the homology of $I_\bullet$. By symmetry, the same is true for $f \circ g$, hence f and g are quasi-isomorphisms. $\qquad\square$

We are now ready to define derived functors. Let $F\colon \mathcal{A} \to \mathcal{B}$ be an additive functor between Abelian categories, and assume that $\mathcal{A}$ has enough injectives. Given an object $A \in \mathcal{A}$, choose an injective resolution $A \to I_\bullet$. Applying F to $I_\bullet$ yields another complex $F(I_\bullet)$:

$$0 \longrightarrow F(I_0) \longrightarrow F(I_1) \longrightarrow F(I_2) \longrightarrow \cdots,$$

which is not necessarily exact. We then define $R^i F(A) := H_i(F(I_\bullet))$.

To check that this is well defined, let $A \to J_\bullet$ be a different injective resolution. Corollary 3.2.5 gives a quasi-isomorphism $f\colon I_\bullet \to J_\bullet$, and F preserves this quasi-isomorphism, therefore inducing an isomorphism $H_i(F(I_\bullet)) \to H_i(F(J_\bullet))$. Moreover, a different quasi-isomorphism $I_\bullet \to J_\bullet$ is chain homotopic to f. The additive functor F preserves the chain homotopy, hence the induced isomorphism $H_i(F(I_\bullet)) \cong H_i(F(J_\bullet))$ does not depend on the choice of the quasi-isomorphism.

To make $R^i F$ into a functor, we need to define $R^i F(A)$ simultaneously for all objects of $\mathcal{A}$. Recall from our discussion in Section 1.2 that classes can be defined either in NBG set theory, or as sets in a model $\mathcal{U}$ of ZFC—in this second approach, we redefine sets to be members of $\mathcal{U}_\kappa$ for a strongly inaccessible cardinal κ of $\mathcal{U}$. In both approaches, the axiom of choice holds for classes.

To define $R^i F$, use the axiom of choice to select an injective resolution for every object A of $\mathcal{A}$, and define $R^i F(A)$ using the above procedure. To define $R^i F$ for a morphism $s\colon A \to B$, let $I_\bullet, J_\bullet$ be the chosen resolutions of A and B respectively. By Proposition 3.2.4, we obtain a map of complexes $I_\bullet \to J_\bullet$, which is unique up to homotopy. This defines a morphism $F(I_\bullet) \to F(J_\bullet)$, again unique up to homotopy, hence a map $R^i F(A) \to R^i F(B)$. The uniqueness up to homotopy immediately implies that if $t\colon B \to C$ is another morphism, then $R^i F(t \circ s) = R^i F(t) \circ R^i F(s)$, so that $R^i F$ is indeed a functor.

The functor $R^i F$ as defined above depends on the arbitrary choice of injective resolutions for all objects of $\mathcal{A}$. If we choose different resolutions, we obtain a different functor $R^i_* F$, but the above discussion gives a natural isomorphism between $R^i F$ and $R^i_* F$. We usually don't bother to insist on this point and denote by $R^i F$ any functor obtained by this procedure, even though $R^i F$ is only defined up to natural isomorphism.

Definition 3.2.6. Let $F\colon \mathcal{A} \to \mathcal{B}$ be an additive functor between Abelian categories, and assume that $\mathcal{A}$ has enough injectives. The functor $R^i F$ defined above is called the i-th *right derived* functor of F.

The basic properties of derived functors are expressed in the following results.

Proposition 3.2.7. *Let $F\colon \mathcal{A} \to \mathcal{B}$ be left exact. Then there is a natural isomorphism $F \cong R^0 F$.*

Proof. Choose an object A and an injective resolution $A \to I_\bullet$. Then

$$0 \longrightarrow F(A) \longrightarrow F(I_0) \longrightarrow F(I_1)$$

is exact, giving an isomorphism $F(A) \cong H_0(F(I_\bullet))$. $\qquad\square$

Proposition 3.2.8. *If A is injective and $F\colon A \to B$ is an additive functor, then $R^i F(A) = 0$ for $i > 0$.*

Proof. Choose the injective resolution $0 \to A \to A \to 0$. $\qquad\square$

Proposition 3.2.9. *Let*

$$0 \longrightarrow A \longrightarrow B \longrightarrow C \longrightarrow 0$$

be a short exact sequence in the Abelian category $\mathcal{A}$, and let $F\colon \mathcal{A} \to \mathcal{B}$ be an additive functor. Then there is a long exact sequence

$$0 \longrightarrow R^0 F(A) \longrightarrow R^0 F(B) \longrightarrow R^0 F(C)$$
$$\longrightarrow R^1 F(A) \longrightarrow R^1 F(B) \longrightarrow R^1 F(C)$$
$$\longrightarrow R^2 F(A) \longrightarrow R^2 F(B) \longrightarrow R^2 F(C) \longrightarrow \cdots .$$

Proof. Choose injective resolutions $A \to I_\bullet^A$, and $C \to I_\bullet^C$, and define $I^B := I^A \oplus I^C$. We claim that I^B is a resolution of B. Since I_0^A is injective, there is map $B \to I_0^A$ extending $A \to I_0^A$, and there is a map $B \to C \to I_0^C$. Putting them together gives a diagram

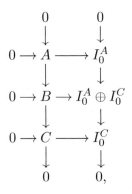

where the central row is exact at B by the four lemma (Lemma 2.5.2). Define $A_0 = \operatorname{coker}(A \to I_0^A)$, and similarly for B_0 and C_0. Then

$$0 \longrightarrow A_0 \longrightarrow B_0 \longrightarrow C_0 \longrightarrow 0$$

is exact by the nine lemma (Lemma 2.5.17). We can repeat the same reasoning to prove that $B_0 \to I_1^B$ is a monomorphism, and so on. By induction, we conclude that I^B is a resolution of B.

We can now use the resolutions I^A, I^B, I^C to compute the derived functors $R^i F$ at A, B, and C. By construction the exact sequence

$$0 \longrightarrow I^A \longrightarrow I^B \longrightarrow I^C \longrightarrow 0$$

is split. Since an additive functor preserves split exact sequences, we get the exact sequence

$$0 \longrightarrow F(I^A) \longrightarrow F(I^B) \longrightarrow F(I^C) \longrightarrow 0 \ .$$

The conclusion now follows by the zig-zag lemma, Lemma 3.1.9.   □

We can obtain dual statements by working with projective resolutions. Let $\mathcal{A}$ be a category with enough projectives, and $F \colon \mathcal{A} \to \mathcal{B}$ an additive functor. Given an object $A \in \mathcal{A}$, we can choose a projective resolution $P_\bullet \to A$, and define the *left-derived functors* $L^i F(A) := H_{-i}(F(P_\bullet))$.

Just as we did for right-derived functors, we can prove that the $L^i F$ are in fact functors, and are well defined up to natural isomorphism. There are obvious analogues to Propositions 3.2.7, 3.2.8, and 3.2.9. In particular, if F is right exact, we have the long exact sequence

$$\cdots \longrightarrow L^2 F(A) \longrightarrow L^2 F(B) \longrightarrow L^2 F(C)$$
$$\longrightarrow L^1 F(A) \longrightarrow L^1 F(B) \longrightarrow L^1 F(C)$$
$$\longrightarrow F(A) \longrightarrow F(B) \longrightarrow F(C) \longrightarrow 0.$$

Remark 3.2.10. The construction of left derived functors does not just follow by duality from the right derived functors, because it applies to covariant functors. If we take duality into account, we get *more* derived functors.

Let $F \colon \mathcal{A}^{op} \to \mathcal{B}$ be a contravariant additive functor. Assuming that $\mathcal{A}$ has enough projectives, $\mathcal{A}^{op}$ has enough injectives, and we can define the right-derived functors $R^i F$. Symmetrically, if $\mathcal{A}$ has enough injectives, we can define the left derived functors $L^i F$. The associated long exact sequences follow from the covariant case.

In particular, if

$$0 \longrightarrow A \longrightarrow B \longrightarrow C \longrightarrow 0$$

is a short exact sequence in $\mathcal{A}$, $\mathcal{A}$ has enough projectives and F is left exact, then we have an associated long exact sequence

$$0 \longrightarrow F(C) \longrightarrow F(B) \longrightarrow F(A)$$
$$\longrightarrow R^1 F(C) \longrightarrow R^1 F(B) \longrightarrow R^1 F(A)$$
$$\longrightarrow R^2 F(C) \longrightarrow R^2 F(B) \longrightarrow R^2 F(A) \longrightarrow \cdots .$$

If instead $\mathcal{A}$ has enough injectives and F is right exact, we have the associated long exact sequences

$$\cdots \longrightarrow L^2 F(C) \longrightarrow L^2 F(B) \longrightarrow L^2 F(A)$$
$$\longrightarrow L^1 F(C) \longrightarrow L^1 F(B) \longrightarrow L^1 F(A)$$
$$\longrightarrow F(C) \longrightarrow F(B) \longrightarrow F(A) \longrightarrow 0.$$

Remark 3.2.11. The construction of derived functors has a naturality property, which we spell out only in the case of covariant functors and right derived functors. Let $F, G \colon \mathcal{A} \to \mathcal{B}$ be additive functors and assume that $\mathcal{A}$ has enough injectives. Let $A \in \mathcal{A}$, and take an injective resolution $A \to I_\bullet$. If $t \colon F \to G$ is a natural transformation, we get a morphism of complexes $t_{I_\bullet} \colon F(I_\bullet) \to G(I_\bullet)$, which induces maps $R^i F(A) \to R^i G(A)$.

It is immediate to check that this defines a natural transformation $R^i F \to R^i G$. In particular, if $\mathcal{A}$ is small, R^i defines a functor $\mathrm{Add}(\mathcal{A}, \mathcal{B}) \to \mathrm{Add}(\mathcal{A}, \mathcal{B})$.

Remark 3.2.12. Let $\mathcal{A}$ be an Abelian category with enough injectives and $F \colon \mathcal{A} \to \mathrm{Ab}$ an additive functor. Then $R^0 F$ is left exact by Proposition 3.2.9. Moreover, if F is already left exact, $R^0 F$ is isomorphic to F by Proposition 3.2.7.

If in addition $\mathcal{A}$ is small, we get an additive functor

$$R^0 \colon \mathrm{Add}(\mathcal{A}, \mathrm{Ab}) \to \mathcal{L}(\mathcal{A}, \mathrm{Ab})$$

such that $R^0 \circ \iota \cong \mathrm{id}$, where $\iota\colon \mathcal{L}(\mathcal{A}, \mathrm{Ab}) \to \mathrm{Add}(\mathcal{A}, \mathrm{Ab})$ is the natural inclusion ($\mathcal{L}(\mathcal{A}, \mathrm{Ab})$ is the category of left exact functors from Definition 2.8.3). In fact, R^0 is left adjoint to ι. To see this, let $F, G\colon \mathcal{A} \to \mathrm{Ab}$ be additive, with G left exact. Then we have the map

$$\mathrm{Hom}(F, \iota G) \xrightarrow{\ R^0\ } \mathrm{Hom}(R^0 F, R^0 \iota G) \cong \mathrm{Hom}(R^0 F, G),$$

and in the other direction

$$\mathrm{Hom}(R^0 F, G) \longrightarrow \mathrm{Hom}(F, \iota G)$$

given by composition with the natural morphism $F \to R^0 F$. One easily checks that they are mutually inverse, giving a natural isomorphism $\mathrm{Hom}(R^0 F, G) \cong \mathrm{Hom}(F, \iota G)$.

The adjunction between $\mathcal{L}(\mathcal{A}, \mathrm{Ab})$ and $\mathrm{Add}(\mathcal{A}, \mathrm{Ab})$ is the crucial step in Theorem 2.8.8 that proves that $\mathcal{L}(\mathcal{A}, \mathrm{Ab})$ is Abelian. Thus, derived functors provide an alternate way to this result, albeit only under the assumption that $\mathcal{A}$ has enough injectives. It is worth noting that even this restriction can be removed—for instance, [**Rus16**] defines 0-th derived functors without assumptions on the underlying Abelian categories.

Example 3.2.13.

(a) Let $\mathcal{A}$ be any Abelian category. By Proposition 2.2.17, the functors h_A and h^A are left exact for all $A \in \mathcal{A}$.

If $\mathcal{A}$ has enough injectives, the right-derived functors of h^A are defined, and they are covariant functors $\mathcal{A} \to \mathrm{Ab}$. They are called the *Ext functors* and are defined by

$$\mathrm{Ext}^i_{\mathcal{A}}(A, B) := R^i h^A(B).$$

Since $h^\bullet$ is a (contravariant) functor $\mathcal{A} \to \mathrm{Add}(\mathcal{A}, \mathrm{Ab})$, by Remark 3.2.11, the Ext functors are also functorial in A. Hence $\mathrm{Ext}^i_{\mathcal{A}}(A, B)$ is a *bifunctor*, contravariant in A, and covariant in B, just like $\mathrm{Hom}_{\mathcal{A}}(A, B)$. When the category $\mathcal{A}$ is clear from the context, we will simply write $\mathrm{Ext}^i(A, B) = \mathrm{Ext}^i_{\mathcal{A}}(A, B)$.

If $\mathcal{A}$ has enough projectives, we can define the right-derived functors of h_B, which are *contravariant* functors $\mathcal{A} \to \mathrm{Ab}$. We then get an alternative definition of the Ext functors as

$$\mathrm{Ext}^i_{\mathcal{A}}(A, B) := R^i h_B(A).$$

For the same reasons as above, this is a bifunctor, contravariant in A and covariant in B. We will see in Section 3.4 that the two definitions agree whenever $\mathcal{A}$ has both enough injectives and enough projectives.

(b) Let A be a commutative ring (most of the following works also in the noncommutative case, see Exercise 11), and M and A-module. By Corollary 2.2.20, the functor

$$T_M := M \otimes - \colon \operatorname{Mod}_A \to \operatorname{Mod}_A$$

is right exact.

Since Mod_A has enough projectives, we can define the left derived functors of T_M, which will be called the *Tor functors*. Explicitly,

$$\operatorname{Tor}_i^A(M, N) := L^i T_M(N).$$

When the ring A is clear from the context, we simply write $\operatorname{Tor}_i(M, N) = \operatorname{Tor}_i^A(M, N)$. It is immediate to check that the association $M \to T_M$ is functorial in M, hence—as it was the case for Ext—the functors $\operatorname{Tor}_i(M, N)$ are bifunctors, covariant both in M and N, just like $M \otimes N$.

Symmetrically, we can also consider the functor

$$T^N := - \otimes N \colon \operatorname{Mod}_A \to \operatorname{Mod}_A,$$

which is also right exact. We then get an alternative definition of the Tor functors as

$$\operatorname{Tor}_A^i(M, N) := L^i T^N(M).$$

For the same reasons as above, this is a bifunctor, covariant in M and N. We will see in Section 3.4 that the two definitions actually give the same functors, justifying our notation.

(c) Let $\mathcal{A}$ be a complete Abelian category, and $\mathcal{I}$ a small category. By Example 1.5.2(h), we know that there is a well-defined limit functor

$$\lim \colon \operatorname{Fun}(\mathcal{I}, \mathcal{A}) \to \mathcal{A},$$

which assigns to every diagram $F \colon \mathcal{I} \to \mathcal{A}$ its limit $\lim F \in \mathcal{A}$. Moreover, we have seen that $\lim$ is a right adjoint, hence by Proposition 1.5.4, it preserves limits.

By Example 2.2.4(d), the category $\operatorname{Fun}(\mathcal{I}, \mathcal{A})$ is itself an Abelian category, and $\lim$ is clearly an additive functor. Since it preserves limits, it is left exact (Exercise 34 in Chapter 2).

Now assume that $\mathcal{A}$ has enough injectives. Then $\operatorname{Fun}(\mathcal{I}, \mathcal{A})$ has enough injectives by Proposition 2.6.20, hence we can define the right derived functors

$$\lim{}^n := R^n \lim.$$

These are called *derived limit functors*.

(d) Let X be a topological space, and consider the functor

$$\Gamma \colon \mathrm{Sh}_{\mathrm{Ab}}(X) \to \mathrm{Ab}$$

that assigns to the sheaf S its global sections $\Gamma(S) := S(X)$. By Proposition 2.3.17, Γ is left exact. Since $\mathrm{Sh}_{\mathrm{Ab}}(X)$ has enough injectives (Corollary 2.6.22), we can define the right derived functors

$$H^n(X, S) := R^n\Gamma(S).$$

The groups $H^n(X, S)$ is called the *n-th cohomology group of X with values in S*. In particular, if G is an Abelian group, we can define $H^n(X, G) := H^n(X, \underline{G})$, where $\underline{G}$ is the associated locally constant sheaf over X. As a notation, it is common to denote $\Gamma(S)$ by $H^0(X, S)$, especially when multiple topological spaces are involved.

(e) Let $\mathcal{A}$ be an Abelian category with enough injectives, and let $\mathrm{Kom}_{\geq 0}(\mathcal{A})$ be the category of complexes concentrated in degree ≥ 0. The functor

$$H_0 \colon \mathrm{Kom}_{\geq 0}(\mathcal{A}) \to \mathcal{A}$$

is left exact by the zig-zag lemma (Lemma 3.1.9). By Proposition 2.6.20, the category $\mathrm{Kom}_{\geq 0}(\mathcal{A})$ has enough injectives, so we can define the right derived functors $R^i H_0$. We claim that these are the same as H_i—in particular, in this case homology itself can be seen as a derived functor, and the zig-zag lemma just becomes the long exact sequence of Proposition 3.2.9.

To see this, take a complex $A \in \mathrm{Kom}_{\geq 0}(\mathcal{A})$ and an injective resolution $A \to I_\bullet$. We can split this as a suite of short exact sequences

$$0 \longrightarrow A \longrightarrow I_0 \longrightarrow K_0 \longrightarrow 0,$$

$$0 \longrightarrow K_0 \longrightarrow I_1 \longrightarrow K_1 \longrightarrow 0,$$

and so on. Since the derived functors vanish for the injective complexes I_k, we find by Proposition 3.2.9 that

$$R^{i+1}H_0(A) = R^i H_0(K_0) \text{ and } R^{i+1}H_0(K_j) = R^i H_0(K_{j+1}),$$

hence by induction $R^i H_0(A) = H_0(K_i)$. On the other hand, applying the zig-zag lemma and using the fact that the *objects* in each complex I_k are injective (Exercise 2), we find

$$H_{i+1}(A) = H_i(K_0) \text{ and } H_{i+1}(K_j) = H_i(K_{j+1}),$$

whence by induction $H_i(A) = H_0(K_i)$, proving our claim.

3.3. Computing derived functors

This section is dedicated to a couple of fairly simple results that allow us to simplify the computation of derived functors, which would otherwise be cumbersome. As a consequence, we obtain some vanishing results for the cohomology of sheaves on topological spaces.

Definition 3.3.1. Let $\mathcal{A}, \mathcal{B}$ be two Abelian categories, F a left exact functor from $\mathcal{A}$ to $\mathcal{B}$ (covariant or contravariant). We say that an object $A \in \mathcal{A}$ is *acyclic* for F if $R^i F(A) = 0$ for all $i > 0$. Similarly, if F is right exact, we say that $A \in \mathcal{A}$ is acyclic for F if $L^i F(A) = 0$ for all $i > 0$.

The advantage of this definition is that acyclic objects are fairly common, and they can be used to compute the value of derived functors. For brevity, we state the results for covariant, left exact functors, but the reader should spell out the obvious generalizations.

The following result is known as *dimension shifting*.

Proposition 3.3.2. *Let $F\colon \mathcal{A} \to \mathcal{B}$ be a left exact functor, $A \in \mathcal{A}$, and take a right resolution*

$$0 \longrightarrow A \longrightarrow X_0 \longrightarrow \cdots \longrightarrow X_{n-1} \longrightarrow B \longrightarrow 0,$$

where the objects X_i are acyclic for F. Then there is an isomorphism $R^{n+k} F(A) \cong R^k F(B)$ for $k > 0$ and

$$R^n F(A) \cong \mathrm{coker}(F(X_{n-1}) \to F(B)).$$

Proof. Split the resolution of A into exact sequences

$$0 \longrightarrow A \longrightarrow X_0 \longrightarrow K_1 \longrightarrow 0$$

$$0 \longrightarrow K_1 \longrightarrow X_1 \longrightarrow C_2 \longrightarrow 0$$

$$\vdots$$

where $K_n = B$. Applying Proposition 3.2.9 with the hypothesis that X_i is acyclic gives isomorphisms

$$R^{n+k} F(A) \cong R^{n+k-1} F(K_1) \cong \cdots \cong R^k F(K_n) = R^k F(B)$$

for $k > 0$. When $k = 0$, we get $R^n F(A) \cong R^1 F(K_{n-1})$, but the last step is a little different. In this case, the long exact sequence associated to

$$0 \longrightarrow K_{n-1} \longrightarrow X_{n-1} \longrightarrow B \longrightarrow 0$$

reads

$$0 \longrightarrow F(K_{n-1}) \longrightarrow F(X_{n-1}) \longrightarrow F(B) \longrightarrow R^1 F(K_{n-1}) \longrightarrow 0,$$

whence $R^1 F(K_{n-1}) \cong \mathrm{coker}(F(X_{n-1}) \to F(B))$. $\qquad\square$

Corollary 3.3.3. *Let* $F\colon \mathcal{A} \to \mathcal{B}$ *be a left exact functor,* $A \in \mathcal{A}$*, and take a right resolution* $A \to X_\bullet$*, where the objects* X_i *are acyclic for* F*. Then there is an isomorphism* $R^k F(A) \cong H_k(F(X_\bullet))$*.*

Proof. Letting $B = \operatorname{coker}(X_{k-2} \to X_{k-1})$, Proposition 3.3.2 gives an isomorphism

$$R^k F(A) \cong \operatorname{coker}(F(X_{k-1}) \to F(B)).$$

Since $B \cong \ker(X_{k-1} \to X_k)$ and F is left exact, we can identify $F(B)$ with $\ker(F(X_{k-1}) \to F(X_k))$, so $\operatorname{coker}(F(X_{k-1}) \to F(B))$ is precisely $H_k(F(X_\bullet))$. $\qquad\qquad\square$

Because of the above result, it is often convenient to find examples of acyclic objects for various functors. As a special case, there are various classes of sheaves that are acyclic for the functor of global sections (or simply acyclic).

Definition 3.3.4. Let X be a topological space, S a sheaf of Abelian groups on X. The sheaf S is called *flasque* (or *flabby*) if for every open set U the restriction $S(X) \to S(U)$ is an epimorphism. In other words, a section over an open set extends to a section over the entire X.

We will see that flasque sheaves are in fact acyclic. For this, we need a couple of preliminary results.

Lemma 3.3.5. *An injective sheaf* $I \in \mathrm{Sh}_{\mathrm{Ab}}(X)$ *is flasque.*

Proof. Let $U \subset X$ be open and consider the sheaf $\mathbb{Z}_U$, which is the sheafification of the presheaf

$$V \mapsto \begin{cases} \mathbb{Z}(V) & \text{if } V \subset U \\ 0 & \text{otherwise.} \end{cases}$$

Here $\underline{\mathbb{Z}}$ is the locally constant sheaf with stalk $\mathbb{Z}$. By construction, if $V \subset U$ are open sets, there is a monomorphism $\mathbb{Z}_V \to \mathbb{Z}_U$. Since I is injective, this gives a surjection $\operatorname{Hom}(\mathbb{Z}_U, I) \to \operatorname{Hom}(\mathbb{Z}_V, I)$. The conclusion follows from the observation that $\operatorname{Hom}(\mathbb{Z}_U, I) \cong I(U)$. $\qquad\square$

Lemma 3.3.6. *Let*

$$0 \longrightarrow A \xrightarrow{\ f\ } B \xrightarrow{\ g\ } C \longrightarrow 0$$

be an exact sequence of sheaves on the topological space X*, with* A *flasque. Then the sequence*

$$0 \longrightarrow H^0(X, A) \xrightarrow{\ f\ } H^0(X, B) \xrightarrow{\ g\ } H^0(X, C) \longrightarrow 0$$

is exact.

Proof. Let $c \in H^0(X, C)$ be a global section. To see that $c = g(b)$ for some $b \in H^0(X, B)$, consider all pairs (U, u) where $U \subset X$ is open, $u \in H^0(U, B)$ and $f(u) = c|_U$. Such pairs are partially ordered by inclusion: $(V, v) \leq (U, u)$ if $V \subset U$ and $u|_V = v$. By the Zorn lemma and the sheaf condition, there is a maximal such pair (U, u), and we need only to prove that $U = X$.

If not, let $x \in X \setminus U$, and take a small neighborhood V of x with a section $v \in H^0(V, B)$ such that $g(v) = c|_V$. On the intersection $W = U \cap V$, we have $v|_W - u|_W = f(a)$ for some $a \in H^0(W, A)$. Since A is flasque, we can extend a to a section defined over the whole X. Subtracting this global section from v, we can assume that u and v agree on W, hence they define a section on $U \cup W$, contradicting the maximality of U. $\qquad\square$

Proposition 3.3.7. *Let S be a flasque sheaf over X. Then $H^k(X, S) = 0$ for $k > 0$, that is, S is acyclic.*

Proof. Take a monomorphism $\iota \colon S \to I$, where I is injective, and let $C = \operatorname{coker} \iota$. By Lemma 3.3.6, the sequence

$$0 \longrightarrow H^0(X, S) \longrightarrow H^0(I, S) \longrightarrow H^0(C, S) \longrightarrow 0$$

is exact. Since I is acyclic, by the long exact sequence in cohomology we obtain that $H^1(X, S) = 0$ and $H^{i+1}(X, S) \cong H^i(X, C)$ for $i > 0$.

Let $U \subset X$ be an open set. In the commutative diagram

$$
\begin{array}{ccccccccc}
0 & \longrightarrow & H^0(X, S) & \longrightarrow & H^0(X, I) & \longrightarrow & H^0(X, C) & \longrightarrow & 0 \\
 & & \downarrow & & \downarrow & & \downarrow & & \\
0 & \longrightarrow & H^0(U, S) & \longrightarrow & H^0(U, I) & \longrightarrow & H^0(U, C) & \longrightarrow & 0,
\end{array}
$$

the rows are exact (all sheaves on X can be seen as sheaves on U by restriction). The first two vertical arrows are surjective, since S and I are flasque by Lemma 3.3.5. By the five lemma, it follows that $H^0(X, C) \to H^0(U, C)$ is surjective. Hence, C is flasque and by induction we get the result. $\qquad\square$

Proposition 3.3.7 is useful because there is a canonical resolution of any sheaf S by flasque sheaves. Namely, for an open set $U \subset X$, define

$$\mathcal{G}^0(S)(U) := \prod_{x \in U} S_x.$$

It is immediate to check that $\mathcal{G}^0(S)$ is a flasque sheaf, and it comes with a natural monomorphism $S \to \mathcal{G}^0(S)$. We can iterate this construction by letting $S_0 = S$ and $S_{i+1} = \operatorname{coker}(S_i \to \mathcal{G}^0(S_i))$. Letting $\mathcal{G}^i(S) = \mathcal{G}^0(S_i)$, we get the right resolution

$$(3.3.1) \qquad 0 \longrightarrow S \longrightarrow \mathcal{G}^0(S) \longrightarrow \mathcal{G}^1(S) \longrightarrow \mathcal{G}^2(S) \longrightarrow \cdots.$$

Definition 3.3.8. Resolution (3.3.1) is called the *Godement resolution* of S.

Since (3.3.1) is a resolution by flasque sheaves, it can be used to compute the cohomology of S by Corollary 3.3.3 and Proposition 3.3.7.

Remark 3.3.9. Flasque sheaves can be used to clarify a subtle point on the cohomology of sheaves. Let X be a topological space endowed with a sheaf $\mathcal{O}_X$ of rings. For a sheaf $\mathcal{M}$ of $\mathcal{O}_X$-modules, one can consider two global section functors: $\Gamma\colon \mathrm{Sh}_{\mathcal{O}_X}(X) \to \mathrm{Ab}$ or $\Gamma'\colon \mathrm{Sh}_{\mathrm{Ab}}(X) \to \mathrm{Ab}$, where one simply forgets the $\mathcal{O}_X$ module structure on $\mathcal{M}$. Are the derived functors of Γ and Γ' the same (up to composition with the forgetful functor)?

A priori this is not obvious: even though the forgetful functor $\mathrm{Sh}_{\mathcal{O}_X}(X) \to \mathrm{Sh}_{\mathrm{Ab}}(X)$ is exact, the two categories do not have the same injectives. But if $\mathcal{I}$ is an injective sheaf in $\mathrm{Sh}_{\mathcal{O}_X}(X)$, a simple modification of Lemma 3.3.5 shows that $\mathcal{I}$ is flasque. Hence, an injective resolution of $\mathcal{M}$ in $\mathrm{Sh}_{\mathcal{O}_X}(X)$ is acyclic in $\mathrm{Sh}_{\mathrm{Ab}}(X)$. It follows that the derived functors of Γ and Γ', computed on $\mathcal{M}$, are isomorphic, hence sheaf cohomology does not depend on the structure of $\mathcal{O}_X$-module.

A different type of acyclic sheaves is common in differential geometry.

Definition 3.3.10. Let X be a topological space, S a sheaf of Abelian groups on X. The sheaf S is called *fine* if for every open cover $\{U_i\}$ of an open set $U \subset X$ there are homomorphisms $f_i\colon S(U_i) \to S(U)$ such that for all sections $s \in S(U)$

(1) $\mathrm{Supp}\, f_i\,(s|_{U_i}) \subset U_i$, where the support is defined by

$$\mathrm{Supp}(\sigma) := \{x \in U \mid \sigma_x \neq 0\}$$

and σ_x is the image of $\sigma \in S(U)$ in the stalk S_x;

(2) for all $x \in U$, $f_i\,(s|_{U_i}) \neq 0$ only for finitely many i;

(3) we can write s as a locally finite sum $s = \sum_i f_i\,(s|_{U_i})$.

This definition may sound contrived, but in fact is quite natural. If X is a paracompact, Hausdorff space, every sheaf of modules over the sheaf C_X of continuous functions is fine. In fact, any open set $U \subset X$ admits a partition of unity $1 = \sum a_i$ for continuous functions a_i with support in U_i, and one can just take $f_i(s) = a_i \cdot s$. Similarly, if M is a smooth manifold, every sheaf of modules over C_M^∞ is fine. The following lemma is the analogue of 3.3.6

Lemma 3.3.11. *Let*

$$0 \longrightarrow A \stackrel{f}{\longrightarrow} B \stackrel{g}{\longrightarrow} C \longrightarrow 0$$

be an exact sequence of sheaves on the topological space X, with A fine. Then the sequence

$$0 \longrightarrow H^0(X, A) \xrightarrow{f} H^0(X, B) \xrightarrow{g} H^0(X, C) \longrightarrow 0$$

is exact.

Proof. Let $c \in H^0(X, C)$ be a section. We can find a covering $\{U_i\}$ of X and sections $b_i \in H^0(U_i, B)$ such that $g(b_i) = c|_{U_i}$. On the intersections $U_{ij} = U_i \cap U_j$, we have $g(b_i|_{U_{ij}} - b_j|_{U_{ij}}) = 0$, hence we find $a_{ij} \in H^0(U_{ij}, A)$ such that $f(a_{ij}) = b_i|_{U_{ij}} - b_j|_{U_{ij}}$. On the triple intersections $U_{ijk} = U_i \cap U_j \cap U_k$, these satisfy

$$(3.3.2) \qquad a_{ij}|_{U_{ijk}} + a_{jk}|_{U_{ijk}} + a_{ki}|_{U_{ijk}} = 0.$$

By the fact that A is fine, we get homomorphisms $f_{ij} \colon S(U_{ij}) \to S(U_i)$ such that $a = \sum_j f_{ij}(a|_{U_{ij}})$ for $a \in H^0(U_i, A)$. We now let $a_i := \sum_j f_{ij}(a_{ij}) \in H^0(U_i, A)$. Using (3.3.2), it is easy to compute (do this!) that on U_{ij}

$$f\left(a_i|_{U_{ij}} - a_j|_{U_{ij}}\right) = b_i - b_j.$$

Letting $b_i' = b_i - f(a_i)$, we conclude that $b_i' = b_j'$ on U_{ij}. Hence the $\{b_i'\}$ glue to give a section b' on the whole X such that $g(b) = c$. $\qquad \square$

Proposition 3.3.12. *Let S be a fine sheaf over X. Then $H^k(X, S) = 0$ for $k > 0$, that is, S is acyclic.*

Proof. Let $S \to G_\bullet$ be the Godement resolution of S, and define $C = \operatorname{coker}(S \to G_0)$. Using Lemma 3.3.11 and arguing as in Proposition 3.3.7, we find that C is flasque. By Proposition 3.3.2, $H^i(X, S) = H^{i-1}(X, C) = 0$ for $i \geq 2$. The vanishing of $H^1(X, S)$ follows from Lemma 3.3.11 and $H^1(X, G_0) = 0$. $\qquad \square$

As an example of an application of this, one obtains a form of the De Rham theorem. In fact, let M be a smooth manifold of dimension n. For all open sets $U \subset M$ one has the vector spaces $\Lambda^k(U)$ of k-differential forms on U. One can bundle all of these together into a fine sheaf Λ_M^k on M. The differential operator d gives rise to a complex of sheaves

$$(3.3.3) \qquad 0 \longrightarrow \mathbb{R} \longrightarrow C_M^\infty \longrightarrow \Lambda_M^1 \longrightarrow \cdots \longrightarrow \Lambda_M^n \longrightarrow 0.$$

By the Poincaré lemma [**BT95**, Corollary 4.1.1], this complex is exact on the stalks, hence it is exact by Exercise 6 of Chapter 2. By Proposition 3.3.12, it is a resolution of $\mathbb{R}$ by acyclic sheaves. It follows that

$$H^k(X, \mathbb{R}) \cong H_k(\Gamma(\Lambda_M^\bullet)) = H_{DR}^k(M).$$

3.4. The Ext and Tor functors

In this section, we are going to study in more details the functors Ext and Tor. In particular, we are going to show that they are well defined: whenever the underlying category $\mathcal{A}$ has enough injectives and projectives, $\text{Ext}^i(A, B)$ can be defined via an injective resolution of B or a projective resolution of A. When such an ambiguity is present, we will see that the two groups that we obtain are actually isomorphic. The same holds for $\text{Tor}_i(M, N)$, which can be computed via a projective resolution of either M or N. As a consequence of these results, we show that Tor is symmetric, and that Ext admits a multiplicative structure. In the second part of the section, we give some examples, and finally we give an alternative characterization of the Ext groups using extensions.

The key to showing the symmetry properties of Ext and Tor is a simple fact on double complexes.

Definition 3.4.1. Let $\mathcal{A}$ be an Abelian category. A *double complex* in $\mathcal{A}$ is a commutative diagram

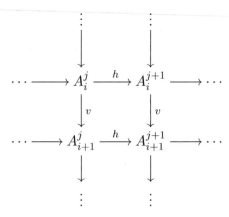

with objects A_i^j and morphisms $h = h_i^j \colon A_i^j \to A_i^{j+1}$ and $v = v_i^j \colon A_i^j \to A_{i+1}^j$ such that the rows and the columns are complexes.

Assume that for all k there is only a finite number of pairs i, j with $i + j = k$ and $A_i^j \neq 0$. The *total complex* associated to the double complex $A_\bullet^\bullet$ is the complex $T_\bullet = \text{Tot}(A_\bullet^\bullet)$ with

$$T_k := \bigoplus_{i+j=k} A_i^j$$

and the differential $d \colon T_k \to T_{k+1}$ defined by $d = h_i^j + (-1)^j v_i^j$ on the summand A_i^j.

It is an easy verification that the total complex of a double complex is actually a complex. The minus sign is there precisely to make this true.

Lemma 3.4.2. *Let $A_\bullet^\bullet$ be a double complex such that $A_i^j = 0$ if either $i < 0$ or $j < 0$. We say that $A_\bullet^\bullet$ is a first quadrant complex. If either the rows or the columns of $A_\bullet^\bullet$ are exact, then $\text{Tot}(A_\bullet^\bullet)$ is acyclic.*

Proof. We prove the result assuming that the rows are exact. Using the Freyd–Mitchell theorem, we can reduce to the case of a category of modules, in which we can perform diagram chasing. Let $T_\bullet = \text{Tot}(A_\bullet^\bullet)$, and fix k. An element of T_k is a sum $t = \sum_{i+j=k} a_i^j$ with $a_i^j \in A_i^j$. Then

$$dt = ha_0^k + ((-1)^k va_0^k + ha_1^{k-1}) + \cdots + va_k^0.$$

If $dt = 0$, each of the terms in the sum must be 0. Since the rows are exact and $ha_0^k = 0$, we find $a_0^{k-1} \in A_0^{k-1}$ such that $ha_0^{k-1} = a_0^k$. By commutativity, $(-1)^k va_0^k = (-1)^k hva_0^{k-1}$, hence

$$h\left((-1)^k va_0^{k-1} + a_1^{k-1}\right) = 0.$$

By exactness again, we find $a_1^{k-2} \in A_1^{k-2}$ such that

$$ha_1^{k-2} = (-1)^k va_0^{k-1} + a_1^{k-1}.$$

We proceed recursively to find elements $a_i^j \in A_i^j$ for $i + j = k - 1$ such that

$$t = d\left(a_0^{k-1} + \cdots + a_{k-1}^0\right),$$

proving that $T_\bullet$ is acyclic. □

Corollary 3.4.3. *Let $A_\bullet^\bullet$ be a first quadrant complex, $B_\bullet^\bullet$ the double complex defined by $B_i^j = A_{i+1}^j$ for $i, j \geq 0$, and 0 otherwise—in other words, B is obtained from A by removing the first row and shifting down. If A has exact columns, then there are isomorphisms between $H_k(A_0^\bullet)$ and $H_k(\text{Tot}(B_\bullet^\bullet))$ for all $k \geq 0$.*

Proof. For a complex $C = \{C_i\}$, let us denote $C[1]$ the complex obtained by shifting indices by 1, that is, $C[1]_i = C_{i+1}$. The complex $\text{Tot}(B_\bullet^\bullet)[1]$ appears as a subcomplex of $\text{Tot}(A_\bullet^\bullet)$, and the quotient is the complex $A_0^\bullet$. The conclusion follows from the zig-zag lemma (Lemma 3.1.9), and the fact that $\text{Tot}(A_\bullet^\bullet)$ is acyclic follows by Lemma 3.4.2. □

Remark 3.4.4. We have stated these results for double complexes in the first quadrant for simplicity, but of course the proof works for double complexes in any quadrant. To state this, it is convenient to invent some notation for the shifts of a double complex. Let $A_\bullet^\bullet$ be a double complex. If $A_i^j = 0$ for $i < 0$, let us denote $\downarrow A_\bullet^\bullet$ the complex obtained by removing the first row of $A_\bullet^\bullet$ and shifting down. A similar notation can be used to define $\leftarrow A_\bullet^\bullet$ when $A_i^j = 0$ for $j < 0$, and so on.

Then Corollary 3.4.3 generalizes in various other forms. For instance, let A be a third-quadrant complex—that is $A_i^j = 0$ for $i > 0$ or $j > 0$. If A has exact columns, then

$$H_k(\text{Tot}(_\uparrow A)) \cong H_k(A_0^\bullet),$$

while if A has exact rows

$$H_k(\text{Tot}(_\to A)) \cong H_k(A_\bullet^0).$$

We can now prove that Tor is well defined independently of the side we take the projective resolution. Recall the notation of functors T_M and T^N from Example 3.2.13(b).

Theorem 3.4.5. *Let A be a ring, M, N two A-modules. Then there is a canonical isomorphism $L^i T_M(N) \cong L^i T^N(M)$.*

Proof. Take projective resolutions $P_\bullet \to M$ and $Q_\bullet \to N$. We can form a double complex C

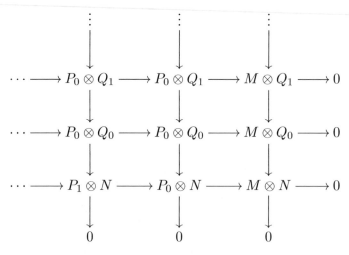

by tensoring the two resolutions. The rows of C, except for the last one, are exact, since they are obtained by tensoring an exact resolution by a projective module Q_i. Since Q_i is projective, $L^1 T^{Q_i} = 0$, hence T^{Q_i} is an exact functor by the long exact sequence in Proposition 3.2.9. For a similar reason, the columns of C, save from the last one, are exact.

By omitting the last column of C we obtain a diagram with exact columns. By Corollary 3.4.3 and the remark after it, we find an isomorphism $L^i T_M(N) \cong H_i(\text{Tot}(P_\bullet \otimes Q_\bullet))$. By omitting the last row, we obtain a similar isomorphism $L^i T^N(M) \cong H_i(\text{Tot}(P_\bullet \otimes Q_\bullet))$, which proves the theorem. $\square$

Corollary 3.4.6. *Let A be a ring. Then there is an isomorphism*

$$\mathrm{Tor}_i(M, N) \cong \mathrm{Tor}_i(N, M)$$

for A-modules M, N, natural in M and N.

Proof. Let $P_\bullet \to M$ be a projective resolution of M. Then the symmetry of tensor product gives an isomorphism of complexes $N \otimes P_\bullet \cong P_\bullet \otimes N$, and the conclusion follows by taking homology and applying Theorem 3.4.5. $\square$

With exactly the same proof of Theorem 3.4.5, we get

Theorem 3.4.7. *Let $\mathcal{A}$ be an Abelian category with enough injectives and projectives, $A, B \in \mathcal{A}$. Then there is a canonical isomorphism $R^i h^A(B) \cong R^i h_B(A)$.*

As a consequence, we are justified in using the notation $\mathrm{Ext}^i(A, B)$ without specifying whether we are computing it by an injective resolution of B or a projective resolution of A. Similarly, the group $\mathrm{Tor}_i(M, N)$ can be computed via a projective resolution of either M or N.

Remark 3.4.8. In the course of the proof of Theorem 3.4.5, we have seen that we can compute $\mathrm{Tor}_i(M, N)$ as $H_i(\mathrm{Tot}(P_\bullet \otimes Q_\bullet))$, where $P_\bullet \to M$ and $Q_\bullet \to N$ are projective resolutions of M and N respectively. The A-module $H_i(\mathrm{Tot}(P_\bullet \otimes Q_\bullet))$ is called the *i-th balanced* Tor of M and N.

Similarly, if $\mathcal{A}$ has enough injectives and projectives, we can compute $\mathrm{Ext}^i(A, B)$ as $H_i(\mathrm{Tot}(\mathrm{Hom}(P_\bullet, I_\bullet)))$, where $P_\bullet \to A$ is a projective resolution of A and $B \to I_\bullet$ is an injective resolution of B. The group $H_i(\mathrm{Tot}(\mathrm{Hom}(P_\bullet, I_\bullet)))$ is called the *i-th balanced* Ext of A and B.

Remark 3.4.9. Let A be a commutative ring, M and A-module. Then the functor $M \otimes -$ takes values in Mod_A, so the same is true for its derived functors. It follows that $\mathrm{Tor}_i(M, N)$ is an A-module for all A-modules M, N.

For a general Abelian category $\mathcal{A}$, the functors $\mathrm{Ext}^i_{\mathcal{A}}(-, -)$ a priori take values in Ab. When $\mathcal{A} = \mathrm{Mod}_A$, though, the groups $\mathrm{Hom}(M, N)$ are actually A-modules themselves. Hence, the functors h^M and h_N are functors with values in Mod_A, and the same holds true for the derived functors $\mathrm{Ext}^i(M, N)$. It follows that the Ext groups have a natural structure of A-module.

Theorem 3.4.7 can also be used to put a multiplicative structure on the Ext groups, called the *Yoneda product*.

Theorem 3.4.10. *Let $\mathcal{A}$ be an Abelian category with enough projectives and injectives. Then for three objects $A, B, C \in \mathcal{A}$ there is a bilinear map*

$$\mathrm{Ext}^i(A, B) \times \mathrm{Ext}^j(B, C) \to \mathrm{Ext}^{i+j}(A, C).$$

In particular, the direct sum $\mathrm{Ext}^*(A, A) = \bigoplus_i \mathrm{Ext}^i(A, A)$ *has a (noncommutative) ring structure.*

Proof. Let $P_\bullet \to A$ be a projective resolution of A, $C \to I_\bullet$ an injective resolution of C. We will use the former to compute $\mathrm{Ext}^i(A, B)$ and the latter to compute $\mathrm{Ext}^j(B, C)$. In particular, $\mathrm{Ext}_i(A, B)$ is a subquotient of $\mathrm{Hom}(P_i, B)$ and $\mathrm{Ext}^j(B, C)$ is a subquotient of $\mathrm{Hom}(B, I_j)$. The composition gives a bilinear map

$$\mathrm{Hom}(P_i, B) \times \mathrm{Hom}(B, I_j) \to \mathrm{Hom}(P_i, I_j)$$

and $\mathrm{Hom}(P_i, I_j)$ appears in the double complex which computes $\mathrm{Ext}^{i+j}(A, C)$ via the balanced Ext. It is a routine verification that this map descends to homology, giving the desired product. $\qquad\square$

We now turn to some examples. As it turns out, the Ext functors are a very general construction, and they subsume many classical cohomological concepts.

Example 3.4.11.

(a) Let X be a topological space endowed with a sheaf of rings $\mathcal{O}_X$. If $\mathcal{F}$ is a sheaf of $\mathcal{O}_X$-modules, there is a natural isomorphism $\Gamma(\mathcal{F}) \cong \mathrm{Hom}(\mathcal{O}_X, \mathcal{F})$. Namely, a section $s \in \Gamma(\mathcal{F})$ determines a morphism $f_s \colon \mathcal{O}_X \to \mathcal{F}$ defined on the open set $U \subset X$ by $f_s(t) = t \cdot s|_U$. Vice versa, a morphism $f \colon \mathcal{O}_X \to \mathcal{F}$ determines the global section $f_X(1) \in \Gamma(\mathcal{F})$. It follows that we have a natural isomorphism $H^k(X, \mathcal{F}) \cong \mathrm{Ext}^k(\mathcal{O}_X, \mathcal{F})$, showing that sheaf cohomology can be expressed using Ext functors.

(b) Let G be a finite group, and let $\mathbb{Z}[G]$ be the group ring defined in Example 1.5.2(e). Elements of $\mathbb{Z}[G]$ are formal linear combinations $\sum a_i g_i$ with $a_i \in \mathbb{Z}$ and $g_i \in G$, and the multiplication is inherited from the group operation in G. This makes $\mathbb{Z}[G]$ into a ring, which is commutative if and only if G is.

We have an associated category $\mathrm{Mod}_G := \mathrm{Mod}_{\mathbb{Z}[G]}$ of left $\mathbb{Z}[G]$-modules. An object in Mod_G is just an Abelian group endowed with a left action of G, and morphisms in Mod_G are group homomorphisms that are equivariant with respect to the group action. Let us regard $\mathbb{Z}$ as a G-module with the trivial action. For all $M \in \mathrm{Mod}_G$,

$$\mathrm{Hom}_{\mathbb{Z}[G]}(\mathbb{Z}, M) \cong M^G := \{m \in M \mid gm = m \text{ for all } g \in G\},$$

the group of invariant elements of M.

It follows that the functor of invariants $-^G$ is left exact. The *cohomology groups* of M are defined by

$$H^i(G, M) = R^i -^G (M),$$

and by the above discussion, we can identify

$$H^i(G, M) = \operatorname{Ext}^i_{\mathbb{Z}[G]}(\mathbb{Z}, M).$$

(c) Let A be a ring, $a \in A$ not a zero divisor. Then the exact sequence

$$0 \longrightarrow A \xrightarrow{\cdot a} A \longrightarrow A/(a) \longrightarrow 0$$

is a free resolution of $A/(a)$. If M is any A-module, we can compute $\operatorname{Ext}^i(A/(a), M)$ by removing the last element of the sequence, applying $\operatorname{Hom}(-, M)$ and taking homology. The complex that we obtain is just

$$0 \longrightarrow M \xrightarrow{\cdot a} M \longrightarrow 0,$$

hence

$$\operatorname{Ext}^0(A/(a), M) = (0 :_M a) := \{m \in M \mid am = 0\},$$

the submodule of a-torsion elements of M, which of course agrees with $\operatorname{Hom}(A/(a), M)$. Similarly, $\operatorname{Ext}^1(A/(a), M) = M/aM$.

We can do a similar computation for Tor.

Example 3.4.12.

(a) Let A be a ring, $a \in A$ not a zero divisor. By the same reasoning as above, we can compute $\operatorname{Tor}(A/(a), M)$ by the homology of the complex

$$0 \longrightarrow M \xrightarrow{\cdot a} M \longrightarrow 0.$$

We deduce that $\operatorname{Tor}_0(A/(a), M) = M/aM \cong M \otimes A/(a)$, as expected, and

$$\operatorname{Tor}_1(A/(a), M) = (0 :_M a) := \{m \in M \mid am = 0\},$$

the submodule of a-torsion elements of M. This is the reason that underlies the name Tor.

(b) In the particular case of $A = \mathbb{Z}$, we can also recover the whole torsion submodule as a Tor group. Namely, let $M = \mathbb{Q}/\mathbb{Z}$. By Exercise 7, Tor commutes with direct limits. The Abelian group M is the direct limit of its finite subgroups, and each of them is isomorphic to $\mathbb{Z}/n\mathbb{Z}$ for some n. Hence, for all Abelian groups N,

$$\operatorname{Tor}_1(M, N) \cong \varinjlim \operatorname{Tor}_1(\mathbb{Z}/n\mathbb{Z}, N) \cong \varinjlim(0 :_N n) = \operatorname{T}(N),$$

the torsion subgroup of N.

In the remaining part of the section, we will see an alternative characterization of the Ext functor. Let $\mathcal{A}$ be an Abelian category. Following Yoneda, we are going to characterize $\mathrm{Ext}^i(A, B)$ as the extensions of A by B of length i, modulo a suitable equivalence relation.

Definition 3.4.13. Let A, B be objects in an Abelian category $\mathcal{A}$. An *extension* of A by B is an exact sequence

$$0 \longrightarrow B \longrightarrow E_1 \longrightarrow \cdots \longrightarrow E_n \longrightarrow A \longrightarrow 0.$$

We denote such an extension by $E_\bullet$, and say that $E_\bullet$ has length n.

Extensions of A by B of length n form a category of their own, which we temporarily denote by $E^n(A, B)$. A morphism $E_\bullet \to E'_\bullet$ is a sequence of morphisms $E_i \to E'_i$ such that the diagram

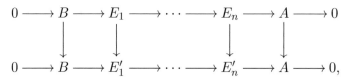

commutes. In particular, we can speak of isomorphic extensions if all maps $E_i \to E'_i$ are isomorphisms. We will need a coarser equivalence relation.

Definition 3.4.14. The equivalence relation $\sim$ for extensions in $E^n(A, B)$ is generated by the relations $E_\bullet \sim F_\bullet$ if there is a morphism $E_\bullet \to F_\bullet$.

Remark 3.4.15. In the case of length 1, every morphism of extensions is an isomorphism by the five lemma (Lemma 2.5.1), so equivalent extensions are isomorphic.

A priori, the extensions of A by B form a proper class, even up to equivalence (see Exercise 29).

Definition 3.4.16. Let $\mathcal{A}$ be an Abelian category. We say that $\mathcal{A}$ is *extension small* if for all objects $A, B \in \mathcal{A}$ and all n there is a set S of extensions such that every extension of A by B of length n is equivalent to an extension in S.

It is worth noticing that the terminology above is not standard, but it will be convenient in the following discussion. Most Abelian categories of interest are extension small: this is clearly true if $\mathcal{A}$ is small, and we will see that it is also true when $\mathcal{A}$ has enough injectives or enough projectives.

Let $\mathcal{A}$ be an extension small Abelian category, $A, B \in \mathcal{A}$. Then there is a set of equivalence classes of length n extensions of A by B. We will denote this set by $\mathrm{Ext}^i_Y(A, B)$ (for Yoneda). Assuming that $\mathcal{A}$ has enough injectives or enough projectives, we will describe a bijective correspondence $\mathrm{Ext}^i_Y(A, B) \cong \mathrm{Ext}^i(A, B)$.

Lemma 3.4.17. *Let $\mathcal{A}$ be an extension small Abelian category, $i \geq 0$. Then $\mathrm{Ext}_Y^i(-,-)$ is a bifunctor with values in* Set, *contravariant in the first variable and covariant in the second.*

Proof. Let $A' \to A$ be a morphism, and let $E_\bullet$ be an extension of A by B. Then we can form an extension

$$
\begin{array}{ccccccccc}
0 & \longrightarrow & B & \longrightarrow & E_1' & \longrightarrow & \cdots & \longrightarrow & E_n' & \longrightarrow & A' & \longrightarrow & 0 \\
& & \downarrow & & \downarrow & & & & \downarrow & & \downarrow & & \\
0 & \longrightarrow & B & \longrightarrow & E_1 & \longrightarrow & \cdots & \longrightarrow & E_n & \longrightarrow & A & \longrightarrow & 0,
\end{array}
$$

where $E_n' := E_n \times_A A'$ and $E_i' := E_i \times_A E_{i+1}'$ for $i < n$. The fact that at the end of the sequence we obtain B follows immediately by induction from Corollary 2.5.19. In a similar way, a morphism $B \to B'$ induces an extension of B', as can be seen by dualizing the above construction. $\square$

Theorem 3.4.18. *Let $\mathcal{A}$ be a category with enough injectives or enough projectives. Then $\mathcal{A}$ is extension small, and there is an isomorphism of bifunctors $\mathrm{Ext}_Y^i(-,-) \cong \mathrm{Ext}^i(-,-)$.*

Remark 3.4.19. In the above theorem, we regard Ext^i as a bifunctor with values in Set, hence the result amounts to a natural bijective correspondence. Actually, one can also define an operation that makes $\mathrm{Ext}_Y^i(A, B)$ into a commutative group, and check that the bijective correspondence is actually a group isomorphism—we will leave this to Exercises 25 and 26.

Proof. We assume that $\mathcal{A}$ has enough injectives, and fix an injective resolution $B \to I_\bullet$ of B, where for uniformity we start numbering by 1. We can truncate it after I_n and complete it with $C := \mathrm{coker}\, I_n \to I_{n+1}$ to obtain the exact sequence

$$
0 \longrightarrow B \longrightarrow I_1 \longrightarrow \cdots \longrightarrow I_n \longrightarrow C \longrightarrow 0.
$$

Let $E_\bullet$ be an extension of A by B of length n. Proposition 3.2.4 gives us a lift of id_B to a commutative diagram

(3.4.1)
$$
\begin{array}{ccccccccccccc}
0 & \longrightarrow & B & \longrightarrow & E_1 & \longrightarrow & \cdots & \longrightarrow & E_n & \longrightarrow & A & \longrightarrow & 0 \\
& & \downarrow & & \downarrow & & & & \downarrow & & \downarrow{\scriptstyle e} & & \\
0 & \longrightarrow & B & \longrightarrow & I_1 & \longrightarrow & \cdots & \longrightarrow & I_n & \longrightarrow & C & \longrightarrow & 0,
\end{array}
$$

where the last map $e \colon A \to C$ comes from the universal property of cokernels. The map e does not depend only on $E_\bullet$, but also on the choices of intermediate lifts. These are determined up to a homotopy, therefore e is the well defined modulo map $A \to I_n$. This construction defines the map

$$
s \colon \mathrm{Ext}_Y^n(A, B) \to \mathrm{coker}(\mathrm{Hom}(A, I_n) \to \mathrm{Hom}(A, C)),
$$

which we will show to be bijective. In fact we don't know yet that $\operatorname{Ext}_Y^n(A, B)$ is a set, but this will follow from injectivity of s. Finally, using Proposition 3.3.2, we obtain an isomorphism

$$\operatorname{coker}(\operatorname{Hom}(A, I_n) \to \operatorname{Hom}(A, C)) \cong \operatorname{Ext}^n(A, B),$$

which concludes the proof.

To show that s is surjective, we start from $e \colon A \to C$ and construct an extension $E_\bullet$ such that $s(E_\bullet) = e$. To do so, just define $E_n := A \times_C I_n$ and $E_i := E_{i+1} \times_{I_{i+1}} I_i$ for $i < n$. Using Corollary 2.5.19 and induction, we get that $\ker E_1 \to E_2 \cong \ker I_1 \to I_2 = B$. Hence, the sequence $E_\bullet$ is an extension of A by B that fits in the commutative diagram (3.4.1), which means that $s(E_\bullet) = e$.

To show that s is injective, we let $F_\bullet$ be any extension such that $s(F_\bullet) = e$, and show that there is a morphism $F_\bullet \to E_\bullet$, where $E_\bullet$ is the extension defined above using fiber products. By hypothesis, $F_\bullet$ fits into a commutative diagram

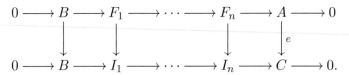

We can then define maps $F_k \to E_k$ by descending induction, using the universal property of the fibered product (check this in detail!). $\qquad\square$

3.5. The $\lim^n$ functors

Let $\mathcal{A}$ be a complete Abelian category with enough injectives, and $\mathcal{I}$ a small category. Then for every functor $F \colon \mathcal{I} \to \mathcal{A}$, the limit $\lim F$ exists, and we can fit all these limits into a limit functor $\lim \colon \operatorname{Fun}(\mathcal{I}, \mathcal{A}) \to \mathcal{A}$ (Example 1.5.2(h)). By Proposition 2.6.20, the category $\operatorname{Fun}(\mathcal{I}, \mathcal{A})$ has enough injectives. As explained in Example 3.2.13(c), the functor $\lim$ is left exact, and we can take its right derived functors $\lim^n = R^n \lim$.

In the present section, we want to specialize this construction to a particular case that is easier to understand: the inverse limit of a linearly ordered family of objects of $\mathcal{A}$. In this setting, we are going to give an alternative characterization of $\lim^1$ and use it to prove some vanishing results.

Remark 3.5.1. The reader may wonder why we are not also considering the derived functors of *direct* limits. The reason is that $\varinjlim$ is always exact in concrete categories, such as Mod_A. To see this, notice that $\varinjlim$ is a filtered colimit, and as such it commutes with finite colimits (obviously) and finite limits (by Theorem 1.4.16). In particular, it commutes with taking kernels and cokernels. Of course, this relies on the category being concrete; see Remark 1.4.18.

Henceforth, let us take $\mathcal{I} = \mathrm{Ord}(\mathbb{N})^{op}$, the order category associated to $\mathbb{N}$ with reversed ordering. A functor $F \colon \mathcal{I} \to \mathcal{A}$ will be equivalent to a sequence

$$\cdots \longrightarrow A_2 \xrightarrow{f_1} A_1 \xrightarrow{f_0} A_0 \longrightarrow 0$$

of objects of $\mathcal{A}$ and morphisms between them. Such a diagram is usually called a *tower* in $\mathcal{A}$.

Since $\mathcal{A}$ is assumed complete, for each tower $A_\bullet$ the product $\prod_{i=0}^\infty A_i$ exists. Moreover, we can define a map

$$\delta \colon \prod_{i=0}^\infty A_i \to \prod_{i=0}^\infty A_i$$

by the universal property from the maps $\delta_j \colon \prod_i A_i \to A_j$ defined by $\delta_j = \pi_j - f_j \circ \pi_{j+1}$, where $\pi_j \colon \prod_i A_i \to A_j$ is the projection on the j-th coordinate.

To express this via the universal property is a little contrived, but when $\mathcal{A}$ is a category of modules, this is just the map defined by

$$\delta(\cdots, a_2, a_1, a_0) = (\cdots, a_2 - f_2(a_3), a_1 - f_1(a_2), a_0 - f_0(a_1)).$$

In this situation, $\ker \delta$ is just the subset of $\prod_i A_i$ of compatible sequences, which is an explicit construction of $\varprojlim A_i$. As it turns out, this is true in general.

Lemma 3.5.2. *Let $\mathcal{A}$ be a complete Abelian category, $A_\bullet$ a tower in $\mathcal{A}$, and $\delta \colon \prod A_i \to \prod A_i$ as defined above. Then $\varprojlim A_i \cong \ker \delta$.*

Proof. It is enough to prove that $\ker \delta$ is a universal cone. Clearly, $\ker \delta$ is a cone over $A_\bullet$ via the monomorphism $\ker \delta \to \prod A_i$. If C is any other cone, we have, in particular a map $C \to A_i$ for all i, which yields a map $c \colon C \to \prod A_i$. The fact that C is a cone says exactly that $\delta \circ c = 0$, which means that c factors through $\ker \delta$. $\square$

Our first aim in this section is to show that—under a suitable condition on $\mathcal{A}$—$\lim^1 A_i \cong \mathrm{coker}\,\delta$, and that higher derived limits vanish.

Definition 3.5.3. Let $\mathcal{A}$ be a cocomplete Abelian category. We say that $\mathcal{A}$ satisfies axiom AB4, or is an AB4 category, if coproducts are exact. This means that if $\{A_i\}$, $\{B_i\}$, $\{C_i\}$ are sequences of objects in $\mathcal{A}$ with short exact sequences

$$0 \longrightarrow A_i \longrightarrow B_i \longrightarrow C_i \longrightarrow 0$$

for all $i \in \mathbb{N}$, then the sequence of coproducts

$$0 \longrightarrow \coprod_i A_i \longrightarrow \coprod_i B_i \longrightarrow \coprod_i C_i \longrightarrow 0$$

is exact. Dually, if $\mathcal{A}$ is complete, we say that $\mathcal{A}$ satisfies axiom AB4*, or is an AB4* category, if products are exact.

As the names suggest, AB4 and AB4* are just two out of a list of axioms for additive categories, introduced in [**Gro57**] and conventionally used ever since (the first two are just our axioms for Abelian categories).

Given a tower $A_{\bullet}$ in $\mathcal{A}$ with the associated morphism $\delta \colon \prod A_i \to \prod A_i$, let us temporarily denote $L^0 A_{\bullet} := \ker \delta$, $L^1 A_{\bullet} := \operatorname{coker} \delta$ and $L^k A_{\bullet} = 0$ for $k \geq 2$. We want to show that these objects satisfy the same properties as derived functors (in the language introduced in the exercises, they form a δ-functor).

Lemma 3.5.4. *Let $\mathcal{A}$ be a complete Abelian category satisfying AB4*,*

$$0 \longrightarrow A_{\bullet} \longrightarrow B_{\bullet} \longrightarrow C_{\bullet} \longrightarrow 0$$

a short exact sequence of towers in $\mathcal{A}$. Then there is a six-term exact sequence

$$0 \longrightarrow L^0 A_{\bullet} \longrightarrow L^0 B_{\bullet} \longrightarrow L^0 C_{\bullet}$$
$$\longrightarrow L^1 A_{\bullet} \longrightarrow L^1 B_{\bullet} \longrightarrow L^1 C_{\bullet} \longrightarrow 0.$$

Proof. By AB4* there is a diagram with exact rows

$$
\begin{array}{ccccccccc}
0 & \longrightarrow & \prod_i A_i & \longrightarrow & \prod_i B_i & \longrightarrow & \prod_i C_i & \longrightarrow & 0 \\
 & & \downarrow{\scriptstyle\delta} & & \downarrow{\scriptstyle\delta} & & \downarrow{\scriptstyle\delta} & & \\
0 & \longrightarrow & \prod_i A_i & \longrightarrow & \prod_i B_i & \longrightarrow & \prod_i C_i & \longrightarrow & 0,
\end{array}
$$

and the conclusion follows from the snake lemma (Lemma 2.5.20). $\qquad\square$

The next step will be to prove that L^1 vanishes on enough injectives.

Lemma 3.5.5. *Let $\mathcal{A}$ be a complete Abelian category satisfying AB4* and having enough injectives. Then for all towers $A_{\bullet}$ in $\mathcal{A}$, there exists an injective tower $I_{\bullet}$ and a monomorphism $A_{\bullet} \to I_{\bullet}$ such that $L^1 I_{\bullet} = 0$.*

Proof. Given an object $X \in \mathcal{A}$, denote by $T_k(X)$ the tower

$$\cdots \longrightarrow X \xrightarrow{\operatorname{id}_X} X \longrightarrow 0 \longrightarrow \cdots \longrightarrow 0,$$

where all nonzero maps are identities and $T_k(X)_i = X$ for $i \geq k$, 0 otherwise. First, we claim that $L^1 T_k(X) = 0$ for all X and k. To see this, let $T = T_k(X)$ and consider the complex C_1 given by

$$0 \longrightarrow \prod_i T_i \overset{\delta}{\longrightarrow} \prod_i T_i \longrightarrow 0,$$

whose homology computes exactly the objects $L^i(T)$. There is another complex C_2 which consists of just the object $\prod_i T_i$ in degree 0, which has a natural inclusion $C_2 \to C_1$ and a projection $C_1 \to C_2$. The definition of δ, together with the explicit shape of $T = T_k(X)$, implies that these two maps are homotopy equivalences. By Corollary 3.1.8, C_1 is quasi-isomorphic to C_2, hence $L^1(T) = H_1(C_1) = 0$.

The same vanishing result holds for products of such towers. In fact, by AB4*, products commute with cokernels, hence with L^1.

Given any tower $A_\bullet$, for each i we can find a monomorphism $A_i \to I_i$, where I_i is injective. Then it is easy to check that $A_\bullet$ admits a monomorphism into $J_\bullet := \prod_i T_i(I_i)$. By the first part of the proof, $L^1 J_\bullet = 0$.

To conclude the proof, we only need to show that $J_\bullet$ is injective. Given the diagram of towers

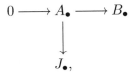

we have for each k a morphism $A_k \to J_k = \bigoplus_{i=0}^{k} I_i$. By induction, we can lift this to a compatible morphism $B_k \to J_k$. The inductive hypothesis gives us a morphism $B_k \to B_{k-1} \to J_{k-1}$, and we can combine this with a morphism $B_k \to I_k$ that extends $A_k \to I_k$. $\qquad \square$

The two results are in fact enough to identify $\lim^i$ with L^i for all i.

Theorem 3.5.6. *Let $\mathcal{A}$ be a complete Abelian category. Assume that $\mathcal{A}$ has enough injectives and satisfies* AB4*. *Let $\mathcal{I} = \mathrm{Ord}(\mathbb{N})^{op}$, so that objects in* $\mathrm{Fun}(\mathcal{I}, \mathcal{A})$ *are towers in $\mathcal{A}$, and consider the limit functor* $\lim \colon \mathrm{Fun}(\mathcal{I}, \mathcal{A}) \to \mathcal{A}$. *Then $\lim^1 = L^1$ and $\lim^i = 0$ for $i > 0$.*

Proof. Lemma 3.5.4 says that the functors L^i satisfy a property that is formally analogous to the long exact sequence for derived functors. Using the same reasoning as in Corollary 3.3.3, we obtain the following fact: if $A \to X_\bullet$ is a right resolution of the tower A with towers X_k such that $L^i(X_k) = 0$ for $i > 0$, then $L^i A \cong H_i(\lim X_\bullet)$. But Lemma 3.5.5 guarantees that we can take such a resolution using *injective* towers, and then this recipe becomes exactly the recipe to compute the right derived functors of $\lim$. $\qquad \square$

Remark 3.5.7. [**Nee01**, Section A.3] gives an explicit description of the functors $\lim^i$ over categories of indices more complex than $\mathrm{Ord}(\mathbb{N})^{op}$. It turns out that these derived functors can be computed as the homology groups of an explicit complex, but that complex has a more complicated shape and has more than two terms.

We now give a useful criteria that guarantees that $\lim^1$ vanishes as well.

Definition 3.5.8. Let $A_\bullet$ be a tower in an Abelian category. We say that $A_\bullet$ satisfies the *Mittag-Leffler condition*[1] if for all n, the descending chain of subobjects $\mathrm{im}(A_{n+k} \to A_n)$ stabilizes, that is, $\mathrm{im}(A_{n+k} \to A_n)$ does not depend on k for $k \gg 0$.

Theorem 3.5.9 (Roos). *Let $\mathcal{A}$ be a complete and cocomplete Abelian category that satisfies* AB4*. If $\mathcal{A}$ has enough projectives, then $L^1 A_\bullet = 0$ for all towers $A_\bullet$ that satisfy the Mittag-Leffler condition.*

In fact, the theorem holds under the much milder hypothesis that $\mathcal{A}$ does not necessarily have enough projectives, but has a generator, see [**Roo06**].

Remark 3.5.10. This result has an interesting history. It was first published in [**Roo61**] without the hypothesis that $\mathcal{A}$ has enough projectives or admits a generator. The error went unnoticed for 40 years, until a counterexample was announced in [**Nee02**]. Shortly after, a revised version of the theorem was published in [**Roo06**].

Proof. First, we are going to prove the result for a category of modules, where one can use elements. Under this assumption, we will reduce to two extreme cases where the result is simple to prove. Let $B_i = \mathrm{im}(A_{i+k} \to A_i)$ for $k \gg 0$. Then there are induced maps $B_{i+1} \to B_i$ that are epimorphisms by construction. By the six-term exact sequence of Lemma 3.5.4, it is enough to establish the vanishing result for the towers $B_\bullet$ and $(A/B)_\bullet$. Thus, it is enough to prove the result in two simple cases: either all maps $f_i \colon A_{i+1} \to A_i$ are surjective (case 1), or, for all i, $\mathrm{im}(A_k \to A_i) = 0$ for $k \gg 0$ (case 2).

As usual, let $\delta \colon \prod A_i \to \prod A_i$ be the map defined by $\delta((a_i)) = ((a_i - f_i(a_{i+1})))$. Given $(b_i) \in \prod_i A_i$, we want to find elements $(a_i) \in \prod A_i$ such that $\delta((a_i)) = (b_i)$, which means that

$$(3.5.1) \qquad\qquad b_i = a_i - f_i(a_{i+1})$$

---

[1]The name comes from the classical Mittag-Leffler theorem in complex analysis, about the existence of a meromorphic function with prescribed singular parts on a discrete set of poles. The connection is not obvious, and goes through an abstract form of this result introduced by Bourbaki; see [**Run07**, Appendix A] for the full story.

for all $i \geq 0$. In case 1, all maps f_i are surjective, and (3.5.1) can be solved by induction starting from any b_0. In case 2, we can allow

$$a_i = \sum_{k=0}^{\infty} \overline{b_{i+k}},$$

where $\overline{b_{i+k}}$ is the image of b_{i+k} inside A_i. The sum is actually finite by hypothesis, and it is easy to check that it solves the system (3.5.1).

To prove the result for a more general category, let P be any projective in $\mathcal{A}$. Then we have a tower in Ab

$$\cdots \longrightarrow \operatorname{Hom}(P, A_2) \longrightarrow \operatorname{Hom}(P, A_1) \longrightarrow \operatorname{Hom}(P, A_0) \longrightarrow 0,$$

and we can check that this satisfies the Mittag-Leffler condition as well. In fact, given i, fix $N \gg 0$ such that $\operatorname{im}(A_k \to A_i)$ does not depend on k for $k \geq N$. For any such k, the map

$$\operatorname{im}(A_{k+1} \to A_i) \to \operatorname{im}(A_k \to A_i)$$

is an isomorphism. Given a morphism $P \to A_k$, the composition $P \to A_k \to A_i$ lands in $\operatorname{im}(A_{k+1} \to A_i)$. Since $A_{k+1} \to \operatorname{im}(A_{k+1} \to A_i)$ is an epimorphism, this can be lifted to a morphism $P \to A_{k+1}$.

Since this is a tower in Ab, by the first half of the proof, the map

$$\prod_i \operatorname{Hom}(P, A_i) \xrightarrow{\delta_P} \prod_i \operatorname{Hom}(P, A_i)$$

is an epimorphism. By definition of product, we can identify $\prod_i \operatorname{Hom}(P, A_i)$ with $\operatorname{Hom}(P, \prod_i A_i)$. It is a simple check that, under this identification, we have $h^P(\delta) = \delta_P$. In other words, every morphism $P \to \prod_i A_i$ can be lifted through δ.

Now let $C = \operatorname{coker} \delta$. Since $\mathcal{A}$ has enough projectives, we can choose an epimorphism $p \colon P \to \prod A_i$, obtaining the diagram

$$
\begin{array}{ccccccc}
& & & P & & & \\
& \swarrow & \swarrow & \downarrow{\scriptstyle p} & & & \\
\prod A_i & \xrightarrow{\ \delta\ } & \prod A_i & \xrightarrow{\ \pi\ } & C & \longrightarrow & 0.
\end{array}
$$

The composition $\pi \circ p$ is an epimorphism, because π and p are. What we have proved provides us the dashed arrow, which by exactness implies that $\pi \circ p = 0$. It follows that $C = 0$ and $L^1 A_\bullet = 0$. $\qquad\square$

3.6. Exercises

1. Let $\mathcal{A}$ be an Abelian category, and for $n \in \mathbb{Z}$ consider the functors $(-)_n \colon \operatorname{Kom}(\mathcal{A}) \to \mathcal{A}$ that sends $C_\bullet$ to C_n and $I_n \colon \mathcal{A} \to \operatorname{Kom}(\mathcal{A})$ that sends

$A \in \mathcal{A}$ to the complex

$$\cdots \longrightarrow 0 \longrightarrow A \longrightarrow A \longrightarrow 0 \longrightarrow \cdots,$$

concentrated in degrees $n-1$ and n. Prove that $(-)_n$ is the left adjoint of I_n.

2. Let $\mathcal{A}$ be an Abelian category. Prove that a complex $P_\bullet \in \mathrm{Kom}(\mathcal{A})$ is projective if an only if each P_i is projective and $P_\bullet$ is exact (use the previous exercise, Lemma 2.6.16, and find an epimorphism $Q_\bullet \to P_\bullet$, where $Q_\bullet$ is exact).

3. Let $\mathcal{A}$ be an Abelian category with enough injectives. Prove that the categories $\mathrm{Kom}(\mathcal{A})$, $\mathrm{Kom}^+(\mathcal{A})$, $\mathrm{Kom}^-(\mathcal{A})$ and $\mathrm{Kom}^b(\mathcal{A})$ have enough injectives.

4. Let $\mathcal{A}$ be an Abelian category with enough injectives and consider the homology functors $H_i \colon \mathrm{Kom}_{\geq 0}(\mathcal{A}) \to \mathcal{A}$. Use Corollary 3.4.3 to show directly that $R^i H_0 \cong H_i$.

5. Let $\mathcal{A}$ be a category with either enough injectives or enough projectives and $\{A_i\}_{i \in I}$ a family of objects of $\mathcal{A}$, $B \in \mathcal{A}$. Prove that if $\coprod_i A_i$ exists, then $\mathrm{Ext}^k(\coprod_i A_i, B) \cong \prod_i \mathrm{Ext}^k(A_i, B)$. Similarly, if $\prod_i A_i$ exists, $\mathrm{Ext}^k(B, \prod_i A_i) \cong \prod_i \mathrm{Ext}^k(B, A_i)$.

6. Let A be a ring, $\{M_i\}_{i \in I}$ a family of A-modules, N an A-module. Prove that $\mathrm{Tor}_k(\bigoplus_i M_i, N) \cong \bigoplus_i \mathrm{Tor}_k(M_i, N)$.

7. Let A be a ring, $F \colon \mathcal{I} \to \mathrm{Mod}_A$ a functor where $\mathcal{I}$ is filtered. Prove that $\mathrm{Tor}_k(\mathrm{colim}\, F, N) \cong \mathrm{colim}_{i \in \mathcal{I}} \mathrm{Tor}_k(F(i), N)$.

8. Let A be a ring, and let M, M', N, N' be A-modules. Show that there is a bilinear product

$$\mathrm{Tor}_i(M, N) \times \mathrm{Tor}_j(M', N') \to \mathrm{Tor}_{i+j}(M \otimes_A M', N \otimes_A N').$$

9. Let A be a ring and B_1, B_2 two A-algebras. Show that the direct sum

$$\mathrm{Tor}_*^A(B_1, B_2) := \bigoplus_i \mathrm{Tor}_i^A(B_1, B_2)$$

has a structure of graded A-algebra.

10. Let $\mathcal{A}$ be an Abelian category with enough projectives, $A_\bullet$ a tower in $\mathcal{A}$, where all maps $A_{i+1} \to A_i$ are epimorphisms. If the inverse limit $\varprojlim A_i$ exists, prove that the map $\varprojlim A_i \to A_0$ is an epimorphism.

11. Let A be a noncommutative ring, Mod_A the category of *right* A-modules, and fix a *left* A-module N. Show that the functor $T_N \colon \mathrm{Mod}_A \to \mathrm{Ab}$ defined by $T_N(M) := M \otimes N$ is well defined and right exact. Conclude that it admits left derived functors $\mathrm{Tor}_i^A(M, N) := (L^i T_N)(M)$. Symmetrically, show that one can define $T'_M \colon {}_A\mathrm{Mod} \to \mathrm{Ab}$, where ${}_A\mathrm{Mod}$ is the category of *left* A-modules, and M is a fixed *right* A-module, by declaring $T'_M(N) := M \otimes N$. Show that $(L^i T_N)(M) \cong (L_i T'_M)(N)$ as Abelian groups, so that the Tor functors can as well be defined by $\mathrm{Tor}^A_i(M, N) := (L^i T'_M)(N)$.

12. Let X be a topological space, endowed with a sheaf of rings $\mathcal{O}$. Given two sheaves of $\mathcal{O}$-modules S and T, show that one can define a sheaf $\mathcal{H}om(S, T)$ by

$$\mathcal{H}om(S, T)(U) = \mathrm{Hom}_{\mathcal{O}(U)}(S(U), T(U)).$$

Show that one can define Ext *sheaves* by

$$\mathcal{E}xt^i(S, T) = (R^i \mathcal{H}om(S, -))(T),$$

and that they satisfy a suitable long exact sequence.

For the next two exercises, we will need the notion of δ-*functor*. Let $\mathcal{A}, \mathcal{B}$ be two Abelian categories. A δ-functor between $\mathcal{A}$ and $\mathcal{B}$ is a sequence of additive functors $T = \{T^n\}_{n \geq 0}$ between $\mathcal{A}$ and $\mathcal{B}$ endowed with morphisms $\delta^n \colon T^n(C) \to T^{n+1}(A)$ for each short exact sequence

$$0 \longrightarrow A \longrightarrow B \longrightarrow C \longrightarrow 0$$

of objects in $\mathcal{A}$ such that

(1) each short exact sequence as above induces a long exact sequence

in $\mathcal{B}$; and

(2) given a morphism of short exact sequences

the induced δ^n morphisms fit into a commutative square

$$
\begin{array}{ccc}
T^n(C) & \xrightarrow{\ \delta^n\ } & T^{n+1}(A) \\
\downarrow & & \downarrow \\
T^n(C') & \xrightarrow{\ \delta^n\ } & T^{n+1}(A').
\end{array}
$$

There is also a notion of morphism of δ-functors. A morphism $S \to T$ between the δ-functors S and T is a collection of natural transformations $F^n \colon S^n \to T^n$ such that, for all short exact sequences

$$
0 \longrightarrow A \longrightarrow B \longrightarrow C \longrightarrow 0
$$

in $\mathcal{A}$, the induced maps between the long exact sequences of S and T are a morphism of complexes. The δ-functor T is called *universal* if for any δ-functor S, all natural transformation $S^0 \to T^0$ extend uniquely to a morphism of δ-functors. Clearly, if S and T are universal δ-functors with $S^0 \cong T^0$, they are isomorphic themselves.

13. Assume that $\mathcal{A}$ has enough injectives, and let $F \colon \mathcal{A} \to \mathcal{B}$ be an additive functor. Show that the derived functors $\{R^i F\}$ are a universal δ-functor between $\mathcal{A}$ and $\mathcal{B}$.

14. Let $\mathcal{A}$ be an Abelian category with enough injectives and consider the homology functors $H_i \colon \mathrm{Kom}_{\geq 0}(\mathcal{A}) \to \mathcal{A}$. Show that $\{H_i\}$ is a universal δ-functor, and deduce again that $H_i \cong R^i H_0$.

15 (**[Bet20]**). Show that an infinite direct sum of flabby sheaves is not necessarily flabby. (On the space $X = \{0\} \cup \{1/n \mid n \in \mathbb{Z}_+\}$ the sheaf $\mathcal{F}$ of all functions with values in $\mathbb{Z}/2\mathbb{Z}$ has a section $s_i \in \mathcal{F}(X \setminus \{0\})$ taking value 1 on $1/i$ and 0 elsewhere).

16. Let G be a finite group, M a G-module. A *crossed homomorphism* is a function $f \colon G \to M$ such that $f(gh) = f(g) + gf(h)$ for all $g, h \in G$. An element $m \in M$ defines a crossed homomorphism, called *principal*, by $f(g) = gm - m$. Prove that $H^1(G, M)$ can be identified with the group of crossed homomorphisms modulo the subgroup of principal crossed homomorphisms.

In [**Fer20**, Section A.8], we developed the Kummer theory of Abelian extension in an elementary way. It turns out that the proofs can be simplified using cohomological methods. Exercises 17–21 develop such an approach.

17. Prove the following cohomological version of Hilbert's theorem 90. Let L/K be a finite Galois extension of fields, $G = Gal(L/K)$ its Galois group. Prove that $H^1(G, L^*) = 0$. (Adapt the proof of Hilbert theorem 90 from [**Fer20**, A.8.5]). Extend the statement to infinite Galois extensions.

18. Let L/K be a finite Galois extension with Galois group G. Let $\mu_n < \overline{K}^*$ be the subgroup of n-th roots of 1, and assume that $\mu_n \subset K^*$ and char K does not divide n. Use the exact sequence

$$0 \longrightarrow \mu_n \longrightarrow L^* \longrightarrow (L^*)^n \longrightarrow 0 ,$$

together with Exercise 17, to derive an isomorphism

$$\frac{K^* \cap (L^*)^n}{(K^*)^n} \cong \mathrm{Hom}(G, \mu_n).$$

19. Let K be a field with algebraic closure $\overline{K}$, $\mu_n < \overline{K}^*$ be the subgroup of n-th roots of 1, and assume that $\mu_n \subset K^*$ and char K does not divide n. Prove that there is an isomorphism $K^*/(K^*)^n \cong \mathrm{Hom}(G, \mu_n)$, where $G = Gal(\overline{K}/K)$ is the absolute Galois group of K.

20. Keep notation as in Exercise 19. An element $\phi \in \mathrm{Hom}(G, \mu_n)$ determines an open subgroup $\ker \phi$ of G, hence a Galois extension L_ϕ/K. Prove that L_ϕ/K is a cyclic extension whose degree divides n. Moreover, $L_\phi = L_{\phi^k}$ for a number k prime with n. Conclude that there is a bijection between the set of cyclic subgroups of $K^*/(K^*)^n$ and the set of cyclic extensions of K of degree dividing n.

21. Extend the previous exercise to prove that there is a bijection between the set of finite Abelian subgroups $K^*/(K^*)^n$ and the set of finite Abelian extensions of K of degree dividing n.

22. Let $A_\bullet$ be a complex in an Abelian category. Show that the mapping cone $C(\mathrm{id}_A)$ of the identity is exact, and in fact the identity of $C(\mathrm{id}_A)$ is homotopic to 0.

23 (**[Ric18]**). Let $\mathcal{A}$ be the Abelian category of countable Abelian groups. Show that $\mathcal{A}$ has enough injectives and projectives, but the subcategories of injective and projective objects in $\mathcal{A}$ are not dual to each other. (There exists an injective group G in $\mathcal{A}$ such that $\mathrm{Hom}(G, G) \cong \mathbb{Q}$).

24. Show that the Mittag-Leffler condition is not necessary for the vanishing of $\lim^1$ (consider the k-vector spaces $x^i k[[x]]$).

25. Let $\mathcal{A}$ be an Abelian category, $A, B \in \mathcal{A}$. Given two extensions $E_\bullet$ and $F_\bullet$ of A by B of length $n \geq 2$, let G_n be the fibered product

and G_1 the fibered coproduct

Show that

$$0 \longrightarrow B \longrightarrow G_1 \longrightarrow E_2 \oplus F_2$$

$$\longrightarrow \cdots \longrightarrow E_{n-1} \oplus F_{n-1} \longrightarrow G_n \longrightarrow A \longrightarrow 0$$

is another extension of A by B, called the *Baer sum* of $E_\bullet$ and $F_\bullet$.

Assuming that $\mathcal{A}$ is extension small, show that the Baer sum gives a commutative group structure to $\mathrm{Ext}^n_Y(A, B)$, and that this makes $\mathrm{Ext}^n_Y(-, -)$ into a bifunctor with values in Ab. What modifications need to be done to define the Baer sum for $n = 1$?

26. Regard $\mathrm{Ext}^n_Y(A, B)$ as a commutative group under the Baer sum. Show that the natural isomorphism $\mathrm{Ext}^n_Y(A, B) \cong \mathrm{Ext}^n(A, B)$ is a group homomorphism, hence an isomorphism of Ab-valued functors.

27. Show that the notion of equivalence is needed when defining the Yoneda Ext functors $\mathrm{Ext}^i_Y(A, B)$. Namely, given an Abelian category $\mathcal{A}$ with enough injectives, two objects $A, B \in \mathcal{A}$ and an extension $E_\bullet$ of A by B of length $i > 1$, show that there is another extension $F_\bullet$ of A by B, not isomorphic to $E_\bullet$, such that $E_\bullet$ and $F_\bullet$ define the same element of $\mathrm{Ext}^i(A, B)$.

28. Let $\mathcal{A}$ be an extension small Abelian category. Show that there is a bilinear product

$$\mathrm{Ext}^i_Y(A, B) \times \mathrm{Ext}^j_Y(B, C) \to \mathrm{Ext}^{i+j}_Y(A, C),$$

defined by joining an extension of A by B to an extension of B by C. Show if $\mathcal{A}$ has enough injectives or enough projectives, this product is compatible with the isomorphisms $\mathrm{Ext}^i_Y(A, B) \cong \mathrm{Ext}^i(A, B)$ and the product of Theorem 3.4.10.

29 (**[Wof16]**). Consider a ring A obtained by extending $\mathbb{Z}$ with a variable x_α for each ordinal α. Technically, A is not a ring, because it is not even a set. Still, we can define a category Mod^*_A of *restricted* modules over A: an element of Mod^*_A will be an Abelian group M, together with an action of x_α on M for $\alpha \in S$, a *set* of ordinals. For $\alpha \notin S$, we will let x_α act trivially on M, i.e., $x_\alpha \cdot m = 0$ for all $m \in M$.

First, check that Mod^*_A is actually a well-defined category. Regard $\mathbb{Z}$ as an element of Mod^*_A, by letting all variables act trivially. For an ordinal α, let $M_\alpha = \mathbb{Z}[x_\alpha]$, with all variables but x_α acting trivially. Prove that there are extensions

$$0 \longrightarrow \mathbb{Z} \longrightarrow M_\alpha \longrightarrow \mathbb{Z} \longrightarrow 0$$

which are pairwise nonisomorphic, hence nonequivalent. In this case, $\mathrm{Ext}^1_{\mathrm{Mod}^*_A, Y}(\mathbb{Z}, \mathbb{Z})$ fails to be a set.

Chapter 4

Spectral Sequences

Spectral sequences are a kind of complicated, 3-dimensional algebraic structure. Part of the difficulty of working with them is that, unlike complexes, which can be drawn linearly, and even double complexes, spectral sequences involve a web of relations between objects indexed by three parameters—something that makes them hard to draw and hard to understand. Still, they provide an indispensable computational tool in algebraic topology, algebraic geometry and homological algebra. As every author of a primer on spectral sequences remarks, it is pretty inevitable that in order to master the subject, one has to manually work out many examples.

To give a little motivation, let us start with the statement of Corollary 3.4.3. There we have a double complex with exact columns and a kind of nontrivial relation between the homology of a row and of a total complex. We would like to understand what happens when the columns are not necessarily exact. One can imagine that there would still be a relation between the homology computed on the rows, on the columns, or on the total complex, but expressing this relation requires that we develop the machinery of spectral sequences.

This problem is not entirely theoretical, since interesting double complexes arise in many situations. For one thing, many of the diagrams appearing in the lemmas that we have encountered about Abelian categories (for instance those in Section 2.5) can be embedded into a suitable double complex, and then the result usually follows by applying the techniques of this chapter. In this sense, the theorems on spectral sequences are a kind of ultimate computation in Abelian categories, as most other results can be derived from them. More importantly, double complexes arise naturally when trying to compute the derived functors of a composition, an observation that directly leads to the Grothendieck spectral sequence for the composition of derived functors.

As it turns out, spectral sequences do not require a double complex, but are more general. In fact, they arise from a very innocent-looking gadget called an exact couple, and essentially all known spectral sequences can be derived in this setting.

Our approach works from the general to the particular. In Section 4.1, we introduce exact couples and show that they give rise to a strange-looking object, which we will dub a spectral sequence. In Section 4.2, we introduce filtered complexes and prove that they always have an associated exact couple, from which we derive a spectral sequence. Section 4.3 specializes this further to the case of a double complex. In this case, since we will have two natural filtrations, we will obtain two spectral sequences, with a relation between them—something very useful for computations. In Section 4.4 we use the general machinery developed so far to give alternative proofs of some results from Section 2.5 and other facts along the same lines. We go back to the theory in Section 4.5, where we specialize the spectral sequence of a double complex to obtain the Grothendieck spectral sequence for the composition of two functors. By specialization, we obtain some very useful spectral sequences for the composition of various Ext and Tor functors. Finally, Section 4.6 introduces two more spectral sequence due to Ischebeck, that relate the Ext and Tor functors for a pair of modules having either finite projective or finite injective resolutions.

4.1. Exact couples

Spectral sequence were invented by Jean Leray while he was interned in Edelbach as a prisoner of war, as a means to study double complexes. While they were soon recognized as a powerful tool, the earliest expositions were fairly technical and formula-ridden. It was Massey in [**Mas52**] that streamlined the presentation, by introducing the concept of an exact couple. In [**EH66**], Eckmann and Hilton extended the notion of exact couple from the category of modules to more general Abelian categories. Our presentation follows a mix of [**May99**] and [**BT95**, Section 14], suitably rephrased for the context of Abelian categories.

Definition 4.1.1. Let $\mathcal{A}$ be an Abelian category. An *exact couple* in $\mathcal{A}$ is the datum of two objects $D, E \in \mathcal{A}$ and three morphisms $i\colon D \to D$,

$j\colon D \to E$ and $k\colon E \to D$ such that the triangle

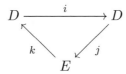

is exact at all vertices (that is, $\operatorname{im} i = \ker j$, $\operatorname{im} j = \ker k$ and $\operatorname{im} k = \ker i$).

The definition looks a little arbitrary right now. To understand it better, notice that we can compose j and k to get other morphisms. By definition, $k \circ j = 0$, so we look at $d := j \circ k \colon E \to E$. We observe that $d^2 = j \circ (k \circ j) \circ k = 0$, so we hope that d can give some interesting homological information about E. If we let $E' := \ker d / \operatorname{im} d$, we have an induced morphism $E' \to D$, and by definition of E' this lands in $\ker j = \operatorname{im} i$.

We are then prompted to define $D' := \ker j = \operatorname{im} i$, so that we have a morphism $k' \colon E' \to D'$. D' is a subobject of D, and by restriction i defines a map $i' \colon D' \to D'$. Defining a morphism $j' \colon D' \to E'$ is trickier: we cannot just take the restriction of j, since that vanishes on D' by definition. On the other hand, one can consider the diagram

$$(4.1.1) \qquad 0 \longrightarrow \ker i \longrightarrow D \xrightarrow{\;i\;} D' \longrightarrow 0$$

$$\begin{array}{ccc} & & \downarrow j \quad \searrow \quad \vdots\, j' \\ & & \ker d \longrightarrow E' \end{array}$$

and notice that $\ker i = \operatorname{im} k$ is a subobject of $\ker(D \to E')$. This is because $D \to E'$ is the composition of j with the quotient map $\ker d \to E'$ that factors out exactly $\operatorname{im} d = \operatorname{im} j \circ k$. Hence, the arrow $D \to E'$ induces a morphism $j' \colon D' \to E'$. The case of modules is possibly a little more explicit: when $\mathcal{A}$ is a category of modules, j' can be defined on elements via $j'(i(x)) = j(x)$, which is well defined thanks to the diagram (4.1.1).

We have then managed to reproduce another triangle

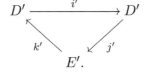

As the reader may now guess,

Lemma 4.1.2. *The objects D', E' and morphisms i', j', k' constructed above form an exact couple.*

Proof. By appealing to the Freyd–Mitchell theorem, Theorem 2.8.1, one can assume that $\mathcal{A} = \operatorname{Mod}_A$ for some noncommutative ring A, and then prove the result by diagram chasing. Alternatively, it is not difficult to prove this using the universal properties of kernels. Both ways are straightforward but tedious, and we leave them to the reader.   □

Definition 4.1.3. The exact couple (D', E', i', j', k') is called the *derived couple* of (D, E, i, j, k).

It is clear that this construction can be iterated to define an infinite sequence of exact couples, letting $(D^1, E^1, i^{(1)}, j^{(1)}, k^{(1)}) = (D, E, i, j, k)$ and $(D^{r+1}, E^{r+1}, i^{(r+1)}, j^{(r+1)}, k^{(r+1)})$ the derived couple of $(D^r, E^r, i^{(r)}, j^{(r)}, k^{(r)})$.

Remark 4.1.4. We have put parenthesis in the maps $i^{(r)}$, $j^{(r)}$, $k^{(r)}$ to distinguish them from the iterates of i, j, k (the latter two do not even compose with themselves).

To study the convergence of this construction, we need something more. Assume that $\mathcal{A}$ admits countable products. Then, by an obvious generalization of Proposition 2.2.7, it also admits all limits of functors $\mathcal{I} \to \mathcal{A}$, where $\mathcal{I}$ is a *countable* category, that is, $\mathcal{I}$ has countably many objects and morphisms. In this case, we shall say that $\mathcal{A}$ is *countably complete*. Dually, if $\mathcal{A}$ admits countable coproducts, it is *countably cocomplete*.

In the sequel, we are going to assume that $\mathcal{A}$ is countably complete and cocomplete. If A is any object of $\mathcal{A}$ and $\{U_i\}_{i \in \mathbb{N}}$ is a sequence of subobjects of A, we can define their intersection as

$$\bigcap_{i \in \mathbb{N}} U_i := \varprojlim_i (U_1 \cap \cdots \cap U_i).$$

Similarly, their sum is defined as

$$\sum_{i \in \mathbb{N}} U_i := \varinjlim_i (U_1 + \cdots + U_i).$$

In the latter case, if it also happens that $U_i \subset U_{i+1}$ for all i, we will write $\bigcup_{i \in \mathbb{N}} U_i = \sum_{i \in \mathbb{N}} U_i$. Recall that the operations of sum and intersection on subobjects are defined in Section 2.5.

Returning to our construction, let (D, E, i, j, k) be an exact couple in $\mathcal{A}$. By construction, $E' = \ker d / \operatorname{im} d$ is a subquotient of E. We let $Z^1 = \ker d$ and $B^1 = \operatorname{im} d$. By induction, we find subobjects $B^i \subset Z^i$ for all i, such that $Z^i \supset Z^{i+1}$, $B^i \subset B^{i+1}$, and $E^i \cong Z^i / B^i$. We can then define

$$Z^\infty := \bigcap_i Z^i$$

$$B^\infty := \bigcup_i B^i,$$

and let $E^\infty := Z^\infty / B^\infty$.

In all common cases, exact couples are endowed with a grading. Henceforth, let us assume that $\mathcal{A}$ consists of bigraded objects over a different Abelian category $\mathcal{B}$. An object of $\mathcal{A}$ is a collection $\{A_{p,q}\}_{(p,q)\in\mathbb{Z}\times\mathbb{Z}}$ of objects of $\mathcal{B}$, indexed by $\mathbb{Z}\times\mathbb{Z}$. A morphism f between $A_{\bullet,\bullet}$ and $B_{\bullet,\bullet}$ is a family of morphisms $f_{p,q}\colon A_{p,q}\to B_{p+s,q+t}$ in $\mathcal{B}$ for some $(s,t)\in\mathbb{Z}\times\mathbb{Z}$. The pair (s,t), which does not depend on (p,q), is called the *degree* of f. It is clear that $\mathcal{A}$ is itself an Abelian category, and is countably complete or cocomplete when $\mathcal{B}$ is.

If (D,E,i,j,k) is an exact couple in $\mathcal{A}$, we want to relate the degree of $d_r = j^{(r)}k^{(r)}\colon E^r \to E^r$ to the degrees of i, j, and k. The first observation is that $\deg i' = \deg i$ and $\deg k' = \deg k$, since i is induced by i and k' is induced by k. But the construction of j' is different, and a brief look at diagram (4.1.1) should convince the reader that $\deg j' = \deg j - \deg i$. By induction, we find

Lemma 4.1.5. *Let (E,D,i,j,k) be an exact couple of bigraded objects. Then $\deg i^{(r)} = \deg i$, $\deg j^{(r)} = \deg j - (r-1)\deg i$ and $\deg k^{(r)} = \deg k$. As a consequence,*

$$\deg d_r = \deg j^{(r)} + \deg k^{(r)} = \deg j - (r-1)\deg i + \deg k.$$

In the most common case, we will assume that $\deg i = (-1,1)$, $\deg j = (0,0)$ and $\deg k = (1,0)$, hence $\deg d_r = (r,-r+1)$. In other words, for each r we will have morphisms

$$d_r \colon E^r_{p,q} \to E^r_{p+r,q-r+1}.$$

As it turns out, the algebraic structure that we have produced, even if a little convoluted, is very useful both for theoretical reasons and for computation.

Definition 4.1.6. Let $\mathcal{B}$ be an Abelian category. A *spectral sequence* with values in $\mathcal{B}$ consists of a collection $E^r_{p,q}$ of objects in $\mathcal{B}$, indexed by $r \in \mathbb{Z}_+$ and $p,q \in \mathbb{Z}\times\mathbb{Z}$, endowed with morphisms $d_r\colon E^r_{p,q} \to E^r_{p+r,q-r+1}$ such that $d_r^2 = 0$ whenever the composition is defined. In other words, each diagonal $\left\{E^r_{p+kr,q-k(r-1)}\right\}_{k\in\mathbb{Z}}$ is made into a complex by d_r. Finally, the datum of a spectral sequence includes isomorphisms

(4.1.2) $$E^{r+1}_{p,q} \cong \frac{\ker\left(E^r_{p,q} \to E^r_{p+r,q-r+1}\right)}{\operatorname{im}\left(E^r_{p-r,q+r-1} \to E^r_{p,q}\right)}.$$

The bigraded object $E^r_{\bullet,\bullet}$ is called the r-th *sheet* of the spectral sequence.

Sometimes—for instance in Section 4.2—we also allow a 0-th sheet $E^0_{\bullet,\bullet}$ in a spectral sequence. More generally, it can sometime happen that we are able to produce the sheets of a spectral sequence only starting from some arbitrary $r \in \mathbb{Z}$.

Usually, spectral sequences are drawn sheet by sheet. A typical sheet in a spectral sequence is a diagram of the form

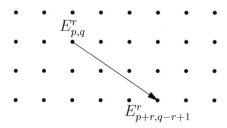

where all arrows share the same slope.

Remark 4.1.7. What we have called spectral sequences are usually called *cohomological* spectral sequences. In the literature one often finds *homological* spectral sequences: these are exactly the same objects, but endowed with morphisms

$$d_r \colon E^r_{p,q} \to E^r_{p-r,q+r-1}.$$

One can switch between the two conventions by changing the signs of p and q—for this reason, we will stick to cohomological spectral sequences and leave to the reader to formulate suitable statements about homological ones.

Let $E^r_{p,q}$ be a spectral sequence in $\mathcal{B}$. By definition, $E^{r+1}_{p,q}$ is a subquotient of $E^r_{p,q}$. If we assume that $\mathcal{B}$ is countably complete and cocomplete, it follows that the objects $E^r_{p,q}$ for fixed p and q converge to an object of $\mathcal{B}$, which we will denote $E^\infty_{p,q}$. In this way, we are able to define a limit sheet $E^\infty_{\bullet,\bullet}$. We will denote this situation by the notation $E^r_{p,q} \implies E^\infty_{p,q}$. In many cases, we will be able to obtain a spectral sequence where E^∞ represents some object of interest, while we have information about some finite sheet, typically E^1 or E^2.

Remark 4.1.8. The isomorphisms in (4.1.2) allow us to compute the sheet $E^{r+1}_{\bullet,\bullet}$ if we know the sheet $E^r_{\bullet,\bullet}$, including its differentials. Still, we only get to know the objects $E^{r+1}_{p,q}$ and not the morphisms between them. Thus, in general a spectral sequence contains nontrivial information in all sheets, and cannot be determined by, say, knowing E^1 alone. Still, in many cases the objects in a sheet are simple enough that the knowledge of the objects themselves is enough to say something about the next sheet. The power of spectral sequences lies in the fact that in such situations we are able to discern a lot of information about the limit E^∞ just by knowing E^1 or E^2.

We subsume the discussion of this section in a result that wraps up our construction.

Theorem 4.1.9. *Let $\mathcal{B}$ be an Abelian category and let $\mathcal{A}$ be the Abelian category of bigraded objects in $\mathcal{B}$. Let (D, E, i, j, k) be an exact couple in $\mathcal{A}$, where $\deg i = (-1, 1)$, $\deg j = (0, 0)$ and $\deg k = (1, 0)$, and let $(D^r, E^r, i^{(r)}, j^{(r)}, k^{(r)})$ be its r-th derived couple. Define*

$$d_r = j^{(r)} k^{(r)} \colon E^r \to E^r,$$

so that $\deg d_r = (r, -r + 1)$. Then E^r, d_r is a (cohomological) spectral sequence in $\mathcal{B}$. If moreover $\mathcal{B}$ is countably complete and cocomplete, the spectral sequence converges to a limit sheet E^∞.

In general, the sheet E^∞ can be difficult to compute, since it is defined as the quotient of an infinite intersection and an infinite union. In some cases, though, the process ends at a finite step.

Definition 4.1.10. Let $E_{p,q}^r$ be a spectral sequence. We say that $E_{p,q}^r$ is a *first quadrant* spectral sequence if $E_{p,q}^1 = 0$ for $p < 0$ or $q < 0$.

Since every sheet is a subquotient of the previous one, if this property holds for E^1, it holds for all sheets E^r. For fixed $p, q \geq 0$, the maps $E_{p,q}^r \to E_{p+r,q-r+1}^r$ and $E_{p-r,q+r-1}^r \to E_{p,q}^r$ are both 0 whenever $r > \max\{p, q+1\}$, because either the source or the target of these morphisms are 0. It follows that we have an isomorphism $E_{p,q}^{r+1} \cong E_{p,q}^r$. In other words, all objects $E_{p,q}^r$ are isomorphic for $r \gg 0$, and in this case we simply have $E_{p,q}^\infty = E_{p,q}^r$ for some large r.

Notice that for a first quadrant spectral sequence the limit sheet E^∞ is well defined even if $\mathcal{B}$ is not countably complete and cocomplete, because for each pair p, q the limit is trivial. The point at which $E_{p,q}^r$ stabilizes, though, can be different for every p, q, so it is not necessarily the case that there exists a fixed r such that $E^\infty = E^r$.

Definition 4.1.11. We say that the spectral sequence $E_{p,q}^r$ *abuts* at $E_{p,q}^\infty$ if for all p, q, the differentials $E_{p,q}^r \to E_{p+r,q-r+1}^r$ and $E_{p-r,q+r-1}^r \to E_{p,q}^r$ are both 0 for large r. We say that $E_{p,q}^r$ *degenerates* at the sheet E^r if the differentials d_i are 0 for $i \geq r$. In this case, we have $E^\infty = E^r$.

With this definition, we can rephrase our conclusion.

Proposition 4.1.12. *Let $E_{p,q}^r$ be a first quadrant spectral sequence. Then $E_{p,q}^r$ abuts at $E_{p,q}^\infty$.*

4.2. Filtered complexes

The previous section allows us to construct a spectral sequence associated to an exact couple, but of course the question stands how to obtain an exact couple in the first place. The most common construction starts from a filtered complex.

Definition 4.2.1. Let $\mathcal{B}$ be an Abelian category, $C_\bullet$ a complex of objects of $\mathcal{B}$. A *filtration* of $C_\bullet$ is a sequence of complexes $F_p C_\bullet$ for $p \in \mathbb{Z}$, such that, for all p and q, $F_{p+1} C_q$ is a subobject of $F_p C_q$. A *filtered* complex is just a complex endowed with a filtration. The filtration is called *canonical* if $F_0 C_\bullet = C_\bullet$, in which case $F_p C_\bullet = C_\bullet$ for all $p \leq 0$. The filtration is called *exhaustive* if for all q, $\bigcup_{p \in \mathbb{Z}} F_p C_q = C_q$ and $\bigcap_{p \in \mathbb{Z}} F_p C_q = 0$.

As usual, we will adopt the suggestive notation $F_{p+1} C_q \subset F_p C_q$, even though the reader should keep in mind that we are talking about subobjects and not necessarily set theoretic inclusions. Similarly, the condition $\bigcup_{p \in \mathbb{Z}} F_p C_q = C_q$ should be interpreted as saying that the natural morphism from the direct limit to C_q is an isomorphism. In the sequel, we will only consider exhaustive filtrations without further mention.

Each object C_q in a filtered complex is endowed with the filtration

$$\cdots \supset F_0 C_q \supset F_1 C_q \supset \cdots \supset F_p C_q \supset \cdots .$$

If $\mathcal{B}$ is cocomplete, we can define the *associated graded object*

$$\operatorname{Gr} C_q := \bigoplus_{p=-\infty}^{\infty} \frac{F_p C_q}{F_{p+1} C_q},$$

where of course one can only consider the sum for $p \geq 0$ when the filtration is canonical.

Let us define $D_{p,q} := F_p C_{p+q}$. Then $D_{\bullet,\bullet}$ is a bigraded object and there is a morphism $i \colon D_{p+1,q-1} \to D_{p,q}$, or in other words a morphism $i \colon D \to D$ of degree $(-1, 1)$. If we let

$$E_{p,q} := F_p C_{p+q} / F_{p+1} C_{p+q},$$

then $E_{\bullet,\bullet}$ is also a bigraded object, and the quotient $j \colon D \to E$ has degree $(0,0)$. For fixed p, the short exact sequence of complexes

$$0 \longrightarrow F_{p+1} C_\bullet \longrightarrow F_p C_\bullet \longrightarrow E_{p,\bullet} \longrightarrow 0$$

induces a long exact sequence in homology

$$(4.2.1) \qquad \cdots \longrightarrow H_n(F_{p+1} C_\bullet) \xrightarrow{\ i\ } H_n(F_p C_\bullet) \xrightarrow{\ j\ } H_n(E_{p,\bullet})$$
$$\xrightarrow{\qquad\qquad k \qquad\qquad}$$
$$\longrightarrow H_{n+1}(F_{p+1} C_\bullet) \longrightarrow \cdots ,$$

which we can interpret as an exact couple (here we are abusing notation by still denoting i, j the maps induced in cohomology, while k is the connecting map). Namely, define bigraded objects

$$D_{p,q}^1 := H_{p+q}(F_p C_\bullet)$$

$$E_{p,q}^1 := H_{p+q}\left(\frac{F_p C_\bullet}{F_{p+1} C_\bullet}\right)$$

so that the maps i, j, k in the long exact sequence above have degrees $(-1, 1)$, $(0, 0)$, and $(1, 0)$ respectively. Then (D^1, E^1, i, j, k) form an exact couple, and the degrees are exactly as in Theorem 4.1.9. From this, we conclude that E^1 is the first sheet in a spectral sequence. Since $E^1 = H(E)$, we can also extend the spectral sequence by adding a 0-th sheet $E^0 = E$, so that $E_{p,q}^0 = F_p C_{p+q}/F_{p+1} C_{p+q}$.

We now figure out what the limit sheet E^∞ looks like. This is slightly easier to discuss under the assumption that $\mathcal{B}$ is cocomplete, so that we can form infinite direct sums of objects in $\mathcal{B}$—in particular a bigraded object $\{A_{p,q}\}$ of $\mathcal{A}$ identifies the object $A = \bigoplus_{p,q} A_{p,q}$ of $\mathcal{B}$. This assumption is not really necessary, as one can perform the following reasoning for each degree, but it simplifies the notation somewhat.

To start, notice that the inclusion of complexes $F_p C_\bullet \to C_\bullet$ induces a map in homology

$$f_p \colon H_\bullet(F_p C) \to H_\bullet(C),$$

where $H_\bullet(C) = \bigoplus_n H_n(C)$, and similarly for $H_\bullet(F_p C)$. If we let $F_p H_\bullet(C) := \operatorname{im} f_p$, then $F_{p+1} H_\bullet(C)$ is a subobject of $F_p H_\bullet(C)$, so that $H_\bullet(C)$ inherits a filtration

$$\cdots \supset F_0 H_\bullet(C) \supset F_1 H_\bullet(C) \supset \cdots \supset F_p H_\bullet(C) \supset \cdots.$$

This has an associated graded object

$$\operatorname{Gr} H_\bullet(C) = \bigoplus_{p=-\infty}^{\infty} \frac{F_p H_\bullet(C)}{F_{p+1} H_\bullet(C)}.$$

As we did above, we see $\operatorname{Gr} H_\bullet(C)$ as a bigraded object, where the component of degree p, q is $F_p H_{p+q}(C)/F_{p+1} H_{p+q}(C)$.

Assume that the filtration of C is canonical and finite, hence $F_{r+1} C = 0$ for some r. By our construction of the exact couple, D^1 is

$$D^1 = H_\bullet(F_r C) \oplus H_\bullet(F_{r-1} C) \oplus \cdots \oplus H_\bullet(F_0 C) \oplus H_\bullet(F_{-1} C) \oplus \cdots,$$

where the degree of $H_{p+q}(F_p C)$ is (p, q). That is, in the displayed equation above, all terms are grouped by the first component of the degree. Since $i^{(1)}$ is induced by the monomorphism $F_{p+1} C \to F_p C$, its action shifts the p degree to the right. Hence,

$$D^2 = 0 \oplus i H_\bullet(F_r C) \oplus i H_\bullet(F_{r-1} C) \oplus \cdots \oplus i H_\bullet F_0 C \oplus i H_\bullet F_{-1} C \oplus \cdots,$$

where $iH_\bullet(F_kC)$ is a subobject of $H_\bullet(F_{k-1}C)$ (here we put a 0 at the beginning so that each summand is aligned to the summand of the same p degree for D_1). Since $F_pC = C$ for $p \leq 0$, $iH_\bullet(F_0C) = H_\bullet(F_0C)$, and the same holds for all terms to the right of it. Notice that $iH_\bullet(F_1C)$ is a subobject of $H_\bullet(F_0C)$, and from this point on, i acts as the identity on $iH_\bullet(F_1C)$ (save for shifting the degree).

Every successive term D^k is obtained by applying k times the map i. Hence, after r steps, we end with a sum of subobjects of $H_\bullet(C)$:

$$D^r = i^r H_\bullet(F_rC) \oplus i^{r-1} H_\bullet(F_{r-1}C) \oplus \cdots \oplus H_\bullet(F_0C) \oplus H_\bullet(F_{-1}C) \oplus \cdots .$$

It follows that $i^{(r)} \colon D^r \to D^r$ is a monomorphism. By exactness of the couple, we conclude that $k^{(r)} = 0$, which implies that $d_r = 0$. Similarly, all differentials d_k for $k \geq r$ are all 0, from which we see that $E^r = E^{r+1} = \cdots = E^\infty$.

We can say a little more, by observing that $i^p H_\bullet(F_pC)$ is just the image of f_p, or what we have dubbed as $F_p H_\bullet(C)$. Hence,

$$D^r = F_r H_\bullet(C) \oplus F_{r-1} H_\bullet(C) \oplus \cdots \oplus H_\bullet C \oplus H_\bullet C \oplus \cdots ,$$

and on D^r the morphism $i^{(r)}$ is just the shift map. Hence, we can identify $E^r = D^r / i^{(r)} D^r$ as

$$E^r = \frac{F_{r-1} H_\bullet(C)}{F_r H_\bullet C} \oplus \cdots \oplus \frac{H_\bullet C}{F_1 H_\bullet C} = \operatorname{Gr} H_\bullet(C).$$

In other words, the spectral sequence degenerates at E^r, to the bigraded object $\operatorname{Gr} H_\bullet(C)$.

We now drop the assumption that $F_{r+1}(C) = 0$, but still assume that the filtration is canonical. In this case, we cannot conclude that the spectral sequence degenerates at E^r, but it is still true that each piece $H_\bullet(F_pC)$ is sent to a subobject of $H_\bullet(C)$ after p steps, after which i acts as the identity on $f_p(H_\bullet(F_pC))$. In this case, the spectral sequence still abuts to $\operatorname{Gr} H_\bullet(C)$, but it does not degenerate to it.

We summarize our conclusions in the following

Theorem 4.2.2. *Let $\mathcal{B}$ be an Abelian category, C a complex in $\mathcal{B}$ endowed with a canonical filtration F_pC. Then there is a spectral sequence $E^r_{p,q}$ where*

$$E^0_{p,q} = \frac{F_p C_{p+q}}{F_{p+1} C_{p+q}},$$

$$E^1_{p,q} = H_{p+q}\left(\frac{F_p C_\bullet}{F_{p+1} C_\bullet} \right).$$

Moreover, this spectral sequence abuts to $E^\infty = \operatorname{Gr} H_\bullet(C)$. Finally, if the filtration is finite, hence $F_{r+1}C = 0$ for some r, the spectral sequence degenerates at E^r.

In essence, this result gives us a way to compute the homology of C, together with its grading, starting from a sheet that contains the homology of the successive quotients of the filtration. In many cases of interest, these will be significantly simpler than C itself.

The case where the filtration is not canonical is slightly more complex, since it is no longer true that $i^p F_p H_\bullet(C)$ becomes stable under the application of i. In general, we cannot expect that E^r abuts to E^∞. Still, assuming that $\mathcal{B}$ is countably complete and cocomplete, the limit sheet E^∞ is well defined as $E^\infty = Z^\infty / B^\infty$.

By definition (the reader should really check this!),

$$Z_{p,q}^\infty = \bigcap_{r=0}^{\infty} Z_{p,q}^r = \bigcap_{r=0}^{\infty} \ker\left(F_p C_{p+q} \to \frac{C_{p+q+1}}{F_{p+r} C_{p+q+1}}\right),$$

while

$$B_{p,q}^\infty = \bigcup_{r=0}^{\infty} B_{p,q}^r = \bigcup_{r=0}^{\infty} \operatorname{im}\left(F_{p-r} C_{p+q-1} \to C_{p+q}\right) \cap F_p C_{p+q}.$$

To say something more, we need to exchange the kernel and intersection. This can be done when $\mathcal{B}$ is a category of sets with additional structure, thanks to Theorem 1.4.16. In this case, if the filtration is exhaustive, we conclude that

$$Z_{p,q}^\infty = \ker\left(F_p C_{p+q} \to C_{p+q+1}\right).$$

Similarly,

$$B_{p,q}^\infty = \operatorname{im}\left(C_{p+q-1} \to C_{p+q}\right) \cap F_p C_{p+q}.$$

In this case, we obtain again that $E_{p,q}^\infty$ is the p-th component of $\operatorname{Gr} H_{p+q}(C)$. For simplicity, we state the conclusion for the category of A-modules.

Theorem 4.2.3. *Let A be a (noncommutative) ring, Mod_A the category of right A-modules. Let $C_\bullet$ be a complex in Mod_A endowed with an exhaustive filtration $F_p C_\bullet$. Then there is a spectral sequence $E_{p,q}^r$ where*

$$E_{p,q}^0 = \frac{F_p C_{p+q}}{F_{p+1} C_{p+q}},$$

$$E_{p,q}^1 = H_{p+q}\left(\frac{F_p C_\bullet}{F_{p+1} C_\bullet}\right),$$

which converges (but not necessarily abuts) to $E^\infty = \operatorname{Gr} H_\bullet(C)$.

4.3. Double complexes

Filtered complexes arise in applications in a natural manner, for instance in algebraic topology. For instance, for any coefficient ring A, if X is a CW-complex and X_n is the union of the cells of dimensions at most n, then the complex $S_\bullet(X, A)$ of singular chains on X has increasing filtration constituted by the chains $S_\bullet(X_n, A)$ supported on X_n. Dually, the cochain complex $\mathrm{Hom}(S_\bullet(X, A), A)$ has a decreasing filtration by the subcomplex $\mathrm{Ann}(S_\bullet(X_n, A))$ costisting of cochains which vanish on $S_\bullet(X_n, A)$. This allows us to use spectral sequences to study the cohomology of X (Exercise 1).

The most common application of the theory that we have developed so far, though, is by the filtered complexes associated to a double complex. Recall from Definition 3.4.1 that a double complex is a commutative diagram with objects A_i^j and morphisms $h = h_i^j \colon A_i^j \to A_i^{j+1}$ and $v = v_i^j \colon A_i^j \to A_{i+1}^j$ such that the rows and the columns are complexes. If we assume that on each diagonal $p + q = k$ we have finite number of nonzero objects A_i^j, then one can form the total complex $T_\bullet = \mathrm{Tot}(A_\bullet^\bullet)$ by putting

$$T_k := \bigoplus_{i+j=k} A_i^j.$$

This complex comes with a natural filtration $F_p T_\bullet$ by letting

$$F_p T_k := \bigoplus_{i+j=k, i \geq p} A_i^j,$$

that is, only summing those objects A_i^j with $i \geq p$. Since the differential on $T_\bullet$ is defined as a (signed) sum of the horizontal and vertical map in the double complex, $F_p T_\bullet$ is actually a subcomplex of $T_\bullet$. Moreover, assuming the underlying Abelian category is countably complete and cocomplete, this filtration is exhaustive.

This lets us produce a spectral sequence $E_{p,q}^r$. Let us compute the first sheets of the sequence. By the previous section,

$$E_{p,q}^0 = F_p T_{p+q} / F_{p+1} T_{p+q} \cong A_p^q.$$

This sheet is endowed with a differential of degree $(0, 1)$, that is, a morphism $A_p^q \to A_p^{q+1}$, which is induced by the differential $T_{p+q} \to T_{p+q+1}$. By definition, on the summand A_p^q, this differential is exactly $h_p^q + (-1)^q v_p^q$, where h is the horizontal morphism of the double complex and v the vertical one. Since v_p^q lands in A_{p+1}^q, it follows that the morphism $E_{p,q}^0 \to E_{p,q+1}^0$ is simply h_p^q.

Consequently, $E_{p,q}^1 = H_q(A_p^\bullet)$. This sheet is endowed with a differential of degree $(1, 0)$, that is, a morphism $H_q(A_p^\bullet) \to H_q(A_{p+1}^\bullet)$, which we can compute from the exact sequence (4.2.1). Recall that this differential is the composition $j \circ k$, where j and k are the maps in the exact sequence (4.2.1). The morphism j is just the map induced on homology from the quotient $F_p T_\bullet \to F_p T_\bullet / F_{p+1} T_\bullet$. The connecting morphism k is induced on homology from the morphism $d = h_p^q + (-1)^q v_p^q \colon A_p^q \to T_{p+q+1}$. Since h_p^q induces the 0 map in homology, it follows that the differential $H_q(A_p^\bullet) \to H_q(A_{p+1}^\bullet)$ is induced from $(-1)^q v_p^q$. In other words, the sheet E^2 is obtained by first taking horizontal homology, and then taking vertical homology (the flipped sign does not change the homology objects).

We will need notation to distiguish the two operations. Hence, we will denote by H_p^I the vertical homology of a double complex, and by H_q^{II} its horizontal homology. Our conclusions above can then be rephrased by saying that $E_{p,q}^1 = H_q^{II}(A_p^\bullet)$ and $E_{p,q}^2 = H_p^I(H_q^{II}(A_\bullet^\bullet))$.

Of course there is no preferential direction in a double complex. We can take the vertical homology first, and then the horizontal homology, and we end up with a different spectral sequence, where the roles of p and q are reversed.

Assume that the double complex is first quadrant, hence $A_p^q = 0$ whenever either p or q is less than 0. Then certainly the total complex is defined, and the filtration $F_p T_\bullet$ is canonical. In this case, we can apply Theorem 4.2.2 to conclude that both spectral sequences abut to $E^\infty = \operatorname{Gr} H_\bullet(T_\bullet)$. We summarize this state of affairs in

Theorem 4.3.1. *Let A_p^q be a double complex in an Abelian category, and assume that the total complex $T_\bullet = \operatorname{Tot}(A)_\bullet$ is defined. Then there are two spectral sequences, $^I E$ and $^{II} E$, the first one having*

$$^I E_{p,q}^0 = A_p^q,$$
$$^I E_{p,q}^1 = H_q^{II}(A_p^\bullet),$$
$$^I E_{p,q}^2 = H_p^I(H_q^{II}(A_\bullet^\bullet)),$$

and the second one having

$$^{II} E_{p,q}^0 = A_p^q,$$
$$^{II} E_{p,q}^1 = H_p^I(A_\bullet^q),$$
$$^{II} E_{p,q}^2 = H_q^{II}(H_p^I(A_\bullet^\bullet)).$$

If A is a first quadrant double complex, then both sequences abut to $E^\infty = \operatorname{Gr} H_\bullet(T_\bullet)$. If moreover A is a finite complex, the spectral sequence degenerates at E^r, where r is maximum of $p + q$ such that $A_p^q \neq 0$.

The reader should use Theorem 4.2.3 to derive a slightly more general result when the underlying category is Mod_A for some ring A. In this case, one has convergence without necessarily assuming that the double complex is first quadrant.

4.4. Simple applications

Now that we can apply the machinery of spectral sequences to double complexes, we can give a significantly quicker proof of Corollary 3.4.3.

Second proof of Corollary 3.4.3. We apply Theorem 4.3.1 to B. By hypothesis, the columns of B are exact except in $p = 0$, hence the first sheet $^{II}E^1$ is 0, except in $p = 0$, where it has the complex B_0^q. Hence, in the second sheet, $^{II}E_{0,q}^2 = H_q(B_0^\bullet)$, and all other entries are 0. All differentials in the sheets E^r for $r \geq 2$ are 0, just because they have degree $(r, -r + 1)$. It follows that $^{II}E^2 = E^\infty$, hence E^2 already computes $\operatorname{Gr} H_\bullet \operatorname{Tot}(B)$.

Taking degrees into account, we see that $H_q(\operatorname{Tot}(B))$ has only one graded piece, of degree 0. In other words, $H_q(\operatorname{Tot}(B)) = E_{0,q}^\infty = H_q(A_0^\bullet)$. $\square$

The above proof may not seem a big simplification over the more direct proof in Section 3.4. But it turns out that the same strategy can be applied over and over. For a different instance, we give another proof of the nine lemma.

Second proof of Lemma 2.5.17. We can regard the whole diagram of the nine lemma as a double complex which is 0 outside a 3×3 square. Since the columns are exact, we already have $^{II}E^1 = 0$, hence $E^\infty = 0$.

Assume that the top row is exact. Then $^{I}E^1$ only contains the homology of the bottom row, and all differentials are 0 for reasons of degree. Hence, ^{I}E *also* degenerates at $^{I}E^1$, and since $E^\infty = 0$, the homology of the bottom row must be 0 as well. A symmetric reasoning applies when the bottom row is assumed exact. $\square$

Not surprisingly, the other results from Section 2.5 can be proved by spectral sequence techniques as well. One interesting instance is the snake lemma, since there we also have to produce a connecting morphism.

Second proof of Lemma 2.5.20. It is an easy exercise to check that it is enough to prove the snake lemma when the rows are short exact sequences, in which case the sequence resulting from the snake lemma is exact also at the ends. As in the previous proof, we can regard the whole diagram of the snake lemma as a double complex, and the fact that the rows are exact tells us that $E^\infty =^{I} E^1 = 0$.

On the other hand, we can compute $^{II}E^1$ by taking vertical homology, which results in the sheet

$$(4.4.1) \qquad 0 \longrightarrow \ker \alpha \longrightarrow \ker \beta \longrightarrow \ker \gamma \longrightarrow 0,$$

$$0 \longrightarrow \operatorname{coker} \alpha \longrightarrow \operatorname{coker} \beta \longrightarrow \operatorname{coker} \gamma \longrightarrow 0.$$

We can go on computing $^{II}E^2$, which will only have nonzero objects in these six spots. The sheet $^{II}E^2$ looks like

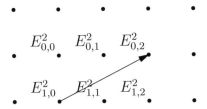

where we have displayed the only possible nonzero arrow (the diagram is upside down with respect to our convention, to preserve the position of the objects in the snake lemma).

Since $E^\infty = 0$, we must have $^{II}E^2_{0,0} =^{II} E^2_{0,1} =^{II} E^2_{1,1} =^{II} E^2_{1,2} = 0$, which means that in (4.4.1) the rows are exact in $\ker \alpha$, $\ker \beta$, $\operatorname{coker} \beta$, and $\operatorname{coker} \gamma$. Moreover, the only arrow in $^{II}E^2$ is an isomorphism, which means that there is an isomorphism between $\ker(\operatorname{coker} \alpha \to \operatorname{coker} \beta)$ and $\operatorname{coker}(\ker \beta \to \ker \gamma)$. The inverse of this last isomorphism is exactly the connecting map that we need to complete the exact sequence

$$
\begin{array}{ccccccc}
0 & \longrightarrow & \ker \alpha & \longrightarrow & \ker \beta & \longrightarrow & \ker \gamma \\
& & & & & \delta & \\
& \operatorname{coker} \alpha & \longrightarrow & \operatorname{coker} \beta & \longrightarrow & \operatorname{coker} \gamma & \longrightarrow 0.
\end{array}
$$

$\square$

To show the versatility of spectral sequences, we are going to use them to prove a handy result that is not yet in our bag of tools.

Proposition 4.4.1 (Schanuel's lemma). *Let $\mathcal{A}$ be an Abelian category,*

$$0 \longrightarrow K_1 \longrightarrow P_1 \longrightarrow A \longrightarrow 0$$

and

$$0 \longrightarrow K_2 \longrightarrow P_2 \longrightarrow A \longrightarrow 0$$

two exact sequences, and assume that P_1 and P_2 are projective. Then $K_1 \oplus P_2 \cong K_2 \oplus P_1$.

Proof. Our proof is taken from [**Pol**]. Since P_1 is projective, we can lift the morphism $P_1 \to A$ to a morphism $P_1 \to P_2$, which gives rise to a morphism of complexes

$$
\begin{array}{ccccccccc}
0 & \longrightarrow & K_1 & \longrightarrow & P_1 & \longrightarrow & A & \longrightarrow & 0 \\
& & \downarrow & & \downarrow & & \downarrow{\scriptstyle \operatorname{id}_A} & & \\
0 & \longrightarrow & K_2 & \longrightarrow & P_2 & \longrightarrow & A & \longrightarrow & 0.
\end{array}
$$

We can regard this as a double complex with exact rows, hence the associated spectral sequence has $E^\infty = {}^I E^1 = 0$. This means that the associated total complex

$$0 \longrightarrow K_1 \longrightarrow P_1 \oplus K_2 \longrightarrow A \oplus P_2 \longrightarrow A \longrightarrow 0$$

is exact. It is easy to check that the kernel of $A \oplus P_2 \to A$ is isomorphic to P_2, hence we can simplify the exact sequence to

$$0 \longrightarrow K_1 \longrightarrow P_1 \oplus K_2 \longrightarrow P_2 \longrightarrow 0.$$

This last exact sequence splits because P_2 is projective, hence the isomorphism $P_1 \oplus K_2 \cong P_2 \oplus K_1$. $\qquad\square$

4.5. The Grothendieck spectral sequence

Spectral sequences can be used to understand the homological properties of a double complex, in particular to relate its horizontal and vertical homology. As a special but fundamental case, double complexes arise when studying the derived functors of a composition. To define the right derived functors of a left exact functor F, we replaced an object A with an injective resolution $I_\bullet$, and then considered the complex $F(I_\bullet)$. If we repeat the same procedure with another left exact functor G, we can expect to obtain a double complex. By applying Theorem 4.3.1, we will then obtain two spectral sequences, and we can try to use them to relate the functors $R^i F$, $R^j G$, and $R^k G \circ F$.

In this section, we will pursue this strategy to obtain the Grothendieck spectral sequence. This is one of the most general spectral sequences, as many other interesting sequences can be obtained from it by specializing to a particular choice of F and G.

To simplify the study of the spectral sequences that will arise from this discussion, we will need a preliminary result.

Lemma 4.5.1. *Let $\mathcal{A}, \mathcal{B}$ be Abelian categories, and $C_\bullet$ a complex in $\mathcal{A}$ such that*

(i) *all objects C_i are injective; and*

(ii) *all kernels $\ker C_i \to C_{i+1}$ are injective.*

Then for each additive functor $G \colon \mathcal{A} \to \mathcal{B}$ we have

$$H_n(G(C_\bullet)) = G(H_n(C_\bullet)).$$

Moreover, $H_n(G(C_\bullet))$ is injective for all n.

Proof. Let $Z_i := \ker C_i \to C_{i+1}$ and $B_i := \operatorname{im} C_{i-1} \to C_i$, so that $H_n(C_\bullet) = Z_n / B_n$. By hypothesis, all C_i and Z_i are injective, and by the exact sequence

$$(4.5.1) \qquad 0 \longrightarrow Z_i \longrightarrow C_i \longrightarrow B_{i+1} \longrightarrow 0 ,$$

all B_i are injective as well. Moreover, (4.5.1) is split. Similarly, since B_i is injective, the exact sequence

$$(4.5.2) \qquad 0 \longrightarrow B_i \longrightarrow Z_i \longrightarrow H_i(C_\bullet) \longrightarrow 0$$

splits. For any additive functor G we then have exact sequences

$$0 \longrightarrow G(Z_i) \longrightarrow G(C_i) \longrightarrow G(B_{i+1}) \longrightarrow 0$$

and

$$0 \longrightarrow G(B_i) \longrightarrow G(Z_i) \longrightarrow G(H_i(C_\bullet)) \longrightarrow 0.$$

From these, we conclude that $G(Z_i) = \ker G(C_i) \to G(C_{i+1})$, $G(B_i) = \operatorname{im} G(C_{i-1}) \to G(C_i)$ and the homology of $G(C_\bullet)$ is just $G(H_n(C_\bullet))$. $\qquad \square$

We can now state the main result of this section.

Theorem 4.5.2. *Let $\mathcal{A}, \mathcal{B}, \mathcal{C}$ be Abelian categories and assume that $\mathcal{A}$ and $\mathcal{B}$ have enough injectives. Let $F \colon \mathcal{A} \to \mathcal{B}$ and $G \colon \mathcal{B} \to \mathcal{C}$ be two left exact functors, and assume that $F(I)$ is G-acyclic for all injective objects I of $\mathcal{A}$. Then for all $A \in \mathcal{A}$ there is a spectral sequence $E_{p,q}^r$ with*

$$E_{p,q}^2 = R^q G(R^p F(A))$$

which abuts to $E^\infty = \operatorname{Gr} R^\bullet (G \circ F)(A)$.

Remark 4.5.3. By working in the opposite category, we obtain another spectral sequence for the composition of left derived functors. Namely, assuming $\mathcal{A}$ and $\mathcal{B}$ have enough projectives and that $F(P)$ is G-acyclic for a projective $P \in \mathcal{A}$, we have a spectral sequence

$$E_{p,q}^2 = L^q G(L^p F(A)) \implies L^{p+q}(G \circ F)(A)$$

for all objects $A \in \mathcal{A}$. We will refer to both versions as Grothendieck spectral sequences.

Proof. Choose an injective resolution $A \to I_\bullet$, and consider the complex $F(I_\bullet)$. As usual, split the complex by letting $Z_p := \ker F(I_p) \to F(I_{p+1})$ and $B_p := \operatorname{im} F(I_{p-1}) \to F(I_p)$. Choose an injective resolution $X_{p,\bullet}$ of Z_p and $Y_{p,\bullet}$ of B_{p+1}. Reasoning as in Proposition 3.2.9 we find that $J_{p,\bullet} := X_{p,\bullet} \oplus Y_{p,\bullet}$ is an injective resolution of $F(I_p)$.

We can put together all objects $J_{p,q}$ to form a double complex. The horizontal maps $J_{p,q} \to J_{p+1,q}$ are defined by induction on q. Assuming $J_{p,q} \to J_{p+1,q}$ has been defined, the composition $J_{p,q} \to J_{p+1,q} \to J_{p+1,q+1}$ induces a map $B_{p,q+1} \to J_{p+1,q+1}$, where $B_{p,q+1} = \operatorname{im} J_{p,q} \to J_{p,q+1}$, as in

the diagram

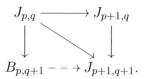

The morphism $B_{p,q+1} \to J_{p+1,q+1}$ extends to $J_{p,q+1}$, by the fact that $J_{p+1,q+1}$ is injective.

Applying G to this, we obtain a double complex $G(J_{\bullet,\bullet})$, and thanks to Theorem 4.3.1 two spectral sequences ${}^{I}E_{p,q}^{r}$ and ${}^{II}E_{p,q}^{r}$. Studying these spectral sequences will give us the desired conclusion.

The sheet ${}^{I}E_{p,q}^{1}$ is obtained by taking vertical homology. The complex $G(J_{p,\bullet})$ is exact, except in degree 0, by the hypothesis that $F(I_p)$ is acyclic for G. Hence the only nonzero terms are ${}^{I}E_{p,0}^{1} = G(F(I_p))$. To compute ${}^{I}E_{p,q}^{2}$ we take horizontal homology, so the only nonzero terms in ${}^{I}E^{2}$ are ${}^{I}E_{p,0}^{2} = R^p(G \circ F)(A)$. After ${}^{I}E^{2}$, all morphisms are 0, hence E^{∞} is the same as ${}^{I}E^{2}$.

In the second spectral sequence, we have ${}^{II}E_{p,q}^{1} = H_p(G(J_{\bullet,q}))$. The complex $J_{\bullet,q}$ satisfies the hypothesis of Lemma 4.5.1, from which we conclude that ${}^{II}E_{p,q}^{1} = G(H_p(J_{\bullet,q}))$. We then take vertical homology with respect to q to compute ${}^{II}E^{2}$. For a fixed p, the complex $H_p(J_{\bullet,q})$ (with respect to q) is a resolution of $H_p(F(I_\bullet)) = R^pF(A)$. To see this, start from the fact that $J_{p,\bullet}$ is a resolution of $F(I_p)$, and use the splittings from the proof of Lemma 4.5.1. The same lemma guarantees that $H_p(J_{\bullet,q})$ is injective, hence the complex $H_p(J_{\bullet,q})$ (with respect to q) is an *injective* resolution of $R^pF(A)$.

Using this to compute R^qG, it follows that ${}^{II}E_{p,q}^{2} = R^qG(R^pF(A))$. Since the double complex is first quadrant, by Theorem 4.3.1 this abuts to E^{∞}, which is what we needed to prove. $\square$

Remark 4.5.4. The proof shows a little more than what is stated in the theorem. Actually, it is enough to assume that there exists a class $\mathcal{I}$ of object of $\mathcal{A}$ such that for all objects $I \in \mathcal{I}$

(i) I is acyclic for F,

(ii) $F(I)$ is acyclic for G, and

(iii) every object A of $\mathcal{A}$ admits a monomorphism $A \to I$ for some $I \in \mathcal{I}$.

For simplicity, we have stated the theorem assuming that $\mathcal{I}$ is the class of all injective objects, since in any case we assume that $\mathcal{A}$ has enough injectives to define the derived functors of F.

Similarly, for the dual version of Remark 4.5.3, one can assume that there exists a class $\mathcal{P}$ of object of $\mathcal{A}$ such that for all objects $P \in \mathcal{P}$

(i) P is a acyclic for F,

(ii) $F(P)$ is acyclic for G, and

(iii) every object A of $\mathcal{A}$ admits an epimorphism $P \to A$ for some $P \in \mathcal{P}$.

In particular, when $\mathcal{A} = \mathrm{Mod}_A$, one can take the class of all free modules, which always satisfies (i) and (refepimorphisms exist. The Grothendieck spectral sequence then holds with the only assumption that $F(P)$ is G-acyclic for all *free* A-modules P.

The Grothendieck spectral sequence has all kinds of consequences, by choosing appropriate functors F and G, some of which we will see in the Exercises.

For our purposes, the most important consequences are the *base change spectral sequences*, which connect the Ext and Tor functors for various rings.

Theorem 4.5.5. *Let* $f\colon A \to B$ *be a morphism of rings,* M *a* B-*module and* N *an* A-*module. Regard* B *and* M *as* A-*modules via* f. *Then there is a spectral sequence*

$$E_{p,q}^2 = \mathrm{Ext}_B^q(M, \mathrm{Ext}_A^p(B, N)) \implies \mathrm{Ext}_A^{p+q}(M, N).$$

Proof. Apply Theorem 4.5.2 with $\mathcal{A} = \mathrm{Mod}_A$, $\mathcal{B} = \mathrm{Mod}_B$ and $\mathcal{C} = \mathrm{Ab}$, where we consider the functors $F = \mathrm{Hom}_A(B, -)\colon \mathrm{Mod}_A \to \mathrm{Mod}_B$ and $G = \mathrm{Hom}_B(M, -)\colon \mathrm{Mod}_B \to \mathrm{Ab}$. Notice that F is the functor of coextension of scalars from Example 1.5.2(g), hence it takes values in Mod_B. Moreover, F is right adjoint to the forgetful functor $\mathrm{Mod}_B \to \mathrm{Mod}_A$. It follows that for an A-module N

$$\mathrm{Hom}_B(M, \mathrm{Hom}_A(B, N)) \cong \mathrm{Hom}_A(M, N).$$

In other words, $G \circ F \cong \mathrm{Hom}_A(M, -)$.

By Lemma 2.6.16, it follows that F preserves injective objects. In particular, if I is an injective A-module, $F(I)$ is acyclic for G. We can then apply Theorem 4.5.2 to obtain the desired conclusion. $\square$

There is a similar spectral sequence for the Tor functors.

Theorem 4.5.6. *Let* $f\colon A \to B$ *be a morphism of rings,* M *a* B-*module and* N *an* A-*module. Regard* B *and* M *as* A-*modules via* f. *Then there is a spectral sequence*

$$E_{p,q}^2 = \mathrm{Tor}_q^B(\mathrm{Tor}_p^A(B, M), N) \implies \mathrm{Tor}_{p+q}^A(M, N).$$

Proof. Again, we apply Theorem 4.5.2 with $\mathcal{A} = \mathrm{Mod}_A$, $\mathcal{B} = \mathrm{Mod}_B$ and $\mathcal{C} = \mathrm{Ab}$. This time, we choose functors $F = B \otimes_A -\colon \mathrm{Mod}_A \to \mathrm{Mod}_B$ and $G = M \otimes_B -\colon \mathrm{Mod}_B \to \mathrm{Ab}$. In this case, F is the functor of *extension* of scalars from Example 1.5.2(f), which is *left* adjoint to the forgetful functor $\mathrm{Mod}_B \to \mathrm{Mod}_A$. It follows that for an A-module N

$$M \otimes_B (B \otimes_A N) \cong M \otimes_A N,$$

that is, $G \circ F \cong M \otimes_A -$.

Since the Tor functors are *left* derived functors, we need to apply the dual version of the Grothendieck spectral sequence, as in Remark 4.5.3. By Remark 4.5.4, it is enough to check that if P is a *free* A-module, then $B \otimes_A P$ is acyclic, which is clear since it is in fact a free B-module. The conclusion is then the Grothendieck spectral sequence. $\square$

Finally, we have a mixed version that involves both Ext and Tor.

Theorem 4.5.7. *Let $f\colon A \to B$ be a morphism of rings, M an A-module and N a B-module. Regard B and N as A-modules via f. Then there is a spectral sequence*

$$E^2_{p,q} = \mathrm{Ext}^q_B(\mathrm{Tor}^A_p(B, M), N) \implies \mathrm{Ext}^{p+q}_A(M, N).$$

Proof. The proof is the same as the previous one, this time taking $F = B \otimes_A -\colon \mathrm{Mod}_A \to \mathrm{Mod}_B$ and $G = \mathrm{Hom}_B(-, N)-\colon \mathrm{Mod}_B \to \mathrm{Ab}$. Since F is left adjoint to the forgetful functor $\mathrm{Mod}_B \to \mathrm{Mod}_A$, we get

$$\mathrm{Hom}_B(B \otimes_A M, N) \cong \mathrm{Hom}_A(M, N)$$

for all A-modules M, that is, $G \circ F \cong \mathrm{Hom}_A(-, N)$. The rest of the proof proceeds as in the spectral sequence for Tor. $\square$

4.6. The Ischebeck spectral sequences

In this section we find more spectral sequences relating the Ext and Tor functors, that cannot be derived simply as applications of the general Grothendieck spectral sequence. They are a little different, in that they require the hypothesis that certain modules admit a finite injective or projective resolution, and as such we are going to give an ad hoc proof. These spectral sequences are due to Ischebeck, and originally appear in [**Isc69**]. The starting point is the following observation.

Remark 4.6.1. Let M, N be A-modules. We have a natural morphism

$$f\colon\ \mathrm{Hom}(M, A) \otimes N \longrightarrow \mathrm{Hom}(M, N)\ .$$

$$\phi \otimes t \longmapsto \phi(-) \cdot t$$

This is clearly an isomorphism if M is finitely generated and free. It is also an isomorphism if M is finitely generated and projective, since in this case M is a direct summand of a free module by Proposition 2.6.4. Moreover, we also find that $\operatorname{Hom}(M, A)$ is still finitely generated projective.

When M is not projective, but has a finite projective resolution, we can replace this isomorphism with a spectral sequence.

Theorem 4.6.2. *Let M, N be A-modules, and assume that M admits a finite resolution by finitely generated projective modules. Then there is a spectral sequence*

$$E^2_{p,q} = \operatorname{Tor}_q(\operatorname{Ext}^p(M, A), N) \implies \operatorname{Ext}^{p-q}(M, N).$$

Proof. Fix projective resolutions $P_\bullet \to M$ and $Q_\bullet \to N$, where $P_\bullet$ is finite and each P_p is finitely generated. This gives a double complex

$$C^q_p = \operatorname{Hom}(P_p, Q_{-q}).$$

Notice the minus sign. We have to be careful with indices—the minus sign ensures that there are maps $C^q_p \to C^q_{p+1}$ and $C^q_p \to C^{q+1}_p$, so that C is indeed a double complex with our conventions.

By the above remark, we can rewrite our double complex as

$$C^q_p \cong \operatorname{Hom}(P_p, A) \otimes Q_{-q}.$$

By our assumption, this is nonzero only in a strip $0 \le p \le N$ for some N. We can then apply Theorem 4.3.1 to find two spectral sequences, and both abut to the same object, since all maps in either ${}^I E^r$ or ${}^{II} E^r$ are 0 for, say, $r > N + 2$.

For the first spectral sequence, we take vertical homology. This leaves us with ${}^I E^1_{p,q} = \operatorname{Tor}_{-q}(\operatorname{Hom}(P_p, A), N)$. We have remarked that $\operatorname{Hom}(P_p, A)$ is projective, so this vanishes except for $q = 0$, where it gives us ${}^I E^1_{p,0} = \operatorname{Hom}(P_p, A) \otimes N = \operatorname{Hom}(P_p, N)$. Taking the horizontal homology, we compute ${}^I E^2_{p,0} = \operatorname{Ext}^p(M, N)$, hence this sequence converges to $\operatorname{Ext}^n(M, N)$, $n = p + q$.

The second spectral sequence comes from horizontal homology, and gives us ${}^{II} E^1_{p,q} = H_p(\operatorname{Hom}(P_\bullet, A) \otimes Q_{-q})$. Since Q_{-q} is projective, the functor $- \otimes Q_{-q}$ is exact, so tensoring with Q_{-q} commutes with homology. This leaves us with

$$^{II} E^1_{p,q} = H_p(\operatorname{Hom}(P_\bullet, A)) \otimes Q_{-q} = \operatorname{Ext}^p(M, A) \otimes Q_{-q}.$$

Taking vertical homology, we find ${}^{II} E^2_{p,q} = \operatorname{Tor}_{-q}(\operatorname{Ext}^p(M, A), N)$. Comparing the two sequences and changing sign to q gives the desired conclusion. $\square$

Remark 4.6.3. If we assume that A is Noetherian, we can simplify the statement a bit. In this case, it is enough to assume that M is finitely generated and admits a finite projective resolution. In this formulation, we can also state a variant. Assume that A is Noetherian, M is finitely generated and N has a finite projective resolution. Then we have the same spectral sequence as in Theorem 4.6.2. The proof is identical, but this time we use the fact that the double complex is zero outside a *horizontal* strip.

With a simple modification of the proof, we obtain another spectral sequence due to Ischebeck, which we will state under the simplifying assumption that A is Noetherian (Exercise 17).

Theorem 4.6.4. *Let A be a Noetherian ring, M, N two A-modules, with M finitely generated. Assume that M admits a finite projective resolution or N admits a finite injective resolution. Then there is a spectral sequence*

$$E_{p,q}^2 = \mathrm{Ext}^q(\mathrm{Ext}^p(M, A), N) \implies \mathrm{Tor}_{q-p}(M, N).$$

4.7. Exercises

1. Let X be a CW-complex, and denote by X_p its p-skeleton. Prove that there is a homological spectral sequence $E_{p,q}^r$ such that $E_{p,q}^2 = H_{p+q}(X_p, X_{p-1})$ (this is relative, singular homology) that converges to $H_{p+q}(X)$. Computing the differentials in $E_{p,q}^1$, show that $E_{p,q}^2 = 0$ for $q \neq 0$ and $E_{p,0}^2$ is the p-th cellular homology $H_p^{cw}(X)$. Conclude that there is an isomorphism $H_p(X) \cong H_p^{cw}(X)$, that is, cellular and singular homology agree.

2. Let E, B be smooth real manifolds, with a map $f \colon E \to B$. Assume that f is a *fiber bundle*, that is, there exists a covering $\{U_i\}$ of B and a real manifold F such that the restriction $f^{-1}(U_i) \to U_i$ is isomorphic to the product $U_i \times F \to U_i$. This means that there exists a diffeomorphism $f^{-1}(U_i) \cong U_i \times F$ that makes the diagram

$$
\begin{array}{ccc}
f^{-1}(U_i) & \longrightarrow & U_i \times F \\
\downarrow & & \downarrow \\
U_i & \longrightarrow & U_i
\end{array}
$$

commute. For a fixed Abelian group G and an open set $U \subset B$, define $h_{G,f}^i(U) := H^i(f^{-1}(U), G)$. Prove that $h_{G,f}^i$ is a sheaf on B, and that $h_{G,f}^i$ is locally isomorphic to the locally constant sheaf associated to the group $H^i(F, G)$. The sheaf $h_{G,f}^i$ is called a *local coefficient system* on B.

3. Let $f \colon E \to B$ be a fiber bundle with fiber F, as in the previous exercise, and let $h_{G,f}^i$ be the associated local coefficient system. Prove that there is a spectral sequence

$$E_{p,q}^2 = H^p(B, h_{G,f}^q) \implies H^{p+q}(E, G),$$

called the *Serre spectral sequence* for f. Further, assuming that B is simply connected, prove that $h_{G,f}^i$ is actually isomorphic to the locally constant sheaf $H^i(F, G)$, in which case the spectral sequence simplifies to

$$E_{p,q}^2 = H^p(B, H^q(F, G)) \implies H^{p+q}(E, G).$$

4. Use a spectral sequence argument to prove the five lemma (Lemma 2.5.1).

5. Use a spectral sequence argument to prove that a short exact sequence of complexes gives rise to a long exact sequence of homology objects.

6. Let $E_{p,q}^r$ be a first quadrant spectral sequence, so that $E_{p,q}^r = 0$ for $p < 0$ or $q < 0$. Prove that there is a morphism from a subobject of $E_{0,p-1}^2$ to a quotient of $E_{p,0}^2$, called the *transgression* map.

7. Let $E_{p,q}^r$ be a first quadrant spectral sequence, arising from a filtered complex $F_p C_\bullet$. Prove that there is an exact sequence

$$0 \longrightarrow E_{1,0}^2 \longrightarrow C_1 \longrightarrow E_{0,1}^2 \longrightarrow E_{2,0}^2 \longrightarrow C_2.$$

8. Let X be a topological space, endowed with a sheaf of rings $\mathcal{O}$. Recall that for two $\mathcal{O}$-modules S, T the sheaf $\mathcal{E}xt^i(S, T)$ was defined in Exercise 12 of Chapter 3. Prove that there is an exact sequence

$$E_2^{p,q} = H^p(X, \mathcal{E}xt^q(S, T)) \implies \operatorname{Ext}_{\mathcal{O}}^{p+q}(S, T).$$

9. Let $f \colon A_\bullet \to B_\bullet$ be a morphism of complexes in an Abelian category. Regarding it as double complex with two rows, show that the mapping cone $C(f)$ is the associated total complex. Then use spectral sequences to derive again the long exact sequence from Proposition 3.1.11.

10. Give a direct proof of Schanuel's lemma, Proposition 4.4.1.

11. Let X be a topological space, with an open covering $U = \{U_i\}$. Given $\mathbf{i} = i_0, \ldots, i_q$, denote $U_{\mathbf{i}} = U_{i_0} \cap \cdots \cap U_{i_q}$ and $\ell(i) = q$. Given a sheaf of Abelian groups S on X, denote

$$C^q(U, S) = \prod_{\ell(\mathbf{i}) = q, U_{\mathbf{i}} \neq \emptyset} S(U_{\mathbf{i}}).$$

Find a natural boundary map $C^q(U, S) \to C^{q+1}(U, S)$ that makes $C^\bullet(U, S)$ into a complex. The homology of this complex is denoted $\check{H}^q(U, S)$ and called the q-th Čech cohomology of S on U.

12. Keep notation as in the previous exercise. Show that there is a spectral sequence
$$E^2_{p,q} = \check{H}^p(U, \mathcal{H}^q(X,S)) \implies H^{p+q}(X,S),$$
where $\mathcal{H}^q(X,S)$ denotes the sheafification of the presheaf $V \mapsto H^q(V,S)$.

13. Let $X = U \cup V$ be a topological space, with U, V open, and S a sheaf of Abelian groups. Use the previous exercise for the covering $\{U, V\}$ to derive an exact sequence relating $H^q(X,S)$, $H^q(U,S)$, $H^q(V,S)$ and $H^q(U \cap V, S)$. This is called the Mayer–Vietoris exact sequence in sheaf cohomology.

14. Let M, N be Abelian groups, where M has no n-torsion, while N is n-torsion (so N is a $\mathbb{Z}/n\mathbb{Z}$-module). Prove that there is an isomorphism
$$\mathrm{Ext}^1_{\mathbb{Z}}(N, M) \cong \mathrm{Hom}_{\mathbb{Z}/n\mathbb{Z}}(N, M/nM).$$

15. Let A be a ring. On A-modules, define an equivalence relation $M \sim N$ whenever there exist projective A-modules P, Q such that $M \oplus P \cong N \oplus Q$. Prove that $\sim$ is indeed an equivalence relation, and that A-modules modulo this relation form an Abelian semigroup $G(A)$, where the only class that has an inverse is 0.

16. Let A be a ring, $G(A)$ the Abelian semigroup defined in the previous exercise, and for a module M write $[M]$ for its class in $G(A)$. Given an A-module M, consider an epimorphism $f\colon P \to M$, where P is projective, and let $K = \ker f$. Prove that $[K] \in G(A)$ is well defined regardless of the choice of P. In fact, prove that $[K]$ depends only on $[M]$, and that the function $\mathcal{P}\colon G(A) \to G(A)$ defined by $\mathcal{P}([M]) = [K]$ is an endomorphism of $G(A)$.

17. Give a full proof for the second Ischebeck spectral sequence of Theorem 4.6.4.

18. Let $\mathcal{A}$ be an Abelian category with enough injectives, $C_\bullet$ a bounded below complex in $\mathcal{A}$. Prove that there exists a double complex $L^\bullet_\bullet$ with $L^j_\bullet = 0$ for $j < 0$ and a morphism $C_\bullet \to L^0_\bullet$ such that

   (i) for all i, the map $C_i \to L^\bullet_i$ is an injective resolution;

   (ii) all boundaries $B^i(L^j_\bullet)$ and cycles $Z^i(L^j_\bullet)$ are injective; and

   (iii) for all j, the exact sequences

$$0 \longrightarrow B^i(L^j_\bullet) \longrightarrow Z^i(L^j_\bullet) \longrightarrow H_i(L^j_\bullet) \longrightarrow 0$$

$$0 \longrightarrow Z^i(L^j_\bullet) \longrightarrow L^j_i \longrightarrow B^{i+1}(L^j_\bullet) \longrightarrow 0$$

   are split.

Such a double complex is called a *Cartan–Eilenberg resolution* of $C_\bullet$, and is what we use in the proof of Theorem 4.5.2.

19. Let $L_\bullet^\bullet$ be a Cartan–Eilenberg resolution of $C_\bullet$. Prove that for all i, $H_i(C_\bullet) \to H_i(L_\bullet^j)$ is an injective resolution.

20. Let $C_\bullet$ and $D_\bullet$ be complexes admitting Cartan–Eilenberg resolutions $L_\bullet^\bullet$ and $M_\bullet^\bullet$ respectively. Show that any morphism of complexes $C_\bullet \to D_\bullet$ induces a morphism between the Cartan–Eilenberg resolutions.

21. Let G be a finite group, M a G-module, $H \lhd G$ a normal subgroup. Prove that there is an action of G/H on M, and a spectral sequence

$$E_{p,q}^2 = H^p(G/H, H^q(H, M)) \implies H^{p+q}(G, M).$$

22. Prove the variant of Theorem 4.6.2 stated in Remark 4.6.3.

Chapter 5

Projective and Injective Modules

While the beginning of the book is a general introduction to homological algebra, in this chapter we start to see the ramifications of homological methods in the realm of commutative algebra. In particular, we begin by understanding the structure of projective and injective *modules*, and use them as a tool to understand modules in general. While usually the theory is developed for noncommutative rings, we will confine ourselves to the commutative case, leaving the general case for some exercises.

The simplest to understand are projective modules. In Section 5.1, we see that they are not far from being the same as locally free modules. This gives us some geometric intuition on the behavior of projective modules. As an application of the techniques of this section, we also prove a theorem of Vasconcelos about finite generation of ideals. In Section 5.2, we define the projective dimension of a module as the length of the shortest projective resolution. This is a fairly well-behaved notion, and we obtain results that allow us to link the projective dimension of modules across various operations, such as quotients, localizations and polynomial rings.

In Section 5.3 we turn to injective modules. These are much less intuitive objects. Our workhorse is Baer's criterion, this allows us to test for injectivity of a module using only monomorphisms of the form $I \to A$, where I is an ideal of the ring A. We use this to prove the celebrated theorem of Bass-Papp, that states that a ring A is Noetherian if and only if injective objects are stable under direct sum. This brings an unexpected link between homological techniques and finiteness conditions. In the rest of the section, we use the notion of injective hull from Section 2.7 to develop a classification theorem for injective modules over Noetherian rings, due to Matlis.

In Section 5.4, we define the notion of injective dimension, which is dual to that of projective dimension. Most of the results are analogous to those of Section 5.2. The main point of departure is the use of Baer's criterion, which allows us to prove a theorem of Kaplansky's that states that the injective dimension of a module M over a local Noetherian ring $A, \mathcal{M}$ can be computed just by looking at the groups $\mathrm{Ext}_A^i(A/\mathcal{M}, M)$.

In Section 5.5 we define the global dimension of a ring as the supremum of the projective dimensions of its modules, or equivalently as the supremum of the injective dimensions of its modules. Many results on global dimension are immediate consequences of the previous sections. The main result of the section is Theorem 5.5.7, which states that the global dimension grows by one when passing from a ring A to the polynomial ring $A[x]$. This is as an abstract version of the famous Hilbert syzygy theorem, which can be seen as the first result of homological algebra.

Hilbert's theorem, though, is concerned with free resolutions (over the ring $k[x_1, \ldots, x_n]$), while the bounds on global dimension only produce projective resolutions. To link the two, one needs to know that a finitely generated projective module over $k[x_1, \ldots, x_n]$) is free. This was conjectured by Serre, who proved a weaker version, and finally proved by Quillen and Suslin. Section 5.7 is devoted to the proof of Serre's theorem, and more generally to the concept of stably free modules. The main result is Theorem 5.7.12, which allows us to produce finite free resolutions in polynomial rings. As a consequence, we also prove Hilbert's syzygy theorem. Of course, Hilbert's original proof did not use all of this machinery, and was in fact quite elementary. We present it in the exercises. Finally, in Section 5.8 we prove the theorem of Quillen and Suslin.

5.1. Projective and free modules

In this section, we investigate the basic properties of projective modules. Recall from Proposition 2.6.4 that an A-module P is projective if and only if it is a direct summand of a free module. In particular, all free modules are projective. The converse is not true, save in simple cases.

Example 5.1.1.

(a) Let A be a principal ideal domain. Then every submodule of a free A-module is free—in particular every projective A-module. This is Theorem A.2.4, and we quickly recall the proof. Let $F = \bigoplus_{s \in S} A_s$ be a free A-module, where each A_s is just a copy of A, and let $N \subset F$ be a submodule. We can prove that N is free by transfinite induction. Namely, take a well-ordering of S and for $s \in S$ define

$$F_{\leq s} := \bigoplus_{t \leq s} F_t$$
$$F_{<s} := \bigoplus_{t<s} F_t.$$

Similarly, let $N_{\leq s} := N \cap F_{\leq s}$ and $N_{<s} := N \cap F_{<s}$. The quotient $N_{\leq s}/N_{<s}$ is a submodule of A_s, so it is either 0 or generated by a single element n_s. If $S' \subset S$ is the set for which this quotient is not 0, we claim that $\{n_s\}_{s \in S'}$ is a basis of N.

To check this, we have to prove that the $\{n_s\}_{s \in S'}$ are linearly independent, and that they generate N. Both of these claims are easy to verify; for details, see [**Fer20**, Theorem 2.1.5].

Conversely, an integral domain having the property that every submodule of a free module is itself free is a principle ideal domain (PID). To see this, let A be an integral domain with this property, and let $I \subset A$ be an ideal. Then I is a free A-module. If I had at least two generators $a, b \in I$, there there would be a nontrivial relation $ab - ba = 0$, hence I can have at most one generator, and A is a PID.

(b) Let A be a Dedekind ring. By Theorem A.5.5, a finitely generated A-module M is projective if and only if it is torsion-free. The details of the argument are fleshed out in the first volume, but since it is an important example, we briefly review them here.

First, take two different nonzero ideals $I, J \subset A$. If $I + J = A$, then $I \cap J = I \cdot J$, and the exact sequence

$$0 \longrightarrow IJ \longrightarrow I \oplus J \longrightarrow A \longrightarrow 0$$

splits, since A is free. It follows that $I \oplus J \cong A \oplus IJ$ as A-modules. The same is true for *any* pair of ideals I, J. To see this, it is enough to switch J for an ideal J' which has the same class in the class group $G(A)$, but having the property that $I + J' = A$.

Given a nonzero ideal $I \subset A$ and $x \in I$, $x \neq 0$, one can find an ideal J such that $I \cdot J = (x)$ by Theorem A.5.4. In this case, we obtain the isomorphism $I \oplus J \cong A \oplus (x) \cong A \oplus A$, showing that I is a direct summand of a free A-module, hence I is projective.

Now let M be a finitely generated A-module that is a direct sum of ideals, and $N \subset M$ a submodule. By induction on the rank of M, one can find ideals $J_k \subset I_k$ and elements $m_1, \ldots, m_r \in M$ such that

$$M \cong I_1 m_1 \oplus \cdots \oplus I_k m_k,$$
$$N \cong J_1 m_1 \oplus \cdots \oplus J_k m_k.$$

It follows that a finitely generated A-module M can be written as $M = T \oplus P$, where T is a direct sum of modules of the form I/J, where $0 \neq J \subset I \subset A$ are ideals, and P is a direct sum of ideals. In the first case, one can find $x \in I \setminus J$ such that $I \cong J + (x)$, and then

$$\frac{I}{J} \cong \frac{J + (x)}{J} \cong \frac{(x)}{(x) \cap J};$$

this is a quotient of (x), so it is isomorphic to a quotient of A. Hence $T \cong A/I_1 \oplus \cdots \oplus A/I_k$ for some ideals $I_1, \ldots, I_k$. In particular, T is torsion. In contrast, P is a direct sum of ideals, hence it is projective.

Thus, for Dedekind rings, torsion is the only obstruction to being projective, at least for finitely generated modules. For modules that are not finitely generated, all of this fails. For instance, $\mathbb{Q}$ is torsion-free, but it is not a free $\mathbb{Z}$-module, hence it is not projective. Notice that in this case, any ideal which is not principal will be a projective A-module that is not free. (Why?)

(c) Actually, projective modules that are not free are quite common. For instance, let A be any ring and consider the $A \times A$-module $A \oplus 0$. This is clearly a direct summand of $A \oplus A$, but it is usually not free. For instance, if A is finite, this is not free for cardinality reasons.

(d) Let X be a topological space, $A = C(X)$ the ring of (real-valued) continuous functions on X. If $E \to X$ is a real vector bundle of dimension n, we can multiply the sections of E by functions in $C(X)$ pointwise, this gives $M = C(E)$ the structure of an A-module. M is in general not free, but by the definition of vector bundle, E is locally isomorphic to a product $\mathbb{R}^n \times U \to U$, where $U \subset X$ is an open set. It follows that $C(E|_U)$ is a free $C(U)$-module of rank n.

Now assume that X is compact and paracompact. Then we can cover X with finitely many such open sets where E is trivial. By extending each of them and taking the direct sum, it follows that E is a subvector bundle of a trivial vector bundle $F \to X$. By taking a Riemannian metric on F, we can write $F = E \oplus E'$, where $E' = E^\perp$ is the orthogonal vector bundle. It follows that $C(E)$ is a *projective* $C(X)$-module. You are asked to flesh out this argument more precisely in Exercise 15.

Taking the last example as a guide, one can get a decent intuition on projective modules by thinking of projective modules as vector bundles, and free modules as trivial vector bundles. This suggests that projective modules should be locally free, in other words, that projective modules over local rings should be free. This turns out to be true.

Theorem 5.1.2 (Kaplansky). *Let A be a local ring, P a projective A-module. Then P is free.*

The above theorem has an easy proof when we assume that P is finitely generated (Exercise 1). The general case is more subtle; our approach follows [**Mat86**, Theorem 2.5]. The proof reduces to the case of countably generated modules via the following

Lemma 5.1.3. *Let A be a ring, M an A-module which is the direct sum of countably generated submodules, N a direct summand of M. Then N is itself a direct sum of countably generated submodules.*

Proof. Write $M = N \oplus P$. By hypothesis, $M = \bigoplus_{i \in I} M_i$, where the M_i are countably generated submodules of M. Given a set $J \subset I$, write $M_J = \bigoplus_{i \in J} M_i$, $N_J = M_J \cap N$, and $P_J = M_J \cap P$.

Consider all sets $J \subset I$ such that $M_J = N_J \oplus P_J$ and N_J, P_J are direct sums of countably generated submodules. By Zorn's lemma, we can take a maximal such J, and we only need to show that $J = I$. If this is not the case, we can find $i_1 \in I \setminus J$. Let $I_1 = \{i_1\}$ and $J_1 = J \cup I_1$. In general, $M_{J_1} \supset N_{J_1} \oplus P_{J_1}$, and the inclusion can be strict. But since M_{i_1} is countably generated, there is a countable set $I_2 \supset I_1$ such that $M_{J_1} \subset N_{J_2} \oplus P_{J_2}$, where $J_2 = J \cup I_2$. Recursively, we can find an increasing chain $I_1 \subset I_2 \subset I_3 \subset \cdots$ of countable sets disjoint from J such that, letting $J_k := J \cup I_k$, we have the inclusions $M_{J_k} \subset N_{J_{k+1}} \oplus P_{J_{k+1}}$ for all k.

If we let $I_\infty = \bigcup_{k=1}^\infty I_k$ and $J_\infty = J \cup I_\infty$, we obtain the equality $M_{J_\infty} = N_{J_\infty} \oplus P_{J_\infty}$. Since N_J is a direct summand of M_J, we can write $M = N_J \oplus M'$, whence $N_{J_\infty} = N_J \oplus (M' \cap N_{J_\infty})$. Let $N'_{J_\infty} = N_{J_\infty} \cap M'$, and define P'_{J_∞} similarly. Using these, we can write

$$M_{J_\infty} = N_J \oplus P_J \oplus N'_{J_\infty} \oplus P'_{J_\infty}.$$

From this we deduce that

$$N'_{J_\infty} \oplus P'_{J_\infty} \cong \frac{M_{J_\infty}}{M_J} \cong M_{I_\infty}.$$

By construction, M_{I_∞} is countably generated, hence the same is true for N'_{J_∞} and P'_{J_∞}, which are its summands. It follows that N_{J_∞} and P_{J_∞} are the direct sum of countably generated modules, and this a contradiction, since J_∞ is strictly larger than J. $\square$

Lemma 5.1.4. *Let $(A, \mathcal{M})$ be a local ring, P a projective A-module, and $x \in P$. Then there exists a free module F which is a direct summand of P, such that $x \in F$.*

Proof. Since P is projective, we can find a free module G such that $G \cong P \oplus Q$. Given a basis $B = \{g_i\}$ of G, we can write $x = \sum_{i=1}^{n} a_i g_i$, where $a_i \in A$. Choose such a basis B and a representation of x such that n is minimal. Writing $g_i = p_i + q_i$, where $p_i \in P$ and $q_i \in Q$, we obtain $x = \sum_{i=1}^{n} a_i p_i$. In turn, we can express

$$p_i = \sum_{j=1}^{n} b_{ij} g_j + c_i,$$

where $b_{ij} \in A$ and c_i is a linear combination of elements of B which are not one of $g_1, \ldots, g_n$. This allows us to write

$$x = \sum_{i=1}^{n} \sum_{j=1}^{n} a_i b_{ij} g_j.$$

Since the expression for x is unique, we obtain relations $a_i = \sum_{j=1}^{n} a_j b_{ji}$ for $i = 1, \ldots n$.

If for any such relation we had an invertible coefficient, we could write one of the a_i as a combination of the other ones, and in turn we could write x using less than n elements of B. By minimality, it follows that $1 - b_{ii} \in \mathcal{M}$ for all i and $b_{ij} \in \mathcal{M}$ for all $i \neq j$. This entails that the determinant of the matrix (b_{ij}) is $\equiv 1 \pmod{\mathcal{M}}$, and so $\det(b_{ij}) \in A^*$. We can then define an inverse of the matrix (b_{ij}) by Cramer's rule, and this inverse has coefficients in A as well. Using the inverse allows us to write $g_1, \ldots, g_n$ as linear combinations of $p_1, \ldots, p_n$ and other elements of B other than $g_1, \ldots, g_n$.

The conclusion of all this is that if we replace $g_1, \ldots, g_n$ by $p_1, \ldots, p_n$ inside B, we obtain another basis of G. We can then take F to be the free module generated by $p_1, \ldots, p_n$: it contains x and it is a direct summand of G, hence of P. $\qquad\square$

Remark 5.1.5. The argument in the lemma is essentially the determinant trick that is used in one proof of Nakayama's lemma; see [**Fer20**, Theorem 5.1.8]. One cannot resort directly to Nakayama's lemma, though, without a hypothesis of finite generation.

Proof of Theorem 5.1.2. Since P is projective, it is a direct summand of a free module, so Lemma 5.1.3 applies, and P is a direct sum of countably generated modules, which are then projective. Hence, it is enough to prove the theorem for a countably generated projective module P. Let $p_1, p_2, \ldots$ be generators of P. Using Lemma 5.1.4 we find a free module $F_1 \ni p_1$ such that $P \cong F_1 \oplus P_1$. We let p_i' be the projection of p_i in P_1 for $i \geq 2$. Since P_1 is again projective, we can find a free module $F_2 \ni p_2'$ such that $P_1 \cong F_2 \oplus P_2$. By continuing in this fashion we find free modules F_i such that $p_i \in F_1 \oplus \cdots \oplus F_i$ for all i, which implies that $P \cong F_1 \oplus F_2 \oplus \cdots$ is free as well. □

Remark 5.1.6. With a little care, the proof of Theorem 5.1.2 also works for noncommutative rings, where a noncommutative ring A is called local if for all $x \in A$ either x or $1 - x$ is invertible [**AF12**, Corollary 26.7].

Kaplansky's theorem reinforces the intuition that projective modules can be seen as an algebraic analogue of vector bundles. Still, one may want to be able to say a little more: that projective modules are actually locally free.

Definition 5.1.7. Let M be an A-module. We say that M is *locally free* if M_P is a free A_P-module for all prime ideals $P \subset A$.

To discuss the relation between projective and locally free modules, we need to understand how projective modules behave under localization. This, in turn, prompts us to understand the behavior of Ext under localization.

Definition 5.1.8. Let M be an A-module. A *presentation* of M is an exact sequence

$$F_1 \longrightarrow F_0 \longrightarrow M \longrightarrow 0,$$

where F_0 and F_1 are free A-modules. If M admits such a presentation with F_0 and F_1 finite free, we say that M is *finitely presented*. More generally, if M admits an exact sequence

$$F_k \longrightarrow F_{k-1} \longrightarrow \cdots \longrightarrow F_0 \longrightarrow M \longrightarrow 0$$

with $F_0, \ldots, F_k$ finite free, we say that M is *k-finitely presented*.

A presentation of M is just a way to give explicit generators $\{m_i\}$ for M and explicit generators for the relations between the $\{m_i\}$.

Remark 5.1.9. A module M is 0-finitely presented if and only if it is finitely generated, and 1-finitely presented if and only if it is finitely presented. If A is Noetherian and M is finitely generated, then M is k-finitely presented for all k (can you see why?).

Proposition 5.1.10. *Let A be a ring, M a k-finitely presented A-module. Then, for all multiplicative subsets $S \subset A$ and all A-modules N,*

$$S^{-1} \operatorname{Ext}_A^i(M, N) \cong \operatorname{Ext}_{S^{-1}A}^i(S^{-1}M, S^{-1}N)$$

for $i < k$.

Proof. Denote, for reasons of space, $A' = S^{-1}A$, $M' = S^{-1}M$ and so on. For $k = 1$ we have a commutative diagram

$$
\begin{array}{ccccccc}
0 & \longrightarrow & \operatorname{Hom}_A(M, N)' & \longrightarrow & \operatorname{Hom}_A(F_0, N)' & \longrightarrow & \operatorname{Hom}_A(F_1, N)' \\
& & \downarrow & & \downarrow & & \downarrow \\
0 & \longrightarrow & \operatorname{Hom}_{A'}(M', N') & \longrightarrow & \operatorname{Hom}_{A'}(F_0', N') & \longrightarrow & \operatorname{Hom}_{A'}(F_1', N'),
\end{array}
$$

which is exact on the left because h^N is left exact and localization is exact. The two vertical arrows on the right are isomorphisms, because F_0 and F_1 are finite free, hence the left arrow is an isomorphism as well by the five lemma (Lemma 2.5.1).

For $k > 1$ the result follows by dimension shifting (Proposition 3.3.2), because the F_i and $S^{-1}F_i$ are acyclic for Hom. $\qquad\square$

By Remark 5.1.9 we immediately obtain:

Corollary 5.1.11. *Let A be a Noetherian ring, M a finitely generated A-module. Then, for all multiplicative subsets $S \subset A$ and all A-modules N,*

$$
S^{-1} \operatorname{Ext}_A^i(M, N) \cong \operatorname{Ext}_{S^{-1}A}^i(S^{-1}M, S^{-1}N)
$$

for all $i \in \mathbb{N}$.

Remark 5.1.12. Since an A-module P is projective if and only if $\operatorname{Ext}^1(P, -)$ vanishes, we also obtain that if P if a finitely generated, projective A-module and A is Noetherian, then $S^{-1}P$ is a projective $S^{-1}A$-module for all multiplicative subsets $S \subset A$. With a slightly different proof, this can be stated in more generality:

Proposition 5.1.13. *Let $f\colon A \to B$ be a ring homomorphism, P a projective A-module. Then $B \otimes_A P$ is a projective B-module.*

Proof. The adjunction of Example 1.5.2(f) gives an isomorphism

$$
\operatorname{Hom}_B(B \otimes_A P, N) \cong \operatorname{Hom}_A(P, N)
$$

for all B-modules N. This means that $h^{B \otimes_A P}$ is the composition of h^P and the forgetful functor $\operatorname{Mod}_B \to \operatorname{Mod}_A$, both of which are exact. $\qquad\square$

In particular, this applies to the localization map $A \to A_P$ for a prime ideal P. Combining this with Kaplansky's theorem, Theorem 5.1.2, we finally get:

Corollary 5.1.14. *Let A be a ring, M a projective A-module. Then M is locally free, that is, for all prime ideals $P \subset A$, the A_P-module M_P is free.*

Under some finiteness conditions, the converse holds. These finiteness conditions are necessary, see Exercise 4.

Proposition 5.1.15. *Let M be a finitely presented A-module. Then M is projective if and only if it is locally free.*

Proof. We only need to prove one implication. Fix a free module F with a surjection $F \to M$, and let $P \subset A$ be a prime. Assuming that M_P is free, $\mathrm{Hom}(M_P, F_P) \to \mathrm{Hom}(M_P, M_P)$ is surjective as well. Since M is finitely presented, Proposition 5.1.10 gives isomorphisms

$$\mathrm{Hom}_{A_P}(M_P, F_P) \cong \mathrm{Hom}_A(M, F)_P \text{ and}$$
$$\mathrm{Hom}_{A_P}(M_P, M_P) \cong \mathrm{Hom}_A(M, M)_P.$$

If this holds for all primes P, Proposition 2.2.21, implies that $\mathrm{Hom}(M, F) \to \mathrm{Hom}(M, M)$ is surjective, and in particular id_M is in the image. This shows that the surjection $F \to M$ splits, hence M is a direct summand of F. $\square$

Remark 5.1.16. A finitely generated projective A-module is finitely presented. In fact, assume that M is projective and that we have a surjection $\phi \colon A^k \to M$. Then ϕ splits, so if we let $K = \ker \phi$, we get an isomorphism $A^k \cong K \oplus M$, showing that K is finitely generated as well.

A nice application of the result of Kaplansky is the following criterion of Vasconcelos, who uses it to derive a finiteness result [**Vas73**].

Proposition 5.1.17 (Vasconcelos). *Let A be a ring, $I \subset A$ a projective ideal. If I is not contained in any minimal prime of A, I is finitely generated.*

The connection with Kaplansky's theorem passes through the following notion.

Definition 5.1.18. Let P be a projective A-module. The *trace* of P is the ideal $\tau_A(P)$ (or simply $\tau(P)$ when the ring is fixed) generated by $f(P)$ for all homomorphisms $f \colon P \to A$. In other words, $\tau(P)$ is the image of the evaluation morphism $\mathrm{Hom}(P, A) \otimes_A P \to A$.

Exercise 18 clarifies the reason for the name "trace." To better understand this ideal, assume that we have a decomposition $F = P \oplus Q$, where F is a free A-module, and that we have chosen a basis $\{v_i\}_{i \in I}$ of F. Then we have corresponding coordinate functions c_i defined by $c_i(\sum_j a_j v_j) = a_i \in A$, and we can restrict them to P.

On the other hand, any homomorphism $f \colon P \to A$ can be extended to F. Consider the free module $F' = F \oplus A$, and let u be a generator of the last summand. Then a basis for F' is $\{u\} \cup \{v_i - f(v_i)u\}_{i \in I}$. With this choice, $f(v_i)$ is the u-coordinate of v_i for all $i \in I$, so that the u-coordinate function agrees with f on F.

We conclude that $\tau(P)$ is generated by the coordinates of elements of P for all decompositions $F = P \oplus Q$ of a free module and all choices of bases for F. Using this characterization, we can prove a change of ring result.

Proposition 5.1.19. *Let $g\colon A \to B$ be a ring homomorphism, M a projective A-module. Then $\tau_B(B \otimes_A M) = g(\tau_A(M)) \cdot B$.*

Notice that $B \otimes_A M$ is in fact a projective B-module by Proposition 5.1.13.

Proof. The diagram

$$
\begin{array}{ccc}
\mathrm{Hom}_A(M, A) \otimes_A M & \longrightarrow & A \\
\downarrow & & \downarrow{\scriptstyle g} \\
\mathrm{Hom}_B(M, B) \otimes_B (B \otimes_A M) & \longrightarrow & B
\end{array}
$$

shows that $g(\tau_A(M)) \subset \tau_B(B \otimes_A M)$.

On the other hand, choose a decomposition of A-modules $F \cong M \oplus N$, where F is free with basis $\{v_i\}$. Then

$$
F_B := B \otimes_A F \cong B \otimes_A M \oplus B \otimes_A N,
$$

and it is enough to check that for any choice of basis of F_B, the corresponding coordinate functions send $B \otimes_A M$ into $g(\tau_A(M)) \cdot B$. Let $\{w_i\}$ be such a basis, and choose $m \in M, b \in B$. With a small abuse of notation, denote g the induced map $F \to F_B$. For each i write

$$
g(v_i) = \sum_j b_{ij} w_j,
$$

where the sum is finite. Then if a_i are the coordinates of m in the basis $\{v_i\}$,

$$
b \otimes m = b \sum_i g(a_i) g(v_i) = \sum_{i,j} b b_{ij} g(a_i) w_j,
$$

so the w_j-coordinate of $b \otimes m$ is $\sum_i b b_{ij} g(a_i) \in g(\tau_A(M)) \cdot B$. $\square$

Corollary 5.1.20. *Let $P \subset A$ be a prime ideal, M a projective A-module, $J = \tau_A(M)$ the trace ideal.*

(i) *If $J \subset P$, $J_P = 0$; and*

(ii) *If $J \not\subset P$, $J_P = A_P$.*

Proof. By the proposition, $J_P = \tau_{A_P}(A_P \otimes_A M) = \tau_{A_P}(M_P)$. The A_P-module M_P is projective, hence free by Theorem 5.1.2. It follows that $J_P = A_P$, unless $M_P = 0$, in which case $J_P = 0$. On the other hand, $J_P = A_P$ happens if and only if $J \not\subset P$. $\square$

Proof of Proposition 5.1.17. Let $J = \tau_A(I)$ be the trace ideal of I. By construction, $I \subset J$. First, assume that J is not all of A, and take a maximal ideal $\mathcal{M} \supset J$. By hypothesis, $\mathcal{M}$ is not a minimal prime, hence by Theorem A.1.2, $\mathcal{M}$ contains a minimal prime P.

In the ring A/P, the ideal $I/PI = A/P \otimes_A I$ has trace

$$\tau_{A/P}(I/PI) = A/P \otimes_A J = (J + P)/P.$$

But the latter is contained in $\mathcal{M}/P$, hence it is not the whole A/P. On the other hand, A/P is an integral domain, hence all localizations of $(J+P)/P$ are different from 0. By Corollary 5.1.20, this means that $(J+P)/P$ is not contained in any prime ideal of A/P, a contradiction.

We conclude that $\tau_A(I) = A$, which means that we can find n homomorphisms $f_i \colon I \to A$ and elements $a_i \in I$ such that $\sum_{i=1}^n f_i(a_i) = 1$. Given any element $x \in I$, we can use this to write

$$x = \sum_{i=1}^n x f_i(a_i) = \sum_{i=1}^n f_i(x) a_i,$$

which shows that $a_1, \ldots, a_n$ generate I. $\qquad\square$

5.2. Projective dimension

In this section, we are going to study A-modules via their projective resolutions. The starting point is the following observation about the length of projective resolutions.

Proposition 5.2.1. *Let $\mathcal{A}$ be an Abelian category with enough projectives, A an object of $\mathcal{A}$. The following are equivalent:*

   (i) *there exists a projective resolution*

$$0 \longrightarrow P_n \longrightarrow \cdots \longrightarrow P_1 \longrightarrow P_0 \longrightarrow A \longrightarrow 0$$

     of length n; and

   (ii) $\operatorname{Ext}^i(A, B) = 0$ *for all $i > n$ and all $B \in \mathcal{A}$.*

Proof. That (i) implies (ii) is clear, since we can use the resolution $P_\bullet$ to compute the Ext groups.

Conversely, assume (ii) and take any projective resolution $P_\bullet \to A$. If $K = \ker(P_{n-1} \to P_{n-2})$, then $\operatorname{Ext}^1(K, -) = 0$ by Proposition 3.3.2. Hence, K is projective and we can substitute K for P_n and get a projective resolution of length n. $\qquad\square$

Definition 5.2.2. Let $\mathcal{A}$ be an Abelian category with enough projectives, $X \in \mathcal{A}$. The minimal length of a projective resolution of X will be called the *projective dimension* of X, and denoted $\operatorname{pd} X$. If X does not admit any projective resolution of finite length, we set $\operatorname{pd} X = \infty$.

When $\mathcal{A} = \operatorname{Mod}_A$ is the category of modules over the ring A, we sometimes denote projective dimension by pd_A for clarity.

By the above result, $\operatorname{pd} X$ is the also minimum n (if any) such that $\operatorname{Ext}^i(X, -) = 0$ for $i > n$.

Example 5.2.3.

(a) Let A be a principal ideal domain. By Example 5.1.1(a), every A-module has projective dimension at most 1. In fact, let M be an A-module, and take an epimorphism $f \colon F \to M$ with F free. Since $K := \ker f$ is free, M admits a free resolution

$$0 \longrightarrow K \longrightarrow F \longrightarrow M \longrightarrow 0$$

with two terms.

(b) Let A be a Dedekind ring, M a finitely generated A-module. By Theorem A.5.5 it follows immediately that $\operatorname{pd}_A M \leq 1$. In fact, the same is true for arbitrary A-modules, as will follow for instance by Proposition 5.5.3.

(c) Let $A = \mathbb{Z}/4\mathbb{Z}$, and consider the A-module $M = \mathbb{Z}/2\mathbb{Z}$. Seeing M as a quotient of A, we get an exact sequence

$$0 \longrightarrow \frac{\mathbb{Z}}{2\mathbb{Z}} \longrightarrow \frac{\mathbb{Z}}{4\mathbb{Z}} \longrightarrow \frac{\mathbb{Z}}{2\mathbb{Z}} \longrightarrow 0 \ .$$

The first term in this sequence can also be seen as quotient of A, and so on. This leads us the the following infinite free resolution of M:

$$\cdots \longrightarrow \frac{\mathbb{Z}}{4\mathbb{Z}} \longrightarrow \frac{\mathbb{Z}}{4\mathbb{Z}} \longrightarrow \frac{\mathbb{Z}}{2\mathbb{Z}} \longrightarrow 0.$$

Applying $\operatorname{Hom}(-, \mathbb{Z}/2\mathbb{Z})$, we can compute that

$$\operatorname{Ext}^i_{\mathbb{Z}/4\mathbb{Z}} \left(\frac{\mathbb{Z}}{2\mathbb{Z}}, \frac{\mathbb{Z}}{2\mathbb{Z}} \right) = \frac{\mathbb{Z}}{2\mathbb{Z}}$$

for all $i \geq 0$. In particular, $\operatorname{pd}_{\mathbb{Z}/4\mathbb{Z}} \mathbb{Z}/2\mathbb{Z} = \infty$.

Remark 5.2.4. Let G be an Abelian group, so that $\operatorname{pd}_{\mathbb{Z}} G \leq 1$. If we assume that $\operatorname{Ext}^1(G, H) = 0$ for all other Abelian groups H, then G is projective, hence free. One may ask whether the condition $\operatorname{Ext}^1(G, \mathbb{Z}) = 0$ suffices. This is known as the Whitehead problem. Shelah proved in [**She74**] that this question is actually independent from the ZFC axioms!

We will be mostly interested in the notion of projective dimension in a category of modules. In particular, we want to understand how the projective dimension of modules changes under various operations. We start by discussing the graded case.

Recall from Definition 1.1.4(d) that when A is a graded ring, we distinguish between the category Mod_A of graded A-modules and the category Mod_A^u of ungraded ones. The functor that forgets the grading is denoted by $U\colon \operatorname{Mod}_A \to \operatorname{Mod}_A^u$.

We can use Proposition 2.6.8 to link the projective dimension in the graded and ungraded case. We state the result in slightly more generality.

Proposition 5.2.5. *Let $\mathcal{A}, \mathcal{B}$ be two Abelian categories with enough projectives, and $U\colon \mathcal{A} \to \mathcal{B}$ an exact functor such that for all $P \in \mathcal{A}$, P is projective if and only if $U(P)$ is. Then $\operatorname{pd} M = \operatorname{pd} U(M)$ for all $M \in \mathcal{A}$. In particular, for a graded A-module the projective dimensions in Mod_A and in Mod_A^u are the same.*

Proof. From a projective resolution

$$0 \longrightarrow P_n \longrightarrow \cdots \longrightarrow P_1 \longrightarrow P_0 \longrightarrow M \longrightarrow 0$$

of M of length n in $\mathcal{A}$, we get a projective resolution

$$0 \longrightarrow U(P_n) \longrightarrow \cdots \longrightarrow U(P_1) \longrightarrow U(P_0) \longrightarrow U(M) \longrightarrow 0$$

of $U(M)$ of length n, hence $\operatorname{pd} U(M) \le \operatorname{pd} M$.

For the opposite inequality, assume that $\operatorname{pd} U(M) \le n$, and consider a truncated resolution

$$P_{n-1} \longrightarrow \cdots \longrightarrow P_1 \longrightarrow P_0 \longrightarrow M \longrightarrow 0$$

of M. Denote $K := \ker P_{n-1} \to P_{n-2}$. Applying U we get a truncated resolution

$$U(P_{n-1}) \longrightarrow \cdots \longrightarrow U(P_1) \longrightarrow U(P_0) \longrightarrow U(M) \longrightarrow 0$$

of $U(M)$, and by Proposition 3.3.2, $U(K)$ is projective. It follows that K is projective as well, hence $\operatorname{pd} M \le n$.

The final statement follows from Proposition 2.6.8. $\qquad\square$

To relate the projective dimension of various modules, we start with a general result, known as the *change of ring* theorem for projective dimension.

Theorem 5.2.6. *Let $f\colon A \to B$ a homomorphism of rings, and M a B-module. Then*

$$\operatorname{pd}_A M \le \operatorname{pd}_A B + \operatorname{pd}_B M.$$

Proof. This follows at once from the base change spectral sequence of Theorem 4.5.5. $\qquad\square$

Taking appropriate choices of A and B, we find many useful specializations of this result, also known as change of ring theorems. In particular, we will investigate what happens under quotients, localization, and polynomial rings. We will follow the approach of [**Wei95**, Section 4.3].

Localization is the easiest case. If $P \subset A$ is a prime ideal, the localization functor $\mathrm{Mod}_A \to \mathrm{Mod}_{A_P}$ is exact, and for any projective A-module M, the localized A-module M_P is again projective by Proposition 5.1.13. The first part of the proof of Proposition 5.2.5 shows that for all A-modules M, $\mathrm{pd}_{A_P} M_P \le \mathrm{pd}_A M$. In fact, we can state a little more under some finite generation assumptions.

Proposition 5.2.7. *Let A be a Noetherian ring, M a finitely generated A-module. Then there exists a maximal ideal $\mathcal{M} \subset A$ such that $\mathrm{pd}_A M = \mathrm{pd}_{A_\mathcal{M}} M_\mathcal{M}$, so*

$$\mathrm{pd}_A M = \sup_{\substack{P \subset A \\ prime}} \mathrm{pd}_{A_P} M_P = \sup_{\substack{\mathcal{M} \subset A \\ maximal}} \mathrm{pd}_{A_\mathcal{M}} M_\mathcal{M}.$$

Proof. Define $d(\mathcal{M}) := \mathrm{pd}_{A_\mathcal{M}} M_\mathcal{M}$. We can assume that $d(\mathcal{M}) < \infty$ for all maximal ideals $\mathcal{M}$, otherwise the result follows from the previous remark.

Consider a projective resolution

$$\cdots \longrightarrow P_2 \longrightarrow P_1 \longrightarrow P_0 \longrightarrow M \longrightarrow 0$$

and let $K_n := \ker P_{n-1} \to P_{n-2}$. Since A is Noetherian, we can choose the resolution so that all P_i are finitely generated. It follows that all K_i are finitely generated as well.

Since localization is exact, we get a projective resolution of $M_\mathcal{M}$ for all $\mathcal{M}$, and moreover $(K_n)_\mathcal{M} = \ker(P_n)_\mathcal{M} \to (P_{n-1})_\mathcal{M}$. By definition of $d(\mathcal{M})$, $(K_{d(\mathcal{M})})_\mathcal{M}$ is projective, hence free by Kaplansky's theorem, Theorem 5.1.2. Let $r(\mathcal{M})$ be the rank of this free $A_\mathcal{M}$-module.

We can choose a homomorphism $f_\mathcal{M} \colon A^{r(\mathcal{M})} \to K_{d(\mathcal{M})}$ such that the localization at $\mathcal{M}$ becomes an isomorphism. In other words, $\ker f_\mathcal{M}$ and $\mathrm{coker} f_\mathcal{M}$ become 0 in the localization at $\mathcal{M}$. Since they are finitely generated, there exists a single element $a_\mathcal{M} \notin \mathcal{M}$ such that $a_\mathcal{M} \cdot \ker f_\mathcal{M} = 0$ and $a_\mathcal{M} \cdot \mathrm{coker} f_\mathcal{M} = 0$.

The set of elements $\{a_\mathcal{M}\}$ generates the whole ring, hence we can choose a *finite* subset $\{a_{\mathcal{M}_1}, \ldots, a_{\mathcal{M}_t}\}$ that generates the whole ring. Let

$$d = \max\{d_{\mathcal{M}_1}, \cdots, d_{\mathcal{M}_t}\}.$$

We claim that $(K_d)_\mathcal{M}$ is free for all maximal ideals $\mathcal{M}$. To see this, choose i such that $a_{\mathcal{M}_i} \notin \mathcal{M}$. Then $f_{\mathcal{M}_i}$ becomes an isomorphism when localized at $\mathcal{M}$, that is to say, $(K_{d(\mathcal{M}_i)})_\mathcal{M}$ is free. But then, since $d \geq d(\mathcal{M}_i)$, $(K_d)_\mathcal{M}$ is projective. (Why?) Again, by Theorem 5.1.2, it is free.

By Proposition 5.1.15, we conclude that K_d is projective. But then $\mathrm{pd}_A M \leq d = d(\mathcal{M}_i)$ for one of the maximal ideals $\mathcal{M}_i$. $\square$

The case of quotients is more subtle, and is handled by the following three results. All of them are proved by induction, based on the following observation.

Remark 5.2.8. Given an A-module M, we can choose a projective module P with a surjection $P \to M$. Let K be the kernel, so that we have the exact sequence

$$0 \longrightarrow K \longrightarrow P \longrightarrow M \longrightarrow 0 .$$

By Proposition 3.3.2, it follows immediately that $\mathrm{pd}_A M = 1 + \mathrm{pd}_A K$.

To state our change of rings results, it is convenient to introduce some terminology.

Definition 5.2.9. Let M be an A-module, $a \in A$. We say that a is *weakly M-regular* if it is not a divisor of 0 in M, that is, if $am = 0$ for some $m \in M$, then $m = 0$. If moreover $aM \subsetneq M$, a is called *M-regular*. In the case where $M = A$, we will simply speak of (weakly) regular elements.

Theorem 5.2.10 (First change of rings theorem). *Let A be a ring, $a \in A$ a regular element, and let M be an $A/(a)$-module. Assume that $\mathrm{pd}_{A/(a)} M < \infty$. Then*

$$\mathrm{pd}_A M = 1 + \mathrm{pd}_{A/(a)} M.$$

Proof. We start with the remark that any module over $A/(a)$ cannot be projective as an A-module. Otherwise, it would be a direct summand of a free A-module, which is a contradiction since multiplication by a is injective on a free A-module.

The exact sequence

$$0 \longrightarrow A \xrightarrow{\cdot a} A \longrightarrow \frac{A}{(a)} \longrightarrow 0$$

gives us $\mathrm{pd}_A A/(a) \leq 1$, so in fact $\mathrm{pd}_A A/(a) = 1$. By Theorem 5.2.6, $\mathrm{pd}_A M \leq 1 + \mathrm{pd}_{A/(a)} M$.

We prove equality by induction on $\mathrm{pd}_{A/(a)} M$. If $\mathrm{pd}_{A/(a)} M = 0$, we have $\mathrm{pd}_A M = 1$, since we have excluded that $\mathrm{pd}_A M = 0$.

For the inductive step, take an epimorphism $P \to M$, where P is a projective $A/(a)$-module, and let K be the kernel. By Remark 5.2.8,

$\mathrm{pd}_{A/(a)} M = 1 + \mathrm{pd}_{A/(a)} K$, and by the inductive hypothesis $\mathrm{pd}_{A/(a)} K = \mathrm{pd}_A K - 1$. We finish the proof if we can show that $\mathrm{pd}_A M = 1 + \mathrm{pd}_A K$. This does *not* follow from Remark 5.2.8 because P is not projective as an A-module.

Still, the exact sequence

$$0 \longrightarrow K \longrightarrow P \longrightarrow M \longrightarrow 0$$

gives us a long exact sequence of $\mathrm{Ext}_A(-, N)$ groups for all A-modules N. Since $\mathrm{pd}_A P = 1$, from that sequence we conclude that $\mathrm{pd}_A M = 1 + \mathrm{pd}_A K$, *unless* $\mathrm{pd}_A M = \mathrm{pd}_A K = 1$ (in which case we know that $\mathrm{pd}_{A/(a)} K = 0$ and $\mathrm{pd}_{A/(a)} M = 1$).

To exclude this case, take another short exact sequence

$$0 \longrightarrow K' \longrightarrow P' \longrightarrow M \longrightarrow 0 \, ,$$

this time of A-modules, with P' projective. If $\mathrm{pd}_A M = 1$, then K' is projective as well. By tensoring with $A/(a)$, we obtain

$$0 \longrightarrow \mathrm{Tor}_1^A(M, A/(a)) \longrightarrow \tfrac{K'}{aK'} \longrightarrow \tfrac{P'}{aP'} \longrightarrow M \longrightarrow 0.$$

Since $\mathrm{pd}_{A/(a)} M = 1$, by dimension shifting (Proposition 3.3.2) we obtain that $\mathrm{Tor}_1^A(M, A/(a))$ is projective. By Example 3.4.12(a),

$$\mathrm{Tor}_1^A(M, A/(a)) = \{m \in M \mid am = 0\} = M,$$

hence M is projective, a contradiction. $\qquad\square$

Our second theorem is similar to the previous one, but starts from an A-module.

Theorem 5.2.11 (Second change of rings theorem)**.** *Let A be a ring, M an A-module and $a \in A$ an M-regular element. Then*

$$\mathrm{pd}_A M \geq \mathrm{pd}_{A/(a)} \frac{M}{aM}.$$

Proof. We can assume that $\mathrm{pd}_A M$ is finite, so the proof will be by induction on $\mathrm{pd}_A M$. The base case is $\mathrm{pd}_A M = 0$, in which case M is projective over A, hence M/aM is projective over $A/(a)$ by Proposition 5.1.13.

For the inductive case, consider an exact sequence

$$0 \longrightarrow K \longrightarrow P \longrightarrow M \longrightarrow 0 \, ,$$

where P is projective, so that $\mathrm{pd}_A K = \mathrm{pd}_A M - 1$. By induction, we know that $\mathrm{pd}_{A/(a)} K/aK \leq \mathrm{pd}_A K$. As in the previous proof, we tensor with $A/(a)$ to find the exact sequence

$$0 \longrightarrow \mathrm{Tor}_1^A(M, A/(a)) \longrightarrow \tfrac{K}{aK} \longrightarrow \tfrac{P}{aP} \longrightarrow \tfrac{M}{aM} \longrightarrow 0.$$

Our hypothesis tells us that a is not torsion on M, hence $\mathrm{Tor}_1^A(M, A/(a)) = 0$. We conclude that either M/aM is projective as an $A/(a)$-module, or

$$\mathrm{pd}_{A/(a)} \frac{M}{aM} = 1 + \mathrm{pd}_{A/(a)} \frac{K}{aK} \leq pd_A M,$$

and in both cases we are done. $\qquad\square$

The last change of rings theorem deals with the case of a local ring.

Lemma 5.2.12. *Let A be a Noetherian ring, M a finitely generated A-module and $I \subset \mathcal{J}(A)$ an ideal. If M/IM is free as an A/I-module, then M is free.*

Proof. Choose a set $S \subset M$ of n elements such that its image is a basis of M/IM. This defines a map $f\colon A^n \to M$, which is surjective by Nakayama's lemma, and injective by Nakayama's lemma applied to $\ker f$. $\qquad\square$

Theorem 5.2.13 (Third change of rings theorem). *Let A be a Noetherian local ring with maximal ideal $\mathcal{M}$, $a \in \mathcal{M}$, and let M be a finitely generated A-module. If a is regular and M-regular, then*

$$\mathrm{pd}_A M = \mathrm{pd}_{A/(a)} \frac{M}{aM}.$$

Proof. If $\mathrm{pd}_{A/(a)} \frac{M}{aM} = \infty$, the equality follows from Theorem 5.2.11. Otherwise, we can prove the result by induction on $\mathrm{pd}_{A/(a)} \frac{M}{aM}$.

The base case is $\mathrm{pd}_{A/(a)} \frac{M}{aM} = 0$, in which case $\frac{M}{aM}$ is projective. Since $A/(a)$ is a local ring, $\frac{M}{aM}$ is free by Theorem 5.1.2. In this case, M is free by Lemma 5.2.12, hence $\mathrm{pd}_A M = 0$ as well.

For the inductive step, consider an exact sequence

$$0 \longrightarrow K \longrightarrow P \longrightarrow M \longrightarrow 0$$

where P is projective, so that $\mathrm{pd}_A K = \mathrm{pd}_A M - 1$. As in the previous proof, we tensor it with $A/(a)$ and use the vanishing of $\mathrm{Tor}_1^A(M, A/(a))$ to obtain the exact sequence

$$0 \longrightarrow \frac{K}{aK} \longrightarrow \frac{P}{aP} \longrightarrow \frac{M}{aM} \longrightarrow 0,$$

from which we get $\mathrm{pd}_{A/(a)} K/aK = \mathrm{pd}_{A/(a)} M/aM - 1$.

We can choose P to be finitely generated (for instance a finitely generated free A-module). Since A is Noetherian, K is finitely generated as well. By induction, $\mathrm{pd}_A K = \mathrm{pd}_{A/(a)} K/aK$, hence $\mathrm{pd}_A M = \mathrm{pd}_{A/(a)} M/aM$. $\qquad\square$

Remark 5.2.14. Since Theorem 5.2.13 requires an element that is both regular and M-regular, it is interesting to understand when such elements exist. Of course, a necessary condition is that there exist, separately, a regular element and an M-regular one. In the local case, this is sufficient as well. Indeed, by Theorem A.3.6, the divisors of zero in M are the union of its associated primes (and similarly for A). If M and A have regular elements, none of these associated primes is the whole $\mathcal{M}$, hence by prime avoidance there is an element of $\mathcal{M}$ that does not lie in any of these associated primes. Such element is regular and M-regular at the same time.

The second change of rings theorem, Theorem 5.2.11, can also be used to link the projective dimension of an A-module M to that of the $A[x]$-module $M[x]$.

Proposition 5.2.15. *Let A be a ring, M an A-module. Then*

$$\operatorname{pd}_A M = \operatorname{pd}_{A[x]} M[x].$$

Proof. The indeterminate $x \in A[x]$ clearly satisfies the hypothesis of Theorem 5.2.11, hence $\operatorname{pd}_{A[x]} M[x] \geq \operatorname{pd}_A M$.

For the converse inequality, notice that a projective resolution $P_\bullet \to M$ as A-modules gives a projective resolution $P_\bullet[x] \to M[x]$ as $A[x]$-modules.
$\square$

5.3. Injective modules

We now turn to the study of injective modules. Recall from Baer's criterion (Theorem 2.6.12) that an A-module M is injective if and only if every morphism $I \to M$, where I is an ideal of A, can be extended to the whole of A. In particular, when A is a PID, an A-module is injective if and only if it is divisible.

An immediate consequence of this is that the category Ab has enough injectives, and we have used this to derive the more general fact that Mod_A has enough injectives for all rings A (Corollary 2.6.18).

Recall from Remark 2.6.2 that a direct product of injective modules is itself injective. A consequence of Baer's criterion is the following surprising connection between direct sums of injective modules and the Noetherian property. The result appeared independently in [**Bas59**] and [**Pap59**]; our proof is taken from [**Cla15**, Section 8.9].

Theorem 5.3.1 (Bass–Papp). *Let A be a ring. Then the following are equivalent:*

  (i) *A is Noetherian,*

 (ii) *any direct limit of injective A-modules is injective, and*

(iii) *the countable direct sum of injective A-modules is injective.*

Proof. Assume that A is Noetherian and let $\{M_i\}$ be a direct system of A-modules, indexed over a directed set, with morphisms $f_{ij}\colon M_i \to M_j$ for $i \leq j$. Let $M = \varinjlim M_i$, with maps $f_i\colon M_i \to M$. To prove (refbass-papp direct limit) by Theorem 2.6.12, we only need to consider a homomorphism $g\colon I \to M$, where $I \subset A$ is an ideal.

Since A is Noetherian, I is finitely generated, hence $g(I)$ is contained in $f_i(M_i)$ for some i. Moreover, there is a finitely generated submodule of M_i, call it N such that $g(I) = f_i(N)$. Letting K be the kernel of $N \to g(I)$, we obtain an exact sequence

$$0 \longrightarrow K \longrightarrow N \longrightarrow g(I) \longrightarrow 0 .$$

Since $K \subset N$ and A is Noetherian, K is finitely generated. Since $f_i(K) = 0$ and the set is directed, there exists an index $j \geq i$ such that $f_{ij}(K) = 0$. If we let $N' = f_{ij}(N)$, it follows that $f_j\colon N' \to g(I)$ is an isomorphism. By inverting it, we get a homomorphism

$$I \to g(I) \to N' \subset M_j,$$

which we can extend to a map $h\colon A \to M_j$. The composition $f_j \circ h\colon A \to M$ is then an extension of g. Since this holds for all morphisms g, M is injective, or in other words (i) implies (ii).

Since (iii) is a special case of (ii), it remains to prove that (iii) implies (i), or better its contrapositive. Assuming that A is not Noetherian, we can find an infinite increasing chain of ideals $I_0 \subset I_1 \subset \cdots$. For each ideal I_k, choose an injective module J_k with an injection $A/I_k \to J_k$.

Consider the ideal $I := \bigcup_k I_k$, and the morphism

$$f_k\colon I \to I/I_k \to J_k.$$

All these morphisms together give a map $f\colon I \to \prod_k J_k$. For an element $a \in I$, $f_k(a) = 0$ except for finitely many k, hence $f(I)$ actually lies in $J := \bigoplus_k J_k$. We claim that f does not extend to a homomorphism $\tilde{f}\colon A \to J$, hence J is not injective.

In fact, assume that the extension $\tilde{f}$ exists. Then

$$f(a) = \tilde{f}(a \cdot 1) = a \cdot \tilde{f}(1)$$

for all $a \in I$. Since $\tilde{f}(1)$ has only finitely many nonzero coordinates, the image $f(I)$ lies in a finite direct sum of the J_k, which is a contradiction since $f_k(a) \neq 0$ for $a \in I \setminus I_k$. $\qquad\square$

Corollary 5.3.2. *Let A be a Noetherian ring, M an A-module. Then M has a maximal injective submodule, and we can decompose $M = I \oplus N$, where N has no nontrivial injective submodules.*

Proof. The union of an increasing chain of injective submodules of M is injective by Theorem 5.3.1, hence the first claim follows by Zorn's lemma. If $I \subset M$ is a maximal injective submodule, then the inclusion $I \to M$ splits, and this implies the second claim.   □

Theorem 5.3.1 should be compared with the following result from [**Cha60**], which looks exactly like its dual, even though the proof is very different.

Theorem (Chase). *Let A be a ring. Then the following are equivalent:*

(i) *A is Artinian,*

(ii) *any inverse limit of projective A-modules is projective, and*

(iii) *the direct product of projective A-modules is projective.*

In the rest of the section, we are going to study A-modules by means of their injective hull. Recall from Definition 2.7.1 that an inclusion of A-modules $M \subset E$ is called an essential extension if $M \cap M' \neq 0$ for all nonzero submodules $M' \subset E$. If moreover E is injective, E is called the injective hull of M. By Theorem 2.7.7, injective hulls exist in Mod_A, and they are unique up to an isomorphism that is the identity on M. The injective hull of M is denoted $E(M)$, or $E_A(M)$ if multiple rings are involved.

Example 5.3.3. Let A be an integral domain, $k = \mathcal{F}(A)$ its fraction field. Then k is injective as an A-module. To see this, we use Theorem 2.6.12. Let $I \subset A$ be an ideal and $f: I \to k$ a homomorphism of A-modules. For any nonzero $a \in I$, let $e = f(a)/a \in k$. It is easy to see that e is independent of the choice of a, so we can extend f to the whole of A by writing $f(1) = e$.

Moreover, k is the injective hull of A, regarded as an A-module over itself. Since k is injective, we know that $A \subset E(A) \subset k$. By Remark 2.6.14, every injective A-module is divisible, so $E(A)$ must be the whole fraction field k.

For injective modules over Noetherian rings, there is a nice structure theory due to Matlis [**Mat58**], based on primary decomposition (see Section A.3 to review).

Definition 5.3.4. Let M be an A-module. If M cannot be written as a direct sum $M = N_1 \oplus N_2$ with $N_1, N_2 \neq 0$, then we say that M is *indecomposable*.

The following result is the reason that primary decomposition appears here.

Proposition 5.3.5. *Let I be an ideal of the ring A, and assume that I has an irredundant decomposition $I = J_1 \cap \cdots \cap J_k$. Assume that $E(A/J_i)$ is* *indecomposable for each i. Then there is a natural isomorphism*

$$E(A/I) \cong E(A/J_1) \oplus \cdots \oplus E(A/J_k).$$

Proof. To start with, the natural map

$$\iota \colon \frac{A}{I} \to \frac{A}{J_1} \oplus \cdots \oplus \frac{A}{J_k}$$

is injective, and this allows us to identify A/I with a submodule of $E(A/J_1) \oplus \cdots \oplus E(A/J_k)$. Since the latter is injective, we only have to check that this is an essential extension.

Since the decomposition is irredundant, for every $i = 1, \ldots, k$ we can find $a \in A$ such that $a \in J_t$ for $t \neq i$ but $a \notin J_i$. It follows that the class of a is nonzero in A/I, and $\iota(\overline{a}) \in A/J_i$. In other words, $A/I \cap A/J_i \neq 0$ when both are regarded as submodules of $E(A/J_1) \oplus \cdots \oplus E(A/J_k)$.

By Proposition 2.7.10, we can regard $E(A/I \cap A/J_i)$ as a submodule of $E(A/J_i)$. The inclusion splits because $E(A/I \cap A/J_i)$ is injective, and the hypothesis that $E(A/J_i)$ is indecomposable implies that $E(A/I \cap A/J_i) = E(A/J_i)$.

Now choose any nonzero element

$$m = (m_1, \ldots, m_k) \in E(A/J_1) \oplus \cdots \oplus E(A/J_k),$$

and assume without loss of generality that $m_1 \neq 0$. Since $E(A/J_1)$ is an essential extension of $A/I \cap A/J_1$, the intersection

$$(A/I \cap A/J_1) \cap Am_1 \neq 0.$$

In other words, $0 \neq am_1 \in A/I$ for some $a \in A$.

We now repeat the procedure with am in place of m, for the first index i such that $am_i \neq 0$, and so on. At the end, we find some $b \in A$ such that $0 \neq bm \in A/I$. It follows that all submodules of $E(A/J_1) \oplus \cdots \oplus E(A/J_k)$ meet A/I—in other words, $E(A/J_1) \oplus \cdots \oplus E(A/J_k)$ is an essential extension of A/I. $\qquad\square$

In view of this result, it is natural to ask for which ideals $J \subset A$, the injective hull $E(A/J)$ is indecomposable. We can answer in slightly more generality.

Let E be an injective A-module. For all $e \in E$, we have a natural monomorphism $A/\operatorname{Ann}(e) \to E$, which we can extend to an embedding $E(A/\operatorname{Ann}(e)) \to E$ by Proposition 2.7.10. If moreover E is indecomposable and $e \neq 0$, we must have the equality $E(A/\operatorname{Ann}(e)) = E$. It follows that all indecomposable injective A-modules appear as the injective hull of a module of the form A/J, where J is an ideal (namely, $J = \operatorname{Ann}(e)$ for any $e \neq 0$). This construction is the basis of the following result.

Theorem 5.3.6. *Let A be a ring, E an A-module. Then E is injective and indecomposable if, and only if, $E \cong E(A/J)$ for some irreducible ideal J.*

Proof. Assume that J is an irreducible ideal. This can be rephrased by saying that any two nontrivial submodules $M_1, M_2 \subset A/J$ have nonzero intersection. If $E(A/J) = N_1 \oplus N_2$ is decomposable, we must have $A/J \cap N_i = 0$ for $i = 1$ or 2, which contradicts the fact that $E(A/J)$ is an essential extension of A/J.

To prove the converse, let E be an indecomposable injective A-module. By the above discussion, we can take $E = E(A/J)$ for some ideal $J \subset A$, and we only need to prove that J is irreducible. If not, we can write $J = J_1 \cap J_2$, where $J \subsetneq J_i$ for $i = 1, 2$, and the decomposition is irredundant.

As in the proof of Proposition 5.3.5, consider A/J as a submodule of $E(A/J_1) \oplus E(A/J_2)$. Following the same proof, we have $A/J \cap A/J_1 \neq 0$. By Lemma 2.7.2, the map $E(A/J \cap A/J_1) \to E(A/J) = E$ is injective, hence an isomorphism because E is indecomposable. Similarly, the map $E(A/J \cap A/J_1) \to E(A/J_1)$ is injective, which is a contradiction since already $A/J \to A/J_1$ is not injective. $\qquad\square$

All results so far apply to any ring. In the Noetherian case, we have more precise results thanks to primary decomposition.

Proposition 5.3.7. *Let A be a Noetherian ring, E an injective A-module. Then E is the direct sum of indecomposable, injective modules.*

Proof. By Zorn's lemma (check that it applies!), there is a maximal submodule $F \subset E$ which is the direct sum of indecomposable injective modules; moreover it is injective by Theorem 5.3.1. If $F \subsetneq E$, write $E = F \oplus G$. If $g \in G$ is any nonzero element, we have an embedding $E(A/\operatorname{Ann}(g)) \hookrightarrow G$ by Lemma 2.7.2.

Let

$$\operatorname{Ann}(g) = Q_1 \cap \cdots \cap Q_k$$

be an irreducible decomposition of $\operatorname{Ann}(g)$. Since each Q_i is irreducible, $E(A/Q_i)$ is indecomposable by Theorem 5.3.6. By Proposition 5.3.5,

$$E(A/\operatorname{Ann}(g)) = E(A/Q_1) \oplus \cdots \oplus E(A/Q_k),$$

hence $F \oplus E(A/\operatorname{Ann}(g))$ is a larger submodule of E which is the direct sum of indecomposable injective modules. $\qquad\square$

In the Noetherian case, we can also sharpen Theorem 5.3.6. An injective indecomposable A-module has the form $E(A/Q)$, where $Q \subset A$ is irreducible, hence primary. We can reduce to the prime case thanks to the following lemma.

Lemma 5.3.8. *Let A be a Noetherian ring, Q a primary ideal of A, $P :=$ $\sqrt{Q}$. Then $E(A/Q) \cong E(A/P)$.*

Proof. Let n be the smallest integer such that $P^n \subset Q$ - we can assume that $n > 1$. Choose any $a \in P^{n-1} \setminus Q$, and let $\overline{a} \neq 0$ be its image in $E(A/Q)$. Since Q is primary, $\operatorname{Ann}(\overline{a}) = P$. The conclusion follows from the discussion before Theorem 5.3.6. $\square$

In conclusion, we can completely classify injective modules over a Noetherian ring.

Theorem 5.3.9. *Let A be a Noetherian ring. Then every injective A-module is the direct sum of indecomposable injective modules. Every indecomposable injective module has the form $E(A/P)$ for some prime ideal $P \subset A$, and all such A-modules are pairwise nonisomorphic.*

Proof. We have already proved all statements but the last. Let P, Q be two prime ideals, and assume that $E = E(A/P) \cong E(A/Q)$. Inside E, we have $A/P \cap A/Q \neq 0$. (Why?) As in the previous proof, for any nonzero $m \in A/P \cap A/Q$ we have $P = \operatorname{Ann}(m) = Q$. $\square$

Remark 5.3.10. Keep the assumption that A is Noetherian. Let $P \subset A$ be a prime, $E = E(A/P)$, and take a nonzero $e \in E$. Letting $J = \operatorname{Ann}(e)$, we know that $E = E(A/J)$, so J is primary by Proposition 5.3.5 and primary decomposition. Letting $Q = \sqrt{J}$, we have $E \cong E(A/Q)$ by Lemma 5.3.8, so ultimately $Q = P$ by Theorem 5.3.9. We deduce that P is the only associated prime of E. Moreover, there exists a k such that $P^k \subset J$ since P is finitely generated. It follows that $P^k e = 0$; in other words, every element of $E(A/P)$ is annihilated by a power of P.

Example 5.3.11. Every injective $\mathbb{Z}$-module is the direct sum of modules of the form $E(\mathbb{Z}/P)$ for some prime ideal P. Let us compute these modules.

When $P = 0$, we get the $\mathbb{Z}$-module $E(\mathbb{Z}) = \mathbb{Q}$. When $P = (p)$ is generated by a prime number, we notice that we can embed

$$\frac{\mathbb{Z}}{p\mathbb{Z}} \cong \frac{\frac{1}{p}\mathbb{Z}}{\mathbb{Z}} \subset \frac{\mathbb{Q}}{\mathbb{Z}}.$$

The subgroup $\frac{1}{p}\mathbb{Z}/\mathbb{Z}$ is divisible by all primes $q \neq p$. To get the injective hull, we only need to guarantee divisibility by powers of p. It follows that

$$E(\mathbb{Z}/p\mathbb{Z}) \cong \frac{\mathbb{Z}[p^{-1}]}{\mathbb{Z}},$$

where $\mathbb{Z}[p^{-1}]$ is the localization of $\mathbb{Z}$ at the powers of p.

5.4. Injective dimension

In this section we introduce the notion of injective dimension of a module, which is the obvious analogue of the projective dimension that we studied in Section 5.2. We start with the analogue of Proposition 5.2.1, whose proof follows immediately from dualization.

Proposition 5.4.1. *Let $\mathcal{A}$ be an Abelian category with enough injectives, and let A be an object of $\mathcal{A}$. The following are equivalent:*

  (i) *there exists an injective resolution*

$$0 \longrightarrow A \longrightarrow I_0 \longrightarrow I_1 \longrightarrow \cdots \longrightarrow I_n \longrightarrow 0$$

  of length n, and

  (ii) $\mathrm{Ext}^i(B, A) = 0$ *for all $i > n$ and all $B \in \mathcal{A}$.*

Definition 5.4.2. Let $\mathcal{A}$ be an Abelian category with enough injectives, $X \in \mathcal{A}$. The minimal length of an injective resolution of X will be called the *injective dimension* of X, and denoted $\mathrm{id}\, X$. If X does not admit any injective resolution of finite length, we set $\mathrm{id}\, X = \infty$. When $\mathcal{A} = \mathrm{Mod}_A$ is the category of modules over the ring A, we sometimes denote injective dimension by id_A for clarity.

By the above result, $\mathrm{id}\, X$ is the also minimum n (if any) such that $\mathrm{Ext}^i(-, X) = 0$ for $i > n$. As in the case of projective dimension, we want to establish some change of ring theorems. Before doing so, we notice one simple consequence of Baer's criterion.

Proposition 5.4.3. *Let A be a ring, M an A-module. Then $\mathrm{id}\, M$ is the minimum n (if any) such that $\mathrm{Ext}^i(A/I, M) = 0$ for $i > n$ and for all ideals $I \subset A$.*

Proof. First assume that $\mathrm{Ext}^1(A/I, M) = 0$ for all ideals I. Then all homomorphisms $I \to M$ can be extended to A by the long exact sequence of Ext, hence M is injective by Theorem 2.6.12.

Assuming $\mathrm{Ext}^{n+1}(A/I, M) = 0$ for all ideals I, take a truncated injective resolution

$$0 \longrightarrow M \longrightarrow I_0 \longrightarrow \cdots \longrightarrow I_{n-1} \longrightarrow N \longrightarrow 0,$$

where $N := \mathrm{coker}\, I_{n-2} \to I_{n-1}$. Then $\mathrm{Ext}^1(A/I, N) = 0$ for all ideals I by Proposition 3.3.2, hence N is injective and $\mathrm{id}\, M \leq n$. $\square$

For local Noetherian rings, we can give a sharper version.

Theorem 5.4.4 (Kaplansky). *Let A be a local Noetherian ring with maximal ideal $\mathcal{M}$, $k = A/\mathcal{M}$, M a finitely generated A-module. Then $\mathrm{id}_A M \leq n$ if and only if $\mathrm{Ext}^i(k, M) = 0$ for all $i > n$.*

Lemma 5.4.5. *Let $A, \mathcal{M}$ be a local Noetherian ring with $k = A/\mathcal{M}$, M a finitely generated A-module, and P a prime ideal of A distinct from $\mathcal{M}$. Assume that $\operatorname{Ext}_A^{i+1}(A/Q, M) = 0$ for all primes Q properly containing P; then $\operatorname{Ext}_A^i(A/P, M) = 0$.*

Proof. Choose any element $a \in \mathcal{M} \setminus P$, and denote $B = A/P$. Notice that $\overline{a}$ is not a zero divisor on B, hence we get the exact sequence

$$0 \longrightarrow B \xrightarrow{\cdot a} B \longrightarrow \frac{B}{\overline{a}B} \longrightarrow 0.$$

Applying the functor $\operatorname{Hom}_A(-, M)$ to it, we get a piece of the associated long exact sequence

$$\operatorname{Ext}_A^i(B, M) \xrightarrow{\cdot a} \operatorname{Ext}_A^i(B, M) \longrightarrow \operatorname{Ext}_A^{i+1}\left(\tfrac{B}{\overline{a}B}, M\right).$$

Since $B/\overline{a}B$ is a finitely generated B-module, by Theorem A.3.7 we get a chain of submodules having quotients isomorphic to B/Q_j' for some primes Q_j' of B. These, in turn, are isomorphic to A/Q_j, where Q_j is the prime of A corresponding to Q_j' in B. By hypothesis, $\operatorname{Ext}_A^{i+1}(A/Q_j, M) = \operatorname{Ext}_A^{i+1}(B/Q_j', M) = 0$ for all j. By induction, $\operatorname{Ext}_A^{i+1}(B/\overline{a}B, M) = 0$.

It follows that multiplication by $\overline{a}$ is surjective on $\operatorname{Ext}_A^i(B, M)$. Since this is finitely generated, it is 0 by Nakayama's lemma. $\square$

Proof of Theorem 5.4.4. By Theorem A.8.5, A has finite dimension. We can then apply the lemma recursively to prove that $\operatorname{Ext}^i(A/P, M) = 0$ for all primes $P \subset A$, for $i > n$. If N is a finitely generated A-module, we can find a chain of submodules of N having quotients isomorphic to A/P for some prime P. By induction, it follows that $\operatorname{Ext}^i(N, M) = 0$ for all finitely generated A-modules N and all $i > n$—in particular, taking $N = A/I$, the conclusion follows from Proposition 5.4.3. $\square$

We end this section with the change of ring theorems for injective dimension. Notice that they are similar, but not identical, to their projective counterparts. For one thing, the general change of rings theorem involves the *projective* dimension of B.

Theorem 5.4.6. *Let $f \colon A \to B$ a homomorphism of rings, and N a B-module. Then*

$$\operatorname{id}_A N \leq \operatorname{pd}_A B + \operatorname{id}_B N.$$

Proof. Apply the base change spectral sequence of Theorem 4.5.7. $\square$

As in the projective case, we specialize this to the important case of quotient rings.

Theorem 5.4.7 (First change of rings theorem). *Let A be a ring, $a \in A$ a regular element, and let M be an $A/(a)$-module. Assume that $\mathrm{id}_{A/(a)} M < \infty$. Then*

$$\mathrm{id}_A M = 1 + \mathrm{id}_{A/(a)} M.$$

Proof. We prove the result by induction on $\mathrm{id}_{A/(a)} M$. First, any $A/(a)$-module cannot be injective as an A-module, for instance, 1 is not divisible by a. If $\mathrm{id}_{A/(a)} M = 0$, then $\mathrm{id}_A M \leq 1$ by Theorem 5.4.6, and in fact $\mathrm{id}_A M = 1$ because M is not injective over A.

For the inductive step, we consider an exact sequence of $A/(a)$-modules

$$0 \longrightarrow M \longrightarrow I \longrightarrow C \longrightarrow 0\,,$$

where I is injective. Then $\mathrm{id}_{A/(a)} C = \mathrm{id}_{A/(a)} M - 1$.

By the previous step, $\mathrm{id}_A I = 1$. If we regard the same exact sequence as a sequence of A-modules and consider the associated long exact sequence of $\mathrm{Ext}(N, -)$ groups for all A-modules N, we conclude that either $\mathrm{id}_A C = \mathrm{id}_A M - 1$—in which case we are done—or $\mathrm{id}_A M = \mathrm{id}_A C = 1$ (in which case, $\mathrm{id}_{A/(a)} C = 0$ and $\mathrm{id}_{A/(a)} M = 1$).

To exclude the latter case, consider an exact sequence of A-modules

$$0 \longrightarrow M \longrightarrow I' \longrightarrow C' \longrightarrow 0\,,$$

with I' and C' injective. We can apply the functor $F = \mathrm{Hom}_A(A/(a), -)$ to get the long exact sequence

$$0 \longrightarrow F(M) \longrightarrow F(I') \longrightarrow F(C') \longrightarrow \mathrm{Ext}_A^1(A/(a), M) \longrightarrow 0,$$

which is actually a sequence of $A/(a)$-modules. Since F preserves injectives (see the proof of Theorem 4.5.5), $F(I')$ and $F(C')$ are injective over $A/(a)$, hence so is $\mathrm{Ext}_A^1(A/(a), M)$.

By Example 3.4.11(c), $\mathrm{Ext}_A^1(A/(a), M) = M/aM = M$, which is a contradiction, since $\mathrm{id}_{A/(a)} M = 1$. $\square$

For the second theorem, we need a lemma that is of independent interest. Recall that $(0 :_M a)$ denotes the elements of the A-module M annihilated by a. This is a module over $A/(a)$ that can be identified with $\mathrm{Hom}_A(A/(a), M)$.

Lemma 5.4.8. *Let A be a ring, M an A-module and $a \in A$ such that $M = aM$. Then*

$$\mathrm{id}_{A/(a)}(0 :_M a) \leq \mathrm{id}_A M.$$

Proof. We can assume that $\mathrm{id}_A M < \infty$, and so prove the result by induction on it. For the case $\mathrm{id}_A M = 0$, we notice that the functor $\mathrm{Hom}_A(A/(a), -)$

preserves injectives. For the inductive step, take an exact sequence of A-modules

$$0 \longrightarrow M \xrightarrow{f} I \xrightarrow{g} C \longrightarrow 0,$$

with I injective, so that $\operatorname{id}_A C = \operatorname{id}_A M - 1$. We claim that

$$0 \longrightarrow (0 :_M a) \longrightarrow (0 :_I a) \longrightarrow (0 :_C a) \longrightarrow 0$$

is exact as well. We can check exactness on the right directly: given $c \in C$ such that $ac = 0$, choose $x \in I$ such that $g(x) = c$. Then $ax \in \ker g = \operatorname{im} f$, so we can write $ax = f(am)$ for some $m \in M$. Then $x - f(m) \in (0 :_I a)$ and maps to c, as required.

Since $(0 :_I a)$ is injective over $A/(a)$, we have

(5.4.1) $$\operatorname{id}_{A/(a)}(0 :_C a) = \operatorname{id}_{A/(a)}(0 :_M a) - 1.$$

Since I is divisible, $aC = C$. By induction, C satisfies the conclusion of the lemma, and hence so does M. $\qquad\square$

Theorem 5.4.9 (Second change of rings theorem). *Let A be a ring, M be an A-module, and let a be weakly regular and weakly M-regular. Then either*

(i) *M is injective and $M/aM = 0$, or*

(ii) *$\operatorname{id}_A M \geq 1 + \operatorname{id}_{A/(a)} \frac{M}{aM}$.*

Proof. We can assume that $\operatorname{id}_A M$ is finite. If $\operatorname{id}_A M = 0$, M is injective. Moreover, the exact sequence

$$0 \longrightarrow M \longrightarrow M \longrightarrow \frac{M}{aM} \longrightarrow 0$$

splits, hence M/aM is a direct summand of M. But multiplication by a is injective on M, hence $M/aM = 0$.

If $\operatorname{id}_A M > 0$, consider an exact sequence of A-modules

$$0 \longrightarrow M \longrightarrow I \longrightarrow C \longrightarrow 0 ,$$

with I injective, so that $\operatorname{id}_A C = \operatorname{id}_A M - 1$. We apply the functor $F = \operatorname{Hom}_A(A/(a), -)$ to get a long exact sequence. Using Example 3.4.11(c) and the fact that a is not a zero divisor on M, this reads

$$0 \longrightarrow \operatorname{Hom}_A(A/(a), I) \longrightarrow \operatorname{Hom}_A(A/(a), C) \longrightarrow \frac{M}{aM} \longrightarrow 0 .$$

Since F preserves injectives, $\operatorname{Hom}_A(A/(a), I)$ is injective over $A/(a)$; moreover $\operatorname{id}_{A/(a)} \operatorname{Hom}_A(A/(a), C) \leq \operatorname{id}_A C$ by Lemma 5.4.8. It follows that

$$\operatorname{id}_{A/(a)} \frac{M}{aM} = \operatorname{id}_{A/(a)} \operatorname{Hom}_A(A/(a), C) \leq \operatorname{id}_A C = \operatorname{id}_A M - 1.$$

$$\square$$

Theorem 5.4.10 (Third change of rings theorem). *Let A be a Noetherian local ring with maximal ideal $\mathcal{M}$, M a finitely generated A-module, and let $a \in A$ be regular and M-regular. Then*

$$\operatorname{id}_A M = \operatorname{id}_A \frac{M}{aM} = 1 + \operatorname{id}_{A/(a)} \frac{M}{aM}.$$

Proof. For the first equality, we consider the exact sequence

$$0 \longrightarrow M \longrightarrow M \longrightarrow \frac{M}{aM} \longrightarrow 0$$

given by multiplication by a. Considering the associated long exact sequence of $\operatorname{Ext}(A/I, -)$ groups, for ideals $I \subset A$, we conclude that $\operatorname{id}_A M = \operatorname{id}_A \frac{M}{aM}$ unless $\operatorname{id}_A M = n$ is finite and

$$\operatorname{Ext}_A^n \left(\tfrac{A}{I}, M \right) \xrightarrow{\ a \cdot\ } \operatorname{Ext}_A^n \left(\tfrac{A}{I}, M \right)$$

is surjective. Since A/I and M are both finitely generated and A is Noetherian, the group $\operatorname{Ext}_A^n(A/I, M)$ is finitely generated as well (why?), hence it is 0 by Nakayama's lemma, a contradiction.

The second equality is Theorem 5.4.7, assuming $\operatorname{id}_{A/(a)} \frac{M}{aM}$ is finite. Otherwise, the equality holds anyway: $\operatorname{id}_A M = \infty$ as well by Theorem 5.4.9. $\square$

5.5. Global dimension of rings

As the reader may imagine, projective and injective dimension are strictly related, and they can be combined to yield an invariant of rings.

Definition 5.5.1. Let A be a ring. The number

$$\sup_{M \in \operatorname{Mod}_A} \operatorname{pd}_A M = \sup_{N \in \operatorname{Mod}_A} \operatorname{id}_A N =$$
$$\sup_{M,N \in \operatorname{Mod}_A} \{ n \mid \operatorname{Ext}_A^n(M, N) \neq 0 \}$$

is called the *global dimension* (or *homological dimension*) of A, and denoted $\operatorname{gl.dim} A$. As it happens for the injective and projective dimensions, this number can also be ∞.

We can easily translate some results about projective and injective dimension. From Kaplansky's theorem (Theorem 5.4.4) we immediately get

Proposition 5.5.2. *Let $A, \mathcal{M}$ be a local Noetherian ring, $k = A/\mathcal{M}$. Then $\operatorname{gl.dim} A = \operatorname{pd}_A k$.*

More generally, Proposition 5.4.3 gives us

Proposition 5.5.3. *For any ring A,*

$$\operatorname{gl.dim} A = \sup_{I \subset A} \operatorname{pd}_A(A/I).$$

We can also relate the global dimension of A to that of its localizations.

Theorem 5.5.4. *For any ring A,*

$$\mathrm{gl.\,dim}\, A \geq \sup_{\substack{P \subset A \\ prime}} \mathrm{gl.\,dim}\, A_P.$$

If moreover A is Noetherian, then we have the equalities

$$\mathrm{gl.\,dim}\, A = \sup_{\substack{P \subset A \\ prime}} \mathrm{gl.\,dim}\, A_P = \sup_{\substack{\mathcal{M} \subset A \\ maximal}} \mathrm{gl.\,dim}\, A_{\mathcal{M}}.$$

Proof. For the first claim, we can assume that $\mathrm{gl.\,dim}\, A < \infty$. Consider any A_P-module M. Regarding M as an A-module, we have $M_P = M$. By the remark before Proposition 5.2.7 and the fact that $\mathrm{pd}_A M < \infty$, we get $\mathrm{pd}_A M \geq \mathrm{pd}_{A_P} M$. We conclude that $\mathrm{gl.\,dim}\, A \geq \mathrm{gl.\,dim}\, A_P$.

Then, assume that A is Noetherian. By Proposition 5.5.3, $\mathrm{gl.\,dim}\, A$ can be computed by considering only cyclic modules A/I. By Proposition 5.2.7, for all finitely generated A-modules M there is a maximal ideal $\mathcal{M}$ such that $\mathrm{pd}_A M = \mathrm{pd}_{A_{\mathcal{M}}} M_{\mathcal{M}}$, and we get the desired conclusion. $\square$

We can develop the same notions in the graded context. Let A be a graded ring, Mod_A its category of graded modules, and Mod_A^u the category of modules over A, when regarded as an ungraded ring. As usual, denote $U \colon \mathrm{Mod}_A \to \mathrm{Mod}_A^u$ the forgetful functor.

Definition 5.5.5. Let A be a graded ring. The number

$$\sup_{M,N \in \mathrm{Mod}_A} \{ n \mid \mathrm{Ext}_A^n(M,N) \neq 0 \}$$

is called the *graded global dimension* of A, and denoted $\mathrm{gr.\,gl.\,dim}\, A$.

Notice that the supremum is taken among graded modules, and that Ext is computed in the category Mod_A of graded modules. It is a simple verification, that, as it happens in the ungraded case,

$$\mathrm{gr.\,gl.\,dim}\, A = \sup_{M \in \mathrm{Mod}_A} \mathrm{pd}_A M.$$

Notice that by Proposition 5.2.5, the projective dimension $\mathrm{pd}_A M$ is the same, whether it is computed in Mod_A or in Mod_A^u. A priori, the ungraded global dimension $\mathrm{gl.\,dim}\, A$ is the supremum of the same quantity, computed over a larger family of modules, and this immediately implies the following fact.

Proposition 5.5.6. *Let A be a graded ring. Then $\mathrm{gr.\,gl.\,dim}\, A \leq \mathrm{gl.\,dim}\, A$.*

We will shortly see in Theorem 5.5.12 that there is a closer relation, and in fact $\mathrm{gl.\,dim}\, A \leq 1 + \mathrm{gr.\,gl.\,dim}\, A$. Before doing so, we need to relate the global dimension of a ring A to that of the polynomial ring $A[x]$.

Theorem 5.5.7. *Let A be a ring. Then* $\mathrm{gl.\,dim}\, A[x] = 1 + \mathrm{gl.\,dim}\, A$. *If moreover A is graded over $\mathbb{Z}$ and $A[x]$ is graded with $\deg x = 1$, then* $\mathrm{gr.\,gl.\,dim}\, A[x] = 1 + \mathrm{gr.\,gl.\,dim}\, A$.

Proof. We prove the case of ungraded rings, and leave it to the reader to check that all arguments generalize to the graded case. If $\mathrm{gl.\,dim}\, A = \infty$, then $\mathrm{gl.\,dim}\, A[x] = \infty$ as well by Proposition 5.2.15, so we can assume that $\mathrm{gl.\,dim}\, A$ is finite.

By Theorem 5.2.10, $\mathrm{pd}_{A[x]} M = 1 + \mathrm{pd}_A M$ for any A-module M such that $\mathrm{pd}_A M < \infty$, hence $\mathrm{gl.\,dim}\, A[x] \geq 1 + \mathrm{gl.\,dim}\, A$.

To get the other inequality, take an $A[x]$-module M, and denote M_A the module M regarded as an A-module. Then we have an exact sequence of $A[x]$-modules

$$(5.5.1) \qquad 0 \longrightarrow M_A[x] \xrightarrow{\ f\ } M_A[x] \xrightarrow{\ g\ } M \longrightarrow 0.$$

To define the maps f and g, it is simpler to think of $M_A[x]$ as $A[x] \otimes_A M_A$. Then

$$f(a \otimes m) := ax \otimes m - a \otimes xm,$$

and $g(a \otimes m) := a \cdot m$, where multiplication by scalars in $A[x]$ happens in M. It is immediate that (5.5.1) is a complex and that g is surjective so we only need to prove that f is injective and $\mathrm{im}\, f = \ker g$. To see that f is injective, write an element $p \neq 0$ of $A[x] \otimes M_A$ as a sum

$$p = x^k \otimes m_k + \cdots + x \otimes m_1 + 1 \otimes m_0,$$

where $m_k \neq 0$, and notice that the only term of degree $k+1$ in $f(p)$ is $x^{k+1} \otimes m_k \neq 0$. We prove that $g(p) = 0$ implies $p \in \mathrm{im}\, f$ by induction on k. If $k = 0$, then $p = 1 \otimes m_0$, so that $g(p) = m_0$. If $g(p) = 0$, then $p = 0$ as well. For $k > 0$, notice that we can find a polynomial $q \in \mathrm{im}\, f$ such that $p - q$ has lower degree, namely $q = f(x^{k-1} \otimes m_k)$. By induction, $p - q \in \mathrm{im}\, f$, so $p \in \mathrm{im}\, f$ as well.

From the exactness of (5.5.1) and Proposition 5.2.15, we get the inequality

$$\mathrm{pd}_{A[x]} M \leq 1 + \mathrm{pd}_{A[x]} M_A[x] = 1 + \mathrm{pd}_A M_A \leq 1 + \mathrm{gl.\,dim}\, A,$$

and the conclusion follows by taking the supremum. $\qquad \square$

Corollary 5.5.8. *If k is a field, then*

$$\mathrm{gr.\,gl.\,dim}\, k[x_1, \ldots, x_n] = \mathrm{gl.\,dim}\, k[x_1, \ldots, x_n] = n.$$

Notice that a similar result holds for power series as well.

Proposition 5.5.9. *If k is a field, then* $\operatorname{gl.dim} k[[x_1, \ldots, x_n]] = n$.

Proof. The ring $A = k[[x_1, \ldots, x_n]]$ is local and Noetherian (Theorem A.2.2), hence $\operatorname{gl.dim} A = \operatorname{pd}_A k$ by Proposition 5.5.2. The equality $\operatorname{pd}_A k = n$ follows by induction from the first change of rings theorem, Theorem 5.2.10. □

The attentive reader has surely noticed that the global dimension of a ring has many properties in common with its Krull dimension. In particular, the above result should be compared with Theorem A.7.2. Moreover, the reader should compare Theorem 5.5.4 with the fact that for all rings A,

$$\dim A = \sup_{\substack{P \subset A \\ \text{prime}}} \dim A_P = \sup_{\substack{\mathcal{M} \subset A \\ \text{maximal}}} \dim A_{\mathcal{M}},$$

which is obvious from the definition. In the following example we compute some global dimensions, showing that the two notions do not always agree.

Example 5.5.10.

(a) Let $A = \mathbb{Z}/4\mathbb{Z}$. By Example 5.2.3(c), we know that $\operatorname{gl.dim} \mathbb{Z}/4\mathbb{Z} = \infty$. On the other hand, $\dim \mathbb{Z}/4\mathbb{Z} = 0$.

(b) More generally, we have $\dim A = 0$ for all Artinian rings. The condition $\operatorname{gl.dim} A = 0$ only holds if all A-modules are projective (in which case A is called *semisimple*). It is easy to see (check this!) that this implies that every A-module is a direct sum of simple modules, that is, modules that have no nontrivial submodules.

Applying this to the ring A itself, we find a decomposition $A = \bigoplus_{i \in I} A_i$ for some simple submodules A_i. The projection $\pi_i \colon A \to A_i$ is surjective, hence we can give A_i a ring structure so that π_i is a ring homomorphism. Since A_i is simple, it has no nontrivial ideals, hence it is a field.

Moreover, every ideal of A is principal (why?), hence A is Noetherian. Since an infinite direct sum of fields is not Noetherian, the index set I must be finite. In conclusion, every semisimple commutative ring is a finite product of fields. This is a very special case of a result for noncommutative rings, called the Artin–Wedderburn theorem [**Lam01**, Theorem 3.5].

In particular, every Artinian ring that is not a direct product of fields has $\dim A = 0$ but $\operatorname{gl.dim} A > 0$.

(c) On the other hand, let A be a principal ideal domain that is not a field. Then $\dim A = 1$, and by Example 5.2.3(a), $\operatorname{gl.dim} A = 1$ as well.

(d) More generally, let A be a Dedekind ring. By Proposition 5.5.3,

$$\text{gl.}\dim A = \sup_{I \subset A} \text{pd } A/I.$$

By Example 5.1.1(b), every ideal $I \subset A$ is projective, and the exact sequence

$$0 \longrightarrow I \longrightarrow A \longrightarrow A/I \longrightarrow 0$$

shows that $\text{pd } A/I \leq 1$, hence $\text{gl.}\dim A = 1 = \dim A$.

Remark 5.5.11. Let A be a Noetherian ring. Then either $\text{gl.}\dim A = \infty$ or else $\text{gl.}\dim A \leq \dim A$. By Theorem 5.5.4, it is enough to prove this in the case where A is local. We will do this in Theorem 8.2.1.

We end the section by using Theorem 5.5.7 to derive a connection between the graded and the ungraded global dimension. This result appears as [**NvO82**, Theorem II.8.2].

Theorem 5.5.12. *Let A be a ring graded over $\mathbb{Z}$. Then* $\text{gl.}\dim A \leq 1 +$ $\text{gr.gl.}\dim A$.

To prove it, we will need the concept of (de)homogeneization.

Definition 5.5.13. Let A be a $\mathbb{Z}$-graded ring, $M = \bigoplus_{i \in \mathbb{Z}} M_i$ a graded A-module. Given a nonzero element $m = \sum_{i=a}^{b} m_i$, where $m_i \in M_i$, and $m_b \neq 0$, the element $m^* \in M[x]$ defined as

$$m^* := \sum_{i=a}^{b} m_i x^{b-i}$$

is called the *external homogeneization* of m. Notice that m^* is always a homogeneous element of $M[x]$. If $N \subset M$ is an ungraded submodule, the graded module $N^* \subset M[x]$ generated by n^* for $n \in N$ is called the *external homogeneization* of N.

In the opposite direction, given a homogeneous element of degree d in $M[x]$, say $m = \sum_{i=a}^{b} m_i x^{d-i}$, where $m_i \in M_i$, the element $m_* \in M$ defined as

$$m_* := \sum_{i=a}^{b} m_i$$

is called the *dehomogeneization* of m. In other words, m_* is obtained by evaluating m at $x = 1$. Given a graded submodule $N \subset M[x]$, the set $N_* = \{0\} \cup \{n_* \mid n \in N\} \subset M$ is an ungraded submodule, called the *dehomogeneization* of N.

We can use these operations to relate graded $A[x]$-modules and ungraded A-modules. Define a functor $E\colon \operatorname{Mod}_{A[x]} \to \operatorname{Mod}_A^u$ by

$$E(M) = \frac{M}{(x-1)M} = \frac{A[x]}{(x-1)} \otimes_{A[x]} M.$$

Notice that $A[x]/(x-1) \cong A$, hence $E(M)$ is an A-module.

Lemma 5.5.14. *The functor E is exact and essentially surjective.*

Proof. The definition of E as a tensor product shows that E is right exact.

Assume that $N \subset M$ is a graded $A[x]$-submodule. Take $n \in N \cap (x-1)M$, say $n = (1-x)m$ with $m = \sum_{i=a}^{b} m_i$ and $m_i \in M_i$. Then

$$n = m_a + (m_{a+1} - xm_a) + \cdots + (m_b - xm_{b-1}) - xm_b.$$

Since N is graded, we find that $m_a \in N$ and recursively $m_i \in N$ for all $i = a, \dots, b$. In other words, $N \cap (x-1)M = (x-1)N$, and this implies that E is exact.

To check that E is essentially surjective, we start with an ungraded A-module M and we need to find a graded module N over $A[x]$ such that $E(N) \cong M$. Take a presentation

$$0 \longrightarrow K \longrightarrow F \longrightarrow M \longrightarrow 0,$$

where F is free. Then we claim that $N = F[x]/K^*$ will do. First notice that for $0 \neq m \in F$, $(m^*)_* = m$. Hence $(K^*)_* = K$, and the result follows by exactness of E and the exact sequence

$$0 \longrightarrow K^* \longrightarrow F[x] \longrightarrow N \longrightarrow 0. \qquad \square$$

Proof of Theorem 5.5.12. Let M be an ungraded A-module. By Lemma 5.5.14, we can find a graded $A[x]$-module N such that $E(N) \cong M$. Let P be a graded projective $A[x]$-module. By Corollary 2.6.8, P is a direct summand of a free $A[x]$-module, hence $E(P)$ is projective as well.

It follows that if $P_\bullet \to N$ is a projective resolution of N, $E(P_\bullet) \to E(M)$ is a projective resolution of M. In particular,

$$\operatorname{pd}_A M \le \operatorname{pd}_{A[x]} N \le \operatorname{gr.gl.dim} A[x] = 1 + \operatorname{gr.gl.dim} A,$$

where the last equality is Theorem 5.5.7. $\qquad \square$

5.6. Free resolutions

Corollary 5.5.8 has a reformulation in classical language, that was already known to Hilbert in 1890 [**Hil90**]. Hilbert formulated his celebrated result in terms of syzygies.

Definition 5.6.1. Let $A = k[x_1, \ldots, x_n]$ be a polynomial ring over a field, M a finitely generated A-module, and choose generators $m_1, \ldots, m_d$. A (first order) *syzygy*[1] between $m_1, \ldots, m_d$ is a relation

$$a_1 m_1 + \cdots + a_d m_d = 0$$

for some $a_1, \ldots, a_d \in A$. The syzygies between $m_1, \ldots, m_d$ form an A-module that we can recognize as the kernel of the map $A^d \to M$ that sends the i-th basis vector e_i to m_i. Recursively, k-th order syzygies are defined as syzygies on the module of $(k-1)$-th order syzygies.

The terminology is more complex than necessary—in our language, Hilbert was looking at presentations

$$F_k \longrightarrow \cdots \longrightarrow F_1 \longrightarrow F_0 \longrightarrow M \longrightarrow 0,$$

where each F_i is a finitely generated, free A-module. The celebrated theorem of Hilbert on syzygies then states:

Theorem 5.6.2 (Hilbert). *Every finitely generated module over the ring $A = k[x_1, \ldots, x_n]$ admits a finite free resolution*

$$0 \longrightarrow F_k \longrightarrow \cdots \longrightarrow F_1 \longrightarrow F_0 \longrightarrow M \longrightarrow 0,$$

of length $k \leq n$.

In other words, if one keeps taking syzygies, at most at the n-th step one ends up with a free A-module. Theorem 5.6.2 is almost equivalent to Corollary 5.5.8, but for one crucial and highly nontrivial fact. Corollary 5.5.8 guarantees that every A-module has a *projective* resolution of length at most n. The missing ingredient is the Quillen–Suslin theorem, stating that every finitely generated projective module over $k[x_1, \ldots, x_n]$ is in fact free. Of course, the original proof of Hilbert did not use neither the Quillen–Suslin theorem, nor the machinery of homological algebra, which did not exist at the time; in fact, Hilbert's theorem can be seen as a forerunner of that machinery. We present Hilbert's elementary proof in Exercise 19. In the next section, we are going to prove Hilbert's syzygy theorem using the theory of stably free modules.

Remark 5.6.3. We can reformulate a special case of Hilbert's theorem as follows. Let k be a field, and $f_1, \ldots, f_k$ be homogeneous polynomials in $A = k[x_1, \ldots, x_n]$, of degrees $d_1, \ldots, d_k$. Then

$$\mathrm{pd}_A A/(f_1, \ldots, f_k) \leq n.$$

---

[1]The word syzygy comes from the Greek, and in astronomy denotes an alignment of three celestial bodies. It was then borrowed for usage in mathematics to denote a more general linear relation.

In [**PS09**], Stillman conjectured that there exists a number p, depending only on $d_1, \ldots, d_k$ such that

$$\mathrm{pd}_A A/(f_1, \ldots, f_k) \leq p.$$

The crucial point is that p does *not* depend on the number n of variables. This is a much deeper result, that was proved by Ananya and Hochster in [**AH19**].

We end this section by studying a particularly simple way to construct free resolutions. Let A be a Noetherian ring, M a finitely generated A-module. By definition, we have a surjective homomorphism $f_0 \colon A^{n_0} \to M$, and we can choose n_0 to be minimal. If we let $K_0 = \ker f_0$, then K_0 is finitely generated as well, and we can repeat the construction with a homomorphism $f_1 \colon A^{n_1} \to K_0$, where n_1 is as small as possible. Continuing in this way, we obtain a (possibly infinite) free resolution of M

$$(5.6.1) \qquad \cdots \longrightarrow A^{n_k} \xrightarrow{f_k} \cdots \xrightarrow{f_1} A^{n_0} \xrightarrow{f_0} M \longrightarrow 0.$$

Definition 5.6.4. A free resolution as in (5.6.1), where each n_i is as small as possible, is called a *minimal* free resolution.

The main result on minimal free resolutions is the following characterization in the local case.

Proposition 5.6.5. *Let $A, \mathcal{M}$ be local Noetherian ring, $k = A/\mathcal{M}$ the residue field, M a finitely generated A-module. Take a free resolution of M as in (5.6.1). The following are equivalent:*

   (i) *the resolution is minimal;*

   (ii) *for each $i \geq 1$, $f_i(A^{n_i}) \subset \mathcal{M} A^{n_{i-1}}$, and $f_0(A^{n_0}) \subset \mathcal{M} M$; and*

   (iii) *for each $i \geq 1$, the matrix representing f_i has coefficients in $\mathcal{M}$, and $f_0(A^{n_0}) \subset \mathcal{M} M$.*

Proof. Denote $F_i = A^{n_i}$, and notice that $n_i = \dim_k F_i \otimes_A k = \dim_k F_i/\mathcal{M} F_i$. Let $K_i = \ker f_i$, and for simplicity of notation let $K_{-1} = M$. When constructing the minimal F_i, we take a surjective map $f_i \colon A^{n_i} \to K_{i-1}$, where n_i is minimal. By Nakayama's lemma, such map is surjective if and only if the map $k^{n_i} \to K_{i-1} \otimes_A k$ is, so the smallest n_i that we can choose is

$$n_i = \dim_k K_{i-1} \otimes_A k.$$

If n_i is chosen in this way and the map of vector spaces $k^{n_i} \to K_{i-1} \otimes_A k$ is surjective, it must be an isomorphism.

Rewrite this as an isomorphism $F_i/\mathcal{M}F_i \cong K_{i-1}/\mathcal{M}K_{i-1}$. It follows that
$$\ker(F_i \to F_{i-1}) = \ker(F_i \to K_{i-1}) \subset \mathcal{M}F_i,$$
thereby showing that (i) implies (ii).

Conversely, if (ii) holds, the map $F_i/\mathcal{M}F_i \cong K_{i-1}/\mathcal{M}K_{i-1}$ is injective. Since it is also surjective, it must be an isomorphism, hence n_i is as small as possible, proving (i).

Finally, (ii) and (iii) are both equivalent to the condition that the tensored complex $F_\bullet \otimes_A k$ has 0 maps, and thus equivalent to each other. □

Corollary 5.6.6. *Let $A, \mathcal{M}$ be a local Noetherian ring, M a finitely generated A-module. Then the numbers n_i in a minimal free resolution (5.6.1) are uniquely determined by M. In fact, $n_i = \dim_k \operatorname{Tor}_i^A(k, M)$, where $k = A/\mathcal{M}$.*

Proof. Let $K_i = \ker f_i$. The number n_i is the dimension of the vector space $K_i/\mathcal{M}K_i$. The result then follows from the fact that if $F_\bullet \to M$ is a minimal resolution, then $F_\bullet \otimes_A k$ has 0 differentials by Proposition 5.6.5(ii). □

Definition 5.6.7. The numbers n_i in (5.6.1) are called the *Betti numbers* of M.

5.7. Stably free modules

In Example 5.1.1(d), we have seen that for a compact and paracompact topological space X, finitely generated projective modules over the ring $A = C(X)$ are the same as locally free modules, and they all arise as the modules of sections of a suitable vector bundle $E \to X$. This led us to investigate the relation between projective and locally free modules for a general ring.

Following this analogy, one can take inspiration from topological K-theory [**Hat17**], which is a cohomology theory that studies spaces by analyzing the category of vector bundles over them. One can hope to translate results in topological K-theory into results about projective modules over general rings. This eventually leads to the study of algebraic K-theory. We will not pursue this in general; instead we will limit ourselves to the simplest example.

If X is a contractible space, all vector bundles on X are actually trivial. In algebraic geometry, one does not have a notion of contractible space, but surely the affine space $\mathbb{A}^n(k)$ over a field k is a natural candidate. This line of reasoning leads to the conjecture that a finitely generated projective module over $k[x_1, \ldots, x_n]$ is in fact free.

This was posed as a problem by Serre in his seminal paper [**Ser55**], and resolved affirmatively by Quillen in [**Qui76**] and Suslin in [**Sus76**] at the same time. The account that we give will follow a simplified proof by Vaserstein in [**VS74**]. For much more on the history of Serre's problem and the development that followed, see the book [**Lam06**].

We start with a result that is slightly weaker, already proved by Serre.

Definition 5.7.1. Let A be a ring, M an A-module. We say that M is *stably free* if there exists a finitely generated free module F such that $M \oplus F$ is free. More generally, two A-modules M, N are *stably isomorphic* if there exist finitely generated free modules F, G such that $M \oplus F \cong N \oplus G$. Hence M is stably free if and only if it is stably isomorphic to a free module.

Remark 5.7.2. A stably free module is projective by Proposition 2.6.4. Moreover, notice that the condition that F is finitely generated is crucial in the definition. Otherwise, according to Exercise 2, *every* projective module would be stably free.

We see that *stably free* is a convenient intermediate notion between *projective* and *free*. These notions are actually distinct, as the next example shows.

Example 5.7.3.

(a) Let $A = \mathbb{Z}/pq\mathbb{Z}$ for two distinct primes p, q, and consider the A-module $M = \mathbb{Z}/p\mathbb{Z}$. Then M is a direct summand of A, hence it is projective, but it is not stably free for cardinality reasons.

(b) Let $S^n \subset \mathbb{R}^{n+1}$ be the n-sphere, and T_{S^n} its tangent bundle. The normal bundle of S^n is trivial, hence there is an exact sequence of vector bundles

$$0 \longrightarrow T_{S^n} \longrightarrow S^n \times \mathbb{R}^{n+1} \longrightarrow S^n \times \mathbb{R} \longrightarrow 0 .$$

Letting $A = C(S^n)$, this gives an exact sequence of the A-modules of sections

$$0 \longrightarrow C(T_{S^n}) \longrightarrow A^{n+1} \longrightarrow A \longrightarrow 0 .$$

This sequence splits, showing that $C(T_{S^n})$ is stably free. On the other hand, $C(T_{S^n})$ is free if and only if T_{S^n} is trivial, which happens only for $n = 1, 3, 7$ by a theorem of Adams [**Hat17**, Theorem 2.16]. Without using the full strength of this result, the hairy ball theorem [**Hat02**, Theorem 2.28] shows that T_{S^n} does not even admit a nonvanishing section for n even, so it is certainly not trivial.

Our aim in the first part of the section is to prove the following result. Our approch follows [**Lan02**, Section $XXI.2$] with some simplifications.

Theorem 5.7.4 (Serre). *Let M be a finitely generated projective module over $A = k[x_1, \ldots, x_n]$, where k is a field or a Dedekind ring. Then M is stably free.*

This will be easy to prove once we have some general results on stably free modules.

Proposition 5.7.5. *Let A be a Noetherian ring, M a finitely generated, projective A-module. Then M is stably free if and only if it admits a finite free resolution.*

Proof. Assume that M has a finite free resolution of length n, say

$$0 \longrightarrow F_n \longrightarrow \cdots \longrightarrow F_1 \longrightarrow F_0 \longrightarrow M \longrightarrow 0.$$

By the Noetherian hypothesis, we can assume that all F_i are finitely generated. We prove that M is stably free by induction on n, the case $n = 0$ being obvious. Letting $K = \ker F_0 \to M$, we get the exact sequence

$$0 \longrightarrow K \longrightarrow F_0 \longrightarrow M \longrightarrow 0 \,,$$

which splits because M is projective. Then K is projective as well, hence stably free by induction, say $K \oplus G \cong H$, where G, H are finitely generated free modules. Hence

$$M \oplus H \cong M \oplus K \oplus G \cong F_0 \oplus G,$$

which is free. The other direction is obvious. $\square$

Notice that this result, together with Hilbert's theorem, Theorem 5.6.2, is already enough to prove Theorem 5.7.4. Of course, this would be circular, so there is still some work to do. Instead, what we are going to do is to prove a weaker version of Hilbert's syzygy theorem, namely that every finitely generated module over $k[x_1, \ldots, x_n]$ has a finite free resolution, which is in any case enough to prove Serre's theorem. We will then obtain Hilbert's syzygy theorem by combining this result with our computation of the global dimension of $k[x_1, \ldots, x_n]$ (Corollary 5.5.8). This allows us to decouple the problem of existence of finite free resolutions from the problem of finding a bound on their length.

We will prove the existence of finite free resolutions by induction. The following result allows the induction step to go through.

Proposition 5.7.6. *Let A be a ring,*

$$0 \longrightarrow M_1 \longrightarrow M_2 \longrightarrow M_3 \longrightarrow 0$$

an exact sequence of A-modules. If two of the modules admit a finite resolution by free finitely generated modules, so does the third.

In order to prove this, we need to introduce the concept of stably free dimension.

Definition 5.7.7. Let A be a ring, M an A-module. The *stably free dimension* of M, denoted sfd M, is the smallest length of a left resolution of M by *finitely generated* stably free modules. If no such resolution exists, we set sfd $M = \infty$.

Remark 5.7.8. Given a stably free resolution of M

$$0 \longrightarrow S_n \longrightarrow \cdots \longrightarrow S_1 \longrightarrow S_0 \longrightarrow M \longrightarrow 0,$$

of length $n > 0$, we can alter it to make it a free resolution by adding suitable pairs of free modules to consecutive terms. In other words, if sfd $M = n > 0$, then M admits a finite free resolution of length n, hence the length of the shortest (finitely generated) free and stably free resolutions of M are the same. It follows that M admits a finite resolution by finitely generated, free modules, if and only if sfd $M < \infty$.

Of course, the above remark does not hold for $n = 0$: a stably free module is not necessarily free. What we need to make the notion of stably free dimension workable is an analogue of Remark 5.2.8. Unfortunately, we cannot leverage the machinery of the Ext functors and dimension shifting, so we will need to produce a more explicit proof.

Lemma 5.7.9. *Let M, N be stably isomorphic A-modules. Then* sfd $M =$ sfd N.

Proof. We can assume that $N = M \oplus F$ for some finitely generated free module F. Clearly, a finite free resolution of M gives rise to a finite free resolution of N of the same length.

Conversely, assume there is a finite resolution

$$0 \longrightarrow S_n \longrightarrow \cdots \longrightarrow S_1 \longrightarrow S_0 \longrightarrow M \oplus F \longrightarrow 0,$$

where the modules S_i are stably free. The surjection $S_0 \to F$ splits, hence we can write $S_0 = R_0 \oplus F$, and $R_0 \to M$ is still surjective. Then

$$\ker(R_0 \to M) = \ker(S_0 \to M \oplus F) = \operatorname{coker}(S_2 \to S_1),$$

and we get a stably free resolution of M of length n. $\qquad\square$

We can now prove the analogue of Remark 5.2.8. Notice that this approach would also work to develop the theory of projective dimension, although using Ext functors is much more convenient.

Proposition 5.7.10. *Let M be an A-module with* sfd $M = n > 0$, *and consider an exact sequence*

$$0 \longrightarrow K \longrightarrow S \longrightarrow M \longrightarrow 0,$$

where S is finitely generated and stably free. Then sfd $K = n - 1$.

Proof. A stably free resolution of K of length k gives a stably free resolution of M of length $k + 1$, hence $\text{sfd } M \leq \text{sfd } K + 1$. For the converse inequality, take a stably free resolution

$$0 \longrightarrow S_n \longrightarrow \cdots \longrightarrow S_1 \longrightarrow S_0 \longrightarrow M \longrightarrow 0$$

of length n. Since S is projective, we can lift the map $S \to M$ to a map $S \to S_0$, giving rise to a diagram

$$
\begin{array}{ccccccccc}
0 & \longrightarrow & K & \longrightarrow & S & \longrightarrow & M & \longrightarrow & 0 \\
& & \downarrow & & \downarrow & & \downarrow{\scriptstyle \text{id}_M} & & \\
0 & \longrightarrow & K_0 & \longrightarrow & S_0 & \longrightarrow & M & \longrightarrow & 0,
\end{array}
$$

where $K_0 = \ker S_0 \to M$. Clearly, $\text{sfd } K_0 = n - 1$.

By Schanuel's lemma (Proposition 4.4.1), there is an isomorphism $K_0 \oplus S \cong K \oplus S_0$. Hence, K and K_0 are stably isomorphic, and the conclusion follows from Lemma 5.7.9. $\qquad \square$

Lemma 5.7.11. *Let*

$$0 \longrightarrow M_1 \longrightarrow M_2 \longrightarrow M_3 \longrightarrow 0$$

be an exact sequence of A-modules, where M_2 is finitely generated and M_3 is finitely presented. Then M_1 is finitely generated.

Proof. First, assume that M_2 is actually free, and take an exact sequence

$$0 \longrightarrow K \longrightarrow F \longrightarrow M_3 \longrightarrow 0 \, ,$$

where K, F are finitely generated and F is free. By Schanuel's lemma (Proposition 4.4.1), $K \oplus M_2 \cong M_1 \oplus F$, from which we conclude that M_1 is finitely generated.

In general, take a finitely generated free module G with a surjection $G \to M_2$, and let $H = \ker G \to M_3$. Then there is a surjection $H \to M_1$, and H is finitely generated by the first part of the proof. $\qquad \square$

Proof of Proposition 5.7.6. There are three cases, according to which pair among (M_1, M_2), (M_1, M_3) and (M_2, M_3) we assume to have finite stably free dimension. In all cases, we will produce a commutative diagram of the form

(5.7.1)

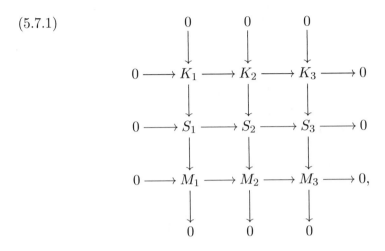

where S_1, S_2, and S_3 are stably free (in fact, finitely generated free).

The first thing to prove is that all of M_1, M_2 and M_3 are finitely generated. This is clear, except in the case where we assume that M_2 and M_3 have finite stably free dimension, where it follows by Lemma 5.7.11.

To obtain diagram (5.7.1), take S_i to be a finitely generated free module with a surjection to M_i for $i = 1, 3$, and let $S_2 = S_1 \oplus S_3$. The map $S_2 \to M_2$ results by letting it restrict to $S_1 \to M_1$ on the summand S_1, and choosing a lift $S_3 \to M_2$ on the other one. Finally, choose $K_i = \ker S_i \to M_i$. By construction, all columns and the bottom two rows are exact, and the top row is exact by the nine lemma.

The proof then splits into two cases. In the first one, we assume that sfd $M_2 < \infty$ and either sfd $M_1 < \infty$ or sfd $M_3 < \infty$—we want to prove that the other one holds. If sfd $M_2 = 0$, M_2 is stably free and the conclusion follows by Proposition 5.7.10. Otherwise, the same Proposition gives us sfd $K_2 = $ sfd $M_2 - 1$, and we finish the proof by induction on sfd M_2.

The remaining case has sfd $M_1 < \infty$ and sfd $M_3 < \infty$. In this case, we argue by induction on max $\{$sfd M_1, sfd $M_3\}$. If both M_1 and M_3 are stably free, the exact sequence between the M_i splits, hence $M_2 = M_1 \oplus M_3$ is stably free as well. Otherwise, Proposition 5.7.10 gives us sfd $K_i = $ sfd $M_i - 1$ for $i = 1, 3$, and we reduce to a smaller case. $\qquad\square$

We can now prove the main result of this section.

Theorem 5.7.12. *Let A be a Noetherian ring such that every finitely generated A-module has a finite free resolution. Then every finitely generated $A[x]$-module has a finite free resolution.*

Notice that we do not specify whether the free modules in the resolution are finitely generated, but over a Noetherian ring, a finitely generated module that has a finite resolution by free modules also has a finite resolution by finitely generated free modules.

Proof. We need to prove that every module M over $A[x]$ has a finite free resolution. By Theorem A.3.7 and Proposition 5.7.6, it is enough to prove the result for $M = A/P$ for some prime ideal P. By another application of Proposition 5.7.6, we can prove the result for $M = P$.

Let $Q = P \cap A$. By hypothesis, Q has a finite free resolution, hence so does $Q[x] = Q \cdot A[x]$. Let $A' = A/Q$, $P' = P/(Q \cdot A[x])$. By Proposition 5.7.6, P has a finite free resolution if and only if P' does.

To find such a resolution for P', we first work over $A'[x]$, where we have the convenience that A' is an integral domain. Let $k = \mathcal{F}(A')$ be its field of fractions. Choose generators $f_1, \ldots, f_d$ for P', and an element $f \in P'$ of minimal degree. Then we can perform the division over k to get $f_i = q_i \cdot f + r_i$, where $q_i, r_i \in k[x]$. By clearing denominators, we find $a' f_i = a' q_i \cdot f + a' r_i$ for some $a' \in A'$ such that $a' q_i \in A'[x]$ and $a' r_i \in A'[x]$. Since f has minimal degree, $a' r_i = 0$, whence $a' \cdot P' \subset (f)$.

Lifting a' to some $a \in A$, we have the exact sequence of $A[x]$-modules

$$0 \longrightarrow a \cdot P' \longrightarrow f \cdot A'[x] \longrightarrow N \longrightarrow 0 \,,$$

where N is the quotient. Since A' is integral, as $A[x]$-modules, $a \cdot P' \cong P'$ and $f \cdot A'[x] \cong A'[x]$, so we read this exaxct sequence as

$$0 \longrightarrow P' \longrightarrow A'[x] \longrightarrow N \longrightarrow 0 \,.$$

Since $A'[x] = A[x]/(Q \cdot A[x])$, by another application of Proposition 5.7.6, we reduce to proving that N has a finite free resolution.

As in the first part of the proof, N has a filtration by $A[x]$-modules of the form $A[x]/R$, for some prime ideals $R \subset A[x]$, and we can reduce to proving that each of them has a finite free resolution. Moreover, by Theorem A.3.7, such primes R are exactly the associated primes of N. We claim that each associated prime R of N is strictly larger than P. Clearly, $P \subset \mathrm{Ann}(n)$ for all $n \in N$. Moreover, $a \in \mathrm{Ann}(n)$ as well, and $a \notin P$, since $a' \neq 0$ in A'.

All of the above works for any prime P. Now, assuming that a prime without a finite free resolution exists at all, we can choose P to be maximal with respect to this condition, since $A[x]$ is Noetherian. With this choice, every associated prime R of N has a finite free resolution, hence the same holds for N, and eventually for P' and P, which is a contradiction. □

As corollaries, we derive Serre's theorem on stably free modules and Hilbert's syzygy theorem.

Proof of Theorem 5.7.4. Combine Theorem 5.7.12 and Proposition 5.7.5.

$\square$

Proof of Theorem 5.6.2. By Corollary 5.5.8, gl. dim $k[x_1, \ldots, x_n] = n$, hence every module over $k[x_1, \ldots, x_n]$ has a projective resolution of length n. If the module is finitely generated, one can choose the projectives to be finitely generated as well. By Theorem 5.7.4, this is a stably free resolution. By Remark 5.7.8, we can alter this resolution to produce a resolution of length n by finitely generated free modules. $\square$

We end this section with a result of Gabel [**Gab72**], showing that stably free modules are interesting only in the finitely generated case. The proof is from [**Con13**].

Theorem 5.7.13 (Gabel). *Let M be a stably free A-module. If M is* not *finitely generated, then M is free.*

Proof. Let $M \oplus A^k \cong F$, with F free, and choose a basis $\mathcal{B}$ for F. Since $F \to A^k$ is surjective, we can choose a finite subset $\mathcal{B}' \subset \mathcal{B}$ such that the free module F' generated by $\mathcal{B}'$ surjects onto A^k.

Then we have $M + F' = F$, hence, letting $N = M \cap F'$,

$$\frac{F}{F'} \cong \frac{M + F'}{F'} \cong \frac{M}{M \cap F'} = \frac{M}{N}.$$

Moreover, the exact sequence

$$0 \longrightarrow N \longrightarrow F' \longrightarrow A^k \longrightarrow 0$$

splits, so $N \oplus A^k \cong F'$. Since F/F' is free of infinite rank, we can write $F/F' \cong A^k \oplus F''$, with F'' free.

Finally, since $M/N \cong F/F'$ is projective,

$$M \cong N \oplus \frac{M}{N} \cong N \oplus \frac{F}{F'} \cong N \oplus A^k \oplus F'' \cong F' \oplus F''. \qquad \square$$

A result in the same spirit is the following from [**Bas63a**].

Theorem (Bass). *Let A be a Noetherian ring, M a projective module, and assume that the only elements $a \in A$ such that $a^2 = a$ are 0 and 1. If M is* not *finitely generated, then M is free.*

Remark 5.7.14. The hypothesis on idempotent elements is necessary to avoid trivial examples, such as the module A over $A \times B$ for any two rings A, B. Less obviously, the Noetherian hypothesis is also necessary; see Exercise 16.

One can also obtain results in the finitely generated case, as long as the rank of the module is big enough with respect to the dimension of the ring. A nice result from [**Yen11**] is

Theorem (Yengui). *Let A be a ring with $\dim A < \infty$, and M a stably free module over $A[x]$, so that $M \oplus A[x]^m \cong A[x]^n$. If $n - m > \dim A$, then M is free.*

5.8. The Quillen–Suslin theorem

We now take the decisive step towards the solution of Serre's problem, passing from stably free to free modules. Our aim is to prove the following result.

Theorem 5.8.1 (Quillen–Suslin). *Let $A = k[x_1, \ldots, x_n]$, where k is a field or a principal ideal domain. If M is a finitely generated projective module over A, then M is free.*

Of course, we have already done the passage from projective to stably free in the previous section. For reasons that will become apparent in a moment, we rephrase the question of freeness of a stably free module in the following way.

Assume that M is a stably free, finitely generated A-module, so that $M \oplus A^m \cong A^n$ for some $m, n \in \mathbb{N}$. Here $m \leq n$, as can be seen by taking the quotient with respect to any maximal ideal. We can see M as the cokernel of a morphism $f \colon A^m \to A^n$, which happens to split. Then f is given by a matrix with coefficients in A. We are free to change the bases for the two free A-modules, which corresponds to multiplication by an element of $\mathrm{GL}_m(A)$ on the left and $\mathrm{GL}_n(A)$ on the right. If, applying these operations, we can choose a matrix for f which represents the standard inclusion $A^m \to A^n$ by the first m coordinates, then the cokernel of f is manifestly free, isomorphic to A^{n-m}. Hence the question can be rephrased by starting from a $m \times n$ matrix that represents a split monomorphism, and checking whether a particular matrix lies in its orbit under the action of these groups.

Now, one can easily see that if this always holds for $m = 1$, then it holds for any m, by a simple induction. We are thus going to study morphisms $f \colon A \to A^n$ such that f is a split monomorphism. Such f is uniquely determined by the vector $v = f(1) \in A^n$.

Definition 5.8.2. The vector $v \in A^n$ is called *unimodular* if its coordinates $v_1, \ldots, v_n$ generate the unit ideal in A.

Lemma 5.8.3. *Let $f \colon A \to A^n$ be a homomorphism. Then f is a split monomorphism if and only if $v = f(1)$ is unimodular.*

Proof. The condition that f is a split monomorphism amounts to saying that there exists a homomorphism $g \colon A^n \to A$ such that $g(v) = 1$. Representing g by a $n \times 1$ matrix with entries $g_1, \ldots, g_n$, this can be written as $\sum_{i=1}^{n} g_i v_i = 1$, which means exactly that v is unimodular. $\quad\square$

In light of the above lemma and the discussion preceding it, we are going to study unimodular vectors in A^n modulo the action of $\mathrm{GL}_n(A)$ on the right. Actually, we will transpose everything, turning our vectors into column vectors, and changing the action of $\mathrm{GL}_n(A)$ to be an action on the left. We will simply say that two unimodular vectors v, w are equivalent if there is a matrix $M \in \mathrm{GL}_n(A)$ such that $M \cdot v = w$, in which case we write $v \sim w$. We aim to find conditions for a ring A that guarantee that every unimodular vector is equivalent to the standard basis vector e_1, or, in other words, that every two unimodular vectors are equivalent.

Remark 5.8.4. Over a PID or a local ring, every projective module is free (the latter is Theorem 5.1.2). Hence, if A is a PID or local, every two unimodular vectors are equivalent.

We will prove the result for polynomial rings by localization. Hence, our point of departure is the following result for local rings.

Theorem 5.8.5 (Horrocks). *Let $A, \mathcal{M}$ be a local ring,*
$$v = (v_1(x), \ldots, v_n(x)) \in A[x]^n$$
a unimodular vector such that one of the elements v_i is monic. Then $v \sim e_1$.

Lemma 5.8.6. *Let A be a ring, $v \in A^n$ a unimodular vector for $n = 1, 2$. Then $v \sim e_1$.*

Proof. The result is obvious for $n = 1$. For $n = 2$, we have an equation $av_1 + bv_2 = 1$. Then the matrix $M = \begin{pmatrix} a & b \\ -v_2 & v_1 \end{pmatrix}$ has $\det M = 1$, so M is invertible over A, and $M \cdot v = e_1$. $\quad\square$

Proof of Theorem 5.8.5. By Lemma 5.8.6, we can assume that $n \geq 3$. Say v_1 is monic, and let $d = \deg v_1$. We will prove the result by induction on d, the case $d = 0$ being obvious (check it!).

By row operations, we can assume that $\deg v_i < d$ for $i \geq 2$. If all coefficients of all v_i lie in $\mathcal{M}$ for $i \geq 2$, v is not unimodular. In fact, reduction modulo $\mathcal{M}$ kills all coordinates but v_1, and v_1 does not generate the unit ideal in $k[x]$, for $k = A/\mathcal{M}$. We can then assume that v_2 has a coefficient outside $\mathcal{M}$, which is thus a unit of A.

We claim that there exists a monic polynomial

(5.8.1) $$h(x) = a(x)v_1(x) + b(x)v_2(x)$$

such that $\deg h < d$. To see this, let $I \subset A$ be the ideal generated by all leading coefficients of polynomials h of the form (5.8.1) where $\deg h < d$. It is easy to see that all coefficients of v_2 belong to I (use descending induction starting from the highest coefficient), hence $I = A$. In particular, $1 \in A$, which is our claim.

We can now use row operations to replace v_3 by $v'_3 = v_3 - c \cdot h$ for some $c \in A[x]$, in such a way that v'_3 is monic and $\deg v'_3 < d$. The proof is then finished by the induction hypothesis. $\square$

Corollary 5.8.7. *Let* $A, \mathcal{M}$ *be a local ring,*

$$v = (v_1(x), \ldots, v_n(x)) \in A[x]^n$$

a unimodular vector such that one of the elements v_i *is monic. Then* $v \sim v(0) \in A^n$.

This is obvious, since both v and $v(0)$ are equivalent to e_1 (the latter because of Remark 5.8.4), but it will turn out to be a convenient reformulation, since it allows induction from A to $A[x]$.

One can also formally derive other relations. Assume that $v \sim v(0)$, so that there is an $n \times n$ matrix $M(x)$ over $A[x]$ such that $M(x) \cdot v(x) = v(0)$. By replacing formally x with $x+y$ we end up with $M(x+y) \cdot v(x+y) = v(0)$. Combining the two relations we get

$$M(x)^{-1} M(x + y) \cdot v(x + y) = v(x),$$

so that $v(x + y) \sim v(x)$ over $A[x, y]$.

Our next task is to generalize Corollary 5.8.7 to arbitrary rings.

Proposition 5.8.8. *Let* A *be a ring,*

$$v = (v_1(x), \ldots, v_n(x)) \in A[x]^n$$

a unimodular vector such that one of the elements v_i *is monic. Then* $v \sim v(0) \in A^n$.

The key is the following localization result.

Lemma 5.8.9. *Let* A *be a ring,* $S \subset A$ *a multiplicative set,* $v \in A[x]^n$, *and assume that* $v \sim v(0)$ *over* $S^{-1}A[x]$. *Then* $v(x) \sim v(x+sy)$ *over* $A[x, y]$, *for some* $s \in S$.

Proof. We have seen that $v(x) \sim v(x + y)$ over $S^{-1}A[x, y]$, so there is a matrix M over $S^{-1}A[x, y]$ such that $M(x, y) \cdot v(x + y) = v(x)$. Replacing y by sy we get

$$M(x, sy) \cdot v(x + sy) = v(x),$$

so it is enough to check that $M(x, sy)$ has coefficients in A for some $s \in S$.

By construction, $M(x,0)$ is the identity, hence $M(x,y) = I + yN$ for some matrix N. Then $M(x,sy) = I + syN$, and it is enough to choose $s \in S$ to clear the denominators in N. $\qquad\square$

Proof of Proposition 5.8.8. Reversing what we have done before, it is enough to show that $v(x+y) \sim v(x)$ over $A[x,y]$, and then substitute $x = 0$ (and rename y to x).

This leads us to define

$$I := \{a \in A \mid v(x+ay) \sim v(x) \text{ over } A[x,y]\}.$$

It will be enough to prove that $1 \in I$. It is easy to check that I is an ideal. For instance, if $a, b \in I$, then $v(x+ay) \sim v(x)$ and $v(x+by) \sim v(x)$. Changing x into $x+ay$ in the latter, we get $v(x+ay+by) \sim v(x+ay) \sim v(x)$. Similarly, I is closed under multiplication for an element of A.

Knowing that I is an ideal, we can rely on our local results. If $I \subsetneq A$, then $I \subset \mathcal{M}$ for some maximal ideal $\mathcal{M}$. Corollary 5.8.7 then guarantees that $v(x) \sim v(0)$ over $A_{\mathcal{M}}[x]$, and by Lemma 5.8.9 there is $s \notin \mathcal{M}$ such that $v(x) \sim v(x+sy)$ over $A[x,y]$, hence $I \not\subset \mathcal{M}$. $\qquad\square$

Proposition 5.8.8 is the crucial inductive step to prove the following formulation of the Quillen–Suslin theorem.

Theorem 5.8.10 (Quillen–Suslin). *Let $A = k[x_1, \ldots, x_n]$, where k is a field or a principal ideal domain. If $v \in A^n$ is unimodular, then $v \sim e_1$.*

Proof. We perform induction on n, the case $n = 0$ being Remark 5.8.4. We look at A as the ring $B[x_n]$, where $B = k[x_1, \ldots, x_{n-1}]$.

We can perform the following invertible change of variables over A. Replace x_i by $x_i - x_n^{M^i}$ for $i < n$, for some $M \gg 0$ (compare the proof of Noether's normalization lemma in [**Fer20**, Theorem 5.3.1]). Then we can guarantee that the entries v_i of v become monic, as polynomials in x_n.

By Proposition 5.8.8, $v \sim v(0) \in A^n$, and since $v(0) \in B^n$, we are done by induction. $\qquad\square$

Proof of Theorem 5.8.1. By Serre's theorem, Theorem 5.7.4, M is stably free. Hence $M \oplus A^m \cong A^n$ for some $m, n \in \mathbb{N}$. By induction over m, we only need to prove the case $m = 1$. In this case, there is a split monomorphism $f \colon A \to A^n$ such that $M \cong \operatorname{coker} f$.

The vector $v = f(1) \in A^n$ is unimodular, and by Theorem 5.8.10, $v \sim e_1$. Hence, we can perform a change of variables in A^n such that f becomes the standard inclusion in the first coordinate, and then $M \cong A^{n-1}$. $\qquad\square$

Remark 5.8.11. A ring A such that all stably free A-modules are free is called a *Hermite ring*. One may be tempted to strengthen the above result to say that if A is Hermite, then $A[x]$ is Hermite. But the proof of Theorem 5.8.10 uses a crucial change of variables that is only available in polynomial rings to achieve the condition that one of the components v_i is monic. In fact, the question of whether being Hermite is inherited from A to $A[x]$ is known as the Hermite ring conjecture by Lam [**Lam78**], and is currently still open.

5.9. Exercises

1. Give a proof of Theorem 5.1.2 under the assumption that P is finitely generated, as follows. Let $\mathcal{M}$ be the maximal ideal of A, and choose a set $\{p_1, \ldots, p_n\} \subset P$ such that the classes $\{\overline{p_1}, \ldots, \overline{p_n}\}$ are a basis of the vector space $P/\mathcal{M}P$. Consider the map $f \colon A^n \to P$ defined by $f(e_i) = p_i$ and apply Nakayama's lemma.

2. Let P be a projective A-module. Show that there exists a free A-module F such that $P \oplus F \cong F$. The construction you will probably come up with is called the Eilenberg swindle.

3. Let A be a Boolean ring, that is, a ring such that $a^2 = 2$ for all $a \in A$. Prove that all localizations of A are isomorphic to $\mathbb{Z}/2\mathbb{Z}$, and conclude that every module over A is locally free.

4. Let A be a direct product of countably many copies of $\mathbb{Z}/2\mathbb{Z}$, and $I \subset A$ their direct sum. Prove that A is Boolean ring (see Exercise 3), and that I is an ideal. Show that I is not principal and conclude that A/I is a locally free, finitely generated A-module that is not projective.

Exercises 5–10, discuss the notion of locally free module in the following strong sense. We say that M is *locally free in the strong sense* if there exist elements $a_1, \ldots, a_n \in A$ that generate the whole ring as an ideal such that M_{a_i} is free for all i. We will provide an example from [**dJea20**, Tag 05WG] of a projective module that is not locally free in the strong sense.

5. Show that if M is locally free in the strong sense, then it is locally free with the usual definition.

6. Let A be a Noetherian ring, M a finitely generated, projective A-module. Show that M is locally free in the strong sense.

7. Let A be a ring, $\{a_i\}_{i \in \mathbb{N}}$ elements of A satisfying $a_i^2 = a_i$, and let $I = (a_1, a_2, \ldots)$ be the ideal that they generate. Show that I is projective. (Up to changing the a_i, one can assume that $(a_i) \subset (a_{i+1})$, so that $I = \mathrm{colim}(a_i)$. Then show that h^I is exact by Theorem 3.5.9.)

8. Let A be a ring, and assume that there is an injective map $f \colon A^n \to M$ for some A-module M. Then M cannot be generated by less that n elements.

(Reduce to the case of M free, and consider the matrix S associated to f. If one of the elements of S is a unit, one can reason by induction. Reduce to this case by localization, unless all elements of S are nilpotent, in which case prove the claim directly.)

9. Let A be a ring $M \subset N$ two A-modules. Show that if N is finitely generated and M is locally free in the strong sense, then M is finitely generated. (Use the previous exercise.)

10. Let $A = \prod_{i=1}^{\infty} \mathbb{Z}/2\mathbb{Z}$ be an infinite product of copies of $\mathbb{Z}/2\mathbb{Z}$. Use the previous exercises to find an ideal $I \subset A$ that is projective but not locally free in the strong sense.

11. Let I be an ideal of A. If A/I is a projective A-module, prove that I is principal.

12. Find a counterexample to Theorem 5.2.10 if we drop the assumption that $\mathrm{pd}_{A/(a)} M < \infty$. Namely, find a ring A, an element $a \in A$, which is not a zero divisor, and a module M over $A/(a)$ such that $\mathrm{pd}_{A/(a)} M = \infty$ but $\mathrm{pd}_A M < \infty$.

13. Show that for a ring A all A-modules are injective if, and only if, all A-modules are projective. In this case, we say that A is *semisimple*.

14. Let M, N be two A-modules such that $M \oplus P \cong N \oplus Q$, with P, Q projective A-modules. Prove that $\mathrm{pd}\, M = \mathrm{pd}\, N$.

15. Let X be a compact and paracompact topological space. Prove that the category of continuous vector bundles over X is equivalent to the category of finitely generated projective modules over the ring $A = C(X)$. This result is known as *Swan's theorem* [**Swa62**].

16. Let $A = C([0,1])$, and consider the ideal $I \subset A$ of functions f such that $f|_{[0,\epsilon]} = 0$ for some $\epsilon > 0$. Show that the only elements $f \in A$ such that $f^2 = f$ are the constants 0 and 1. Prove that I is projective, but I is not free. Conclude that I is *not* finitely generated.

17. Let P, Q be projective A-modules. Prove that $P \otimes_A Q$ is projective. Is the same true for injective modules?

18. Let P be a finitely generated projective A-module. Prove that there is a canonical isomorphism $\mathrm{Hom}(P, A) \otimes P \cong \mathrm{End}_A(P)$. If P is moreover free, prove that the composition

$$\mathrm{End}_A(P) \cong \mathrm{Hom}(P, A) \otimes P \to A,$$

where the latter is the evaluation morphism, is the usual trace (that is, if $f \colon P \to P$ is represented by a matrix M, this is the sum of the diagonal entries of M).

19. Let k be a field, $A = k[x_1, \ldots, x_n]$. Show that if

$$0 \longrightarrow F_0 \longrightarrow F_1 \longrightarrow \cdots \longrightarrow F_n$$

is an exact sequence of graded A-modules and $F_1, \ldots, F_n$ are finitely generated free, then so is F_0. Deduce the graded version of Hilbert's syzygy theorem (Theorem 5.6.2). (Show that the sequence remains exact up to F_{n-1} when taking the quotient with respect to x_n.)

20. Derive from the previous exercise another proof of the existence of the Hilbert polynomial for finitely generated graded modules over $A = k[x_1, \ldots, x_n]$. Namely, if M is such a module, there exists a polynomial $p_M \in \mathbb{Q}[x]$ such that $\dim_k M_i = p_M(i)$ for all i big enough.

21. Let M be an A-module. We say that a morphism $f \colon P \to M$ is a *projective cover* if P is projective, f is surjective, and no proper submodule of P maps surjectively onto M. Show that this is the same as saying $M \to P$ is an injective hull in Mod_A^{op}, and that the projective cover, if it exists, is unique up to isomorphism.

22. Show that $\mathbb{Z}/2\mathbb{Z}$ does not admit a projective cover as a $\mathbb{Z}$-module.

Quite surprisingly, every A-module M admits a *flat cover*, that is, a morphism $F \to M$ as in Exercise 21, where F is only assumed flat (that is, acyclic for Tor, see Chapter 6). This used to be called the *flat cover conjecture*, now a theorem thanks to [**BEBE01**] (see also [**EEE01**]).

23. Let $A, \mathcal{M}$ be a local Artinian ring, with $\mathcal{M}$ principal, and M a finitely generated A-module. Show that M is a direct sum of cyclic modules. (Replace A by $A/\operatorname{Ann} M$, then show that A is injective over itself.)

24. Let $A = \mathbb{R}[x, y, z]/(x^2 + y^2 + z^2 - 1)$, and

$$T = \{(a, b, c) \in A^3 \mid xa + yb + zc = 0\}.$$

Show that T is a stably free A-module, but is not free.

Starting from this ring, Hochster has given an example of two non-isomorphic rings B, C such that $B[t] \cong C[t]$ [**Hoc72**]. We give his construction in Exercises 25–27. We will need some notation: given an A-module M, we denote by $\operatorname{Sym}^n M$ its n-th symmetric power (see Definition 6.5.4). We also denote $\operatorname{Sym} M = \bigoplus_{n=0}^{\infty} \operatorname{Sym}^n M$, which has the structure of a commutative ring.

25. Let A, T be as in Exercise 24, so that $A^3 \cong A \oplus T$. Show that

$$\operatorname{Sym}(A^3) \cong \operatorname{Sym}(T) \otimes \operatorname{Sym}(A) \cong \operatorname{Sym}(T)[t],$$

where t is an indeterminate. Also show that

$$\mathrm{Sym}(T) \cong \frac{A[a, b, c]}{(xa + yb + zc)}.$$

Hence, if we let $B = A[u, v]$ and $C = \mathrm{Sym}(T)$, we have $B[t] \cong C[t]$ as rings.

26. Keep the notation of the previous exercise. Assume that there is an isomorphism $f \colon B \to C$. Show that f is an isomorphism of $\mathbb{R}$-algebras. Moreover, show that the only solutions to $X^2 + Y^2 + Z^2 = 1$ in B actually lie in A. Deduce that the same holds for C, so $f(A) \subset A$ and $f^{-1}(A) \subset A$. Thus, up to composing with an automorphism of A, f is an isomorphism of A-algebras.

27. Keep the notation of the previous exercise. Show that there is no isomorphism of A-algebras $f \colon B \to C$. (Assume that there is one, and let $c = f(u)$ and $d = f(v)$. Obtain from them two generators of T as an A-module and prove that $T \cong A^2$, a contradiction.)

28. By Example 5.3.11 and Theorem 5.3.9, we know that $\mathbb{Q}/\mathbb{Z}$ must be a direct sum of factors, all of which are either $\mathbb{Q}$ or $\mathbb{Z}[p^{-1}]/\mathbb{Z}$ for some prime p. Find explicitly such a decomposition.

29. Let A be a Noetherian ring, $\mathcal{M}$ a maximal ideal. Prove that $\mathrm{gl.\,dim}\, A_{\mathcal{M}} = \mathrm{pd}_A\, A/\mathcal{M}$.

30 ([**AHRT02**]). Prove that injective hulls cannot be made functorial. Namely, assume $E \colon \mathrm{Mod}_A \to \mathrm{Mod}_A$ is a functor with a natural transformation $\iota \colon \mathrm{id} \to E$ such that for all A-modules M, $\iota_M \colon M \to E(M)$ is an injective hull. Prove that all A-modules are injective and E is isomorphic to the identity. (Show that $E(\iota_M)$ is an isomorphism, then use the following diagram

$$
\begin{array}{ccccc}
M & \xrightarrow{\ \iota_M\ } & E(M) & \overset{f}{\underset{g}{\rightrightarrows}} & N \\
{\scriptstyle \iota_M}\big\downarrow & & {\scriptstyle \iota_{E(M)}}\big\downarrow & & \big\downarrow{\scriptstyle \iota_N} \\
E(M) & \xrightarrow{E(\iota_M)} & E(E(M)) & \overset{E(f)}{\underset{E(g)}{\rightrightarrows}} & E(N)
\end{array}
$$

to show that if $f \circ \iota_M = g \circ \iota_M$, then $f = g$.)

31. Let $A, \mathcal{M}$ be a local ring with $\mathrm{gl.\,dim}\, A = 2$. Assuming that $\mathcal{M}$ is principal, show that A is a valuation domain.

32. Let P be a finitely generated projective A-module. Show that for all maximal ideals $\mathcal{M} \subset A$, there exists $a \notin \mathcal{M}$ such that P_a is free over A_a. Deduce that if Q_1, Q_2 are two prime ideals such that $Q_1 + Q_2 \subsetneq A$, then P_{Q_1} and P_{Q_2} are free of the same rank (for readers familiar with the terminology, the rank function is locally constant for the Zariski topology). Show that if the only solutions to $a^2 = a$ in A are 0 and 1, then the rank of P_Q is the same regardless of the prime $Q \subset A$. Show that all these properties can fail for nonprojective modules.

Chapter 6

Flatness

The previous chapter was dedicated to projective and injective modules, which are acyclic for the Ext functors. We now turn our attention to flat modules, that is, modules which are acyclic for the Tor functors. Explicitly, a module M is flat if, for every exact sequence $S_\bullet$, the sequence $S_\bullet \otimes M$ remains exact. This is a key concept, especially in the study of ring homomorphisms. A ring homomorphism $f\colon A \to B$ is called flat if it makes B into a flat A-module.

Grothendieck was the first to recognize the centrality of this algebraic notion in the geometric setting. At an intuitive and fairly imprecise level, the idea is the following. Let $f\colon V \to W$ be a morphism of affine algebraic varieties over a field k. We look at f as a family of varieties. The members of this family are the fibers $V_p := f^{-1}(p)$, parametrized over the base space W. Algebraically, f corresponds to a map $f^*\colon R(W) \to R(V)$ between the coordinate rings. The condition that f^* is flat should amount to saying that the fibers V_p vary continuously as p varies in W. This means for instance, that their dimensions should not jump, and other numerical invariants should be locally constant.

This description has various limitations. The most obvious one is the fact that to make it more rigorous we should talk about schemes instead of varieties, but we have not introduced this language. That said, we should also specify what it does *not* say. For one thing, one should not expect that the fibers of a flat morphism (say over $\mathbb{C}$) are homeomorphic to each other in the Euclidean topology. Moreover, it can even happen that most fibers are smooth while a single fiber is singular. Still (in the context of schemes) the notion of flatness provides a good notion of a continuously varying family.

We start the chapter with the basic properties of flat modules. In particular, it is immediate that projective modules are flat, and in Section 6.1 we see that the converse holds for finitely presented modules. As a consequence, we prove that the weak dimension of an A-module, which is the length of a shortest resolution by flat modules, is the same as the projective dimension when A is Noetherian. In Section 6.2 we specialize to the case of flat morphisms, and in particular we prove the key results that localizations and completions are flat. This is consistent with our intuition for flat morphisms above. Let V be an affine variety with coordinate ring A. In [**Fer20**, Section 8.5], we have argued that localizing A at a maximal ideal $\mathcal{M}$ corresponds to the geometric operation of looking at a small neighborhood of the point $p \in V$ corresponding to $\mathcal{M}$. Similarly, passing to the completion with respect to the $\mathcal{M}$-adic topology would correspond to looking at an infinitesimal neighborhood of p. If our geometric analogy makes any sense, these operations should locally be isomorphisms, and in particular flat.

In Section 6.3, we study the stronger notion of faithfully flat modules and morphisms. We say that M is faithfully flat if a complex $S_{\bullet}$ is exact if *and only if* $S_{\bullet} \otimes M$ is. This is stronger than flatness: in particular, every faithfully flat morphism $A \to B$ is injective, and if $f \colon A \to B$ is a flat inclusion, then it is faithfully flat if and only if every prime of A is the restriction of a prime of B. This leads to an analogue of the going down theorem for integral extensions, which can be used to prove that the dimension of the nonempty fibers of a flat morphism is constant.

Section 6.4 is dedicated to various criteria that ensure that a module is flat. Of particular importance is the local criterion of Theorem 6.4.5, which guarantees that over a local ring $A, \mathcal{M}$ a module M is flat if the single group $\mathrm{Tor}_1(M, A/\mathcal{M})$ vanishes. Finally, Section 6.5 is dedicated to the proof of the Govorov–Lazard theorem, Theorem 6.5.1, which states that a module is flat if and only if it is the direct limit of free modules. This allows us to reduce many questions about flatness to the easy case of free modules. In the final part of the section, we use this strategy to derive the flatness of symmetric and exterior powers.

6.1. Flat modules

In the previous chapter, we investigated modules that are acyclic for the Ext functors (on either side). We now turn to the Tor functors.

Definition 6.1.1. Let A be a ring. An A-module M is called *flat* if the functor $M \otimes - \colon \mathrm{Mod}_A \to \mathrm{Mod}_A$ is exact.

Remark 6.1.2. The condition of M being flat is equivalent to asking that $\mathrm{Tor}_1^A(M, N) = 0$ for all A-modules N. In this case, one also has

$\mathrm{Tor}_i^A(M, N) = 0$ for all $i > 0$. Moreover, since Tor is symmetric, these conditions are also equivalent to $\mathrm{Tor}_1^A(N, M) = 0$ for all A-modules N, or to $\mathrm{Tor}_i^A(M, N) = 0$ for all A-modules N and $i > 0$.

Remark 6.1.3. By Corollary 3.3.3, the functors $\mathrm{Tor}_i^A(M, N)$ can be computed using a left resolution of either M or N by *flat* modules.

Remark 6.1.4. Flatness is stable under various operations. Let $\{M_i\}$ be a family of A-modules, N an A-module. By the natural isomorphism

$$(\bigoplus M_i) \otimes N \cong \bigoplus M_i \otimes N,$$

we immediately conclude that the direct sum $\bigoplus M_i$ is flat if and only if each M_i is flat. More generally, by Exercise 7 of Chapter 3, Tor commutes with filtered colimits. It follows that a filtered colimit (in particular, a direct limit) of flat modules is itself flat. Finally, if M and N are flat, $M \otimes N$ is clearly flat as well.

Remark 6.1.5. Let M be an A-module. Since $M \otimes -$ is always right exact, M is flat if and only if $M \otimes N_1 \to M \otimes N_2$ is injective for all A-modules $N_1 \subset N_2$. If this map is *not* injective, then there are finitely generated submodules $N_1' \subset N_1$ and $N_2' \subset N_2$ such that $M \otimes N_1' \to M \otimes N_2'$ is not injective. (Why?) Hence, to check that M is flat, one only needs to consider exact sequences of finitely generated modules.

Since the Tor_i are left derived functors, they vanish on projective modules. In other words, every projective module is flat. Quite surprisingly, the converse holds under some finiteness hypothesis.

Proposition 6.1.6. *Let M be a finitely presented A-module. Then M is flat if and only if it is projective.*

We will prove this by a local argument, hence our first remark is that flatness plays well with localization.

Proposition 6.1.7. *Let M be a flat A-module, $S \subset A$ a multiplicative set. Then $S^{-1}M$ is a flat $S^{-1}A$-module.*

Proof. It is enough to prove that tensoring with $S^{-1}M$ preserves exact sequences of $S^{-1}A$-modules. But any sequence of $S^{-1}A$-modules can be regarded as a sequence of A-modules as well. If N is any $S^{-1}A$-module, regarding it as an A-module we have

$$S^{-1}M \otimes N \cong M \otimes (S^{-1}A \otimes N) \cong M \otimes N,$$

hence M being flat implies $S^{-1}M$ being flat. $\square$

In fact, the following result shows that flatness is a local condition.

Corollary 6.1.8. *If M is a flat A-module, then M_P is flat over A_P for all primes $P \subset A$. Conversely, if M_P is flat over A_P for all maximal ideals P, then M is A-flat.*

Proof. We only need to prove the second claim. Let $N \to N'$ be a monomorphism of A-modules. For any maximal ideal P, $N_P \to N'_P$ is also injective. Since $M_P \otimes_{A_P} N_P \cong (M \otimes N)_P$ and M_P is flat, we conclude that $(M \otimes N)_P \to (M \otimes N')_P$ is injective. By Proposition 2.2.21, we find that $M \otimes N \to M \otimes N'$ is injective, hence M is flat. $\qquad\square$

We also need a result that simplifies dealing with modules of finite presentation: once we have *one* finite presentation for M, *any* way of presenting M with a finite number of generators also ends with a finitely generated module of relations.

Lemma 6.1.9. *Let M be a finitely presented A-module. If we have an exact sequence*

$$0 \longrightarrow K \longrightarrow A^r \longrightarrow M \longrightarrow 0 \,,$$

then K is finitely generated.

Proof. Take any finite presentation

$$0 \longrightarrow L \longrightarrow A^s \longrightarrow M \longrightarrow 0 \,,$$

so that L is finitely generated. By Schanuel's lemma (Proposition 4.4.1) we get

$$K \oplus A^s \cong L \oplus A^r,$$

so K is finitely generated as well. $\qquad\square$

Proof of Proposition 6.1.6. One implication is always true, so assuming M finitely presented and flat, we will prove that it is projective. By Proposition 5.1.15, it is enough to prove that M is locally free. Moreover, using Corollary 6.1.8, we reduce to the case where A is local, and we need to prove that M is free.

Let $\mathcal{M}$ be the maximal ideal of A, $k = A/\mathcal{M}$ and take a (finite) k-basis of $M/\mathcal{M}M$. Lifting the basis to a set of elements of M, we obtain a map $\phi \colon A^r \to M$ for some r. By Nakayama's lemma, ϕ is surjective. Let $K = \ker \phi$, and consider the exact sequence

$$0 \longrightarrow K \longrightarrow A^r \longrightarrow M \longrightarrow 0 \,.$$

By tensoring with k and using the fact that $\mathrm{Tor}_1(M, k) = 0$, we conclude that $K/\mathcal{M}K = K \otimes k = 0$. By Lemma 6.1.9, K is finitely generated, hence by Nakayama again we get that $K = 0$ and ϕ is an isomorphism. $\qquad\square$

Remark 6.1.10. In Proposition 6.1.6, if moveover A is Noetherian, it is enough to assume that M is finitely generated. There is a far reaching generalization of this: in [**RG71**, Theorem 3.4.6] it is proved that if A has finitely many associated primes (something that certainly holds if A is Noetherian, or if A is an integral domain) and M is a finitely generated flat A-module, then M is finitely presented—hence the result applies and M is projective.

We now turn to examples. Since every projective A-module is flat, we give some examples of flat, nonprojective modules.

Example 6.1.11.

(a) Over $A = \mathbb{Z}$, consider the module $M = \mathbb{Q}$. Tensoring $\mathbb{Z}$-modules by $\mathbb{Q}$ preserves exact sequences, since it is the same a localizing with respect to $S = \mathbb{Z} \setminus \{0\}$. On the other hand, $\mathbb{Q}$ is not free, and since $\mathbb{Z}$ is a principal ideal domain, it is not projective.

(b) Giving a finitely generated example takes more work. Let $A = \prod_{i=0}^{\infty} F_i$ be an infinite product of fields, where each $F_i = F$, a fixed field. Inside A, the direct sum $I = \bigoplus_{i=0}^{\infty} F_i$ is an ideal. We claim that $M = A/I$ is flat over A, but not projective.

If M was projective, we would have a decomposition $A \cong I \oplus M$, which is not possible, since I is essential in A (check it!). If $J \subset A$ is any finitely generated ideal, then there is element $a \in A$ such that $J = (a)$ and $a^2 = a$. This is proved by induction on the number of generators (Exercises 16 and 17). Letting $J' = (1 - a)$, we get the decomposition $A = J \oplus J'$, showing that $J' \cong A/J$ is projective, hence flat. Now I is not finitely generated, but we can write I as the increasing union of finitely generated ideals, say $I = \bigcup J_i$. Correspondingly, $A/I = \varinjlim A/J_i$. Since flatness is preserved under direct limits by Remark 6.1.4, $M = A/I$ is flat.

(c) [**Vas69**, Example 3.2] This is an example of a principal ideal that is flat but not projective. Regard $A_0 = \bigoplus_{i \in \mathbb{N}} \mathbb{Z}/2\mathbb{Z}$ as a ring without identity, and let $A = \mathbb{Z} \oplus A_0$ be the ring obtained by adding the identity to A_0. Multiplication in A is defined by $(a, x) \cdot (b, y) = (ab, xy + ay + bx)$, so that $(1, 0)$ is the multiplicative identity. This makes A_0 into an ideal of A.

Let $a = (2, 0) \in A$, and $I = (a)$ the principal ideal that it generates. Notice that $\text{Ann}(a) = A_0$, and that we have an exact sequence

$$0 \longrightarrow \text{Ann}(a) \longrightarrow A \longrightarrow I \longrightarrow 0 .$$

Since I is finitely generated, but A_0 is not, it follows by Lemma 6.1.9 that I is not finitely presented. By Remark 5.1.16, I is not projective.

We can show that I is locally free as follows. Let $P \subset A$ be any prime. If $A_0 \not\subset P$, then $I_P = 0$. Assume instead that $A_0 \subset P$. Notice that any element $b \in A_0$ satisfies $b(1 - b) = 0$, hence b becomes 0 in the localization A_P. But then the annihilator of a becomes trivial in the localization, hence $I_P = aA_P \cong A_P$ is again free. Since I is locally free, it is flat by Corollary 6.1.8.

We can also relate the notion of flatness to that of torsion freeness.

Proposition 6.1.12. *Let M be a flat A-module, where A is an integral domain. Then M is torsion free. Conversely, if M is torsion free and A is a Dedekind ring, then M is flat.*

Proof. The first claim follows immediately by tensoring the exact sequence

$$0 \longrightarrow A \xrightarrow{\cdot a} A \longrightarrow A/(a) \longrightarrow 0$$

with M, where a is any nonzero element of A.

For the second claim, we reduce to the case of finitely generated A-modules by Remark 6.1.4, since every A-module is the direct limit of finitely generated ones. Moreover, it is enough to prove that M_P is flat over A_P for all primes $P \subset A$. But in this case A_P is a *discrete valuation ring* (DVR), in particular a PID. By the classification theorem for finitely generated modules over a PID (Theorem A.2.5), a finitely generated, torsion-free module over a PID is free, and we are done. $\qquad\square$

At this point, it may be useful to state a result that summarizes all links between the various notions of projective, free, locally free, flat, torsion-free modules. All the claims are either obvious or have already been proved, in which case we cite a reference.

Theorem 6.1.13. *Let A be a ring, M an A-module.*

(i) *If M is free, then M is projective;*

(ii) *if M is projective, then M is flat;*

(iii) *if M is finitely presented, then it is projective if and only if it is locally free (5.1.15), if and only if it is flat (6.1.6);*

(iv) *if A is local and M is projective, then M is free (5.1.2);*

(v) *if A is a PID and M is projective, then M is free (5.1.1);*

(vi) *flatness is a local condition (6.1.8), hence if M is locally free, it is flat;*

(vii) *if A is a domain and M is flat, then M is torsion-free* (6.1.12);

(viii) *if A is a Dedekind ring and M is torsion-free, then M is flat* (6.1.12); *and*

(ix) *if A is a PID and M is torsion-free and finitely generated, then M is free* (A.2.5).

To conclude this section, we mimic what we have done in Sections 5.2 and 5.4 to introduce yet another notion of dimension, this time using the vanishing of Tor functors.

Definition 6.1.14. Let A be a ring, M an A-module. A left resolution of M by flat modules is called a *flat resolution*. The length of the smallest flat resolution of M (if any) is called the *weak dimension* (or Tor dimension) of M, denoted $\mathrm{wd}_A M$ (or simply $\mathrm{wd}\, M$). As usual, this is the same as the smallest n (if any) such that $\mathrm{Tor}_i^A(M, N) = 0$ for all A-modules N and all $i > n$. If no such n exists, we set $\mathrm{wd}_A M = \infty$.

The *global weak dimension* of A, denoted $\mathrm{gl.\,wdim}\, A$ is the least n (if any) such that $\mathrm{Tor}_i^A(M, N) = 0$ for all A-modules M, N and all $i > n$. If no such n exists, we set $\mathrm{gl.\,wdim}\, A = \infty$.

Since a projective resolution is a flat resolution, we always have $\mathrm{wd}\, M \le \mathrm{pd}\, M$. Using this notion and Theorem 4.5.7, we can immediately sharpen Theorem 5.4.6.

Theorem 6.1.15. *Let $f \colon A \to B$ a homomorphism of rings, and N a B-module. Then*

$$\mathrm{id}_A N \le \mathrm{wd}_A B + \mathrm{id}_B N.$$

This is less of an improvement than one might think. As it turns out, the weak dimension is the same as the projective dimension under some finiteness hypothesis.

Theorem 6.1.16. *Let A be a Noetherian ring, M a finitely generated A-module. Then $\mathrm{pd}\, M = \mathrm{wd}\, M$.*

Proof. Assume that $\mathrm{wd}\, M = n$. Choose a projective resolution $P_\bullet \to M$. Since M is finitely generated and A is Noetherian, we can take each P_i finitely generated. By truncating this resolution we obtain

$$0 \longrightarrow K \longrightarrow P_{n-1} \longrightarrow \cdots \longrightarrow P_1 \longrightarrow P_0 \longrightarrow M \longrightarrow 0,$$

where K is itself finitely generated, hence finitely presented. By dimension shifting (Proposition 3.3.2), $\mathrm{Tor}_i(K, N) = 0$ for all $i > 0$ and all A-modules N, that is, K is flat. So by Proposition 6.1.6, K is projective and $\mathrm{pd}\, M \le n$. $\qquad\square$

Corollary 6.1.17. *Let A be a Noetherian ring. Then* gl. wdim $A =$ gl. dim A.

Proof. Since Tor commutes with direct limits and every A-module is a direct limit of finitely generated ones, this follows at once from Theorem 6.1.16. ☐

6.2. Flat morphisms

More interesting than flatness for a single module is the notion of flatness for morphisms of rings. The definition is very simple.

Definition 6.2.1. Let $f \colon A \to B$ be a ring homomorphism. We say that f is flat if B, regarded as an A-module via f, is flat.

In other words, for all short exact sequence

$$0 \longrightarrow M_1 \longrightarrow M_2 \longrightarrow M_3 \longrightarrow 0$$

of A-module, the sequence

$$0 \longrightarrow B \otimes_A M_1 \longrightarrow B \otimes_A M_2 \longrightarrow B \otimes_A M_3 \longrightarrow 0$$

must be exact as well. The thing to notice is that the second sequence is also a sequence of B-modules, hence tensoring with B can be seen as a functor $\mathrm{Mod}_A \to \mathrm{Mod}_B$ (extension of scalars). What we are asking for is that this functor be exact.

Example 6.2.2. Let A be a ring, $B = A[x]$. Then B is free as an A-module, hence the natural map $A \to A[x]$ is flat.

As we argued in the introduction of the chapter, geometric intuition leads us to expect that localization and completion maps are flat. We will see that this is indeed the case.

Proposition 6.2.3. *Let A be a ring, $S \subset A$ a multiplicative set. The localization map $A \to S^{-1}A$ is flat.*

Proof. This follows from the isomorphism, $S^{-1}A \otimes_A M \cong S^{-1}M$, plus the fact that localization is exact. ☐

There is also another version of a result expressing the link between flatness and localization. Before that, we will need some preliminary lemmas about the behavior of Tor under localization.

Lemma 6.2.4. *Let $f \colon A \to B$ be a flat morphism, M an A-module and N a B-module, which we regard as an A-module via f. Then there is an isomorphism of A-modules*

$$\mathrm{Tor}_i^A(M, N) \cong \mathrm{Tor}_i^B(B \otimes_A M, N).$$

and an isomorphism of B-modules

$$B \otimes_A \operatorname{Tor}_i^A(M, N) \cong \operatorname{Tor}_i^B(B \otimes_A M, N).$$

Proof. The first claim follows by the base change spectral sequence of Theorem 4.5.6, and the second one by tensoring with B.   □

Lemma 6.2.5. *Let $f \colon A \to B$ be a morphism of rings, N an A-module, M a B-module. Let $Q \subset B$ be prime and $P = f^{-1}(Q)$. Then there is an isomorphism of B_Q-modules*

$$\operatorname{Tor}_i^A(M, N)_Q \cong \operatorname{Tor}_i^{A_P}(M_Q, N_P).$$

Proof. Let $S = A \setminus P$ and denote $B_P = S^{-1}B$, so that B_Q is a further localization of B_P. By flatness of localization (Proposition 6.2.3), we have an isomorphism

$$\operatorname{Tor}_i^A(M, N)_P \cong \operatorname{Tor}_i^{A_P}(M_P, N_P).$$

Notice that both sides are already B_P-modules (not only A_P-modules), since M is a B-module. The map $B_P \to B_Q$ is a localization, hence also flat. Tensoring by B_Q and using Lemma 6.2.4 we get

$$\operatorname{Tor}_i^A(M, N)_Q \cong B_Q \otimes \operatorname{Tor}_i^{A_P}(M_P, N_P) \cong \operatorname{Tor}_i^{A_P}(M_Q, N_P).$$

□

We can use the previous lemmas to prove a relative version of Corollary 6.1.8.

Proposition 6.2.6. *Let $f \colon A \to B$ be a morphism of rings, N a B-module. Then the following are equivalent:*

(i) *N is flat over A;*

(ii) *N_Q is flat over A_P, where $P = f^{-1}(Q)$, for all prime ideals $Q \subset B$; and*

(iii) *$N_{\mathcal{M}}$ is flat over A_P, where $P = f^{-1}(\mathcal{M})$, for all maximal ideals $\mathcal{M} \subset B$;*

Proof. Assume (i), and fix a prime $Q \subset B$, letting $P = f^{-1}(Q)$. Then N_P is flat over A_P by Proposition 6.1.7. The module N_Q is a further localization of N_P, as an A_P-module, hence it is flat as well, proving (ii). That this implies (iii) is obvious.

Assume (iii), and let M be any A-module. By Lemma 6.2.5,

$$\operatorname{Tor}_1^A(M, N)_{\mathcal{M}} \cong \operatorname{Tor}_1^{A_P}(M_P, N_{\mathcal{M}}) = 0,$$

since $N_{\mathcal{M}}$ is assumed A_P-flat, for all maximal ideals $\mathcal{M} \subset B$. This means that $\operatorname{Tor}_1^A(M, N) = 0$ for all A-modules M, that is, N is flat.   □

Corollary 6.2.7. *Let* $f \colon A \to B$ *be a morphism of rings. Then the following are equivalent:*

  (i) f *is flat;*

  (ii) $A_P \to B_Q$ *is flat, where* $P = f^{-1}(Q)$, *for all prime ideals* $Q \subset B$; *and*

  (iii) $A_P \to B_{\mathcal{M}}$ *is flat, where* $P = f^{-1}(\mathcal{M})$, *for all maximal ideals* $\mathcal{M} \subset B$.

To understand the next results, recall the basic facts around completion from Section A.6. If M is an A-module, one can consider the $\widehat{A}$-modules $\widehat{M}$ and $\widehat{A} \otimes_A M$. By the universal property of inverse limits, there is a natural map $t_M \colon \widehat{A} \otimes_A M \to \widehat{M}$, which need not be an isomorphism in general. But there is a partial result.

Proposition 6.2.8. *Let* M *be a finitely generated* A-module. Then t_M is surjective. If moreover A is Noetherian, t_M is an isomorphism.

Proof. Since M is finitely generated, we can choose a surjection $A^k \to M$, and let K be its kernel. We then get the commutative diagram

$$
\begin{array}{ccccccc}
\widehat{A} \otimes K & \longrightarrow & \widehat{A} \otimes A^k & \longrightarrow & \widehat{A} \otimes M & \longrightarrow & 0 \\
\downarrow{\scriptstyle t_K} & & \downarrow{\scriptstyle t_{A^k}} & & \downarrow{\scriptstyle t_M} & & \\
\widehat{K} & \longrightarrow & \widehat{A^k} & \longrightarrow & \widehat{M} & \longrightarrow & 0.
\end{array}
$$

The map t_{A^k} is an isomorphism, hence t_M is surjective. If A is Noetherian, the bottom row is exact also on the left by Theorem A.6.2. Moreover, K is finitely generated, hence t_K is surjective by the first part of the proof. Hence, t_M is an isomorphism by the five lemma (Lemma 2.5.1). $\qquad\square$

This fact has an important consequence.

Theorem 6.2.9. *Let* A *be a Noetherian ring,* $I \subset A$ *an ideal, and* $\widehat{A}$ *the* I-adic completion of A. Then $A \to \widehat{A}$ is a flat morphism.

Proof. Combining Theorem A.6.2 with Proposition 6.2.8, we see that tensoring with $\widehat{A}$ preserves exact sequences of finitely generated A-modules. But then, it preserves arbitrary exact sequences by Remark 6.1.5. $\qquad\square$

Remark 6.2.10. Notice that we used two different functors in the proof: $\widehat{-}$ and $\widehat{A} \otimes_A -$. The two are not the same, but they agree on finitely generated A-modules. The functor $\widehat{-}$ is exact on finitely generated modules, while at the end we conclude that $\widehat{A} \otimes_A -$ is exact on *arbitrary* modules.

Another useful fact about flat morphisms is that they are stable under base change. The algebraic version of this fact is really just a simple observation.

Proposition 6.2.11. *Let $A \to B$ be a homomorphism of rings, and M a flat A-module. Then $M \otimes_A B$ is a flat B-module.*

Proof. Immediate from the fact that for a B-module N we have

$$(M \otimes_A B) \otimes_B N \cong M \otimes_A N$$

as A-modules. □

Corollary 6.2.12. *Let $f\colon A \to B$ be and $g\colon A \to C$ be homomorphisms of rings, with f flat. Then $f'\colon C \to B \otimes_A C$, where $f'(c) := 1 \otimes c$, is flat as well.*

6.3. Faithful flatness

In this section, we introduce a more restrictive condition that turns out to be very useful for investigating flat morphisms.

Definition 6.3.1. Let M be an A-module. We say that M is *faithfully flat* if the following holds: the sequence of A-modules

$$0 \longrightarrow N_1 \longrightarrow N_2 \longrightarrow N_3 \longrightarrow 0$$

is exact if, and only if,

$$0 \longrightarrow M \otimes N_1 \longrightarrow M \otimes N_2 \longrightarrow M \otimes N_3 \longrightarrow 0$$

is exact. A homomorphism of rings $f\colon A \to B$ is called faithfully flat if it makes B into a faithfully flat A-module.

The following general result explains the terminology.

Proposition 6.3.2. *Let $F\colon \mathcal{A} \to \mathcal{B}$ be an additive functor between Abelian categories. Then the following are equivalent:*

(i) *F is exact and faithful, and*

(ii) *for all sequences*

$$(6.3.1) \qquad A \xrightarrow{\; a \;} B \xrightarrow{\; b \;} C$$

in $\mathcal{A}$, the sequence is exact if, and only if

$$(6.3.2) \qquad F(A) \xrightarrow{\; F(a) \;} F(B) \xrightarrow{\; F(b) \;} F(C)$$

is exact.

Proof. Assume that F is faithful and exact, and assume that (6.3.2) is exact. Since $F(b \circ a) = F(b) \circ F(a) = 0$ and F is faithful, $b \circ a = 0$ and (6.3.1) is a complex. Then, notice that, since F is faithful, $F(X) \neq 0$ whenever $X \neq 0$. By assumption, $\ker F(b)/\operatorname{im} F(a) = 0$, and by exactness this is the same as $F(\ker b/\operatorname{im} a)$. Hence $\ker b/\operatorname{im} a = 0$, and (6.3.1) is exact.

Conversely, if F is not faithful, there are objects $A, B \in \mathcal{A}$, and a nonzero map $f\colon A \to B$ such that $F(f) = 0$. But this means that $A \xrightarrow{\operatorname{id}_A} A \xrightarrow{f} B$ is not exact, while $F(A) \xrightarrow{\operatorname{id}_{F(A)}} F(A) \xrightarrow{F(f)} F(B)$ is exact, which contradicts (ii). $\qquad\square$

This yields the following characterization of faithfully flat modules.

Theorem 6.3.3. *Let M be an A-module. Then the following are equivalent:*

   (i) *M is faithfully flat,*

   (ii) *the functor $M \otimes -$ is exact and faithful,*

   (iii) *M is flat and $M \otimes N \neq 0$ for all non zero A-modules N, and*

   (iv) *M is flat and $M \otimes A/\mathcal{M} \neq 0$ for all maximal ideals $\mathcal{M} \subset A$.*

Proof. The equivalence between (i) and (ii) is Proposition 6.3.2.

Assume (ii), and let $N \neq 0$. Then $M \otimes N \neq 0$, otherwise the nonexact sequence $0 \longrightarrow N \longrightarrow 0$ is sent to an exact sequence by the functor $M \otimes -$. Vice versa, assume (iii). If $f\colon N_1 \to N_2$ is a nonzero map, then $M \otimes N_1 \to M \otimes N_2$ is also nonzero, otherwise $M \otimes \operatorname{im} f = 0$. Hence, $M \otimes -$ is faithful, showing (ii).

Since (iii) implies (iv) trivially, we prove that (iv) implies (iii). Given $N \neq 0$, choose any nonzero $n \in N$, and let $\mathcal{M}$ be a maximal ideal containing $\operatorname{Ann}(n)$. Then $A/\mathcal{M}$ embeds as a submodule of N by the map $a \to an$. Since $M \otimes A/\mathcal{M} \neq 0$, and M is flat, a fortiori $M \otimes N \neq 0$. $\qquad\square$

Corollary 6.3.4. *Let $f\colon A \to B$ be a faithfully flat homomorphism of rings. Then for all A-modules M, the natural map $M \to M \otimes_A B$ is injective, in particular, f itself is injective.*

Proof. If $K = \ker(M \to M \otimes_A B)$, then the functor $- \otimes_A B$ sends K to 0, hence $K = 0$. $\qquad\square$

Corollary 6.3.5. *Let $f\colon A \to B$ be a faithfully flat morphism, which we regard as an inclusion by Corollary 6.3.4. Then for all ideals $I \subset A$, we have $I \cdot B \cap A = I$.*

Proof. Under the hypothesis, we have $B \otimes_A (A/I) = B/IB$. This is faithfully flat over A/I (why?), hence the map $A/I \to B/IB$ is injective, which is what we needed to prove. $\square$

The last characterization in Theorem 6.3.3 is especially useful in the local case. Let $(A, \mathcal{M}_A)$ and $(B, \mathcal{M}_B)$ be local rings. We shall say that a homomorphism $f \colon A \to B$ is a *local morphism* if $f(\mathcal{M}_A) \subset \mathcal{M}_B$.

Corollary 6.3.6. *Let $f \colon A \to B$ be a local morphism of rings, M a finitely generated B-module. Then, as an A-module, M is flat if and only if it is faithfully flat.*

Proof. Assuming M flat, we check condition (iv) of Theorem 6.3.3. The only maximal ideal to consider is $\mathcal{M}_A$. By Nakayama's lemma, $M/\mathcal{M}_B M \neq 0$. By flatness, $M \otimes A/\mathcal{M}_A \cong M/f(\mathcal{M}_A)M \neq 0$ as well. $\square$

In particular, we get the following special cases.

Corollary 6.3.7. *A flat local morphism of rings is faithfully flat.*

Combining this with Theorem 6.2.9 we get:

Corollary 6.3.8. *Let $A, \mathcal{M}$ be a local Noetherian ring, $\widehat{A}$ the completion of A in the $\mathcal{M}$-adic topology. Then the map $A \to \widehat{A}$ is faithfully flat—in particular it is injective.*

The last fact is of course also a consequence of Krull's intersection theorem, Theorem A.6.4. Another useful consequence is:

Corollary 6.3.9. *Let A be a local ring, M a finitely generated flat A-module. Then M is faithfully flat.*

An important consequence of this fact is that flat homomorphisms of rings satisfy the *going down* property, similarly to integral extensions (compare Theorem A.4.4).

Theorem 6.3.10 (Flat going down theorem). *Let $f \colon A \to B$ be a flat homomorphism of rings. Let $P_1 \supsetneq P_2$ be prime ideals of A and Q_1 a prime ideal of B such that $P_1 = f^{-1}(Q_1)$. Then there exists a prime ideal $Q_2 \subset Q_1$ such that $f^{-1}(Q_2) = P_2$.*

The link to faithful flatness passes through the following result.

Proposition 6.3.11. *Let $f \colon A \to B$ be a flat morphism. Then f is faithfully flat if and only if for every prime $P \subset A$ there exists a prime $Q \subset B$ such that $f^{-1}(Q) = P$.*

Proof. Assume that f is faithfully flat, in which case we can regard it as an inclusion of rings by Corollary 6.3.4. Let $P \subset A$ be a prime, and denote $B_P := A_P \otimes_A B$. Then B_P is faithfully flat over A_P, hence $PA_P = PB_P \cap A_P$. Any maximal ideal $\mathcal{M}$ of B_P containing PB_P satisfies $\mathcal{M} \cap A_P = PA_P$. If we let $Q \subset B$ be the prime ideal corresponding to $\mathcal{M}$, then $Q \cap A = P$.

For the converse, we are going to prove property (iv) in Theorem 6.3.3. If $\mathcal{M} \subset A$ is any maximal ideal, then choose a prime $Q \subset B$ such that $f^{-1}(Q) = \mathcal{M}$. Then $B/Q \neq 0$, and a fortiori $B \otimes_A (A/\mathcal{M}) = B/\mathcal{M}B \neq 0$. $\qquad\square$

Proof of Theorem 6.3.10. There is an induced map of local rings $A_{P_1} \to B_{Q_1}$, which is flat by Corollary 6.2.7. Since this is a local morphism, it is faithfully flat by Corollary 6.3.7. The conclusion follows by applying Proposition 6.3.11 to the prime $P_2 \cdot A_{P_1}$. $\qquad\square$

Theorem 6.3.10 allows us to prove a basic result, whose geometric meaning is that the dimension of fibers stays constant in a flat family. We will see in Theorem 9.3.1 that, under suitable conditions, this result has a surprising converse.

Theorem 6.3.12. *Let $f \colon A \to B$ be a flat morphism between Noetherian rings. For a prime $Q \subset B$, let $P = f^{-1}(Q)$. Then*

$$\dim B_Q = \dim A_P + \dim(B/PB)_Q.$$

Recall from Theorem A.7.5 that one inequality is true without any flatness hypothesis.

Proof of Theorem 6.3.12. Since $(B/PB)_Q \cong B_Q/PB_Q$, we can work inside B_Q. Notice that $A_P \to B_Q$ is again flat by Corollary 6.2.7. If $\dim B_Q/PB_Q = s$, then we can find a chain of $s + 1$ prime ideals

$$Q_0 \subsetneq \cdots \subsetneq Q_s = QB_Q$$

in B_Q, such that $PB_Q \subset Q_0$. Then $f^{-1}Q_0 = PA_P$. Let

$$P_0 \subsetneq \cdots \subsetneq P_r = PA_P$$

be a maximal chain of ideals in A_P. By flat going down, this gives rise to a chain of ideals in B_Q of the same length, ending in Q_0. Putting the two chains together, we get $\dim B_Q \geq \dim A_P + \dim(B/PB)_Q$, while the other inequality is Theorem A.7.5. $\qquad\square$

Example 6.3.13. To better understand the geometric picture, let k be an algebraically closed field, $f \colon V \to W$ a morphism between affine varieties over k. For a point $p \in V$, let $q = f(p)$. Then p corresponds to a maximal ideal $Q \subset B := R(V)$ and q to a maximal ideal $P \subset A := R(W)$.

In this situation, $\dim B_P$ is the local dimension of V around p, $\dim A_Q$ is the local dimension of W around q, and B/PB is the ring corresponding to the affine variety $V_q := f^{-1}(q) \subset V$, so that $\dim(B/PB)_Q$ is the local dimension of V_q around p.

Theorem A.7.5 states that the dimension of V in p is at most the dimension of V_q at p, plus the dimension of W at q. Theorem 6.3.12 states that if, moreover, the morphism is flat, then we have equality.

If we assume that V and W are irreducible, things are a little easier, since the local dimension of V in p is just $\dim V$, and similarly for W. The fiber V_q itself may not be irreducible, but in the flat case, we see the dimension of V_q at p is $\dim V - \dim W$, which does not depend on p, that is, V_q has the same dimension around all its points. Moreover, this dimension is independent of q, hence all fibers have the same dimension.

6.4. Criteria for flatness

In Remark 6.1.5, we noticed that, in order to test whether the A-module M is flat, we only need to check that $\operatorname{Tor}_1^A(M, N) = 0$ for A-modules N that are finitely generated. In fact, we can strengthen this.

Proposition 6.4.1. *Let M be an A-module. Then M is flat if and only if $\operatorname{Tor}_1^A(M, A/I) = 0$ for all ideals $I \subset A$.*

Proof. Let $N = \langle n_1, \ldots, n_k \rangle_A$ be a finitely generated A-module. Letting $N_i := \langle n_1, \ldots, n_i \rangle_A$, we have a filtration

$$0 = N_0 \subset \cdots \subset N_k = N,$$

and each successive quotient $N_{i+1}/N_i \cong A/I$ for some ideal I (namely, the map $f : A \to N_{i+1}/N_i$ defined by $f(1) = n_{i+1}$ is surjective). By the long exact sequence for Tor and induction, if $\operatorname{Tor}_1^A(M, A/I) = 0$ for all ideals $I \subset A$, then $\operatorname{Tor}_1^A(M, N) = 0$. Since this holds for all finitely generated A-modules N, M is flat. $\qquad\square$

Using the exact sequence for Tor, we can rephrase this more explicitly.

Corollary 6.4.2. *Let M be an A-module. Then M is flat if and only if the natural map $I \otimes_A M \to M$ is injective for all ideals $I \subset A$. Equivalently, the map $I \otimes_A M \to IM$ is an isomorphism.*

There is yet another reformulation that is even more down to earth. Given an A-module M and elements $m_1, \ldots, m_k \in M$, recall that a *syzygy* between them is a vector $(a_1, \ldots, a_k) \in A^k$ such that $\sum_{i=1}^k a_i m_i = 0$.

Definition 6.4.3. There is an obvious way to produce syzygies: choose a matrix $B = (b_{ij}) \in A^{k \times h}$ and a vector $v = (a_1, \ldots, a_k) \in \ker B^T$. Then for any choice of elements $n_1, \ldots, n_h \in M$, let

$$m_i = \sum_{j=1}^{h} b_{ij} n_j.$$

With this choices, v is a syzygy for $m_1, \ldots, m_k$. Such a syzygy will be called *trivial*.

Corollary 6.4.4 (Equational criterion of flatness). *Let M be an A-module. Then M is flat if and only if every syzygy on M is trivial.*

Proof. Assume that M is flat, and let $a_1, \ldots, a_k$ be a syzygy for $m_1, \ldots, m_k$. Let $I = (a_1, \ldots, a_k) \subset A$, and $f \colon A^k \to I$ be the map defined by

$$f(v_1, \ldots, v_k) = a_1 v_1 + \cdots + a_k v_k.$$

Letting $K := \ker f$, by flatness of M we get the exact sequence

$$0 \longrightarrow K \otimes M \longrightarrow M^k \longrightarrow I \otimes M \longrightarrow 0 \,.$$

By hypothesis, $(m_1, \ldots, m_k) \in M^k$ is an element that goes to 0 in IM, and so it goes to 0 in $I \otimes M$ by Corollary 6.4.2. Hence, it comes from an element $\sum_{j=1}^{h} k_j \otimes n_j \in K \otimes M$. Writing an element $k_j \in K \subset A^k$ as a vector and expanding out shows that $a_1, \ldots, a_k$ is a trivial syzygy.

For the converse implication, let $I \subset A$ be an ideal and take an element $\sum_{i=1}^{k} a_i \otimes m_i \in \ker(I \otimes M \to M)$. This means that $a_1, \ldots, a_k$ is a syzygy for $m_1, \ldots, m_k$. The syzygy is trivial, so we can write $m_i = \sum_{j=1}^{h} b_{ij} n_j$, where $\sum_{i=k}^{h} a_i b_{ij} = 0$, and so

$$\sum_{i=1}^{k} a_i \otimes m_i = \sum_{i,j} a_i \otimes b_{ij} n_j = \sum_{i,j} a_i b_{ij} \otimes n_j = 0.$$

In other words, $I \otimes M \to M$ is injective, hence M is flat by Corollary 6.4.2. $\qquad\square$

The main result of this section is a powerful generalization of Proposition 6.4.1 in the local case.

Theorem 6.4.5 (Local criterion of flatness). *Let $A, \mathcal{M}$ be a local Noetherian ring, $k = A/\mathcal{M}$. The A-module M is flat if and only if $\mathrm{Tor}_1^A(M, k) = 0$.*

In other words, in Proposition 6.4.1, we need only check the maximal ideal $\mathcal{M}$. The reader should compare this statement with Kaplansky's theorem (Theorem 5.4.4), which in particular says that a finitely generated A-module M is injective if and only if for all $i > 0$ we have $\mathrm{Ext}_A^i(M, k) = 0$.

Theorem 6.4.5 is an immediate consequence of Krull's intersection theorem (Theorem A.6.4) and the following more general statement.

Theorem 6.4.6 (Local criterion of flatness, strong form). *Let A be a Noetherian ring, $I \subset A$ and M an A-module. Assume that for all ideals $J \subset A$, $J \otimes_A M$ is Hausdorff in the I-adic topology. Then M is flat if and only if M/IM is flat over A/I and $\mathrm{Tor}_1^A(M, A/I) = 0$.*

Lemma 6.4.7. *Let A be a ring, $I \subset A$ and ideal. If M is an A-module such that $\mathrm{Tor}_1^A(M, N) = 0$ for all A/I-modules N, then $\mathrm{Tor}_1^A(M, N) = 0$ for all A/I^n-modules N and all $n \geq 1$.*

Proof. By hypothesis, $\mathrm{Tor}_1^A(M, I^k N/I^{k+1} N) = 0$, which implies the surjectivity of $\mathrm{Tor}_1^A(M, I^{k+1} N) \to \mathrm{Tor}_1^A(M, I^k N)$. Since $I^n N = 0$, the conclusion follows by descending induction. $\qquad\square$

Proof of Theorem 6.4.6. We need to prove that $\mathrm{Tor}_1(M, A/J) = 0$ for all ideals $J \subset A$. Observe that $J \otimes M \to M$ is injective whenever $J \otimes N \to N$ is injective for all finitely generated submodules $N \subset M$. This is clear, since an element in the kernel of $J \otimes M \to M$ involes only finitely many generators of M. Hence, in the sequel, we will assume that M is finitely generated.

We start from the exact sequence

$$0 \longrightarrow \frac{J}{J \cap I^n} \longrightarrow \frac{A}{I^n} \longrightarrow \frac{A}{J + I^n} \longrightarrow 0 \ .$$

Tensoring it with M and using Lemma 6.4.7, we get

$$(6.4.1) \qquad 0 \longrightarrow \frac{J}{J \cap I^n} \otimes M \longrightarrow \frac{M}{I^n M} \longrightarrow \frac{M}{(J + I^n) M} \longrightarrow 0 \ .$$

Similarly, starting from

$$0 \longrightarrow J \cap I^n \longrightarrow J \longrightarrow \frac{J}{J \cap I^n} \longrightarrow 0$$

we obtain

$$(6.4.2) \qquad (J \cap I^n) \otimes M \longrightarrow J \otimes M \longrightarrow \frac{J}{J \cap I^n} \otimes M \longrightarrow 0,$$

this time without left exactness.

Let $m \colon J \otimes M \to M$ be the multiplication map. Take an element $x \in J \otimes M$ such that $m(x) = 0$. Using (6.4.1) we see that x goes to 0 already in $\frac{J}{J \cap I^n} \otimes M$, and by (6.4.2) we see that x is the image of an element in $(J \cap I^n) \otimes M$. By the Artin–Rees theorem, Theorem A.6.1, the I-adic topology on J is the restriction of the I-adic topology on A. In particular, $x \in (J \cap I^n) \otimes M$ for all n if and only if $x \in I^n (J \otimes M)$ for all n. Since we assumed $J \otimes_A M$ to be Hausdorff, it follows that $x = 0$, and we are done by Corollary 6.4.2. $\qquad\square$

Proof of Theorem 6.4.5. As in the proof of Theorem 6.4.6, we can assume that M is finitely generated. In this case, the result is a special case of Theorem 6.4.6. In fact, the condition that $M/\mathcal{M}M$ is flat over k holds trivially; moreover for every ideal J, $J \otimes M$ is Hausdorff by Theorem A.6.4. $\square$

A variant of this result relates the flatness of M to the flatness of the quotients $M/I^n M$. We can look at it as an infinitesimal criterion, since these are the finite convergents to the completion $\widehat{M}$ with respect to the I-adic topology.

Theorem 6.4.8 (Infinitesimal criterion of flatness, strong form). *Let A be a Noetherian ring, $I \subset A$ an ideal, M an A-module such that $J \otimes M$ is Hausdorff in the I-adic topology for all ideals $J \subset A$. Then M is flat if and only if for all $n \gg 0$, $M/I^n M$ is flat over A/I^n.*

Proof. If M flat, $M/I^n M$ is flat as well by Proposition 6.2.11.

For the converse, consider again the sequences (6.4.1) and (6.4.2). These are still exact (for $n \gg 0$), since the first one is actually a sequence of A/I^n-modules, and we can use flatness of $M/I^n M$. The rest of the argument proceeds as in the proof of Theorem 6.4.5. $\square$

As above, it is useful to write out explicitly the specialization of this result to the local case.

Corollary 6.4.9 (Infinitesimal criterion of flatness). *Let $A, \mathcal{M}$ be a local Noetherian ring, M an A-module. Then M is flat if and only if for all $n \gg 0$, $M/\mathcal{M}^n M$ is flat over $A/\mathcal{M}^n$.*

We can also rephrase the above results in terms of graded rings. If M is an A-module and I an ideal, we can consider $\mathrm{Gr}_I(M) := \bigoplus_{n=0}^{\infty} I^n M/I^{n+1} M$ as a graded module over the ring $\mathrm{Gr}_I(A) := \bigoplus_{n=0}^{\infty} I^n/I^{n+1}$. Notice that $\mathrm{Gr}_I(A)_0 = A/I$ and $\mathrm{Gr}_I(M)_0 = M/IM$. To simplify the notation, we shall denote them by A_0 and M_0 respectively. For each n, we have a multiplication map

$$\frac{I^n}{I^{n+1}} \otimes_{A_0} \frac{M}{IM} \to \frac{I^n M}{I^{n+1} M},$$

which can be assembled to yield a morphism

$$m \colon \mathrm{Gr}_I(A) \otimes_{A_0} M_0 \to \mathrm{Gr}_I(M).$$

Theorem 6.4.10. *Let A be a Noetherian ring, $I \subset A$ an ideal, M an A-module such that $J \otimes M$ is Hausdorff in the I-adic topology for all ideals $J \subset A$. Define $A_0 = A/I$ and $M_0 = M/IM$. Then M is flat if, and only if, M_0 is flat over A_0 and the natural map $m \colon \mathrm{Gr}_I(A) \otimes_{A_0} M_0 \to \mathrm{Gr}_I(M)$ is an isomorphism.*

Proof. Assume that M is flat. Then $I^n \otimes M \to I^n M$ is an isomorphism for all n by Corollary 6.4.2. By tensoring the exact sequence

$$0 \longrightarrow I^{n+1} \longrightarrow I^n \longrightarrow \frac{I^n}{I^{n+1}} \longrightarrow 0$$

with M, we also find that $(I^n/I^{n+1} \otimes M) \to (I^n M/I^{n+1} M)$ is an isomorphism, which means that m is an isomorphism. Moreover, M_0 is flat over A_0 by base change.

For the converse implication, we are going to show that $M/I^n M$ is flat over A/I^n for all n, and use Theorem 6.4.8. By Theorem 6.4.6, it is enough to show that $\operatorname{Tor}_1^{A/I^n}\left(\frac{M}{I^n M}, \frac{A}{I}\right) = 0$. To simplify notation, denote $A_n := A/I^n$ and $M_n := M/I^n M$, so that our task is to prove that $\operatorname{Tor}_1^{A_n}(M_n, A/I) = 0$. For a fixed $i \leq n$, we look at the commutative diagram

$$
\begin{array}{ccccccccc}
0 & \longrightarrow & \operatorname{Tor}_1^{A_n}(A/I^{i+1}, M_n) & \longrightarrow & I^{i+1} A_n \otimes M_n & \longrightarrow & I^{i+1} M_n & \longrightarrow & 0 \\
& & \downarrow & & \downarrow & & \downarrow & & \\
0 & \longrightarrow & \operatorname{Tor}_1^{A_n}(A/I^i, M_n) & \longrightarrow & I^i A_n \otimes M_n & \longrightarrow & I^i M_n & \longrightarrow & 0 \\
& & \downarrow & & \downarrow & & \downarrow & & \\
& & 0 & \longrightarrow & \frac{I^i}{I^{i+1}} \otimes M_n & \xrightarrow{m_i} & \frac{I^i M_n}{I^{i+1} M_n} & \longrightarrow & 0.
\end{array}
$$

The reader can easily check that the rows and columns are exact, and the map m_i in the bottom row is the component of m in degree i, hence an isomorphism. By descending induction, starting from $i = n$, we deduce that $\operatorname{Tor}_1^{A_n}(A/I^i, M_n) = 0$ for all i, and for $i = 1$ we reach the desired conclusion. $\qquad\square$

To understand the consequence of this result, we will make use of the Hilbert polynomial (see Section A.8). We will also need a simple result on the length of a tensor product.

Lemma 6.4.11. *Let M, N be A-modules of finite length. Then*

$$\ell(M \otimes_A N) \leq \ell(M) \cdot \ell(N).$$

If, moreover, N is flat,

$$\ell(M) \leq \ell(M \otimes_A N).$$

Proof. Let $M' \subset M$ be a submodule. We have the exact sequence

$$(6.4.3) \qquad M' \otimes N \longrightarrow M \otimes N \longrightarrow \frac{M}{M'} \otimes N \longrightarrow 0,$$

from which we find that

$$(6.4.4) \qquad \ell(M \otimes N) \leq \ell(M' \otimes N) + \ell\left(\frac{M}{M'} \otimes N\right).$$

From this, we derive the desired inequality by induction over a series of composition for M. For the base case, we need to consider what happens when M is simple. Then, we have a surjection $A \to M$, and tensoring with N we get a surjection $N \to M \otimes N$, showing that $\ell(M \otimes N) \leq \ell(N)$, as desired.

If, morever, N is flat, the sequence in (6.4.3) is also exact on the left and all inequalities in (6.4.4) becomes equalities. The conclusion follows again by induction starting from the obvious fact that $\ell(M \otimes N) \geq 1$ when M is simple. $\square$

Theorem 6.4.12. *Let A be a semilocal Noetherian ring, $I \subset A$ an ideal of definition, and M a finitely generated flat A-module. Then,*

$$\chi_A^I(n) \leq \chi_M^I(n) \leq \ell(M/IM)\chi_A^I(n).$$

In particular, $\dim M = \dim A$.

Proof. For all ideals $J \subset A$, $J \otimes M$ is finitely generated. Hence by Krull intersection Theorem A.6.4, Theorem 6.4.10 applies. We conclude that M_0 is flat and $\mathrm{Gr}_I(A) \otimes_{A_0} M_0 \to \mathrm{Gr}_I(M)$ is an isomorphism, where $A_0 = A/I$ and $M_0 = M/IM$. In particular,

$$\ell(A/I^k) \leq \ell(M/I^k M) = \ell(A/I^k \otimes_{A_0} M_0) \leq \ell(A/I^k) \cdot \ell(M_0)$$

by Lemma 6.4.11. $\square$

We also record a variant of Lemma 6.4.11 that will be useful later.

Lemma 6.4.13. *Let $f \colon A \to B$ be a flat, local homomorphism of rings, $\mathcal{M}_A$ the maximal ideal of A, and let M be an A-module of finite length. Denote by ℓ_A (resp., ℓ_B) the length as an A-module (resp., as a B-module). Then,*

$$\ell_B(M \otimes_A B) = \ell_A(M) \cdot \ell_B(B/\mathcal{M}_A B).$$

Proof. Let $n = \ell_A(M)$. By Theorem A.3.7, M admits a series of composition of length n, where each factor is isomorphic to $A/\mathcal{M}_A$. Tensoring with B, which is flat, gives a decomposition of $M \otimes_A B$ with n factors, each one being isomorphic to $B/\mathcal{M}_A B$. $\square$

6.5. Flatness and freeness

In Theorem 6.1.13 we have summarized various implications that exist between the notions of free, projective, locally free, torsion-free and flat modules. In this section, we are going to see a different kind of link that exists between flatness and freeness. The following result allows one to reduce many results about flatness to the free case.

Theorem 6.5.1 (Govorov–Lazard). *Let M be an A-module. Then M is flat if and only if is the direct limit of a directed system of finitely generated, free A-modules.*

This result appears in [**Gov65**] and [**Laz69**]; the proof we give comes from [**dJea20**, Tag $058G$].

Lemma 6.5.2. *Let M be a flat A-module, $f: A^n \to M$ a homomorphism, and $K \subset \ker f$ a finitely generated submodule. Then there exists $m \in \mathbb{N}$ and a factorization $f = h \circ g$, where $g: A^n \to A^m$, $h: A^m \to M$, and $K \subset \ker g$.*

Proof. By induction on the number of generators of K, we reduce to the case where $K = \langle k \rangle$ has a single generator. Letting $e_1, \ldots, e_n$ be the standard basis of A^n and $m_i = f(e_i)$ for $i = 1, \ldots, n$, the element k amounts to a syzygy between $m_1, \ldots, m_n$. By the equational criterion (Corollary 6.4.4), this must be a trivial syzygy. That is, there exist a matrix $b_{ij} \in A^{n \times m}$ and elements $n_1, \ldots, n_m \in M$ such that we can write

$$m_i = \sum_j b_{ij} n_j$$

and, moreover, $\sum_i k_i b_{ij} = 0$, where $k = (k_1, \ldots, k_n)$. Then the matrix (b_{ij}) gives the desired map $g: A^n \to A^m$, whereas h is defined by letting $n_j = h(e_j)$ for $j = 1, \ldots, m$. $\qquad\square$

Although the above lemma looks a bit technical, it can be reformulated in a more convenient way.

Corollary 6.5.3. *Let M, N be A-modules, where M is flat and N is finitely presented. Given a map $f: N \to M$, there exists a finitely generated free module F and a factorization $f = h \circ g$, where $g: N \to F$ and $h: F \to M$.*

Proof. Take a presentation

$$0 \longrightarrow K \longrightarrow G \longrightarrow N \longrightarrow 0 \,,$$

where G and K are finitely generated, and G is free, and apply Lemma 6.5.2 to the composition $G \to N \to M$. $\qquad\square$

Proof of Theorem 6.5.1. One implication is easy: by Remark 6.1.4, a direct limit of flat modules is flat. The content of the theorem is the other implication.

To start, we give a construction that allows us to write any module as the direct limit of finitely presented ones. Namely, take a free module $F = \bigoplus_{i \in I} A$ with a surjection $f: F \to M$, and let $K \subset F$ be the kernel. For a subset $J \subset I$, let $F_J := \bigoplus_{i \in J} A$, and consider the set of pairs (J, N), where J is a finite subset of I and N a finitely generated submodule of $F_J \cap K$.

By construction, the cokernel of $N \to F_J$ is finitely presented. The set P of such pairs has a natural directed order: $(J_1, N_1) \leq (J_2, N_2)$ when $J_1 \subset J_2$ and $N_1 \subset N_2$. If $p = (J, N)$, we denote $M_p := F_J/N$. Then it is immediate to check (do it!) that

$$M = \varinjlim_{p \in P} M_p.$$

Now assume that M is flat, and make a specific choice of I. Namely, choose $I = M \times \mathbb{Z}$—the map $f \colon F \to M$ is defined on the i-th coordinate by sending 1 to m for $i = (m, k) \in I$. With this choice, we will prove that M_p is in fact free for $p \in P'$, where P' is a cofinal subset of P.

To see this, take any $p = (J, N) \in P$, and consider the complex

$$N \longrightarrow F_J = A^J \longrightarrow M.$$

By Lemma 6.5.2, there exists a factorization of the map $F_J \to M$ as

$$F_J \xrightarrow{g} A^m \xrightarrow{h} M$$

such that $N \subset \ker g$. We will prove that there exists $p' \geq p$ such that $M_{p'} = A^m$. If this holds, M is the direct limit of finitely generated free modules.

To prove the claim, let $\{e_1, \ldots, e_m\}$ be the standard basis of A^m and $\{a_i\}_{i \in I}$ that of F. Notice that our choice of index set I guarantees that every element of M is the image of infinitely many basis elements of F. Let $m_i = h(e_i)$ for $i = 1, \ldots, m$. We can then choose elements $j_1, \ldots, j_m \notin J$ such that $f(a_{j_k}) = m_k$, for $k = 1, \ldots, m$. Let $J' = J \cup \{j_1, \ldots, j_m\}$ and extend the map $F_J \to A^m$ to a map $s \colon F_{J'} \to A^m$, letting $s(a_{j_k}) = e_k$. This makes the square

$$
\begin{array}{ccc}
F_{J'} & \longrightarrow & A^m \\
\downarrow & & \downarrow \\
F & \longrightarrow & M
\end{array}
$$

commute. In addition, s is surjective, say with kernel N'. Since A^m is free, s splits, which implies that N' is finitely generated. It follows that $p' = (J', N') \in P$, and by construction $p' \geq p$ and $M_{p'} = A^m$. $\square$

The Govorov–Lazard theorem allows us to reduce many questions on flat modules to the simple case of finitely generated free modules. The reader can try to apply this principle to some of the results in this chapter. As an example, we will prove flatness of symmetric and alternating powers of flat modules.

Definition 6.5.4. Let M be an A-module. The n-th *symmetric power* of M, denoted $\operatorname{Sym}^n M$, is the quotient $M^{\otimes n}/S$, where $S \subset M^{\otimes n}$ is the submodule generated by tensors of the form

$$m_1 \otimes \cdots \otimes m_n - m_1 \otimes \cdots \otimes m_{i+1} \otimes m_i \otimes \cdots \otimes m_n$$

for elements $m_1, \ldots, m_n \in M$ and $1 \le i < n$. In other words, it is the quotient of the tensor module obtained by imposing the commutative law. Similarly, the n-th *exterior power* of M, denoted $\Lambda^n M$, is the quotient $M^{\otimes n}/T$, where $T \subset M^{\otimes n}$ is the submodule generated by tensors of the form

$$m_1 \otimes \cdots \otimes m_n + m_1 \otimes \cdots \otimes m_{i+1} \otimes m_i \otimes \cdots \otimes m_n$$

for elements $m_1, \ldots, m_n \in M$ and $1 \le i < n$. In other words, it is the quotient of the tensor module obtained by imposing the anticommutative law.

The image of $m_1 \otimes \cdots \otimes m_n$ in $\operatorname{Sym}^n M$ is denoted $m_1 \cdots m_n$, while its image in $\Lambda^n M$ is denoted $m_1 \wedge \cdots \wedge m_n$.

Remark 6.5.5. There exists a multilinear map $s_n \colon M^n \to \operatorname{Sym}^n M$ that is commutative, that is $s_n(m_1, \ldots, m_n)$ does not depend on the order of $m_1, \ldots, m_n$. Moreover, it has the following universal property: for every A-module N and every commutative multilinear map $f \colon M^n \to N$, there exists a unique homomorphism $\operatorname{Sym}^n M \to N$ that makes the diagram

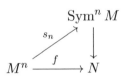

commute. We leave to the reader the formulation of an analogous universal property of the exterior power.

Proposition 6.5.6. *Let M be a flat A-module, $n \in \mathbb{N}$. Then $\operatorname{Sym}^n M$ and $\Lambda^n M$ are flat.*

Proof. First assume that M is free with basis $e_1, \ldots, e_m$. Then, as in the familiar case of vector spaces, one can check that $\operatorname{Sym}^n M$ is free with basis $e_{i_1} \cdots e_{i_n}$ for indices $1 \le i_1 \le \cdots \le i_n \le m$. Similarly, $\Lambda^n M$ is free with basis $e_{i_1} \wedge \cdots \wedge e_{i_n}$ for indices $1 \le i_1 < \cdots < i_n \le m$ if $n \le m$, and 0 otherwise.

If M is flat, use the Govorov–Lazard theorem (Theorem 6.5.1) to write $M = \varinjlim M_i$, where the M_i are finitely generated free. Then the universal properties of direct limits and symmetric and exterior powers easily imply that $\operatorname{Sym}^n M = \varinjlim \operatorname{Sym}^n M_i$ and $\Lambda^n M = \varinjlim \Lambda^n M_i$, and thus are flat. $\qquad\square$

Remark 6.5.7. Notice that $\mathrm{Sym}^n M$ and $\Lambda^n M$ are *not* direct summands of $M^{\otimes n}$ in general (they are, for instance, when A is a field). Hence a direct argument for the flatness of $\mathrm{Sym}^n M$ and $\Lambda^n M$ fails.

Remark 6.5.8. The functors Sym^n and Λ^n are otherwise not very well behaved, for instance they are not either left or right exact. Still, if $M \to N$ is surjective, $\mathrm{Sym}^n M \to \mathrm{Sym}^n N$ is surjective as well. This is an example of a functor that preserves surjections but is not right exact (see Exercise 10). Actually, a moment's thought should convince the reader they are not even additive functors.

6.6. Exercises

Exercises 1–3 prove a result of Miyata [**Miy67**] that apparently is a basic result in module theory, but uses the notion of flatness crucially in its proof.

1. Let $A, \mathcal{M}$ be a local Noetherian ring, M, N two finitely generated A-modules with an inclusion $\iota \colon N \hookrightarrow M$, and assume that, abstractly, $M \cong N \oplus M/N$. Prove that for all A-modules L of finite length, the sequence

$$0 \longrightarrow \mathrm{Hom}\left(\tfrac{M}{N}, L\right) \longrightarrow \mathrm{Hom}(N, L) \longrightarrow \mathrm{Hom}(M, L) \longrightarrow 0$$

induced by ι is exact.

2. Let A, M, N be as in the previous exercise, and denote by $\widehat{A}$ the completion of A with respect to $\mathcal{M}$. Prove that for all finitely generated A-modules L, the sequence

$$0 \longrightarrow \widehat{A} \otimes \mathrm{Hom}\left(\tfrac{M}{N}, L\right) \longrightarrow \widehat{A} \otimes \mathrm{Hom}(N, L)$$
$$\longrightarrow \widehat{A} \otimes \mathrm{Hom}(M, L) \longrightarrow 0$$

induced by ι is exact.

3. Let A be a Noetherian ring, M, N two finitely generated A-modules. Prove that if the sequence

$$0 \longrightarrow M \longrightarrow M \oplus N \longrightarrow N \longrightarrow 0$$

is exact, then it splits.

4. Let $A \subset B$ be integral domains, and assume that B is integral and flat over A. Prove that B is faithfully flat.

5. Let $A \subset B$ be an inclusion of integral domains with the same field of fractions. Assuming that B is faithfully flat over A, prove that $A = B$. Conclude that the integral closure of an integral domain A is never flat over A, unless A is already integrally closed.

6. Let $A \to B$ be a flat morphism. Prove that for all ideals $I, J \subset A$, $(I \cap J) \cdot B = (I \cdot B) \cap (J \cdot B)$.

7. Let $A \to B$ be a flat morphism, $I, J \subset A$ ideals with J finitely generated. Prove that $(I : J) \cdot B = (I \cdot B : J \cdot B)$.

8. Use the equational criterion of flatness to give an alternative proof of Proposition 6.1.6.

9. Let P be a projective A-module. Prove that $\operatorname{Sym}^n P$ and $\Lambda^n P$ are projective.

10. Find an example of an exact sequence of A-modules

$$0 \longrightarrow M_1 \longrightarrow M_2 \longrightarrow M_3 \longrightarrow 0$$

such that

$$\operatorname{Sym}^2 M_1 \longrightarrow \operatorname{Sym}^2 M_2 \longrightarrow \operatorname{Sym}^2 M_3 \longrightarrow 0$$

is *not* exact.

For the next exercise, we need a couple of definitions. Let $f \colon A \to B$ be a ring homomorphism. We say that f is *formally smooth* if the following holds. For every ring C with an ideal $I \subset C$ such that $I^2 = 0$, and every commutative square

$$\begin{array}{ccc} B & \longrightarrow & C/I \\ {\scriptstyle f}\big\uparrow & \diagdown & \big\uparrow \\ A & \longrightarrow & C, \end{array}$$

there exists a map $B \to C$ (the dashed line) that makes the whole diagram commute. In other words, every morphism of A-algebras $B \to C/I$ can be lifted to a morphism $B \to C$. We say that f is *smooth* if it is formally smooth, and moreover there is a finitely generated A-algebra R such that B is a localization of R.

11. Let $f \colon A \to B$ be a smooth morphism or Noetherian rings. Prove that f is flat. (Write $B = R/I$ where R is a localization of a polynomial algebra over A, hence flat. Show that $B = \widehat{R}/\widehat{I}$, where $\widehat{}$ denotes I-adic completion. Show that the map $\widehat{R} \to B$ admits a section, hence B is a direct summand of $\widehat{R}$, and conclude. This quick proof is from [**Con17**]. The result is true with just the assumption that f is formally smooth, see [**Gro64**, 0_{IV}.19.7.1]).

12. Let A be a Noetherian ring. Prove that $A \to A[[x]]$ is a flat morphism.

13. Let $I \subset A$ be an ideal such that $I^2 \neq I$. Prove that A/I is not flat over A.

Exercises 14–19, characterize the rings A such that every A-module is flat. To start, we say that a (possibly noncommutative) ring A is *Von Neumann regular* (VNR) if for every $a \in A$ there exists $b \in A$ such that $a = aba$. In the exercises, we will only consider the commutative case.

14. Let A be a local ring. Show that A is VNR if and only if it is a field.

15. Show that a (commutative) ring A is VNR if and only if $\dim A = 0$ and A has no nilpotents.

16. Show that a (possibly infinite) product of fields is Von Neumann regular.

17. Let A be a ring. Show that A is Von Neumann regular if and only if for every finitely generated ideal $I \subset A$ there exists $a \in A$ such that $I = (a)$ and $a^2 = a$.

18. Let A be a ring. Show that A is VNR if and only if A/I is projective for all ideals I. In turn, prove that this is equivalent to asking that all A-modules are flat. For this reason, VNR rings are also known as *absolutely flat* rings.

19. Let A be a Noetherian VNR ring. Show that A is semisimple, that is, all A-modules are injective and projective,

20. Let M be a projective A-module. Show that M is faithfully flat.

21. Let A be a nonlocal ring, and denote by $\widehat{A}$ the completion of A with respect to a maximal ideal $\mathcal{M}$. Show that $A \to \widehat{A}$ is not faithfully flat, and in particular $\widehat{A}$ is not projective over A.

22. Let $f \colon A \to B$ and $g \colon B \to C$ be ring maps. Assume that $g \circ f$ is flat and g is faithfully flat. Show that f is flat. Can you find an example where g is flat (not faithfully) and f is not?

Chapter 7

Koszul Complexes and Regular Sequences

Up to this point, we have used resolutions as an abstract tool to compute derived functors. In this chapter, we are going to introduce an explicit complex, named the Koszul complex, that in many cases can be proved to be a resolution. This is a complex of modules inspired by the construction of De Rham cohomology, that allows us to compute derived functors in many situations, generalizing the basic Example 3.4.11(c). Short of being only a computational tool, the Koszul complex can also be used to prove theoretical properties of the Ext and Tor functors.

In Section 7.1, we introduce the Koszul complex and prove its basic properties. The key property is Proposition 7.1.10, which expresses the Koszul complex of a sequence as a tensor product of many Koszul complexes, each one associated to a single element, and allows us to prove properties of the Koszul complex by induction. In Section 7.2, we investigate the natural question of understanding when the Koszul complex is actually a resolution. This naturally leads to the introduction of regular sequences: these are sequences of elements $a_1, \ldots, a_n \in A$ such that a_i is not a zero divisor modulo $(a_1, \ldots, a_{i-1})$ for all i (and that do not generate the unit ideal). More generally, we are going to define M-regular sequences with respect to a module M. This definition may seem artificial at first, but it turns out that regular sequences are the linking element between homological and commutative algebra. They are defined purely in terms of zero divisors, but the existence of a regular sequence allows us to define a Koszul resolution and use it to compute derived functors. This link underlies many of the applications of homological methods to commutative algebra.

In Section 7.3, we are going to prove that, under mild hypothesis, the maximal M-regular sequences contained in a given ideal all have the same length. This leads to defining this length as an invariant of ideals and modules, called *depth*. The section is devoted to studying the basic properties of depth, and linking it to other notions of dimension. A crucial result is Rees's theorem (Theorem 7.3.12), that expresses depth in terms of vanishing of the Ext functors.

Section 7.4 is devoted to the proof of the celebrated Auslander–Buchsbaum formula, that gives an unexpected relation between depth and projective dimension, allowing to compute one in terms of the other.

We end the chapter by giving applications of the Koszul complex and regular sequences to familiar topics. In Section 7.5, we recall the concept of multiplicity of a module over a local ring, and prove a result of Auslander and Buchsbaum that expresses the multiplicity in terms of the homology of the Koszul complex. Section 7.6 introduces the Macaulay resultant, a deep generalization of the classical resultant of two polynomials (see [**Fer20**, Section 4.1]), and an important tool in computational algebra and geometry. The theory of Macaulay resultants can be developed in purely elementary terms, as was done by Hurwitz and Macaulay, but it turns out that a key step in the construction is greatly simplified, both technically and conceptually, by using the notions of depth and regular sequences (see Theorem 7.6.8). Finally, Section 7.7 uses the Koszul complex to give an explicit algorithm (due to Chardin) to compute the Macaulay resultant, which is more efficient than the original one introduced by Macaulay.

7.1. The Koszul complex

Let k be a field, and consider the polynomial ring $A = k[x_1, \ldots, x_n]$. By Hilbert's syzygy theorem, (Theorem 5.6.2), we know that every finitely generated module over A admits a free resolution of length at most n. In fact, by repeated use of the first change of rings theorem, Theorem 5.2.10, we see that the shortest free resolution of k as an A-module has length *exactly* n.

In this section we address the question of how to describe such a resolution explicitly. Tracing back the proof of Theorem 5.2.10 is a little messy, since it ultimately relies on the base change spectral sequence for Ext. We will instead take inspiration from De Rham cohomology.

Recall that in Definition 6.5.4 we have defined the exterior powers of an A-module. Moreover we have noticed that if F is a finitely generated free A-module, the exterior powers $\Lambda^k F$ are finitely generated free as well. Notice that $\Lambda^1 F \cong F$, and we can define $\Lambda^0 = A$, generated by 1, the empty product.

Let $\mathcal{M} = (x_1, \ldots, x_n)$ be the maximal ideal of the point 0. This is the image of an A-linear map $s \colon A^n \to A$ given by

$$s(f_1, \ldots, f_n) = x_1 f_1 + \cdots + x_n f_n.$$

In other words, we have the exact sequence

$$A^n \xrightarrow{\ s\ } A \longrightarrow k \longrightarrow 0.$$

Identifying A^n with $\Lambda^1 A^n$ and A with $\Lambda^0 A^n$, we want to extend this sequence in a way that mimics (the dual of) the De Rham complex (3.3.3). In this parallel, k will take the place of $\mathbb{R}$, and A of $C^\infty(M)$.

With a leap of faith, we look at a resolution of k of the form

$$(7.1.1) \qquad 0 \longrightarrow \Lambda^n A^n \xrightarrow{\ d_n\ } \cdots \longrightarrow \Lambda^2 A^n \xrightarrow{\ d_2\ } A^n \xrightarrow{\ d_1 = s\ } A \longrightarrow 0.$$

We need a way to define the differentials d_k to make this into an exact sequence, except at the right.

To understand how to proceed, we begin to highlight some structure. For this purpose, let us now switch to an arbitrary ring A. If F is a free A-module of rank n, the direct sum

$$\Lambda^* F = \bigoplus_{k=0}^n \Lambda^k F$$

is an anticommutative, graded A-algebra, by the wedge product $\wedge$. This just means that $\wedge$ is bilinear over A, and for $x \in \Lambda^k F$ and $y \in \Lambda^h F$ we have the commutation rule $x \wedge y = (-1)^{hk} y \wedge x$. This is easy to verify directly.

Proposition 7.1.1. *Let A be a ring, F a free A-module of rank n and $s \colon F \to A$ a homomorphism. Then there is a unique morphism of A-modules $d \colon \Lambda^* F \to \Lambda^* F$ such that $d|_{\Lambda^1 F} = s$ and d satisfies the graded Leibniz rule*

$$d(x \wedge y) = x \wedge dy + (-1)^k dx \wedge y$$

for $x \in \Lambda^k F$ and $y \in \Lambda^h F$. In fact, $d_k = d|_{\Lambda^k F}$ is given on decomposable forms by the formula

$$(7.1.2) \qquad d_k(v_1 \wedge \cdots \wedge v_k) := \sum_{i=1}^k (-1)^{i+1} s(v_i)(v_1 \wedge \cdots \wedge \widehat{v_i} \wedge \cdots \wedge v_k).$$

Proof. The Leibniz rule, together with the requirement that $d_1 = s$, immediately forces the validity of (7.1.2), hence uniqueness of the differential. For existence, notice that (7.1.2) yields a well-defined morphism $d_k \colon \Lambda^k F \to \Lambda^{k-1} F$ thanks to the universal property of the exterior power. Putting together all d_k, we obtain a map d that satisfies the conclusions. $\square$

Remark 7.1.2. The differential d, together with the grading and the wedge product, makes $\Lambda^* F$ into a so-called *differential graded algebra*.

Using this result, we can now define a suitable complex.

Definition 7.1.3. Let A be a ring, F a free A-module of rank n and $s\colon F \to A$ a homomorphism. The *Koszul complex* associated to s—denoted $K_\bullet(s)$—is the complex

$$(7.1.3) \qquad 0 \longrightarrow \Lambda^n F \xrightarrow{d_n} \cdots \longrightarrow \Lambda^2 F \xrightarrow{d_2} F \xrightarrow{s} A \longrightarrow 0,$$

where d_k is defined by (7.1.2).

It is a straighforward verification (do it!) that (7.1.3) is indeed a complex. When $F = A^n$, every map $s\colon F \to A$ takes the form $s(v) = v \cdot \mathbf{a}$ for some $\mathbf{a} = (a_1, \dots, a_n) \in A^n$. In this case, we will denote $K(a_1, \dots, a_n) = K(\mathbf{a}) := K(s)$.

Finally, let M be an A-module. By tensoring (7.1.3) with M, we obtain another complex, called the Koszul complex associated to s and M, which we denote by $K_\bullet(s; M) = K_\bullet(s) \otimes M$, or $K_\bullet(a_1, \dots, a_n; M)$ when $s(v) = v \cdot \mathbf{a}$. There is also a dual complex $K^\bullet(s; M) = \mathrm{Hom}_A(K_\bullet(s), M)$. Notice that arrows go in the opposite direction, hence we invert the indices and define $K^i(s, M) = \mathrm{Hom}_A(K_{n-i}(s), M)$.

Remark 7.1.4. In Corollary 7.2.3, we will see that when $A = k[x_1, \dots, x_n]$, the complex $K_\bullet(x_1, \dots, x_n)$ is indeed a resolution of k, as promised.

It is not always the case that the complex $K_\bullet(a_1, \dots, a_n; M)$ is acyclic. This depends on the properties of the sequence $a_1, \dots, a_n$ and the module M. It is this fact that allows us to link the homological properties of M with its commutative algebraic ones. Before giving some examples, we need a further definition.

Definition 7.1.5. Let F be a finitely generated free A-module, $s\colon F \to A$, and let M be an A-module. The i-th *Koszul homology* module of M with respect to s is $H_i(K_\bullet(s; M))$. The i-th *Koszul cohomology* module of M with respect to s is $H_i(K^\bullet(s; M))$. When there is no ambiguity, we will shorten the notation to $H_i(s; M) := H_i(K_\bullet(s; M))$ and $H^i(s; M) := H_i(K^\bullet(s; M))$ respectively.

Example 7.1.6.

(a) For a single $a \in A$, the Koszul complex $K(a)$ is just

$$(7.1.4) \qquad\qquad 0 \longrightarrow A \xrightarrow{\cdot a} A \longrightarrow 0.$$

So, for every A-module M the Koszul complexes $K_\bullet(a; M)$ is

$$0 \longrightarrow M \xrightarrow{\cdot a} M \longrightarrow 0.$$

It follows that $H_1(a; M) = (0 :_M a)$ and $H_0(a, M) = M/aM$. Similarly, we see that $H^0(a, M) = (0 :_M a)$ and $H^1(a; M) = M/aM$.

(b) Let $a, b \in A$. To compute the Koszul complex of a, b, let $\{e_1, e_2\}$ be the standard basis for A^2 and notice that $\Lambda^2 A^2 \cong A$, generated by $e_1 \wedge e_2$. The Koszul complex is

$$0 \longrightarrow \Lambda^2 A^2 \xrightarrow{d_2} \Lambda^1 A^2 \xrightarrow{d_1} A \longrightarrow 0,$$

where $d_2(e_1 \wedge e_2) = ae_2 - be_1$, $d_1(e_1) = a$ and $d_1(e_2) = b$. In other words, it is isomorphic to the complex

$$0 \longrightarrow A \xrightarrow{f_2} A^2 \xrightarrow{f_1} A \longrightarrow 0,$$

where the matrix of f_2 is $(-b, a)$ and that of f_1 is $(a, b)^T$. An element of $H_1(K_\bullet(a, b))$ is represented by a pair $(x, y) \in A^2$ such that $xa + yb = 0$, that is, a syzygy between a and b. The pair (x, y) represents 0 in homology if and only if it takes the form $x = -bc$, $y = ac$ for some $c \in A$—in other words it is a trivial syzygy according to Definition 6.4.3. Thus, $H_1(K_\bullet(a, b))$ measures the extent to which syzygies between a, b are trivial.

(c) More generally, let $\mathbf{a} = (a_1, \ldots, a_n)$ be any sequence in A. Let $\{e_1, \ldots, e_n\}$ be the standard basis for A^n, so that $\Lambda^n A^n$ is generated by $e_1 \wedge \cdots \wedge e_n$ and $\Lambda^{n-1} A^n$ is generated by all products of the form $e_1 \wedge \cdots \wedge \widehat{e_i} \wedge \cdots \wedge e_n$. Then we can represent the map $d_n \colon \Lambda^n A^n \to \Lambda^{n-1} A^n$ via the matrix $(a_1, -a_2, \cdots, (-1)^{n+1} a_n)$, whereas the map $d_1 \colon \Lambda^1 A^n \to A$ is given by scalar product with $(a_1, \cdots, a_n)$. In particular

$$H_0(K_\bullet(\mathbf{a})) \cong A/(a_1, \ldots, a_n),$$

while

$$H_n(K_\bullet(\mathbf{a})) \cong \bigcap_{i=1}^n (0 : a_i).$$

More generally, for an A-module M,

$$H_0(\mathbf{a}; M) \cong M/(a_1, \ldots, a_n)M.$$

We now give a different construction for the Koszul complex. The idea is to reduce to the complex for a single element via the tensor product. We start by discussing tensor products of complexes.

Definition 7.1.7. Let $K_\bullet, L_\bullet$ be two complexes of A-modules, with differentials d_K, d_L. We assume that $K_\bullet, L_\bullet$ are bounded below. Define $(K \boxtimes L)_q^p := K_p \otimes L_q$, so that $(K \boxtimes L)_\bullet^\bullet$ is a double complex. The *tensor product* of $K_\bullet$ and $L_\bullet$ is the total complex $K_\bullet \otimes L_\bullet = \mathrm{Tot}(K_\bullet \boxtimes L_\bullet)$. More explicitly,

$$(K \otimes L)_r = \bigoplus_{p+q=r} K_p \otimes L_q,$$

with the differential defined on the $K_p \otimes L_q$ component as $d = (-1)^q d_K + d_L$.

Remark 7.1.8. Due to signs, it is not entirely immediate, but the tensor product of complexes is associative. Namely, there is a natural isomorphism of complexes

$$(K_\bullet \otimes L_\bullet) \otimes M_\bullet \cong K_\bullet \otimes (L_\bullet \otimes M_\bullet).$$

In fact, both complexes have the component of degree k isomorphic to

$$\bigoplus_{p+q+r=k} K_p \otimes L_q \otimes M_r.$$

For the left side, the differential is $(-1)^r((-1)^q d_K + d_L) + d_M$, while for the right-hand side it is $(-1)^{q+r} d_K + ((-1)^r d_L + d_M)$, hence they are the same.

Our main results about the Koszul complex will all follow from the computation of the tensor product of an arbitrary complex with a Koszul complex.

Proposition 7.1.9. *Let $L_\bullet$ be a bounded below complex of A-modules, $a \in A$. Then*

$$(L_\bullet \otimes K_\bullet(a))_i \cong L_i \oplus L_{i-1},$$

where the differential $d_i \colon (L_i \oplus L_{i-1}) \to (L_{i-1} \oplus L_{i-2})$ is given by the matrix

$$(7.1.5) \qquad \begin{pmatrix} d_L & (-1)^i \cdot a \\ 0 & d_L \end{pmatrix}.$$

Proof. We have computed the complex $K(a)$ in (7.1.4). Since the only nonzero component are in degree 0 and 1, both isomorphic to A, the first claim is evident. The second claim also follows by the explicit differential of a tensor product, along with the fact that the map $K_1(a) \to K_0(a)$ is multiplication by a. $\qquad\square$

Corollary 7.1.10. *Let $a_1, \ldots, a_n \in A$. Then there is a natural isomorphism*

$$K_\bullet(a_1, \ldots, a_n) \cong K_\bullet(a_1) \otimes \cdots \otimes K_\bullet(a_n).$$

Proof. By induction, it is enough to prove that

$$K_\bullet(a_1, \ldots, a_n, a_{n+1}) \cong K_\bullet(a_1, \ldots, a_n) \otimes K_\bullet(a_{n+1}).$$

To simplify notation, given indices $i_1 < \cdots < i_k$, we will write $e_{i_1} \wedge \cdots \wedge e_{i_k}$ as e_I, where $I = \{i_1, \ldots, i_k\}$.

A k-element subset I of $\{1, \ldots, n+1\}$ is either a k-element subset of $\{1, \ldots, n\}$, or the union $I' \cup \{n+1\}$, where I' is a $k-1$-element subset of $\{1, \ldots, n\}$. This gives a bijection between a basis of $\Lambda^k A^{n+1}$ and the union of a basis of $\Lambda^k A^n$ and a basis of $\Lambda^{k-1} A^n$. In other words, we get an isomorphism

$$\Lambda^k A^{n+1} \cong \Lambda^k A^n \oplus \Lambda^{k-1} A^n.$$

Let us denote by d the differential in $K_\bullet(a_1, \ldots, a_{n+1})$ and d' that in $K_\bullet(a_1, \ldots, a_n)$. For a basis element e_I of $\Lambda^k A^n$, we have $d e_I = d' e_I$, since

e_{n+1} does not appear. Let I' be a $k-1$-element subset of $\{1, \ldots, n\}$, and $I = I' \cup \{n+1\}$. Then $e_I = e_{I'} \wedge e_{n+1}$, hence

$$de_I = (d'e_{I'}) \wedge e_{n+1} + (-1)^k a_{n+1} e_{I'}.$$

This description agrees with the matrix presentation in (7.1.5). By Proposition 7.1.9, we get the desired isomorphism. $\qquad\square$

Remark 7.1.11. The Koszul complex is built using exterior products, and not just tensor powers. The presence of the -1 sign in the definition of the tensor product of complexes is exactly what is needed to make Corollary 7.1.10 work.

Corollary 7.1.12. *Let $L_\bullet$ be any complex of A-modules, and $a \in A$. Then there is a long exact sequence*

$$\cdots \longrightarrow H_i(L_\bullet) \xrightarrow{\cdot a} H_i(L_\bullet) \longrightarrow H_i(L_\bullet \otimes K_\bullet(a)) \longrightarrow H_{i-1}(L_\bullet) \longrightarrow \cdots .$$

Proof. If $f \colon L_\bullet \to L_\bullet$ is multiplication by a, Proposition 7.1.9 describes $L_\bullet \otimes K_\bullet(a)$ exactly as the mapping cone $C(f)$ (actually, the matrix is transposed with respect to Definition 3.1.10, since we have reversed the direction of the complexes). The conclusion then follows from Proposition 3.1.11. $\quad\square$

Combining Corollaries 7.1.10 and 7.1.12, we immediately get the following special case.

Corollary 7.1.13. *Let $a_1, \ldots, a_{n+1} \in A$ and denote $\mathbf{a} = (a_1, \ldots, a_n)$ and $\mathbf{a}' = (a_1, \ldots, a_{n+1})$. Let M be an A-module. Then there is a long exact sequence*

$$\cdots \longrightarrow H_i(\mathbf{a}; M) \xrightarrow{\cdot a_{n+1}} H_i(\mathbf{a}; M) \longrightarrow H_i(\mathbf{a}'; M) \longrightarrow H_{i-1}(\mathbf{a}; M) \longrightarrow \cdots .$$

There is also a very similar, but slightly different, exact sequence, that we will need in the sequel.

Proposition 7.1.14. *Let $a_1, \ldots, a_n, b \in A$ and denote $\mathbf{a} = a_1, \ldots, a_n$. Let M be an A-module, and assume that b is not a zero divisor on M. Then there is a long exact sequence*

$$\cdots \longrightarrow H_i(\mathbf{a}; M) \xrightarrow{\cdot b} H_i(\mathbf{a}; M) \longrightarrow H_i\left(\mathbf{a}; \tfrac{M}{bM}\right) \longrightarrow H_{i-1}(\mathbf{a}; M) \longrightarrow \cdots .$$

Proof. By hypothesis we have an exact sequence

$$0 \longrightarrow M \xrightarrow{\cdot b} M \longrightarrow \tfrac{M}{bM} \longrightarrow 0.$$

We can take the tensor product with the Koszul complex $K_\bullet(\mathbf{a})$. Since every term in $K_\bullet(\mathbf{a})$ is a free A-module, we obtain a short exact sequence of complexes

$$0 \longrightarrow K_\bullet(\mathbf{a}; M) \xrightarrow{\cdot b} K_\bullet(\mathbf{a}; M) \longrightarrow K_\bullet\left(\mathbf{a}; \tfrac{M}{bM}\right) \longrightarrow 0.$$

The conclusion follows by taking the associated long exact sequence in homology. $\qquad\square$

The results above are about the homology of a Koszul complex. We end this section by proving a result, called *Koszul duality*, that allows us to translate between Koszul homology and cohomology.

Proposition 7.1.15 (Koszul duality). *Let* $s\colon A^n \to A$, *and let* M *be an* A-*module. Then there is a natural isomorphism* $H^i(s; M) \cong H_{n-i}(s; M)$.

Proof. To simplify notation, denote $K_i = K_i(s)$. Identify $\Lambda^n A^n$ with A, generated by the element $e_1 \wedge \cdots \wedge e_n$. This gives a perfect pairing

$$m\colon \quad K_{n-i} \otimes K_i \longrightarrow A,$$

$$x \otimes y \longmapsto x \wedge y$$

and therefore an isomorphism $K_{n-i} \cong \mathrm{Hom}_A(K_i, A)$. This isomorphism makes the diagram

(7.1.6)
$$
\begin{array}{ccc}
K_{n-i} & \longrightarrow & K_{n-i-1} \\
\cong \downarrow & & \downarrow \cong \\
\mathrm{Hom}_A(K_i, A) & \longrightarrow & \mathrm{Hom}_A(K_{i+1}, A)
\end{array}
$$

commute. Since K_i is finitely generated free, we have a natural isomorphism $\mathrm{Hom}_A(K_i, M) \cong \mathrm{Hom}_A(K_i, A) \otimes M$, therefore the squares (7.1.6), tensored with M, fit together to give an isomorphism between $K_\bullet(s; M)$ and $K^\bullet(s; M)$ (recall that the latter has reversed indices). $\qquad\square$

7.2. Regular sequences

Corollary 7.1.13 immediately suggests a way to prove that a Koszul complex is acyclic by induction. To make this precise, we extend Definition 5.2.9 to sequences of elements.

Definition 7.2.1. Let M be an A-module. A sequence $a_1, \ldots, a_n \in A$ is called *weakly M-regular* if a_1 is weakly M-regular and for all $i > 1$, a_i is weakly $M/(a_1, \ldots, a_{i-1})M$-regular. If moreover $M/(a_1, \ldots, a_n)M \neq 0$, we say that $a_1, \ldots, a_n$ is M-*regular*. When $M = A$, we drop the M prefix, and simply say that $a_1, \ldots, a_n$ is a *regular sequence*.

The definition may seem a little arbitrary right now, but it is crucial to the link between homological and commutative algebra. The main reason is the following fact, which is now immediate to prove.

Theorem 7.2.2. *Let M be an A-module, $\mathbf{a} = (a_1, \ldots, a_n)$ a weakly M-regular sequence. Then $H_i(\mathbf{a}; M) = 0$ for $i > 0$.*

Proof. We use induction on n. When $n = 1$, we have $H_1(\mathbf{a}; M) = (0 :_M a_1)$ by Example 7.1.6(a), and this is 0 by assumption. For the inductive step, we use the exact sequence from Corollary 7.1.13. From it, we see that $H_i(\mathbf{a}; M) = 0$ for all $i > 1$. For $i = 1$ we have the exact sequence

$$0 \longrightarrow H_1(\mathbf{a}; M) \longrightarrow H_0(\mathbf{a}'; M) \xrightarrow{\cdot a_n} H_0(\mathbf{a}'; M),$$

where $\mathbf{a}' = (a_1, \ldots, a_{n-1})$. By Example 7.1.6(c), we can identify $H_0(\mathbf{a}'; M)$ with $M/(a_1, \ldots, a_{n-1})M$, and multiplication by a_n is injective by assumption on this module. $\square$

In other words, if $\mathbf{a} = (a_1, \ldots, a_n)$ is weakly M-regular, the Koszul complex $K_\bullet(\mathbf{a}; M)$ is a resolution of $M/(a_1, \ldots, a_n)M$. In particular, if $\mathbf{a}$ is weakly regular, $K_\bullet(\mathbf{a})$ is a free resolution of $A/(a_1, \ldots, a_n)$. This is a powerful criterion, that makes the computation of many derived functors easier.

Corollary 7.2.3. *Let k be a field, $A = k[x_1, \ldots, x_n]$. Then the Koszul complex $K_\bullet(x_1, \ldots, x_n)$ is a free resolution of k as an A-module.*

Proof. It is clear that $x_1, \ldots, x_n$ is a regular sequence. By Theorem 7.2.2, the complex $K_\bullet(x_1, \ldots, x_n)$ is acyclic except at the right, where the cokernel of $A^n \to A$ is k by construction. $\square$

There is actually a partial converse to Theorem 7.2.2.

Theorem 7.2.4. *Let A be a Noetherian ring, M a finitely generated A-module, $\mathbf{a} = (a_1, \ldots, a_n)$ a sequence in the Jacobson radical $\mathcal{J}(A)$. If $H_1(\mathbf{a}; M) = 0$, then $\mathbf{a}$ is M-regular.*

Proof. We first prove that $\mathbf{a}$ is weakly M-regular by induction on n. For $n = 1$, we have $0 = H_1(a_1; M) = (0 :_M a_1)$ by Example 7.1.6(a), so a_1 is not a divisor of 0 on M. For the induction step, denote $\mathbf{a}' = (a_1, \ldots, a_{n-1})$, and consider the exact sequence from Corollary 7.1.13. First, by the vanishing of $H_1(\mathbf{a}; M)$, we get that multiplication by a_n is a surjective map $H_1(\mathbf{a}'; M) \to H_1(\mathbf{a}'; M)$. Since A is Noetherian, $H_1(\mathbf{a}'; M)$ is finitely generated, so by Nakayama's lemma we obtain $H_1(\mathbf{a}'; M) = 0$. By the inductive hypothesis, $\mathbf{a}'$ is weakly M-regular. The tail of the exact sequence is

$$0 \longrightarrow H_0(\mathbf{a}'; M) \xrightarrow{\cdot a_n} H_0(\mathbf{a}'; M) \longrightarrow H_0(\mathbf{a}; M) \longrightarrow 0.$$

By Example 7.1.6(c), this can be rewritten as

$$0 \longrightarrow M/\mathbf{a}'M \xrightarrow{\cdot a_n} M/\mathbf{a}'M \longrightarrow M/\mathbf{a}M \longrightarrow 0,$$

showing that a_n is not a zero divisor on $M/\mathbf{a}'M$, so $\mathbf{a}$ is weakly M-regular as well. Finally, $M/\mathbf{a}M \neq 0$ by Nakayama's lemma, hence $\mathbf{a}$ is M-regular. $\square$

The two theorems above immediately imply the following *exchange property*.

Corollary 7.2.5. *Let A be a Noetherian ring, M a finitely generated A-module, $\mathbf{a} = (a_1, \ldots, a_n)$ a regular sequence in the Jacobson radical $\mathcal{J}(A)$. Then any permutation of $(a_1, \ldots, a_n)$ is also M-regular.*

Proof. Just notice that $H_i(\mathbf{a}; M)$ does not depend on the order of the elements in $\mathbf{a}$ thanks to Corollary 7.1.10. $\square$

This is not obvious at all, since the definition of regular sequence is order dependent.

Example 7.2.6.

(a) The prototypical example of a regular sequence is the sequence $x_1, \ldots, x_n$ in a polynomial ring $A[x_1, \ldots, x_n]$.

(b) Let k be a field, $A = k[x, y]$. Then x^2, xy is not regular, since multiplication by xy is not injective on $k[x, y]/(x^2)$. For a similar reason, xy, x^2 is not regular as well.

(c) Let k be a field, $A = k[x, y, z]$. The sequence $x, y(1-x), z(1-x)$ is regular: indeed, once we quotient by x, we are left with the sequence y, z in the ring $k[y, z]$. But the sequence $y(1-x), z(1-x), x$ is *not* regular: $z(1-x)$ is a divisor of 0 in $k[x, y, z]/(y(1-x))$, as can be seen multiplying it by y.

Let $\mathbf{a} = a_1, \ldots, a_n$ be a regular sequence in the ring A. As we have seen, the Koszul complex $K_\bullet(\mathbf{a})$ is a free resolution of $A/(a_1, \ldots, a_n)$. This allows us to compute derived functors. In fact, it is immediate that

$$H_i(\mathbf{a}; M) \cong \mathrm{Tor}_i^A(A/(\mathbf{a}), M) \text{ and}$$
$$H^i(\mathbf{a}; M) \cong \mathrm{Ext}_A^i(A/(\mathbf{a}), M)$$

for all i and all A-modules M. This is just because the complexes $K_\bullet(\mathbf{a}; M)$ and $K^\bullet(\mathbf{a}; M)$ are obtained from $K_\bullet(\mathbf{a})$ by applying the functors $- \otimes M$ and $\mathrm{Hom}(-, M)$ respectively.

More generally, let $\mathbf{a}$ be any sequence, not necessarily regular. Choose a projective resolution $P_\bullet \to A/\mathbf{a}$. Since the modules $K_i(\mathbf{a})$ are free, applying (the dual of) Proposition 3.2.4 to the identity of $A/\mathbf{a}$, we obtain a map of complexes $K_\bullet(\mathbf{a}) \to P_\bullet$, which is unique up to chain homotopy. Applying the functor $- \otimes M$ and taking homology, we obtain a well-defined natural map $H_i(\mathbf{a}; M) \to \mathrm{Tor}_i^A(A/(\mathbf{a}), M)$. If we apply $\mathrm{Hom}(-, M)$ instead, and then take homology, we obtain a well-defined natural map $\mathrm{Ext}_A^i(A/(\mathbf{a}), M) \to H^i(\mathbf{a}; M)$. We summarize the discussion so far.

Proposition 7.2.7. *Let $\mathbf{a}$ be a sequence in a ring A and M be any module. Then there are natural homomorphisms $H_i(\mathbf{a}; M) \to \mathrm{Tor}_i^A(A/(\mathbf{a}), M)$ and $\mathrm{Ext}_A^i(A/(\mathbf{a}), M) \to H^i(\mathbf{a}; M)$. These maps are isomorphisms if $\mathbf{a}$ is regular.*

Since regular sequences are useful in computing derived functors, it is convenient to understand what kind of operations preserve them.

Proposition 7.2.8. *Let $f\colon A \to B$ be a flat homomorphism of rings, M an A-module and $\mathbf{a}$ a weakly M-regular sequence in A. Let $M_B = M \otimes_A B$. Then $f(\mathbf{a})$ is weakly M_B-regular. If moreover f is faithfully flat and $\mathbf{a}$ is M-regular, then $f(\mathbf{a})$ is M_B-regular.*

Proof. Writing $\mathbf{a} = (a_1, \ldots, a_n)$, we prove the first statement by induction on n. For $n = 1$, we know that multiplication by a_1 is an injective map $M \to M$. Hence it remains injective by tensoring with B. But the tensored map is just multiplication by $f(a_1)$.

For the inductive step, notice that the map $A/(a_1) \to B/(f(a_1))$ is obtained from f by tensoring with $A/(a_1)$, hence it is flat by Corollary 6.2.12. By inductive hypothesis, $(f(a_2), \ldots, f(a_n))$ is weakly regular in $M_B/(f(a_1))$, and putting together with the case $n = 1$ we conclude that $(f(a_1), \ldots, f(a_n))$ is weakly M_B-regular.

For the second claim, we have $M/(\mathbf{a})M \neq 0$ by hypothesis. Hence this module remains nonzero by tensoring with B, by Theorem 6.3.3(iii). $\qquad\square$

Regular sequences can also be used to prove a kind of base change theorem for Tor.

Proposition 7.2.9. *Let M be an A-module, $\mathbf{a}$ a regular and M-regular sequence, and let N be a B-module, where $B = A/(\mathbf{a})$. Then there is an isomorphism $\mathrm{Tor}_i^A(M, N) \cong \mathrm{Tor}_i^B(M/\mathbf{a}M, N)$ for all i.*

Proof. By induction, it is enough to prove the case where $\mathbf{a} = (a)$ consists of a single element. Since a is regular, we have the exact sequence

$$0 \longrightarrow A \overset{\cdot a}{\longrightarrow} A \longrightarrow B \longrightarrow 0.$$

Taking the associated exact sequence of $\mathrm{Tor}_i^A(-, M)$, we get $\mathrm{Tor}_i^A(B, M) = 0$ for $i > 1$. The tail of this sequence looks like

$$0 \longrightarrow \mathrm{Tor}_1^A(B, M) \longrightarrow M \overset{\cdot a}{\longrightarrow} M \longrightarrow B \otimes_A M \longrightarrow 0.$$

Since a is M-regular, the second map is injective, and we deduce that $\operatorname{Tor}_1^A(B, M) = 0$ as well.

On the other hand, $\operatorname{Tor}_i^A(B, M)$ can also be computed by means of a free resolution $F_\bullet$ of M. The vanishing of these groups implies that $F_\bullet \otimes_A B$ is still a resolution of $M \otimes_A B = M/aM$. Moreover, the modules $F_i \otimes_A B$ are free over B, hence $G_\bullet = F_\bullet \otimes_A B$ is a free resolution of M/aM as B-modules. It follows that

$$\operatorname{Tor}_i^B(M, N) \cong H_i(G_\bullet \otimes_B N) \cong H_i(F_\bullet \otimes_A N) \cong \operatorname{Tor}_i^A(M, N). \qquad \square$$

7.3. Depth

Our aim in this section is to use regular sequences to define an invariant of modules.

Definition 7.3.1. Let M be an A-module, $I \subset A$ an ideal. The *depth* (or *grade*) of M with respect to I, denoted $\operatorname{depth}_I M$, is the supremum of the lengths of M-regular sequences contained in I (if $IM \neq M$), or ∞ otherwise. If $A, \mathcal{M}$ is a local ring, we denote $\operatorname{depth} M := \operatorname{depth}_{\mathcal{M}} M$.

Remark 7.3.2. It will turn out that in the Noetherian case, *all* maximal M-regular sequence in I have the same length (Corollary 7.3.11), which is $\operatorname{depth}_I M$. This gives a reason to single out this invariant, even though to prove the result we need some preliminary work.

Remark 7.3.3. Let M be an A-module, $I, P \subset A$ ideals, with P prime. Since localization is flat, by Proposition 7.2.8 we immediately get the inequality

$$\operatorname{depth}_I M \leq \operatorname{depth}_{I_P} M_P,$$

provided $I_P M_P \neq M_P$.

A consequence of this is that to prove that $\operatorname{depth}_I M$ is finite one can reduce to a local situation. But in general, even if a ring is local, we cannot expect that the depth is finite.

Example 7.3.4. Let $A = k[x_1, x_2, \dots]$, where k is a field, and $\mathcal{M} = (x_1, x_2, \dots)$ the ideal of 0. The sequence $x_1, x_2, \dots, x_n$ is regular for all n, hence $\operatorname{depth}_{\mathcal{M}} A = \infty$. If we localize at $\mathcal{M}$, we obtain a local ring of infinite depth.

Using Definition 7.3.1, it is not obvious how depth can be computed, or even how to detect whether it is finite. Hence in this section, we aim for two developments. First, we prove that depth is indeed finite under some finite generation hypothesis. Second, we are going to prove the *depth sensitivity* of some functors, that is, that the vanishing of some homology groups measures the depth of some modules. To get our finiteness result, we will use some results on associated primes from Section A.3.

Theorem 7.3.5. *Let A be a Noetherian ring, M a finitely generated A-module. Let $I \subset A$ be an ideal with $IM \neq M$. Then $\operatorname{depth}_I M < \infty$.*

Proof. First, assume that A is local and that $I = \mathcal{M}$ is the maximal ideal. By Theorem A.8.5, A has finite dimension, so we can take a maximal chain

$$P_0 \subsetneq P_1 \subsetneq \cdots \subsetneq P_n$$

such that $P_i \in \operatorname{Supp}(M)$ for all i. We claim that $\operatorname{depth} M \leq n$, and prove this by induction on n.

Notice that by Nakayama's lemma, $\mathcal{M} \in \operatorname{Supp}(M)$. If $n = 0$, this means that $\operatorname{Supp}(M) = \{\mathcal{M}\}$, hence $\operatorname{Ass}(M) = \{\mathcal{M}\}$ as well by Theorem A.3.4. Write $\mathcal{M} = \operatorname{Ann}(m)$ for some $m \in M$. Then no element of $\mathcal{M}$ can be M-regular, so $\operatorname{depth} M = 0$.

For the inductive step, assume that $n > 0$, and let $a_1, \ldots, a_d$ an M-regular sequence. Letting $M' = M/a_1 M$, we notice that $\operatorname{Supp}(M') \subset \operatorname{Supp}(M)$. Moreover, since a_1 is not a divisor of 0 on M, a_1 does not belong to any minimal prime of M by Theorem A.3.6. It follows that $\dim \operatorname{Supp}(M') < n$, hence by induction $d - 1 \leq \dim \operatorname{Supp}(M')$, and we are done.

In the general case, using Remark 7.3.3, it is enough to prove that there is a prime P such that $I_P M_P \neq M_P$. This is equivalent to $(M/IM)_P \neq 0$, or $P \in \operatorname{Supp}(M/IM)$. By hypothesis, $M/IM \neq 0$, hence its support is nonempty by Theorem A.3.4. $\qquad\square$

We now turn to methods for computing depth. Our first result on depth sensitivity uses the Koszul complex. To get there, we are going to investigate the relation between the Koszul complex of a sequence $\mathbf{a} = a_1, \ldots, a_n$ and the ideal $I = (a_1, \ldots, a_n)$ generated by the same sequence. We start by noticing that I annihilates the Koszul homology of $\mathbf{a}$.

Proposition 7.3.6. *Let M be an A-module, $\mathbf{a} = a_1, \ldots, a_n$ a sequence in A, $I = (\mathbf{a})$ the ideal it generates. Then for all $i \in \mathbb{N}$ we have $I \subset \operatorname{Ann} H_i(\mathbf{a}; M)$.*

Proof. First assume that $A = B[x_1, \ldots, x_n]$ is a polynomial ring and that $a_i = x_i$. In this case, the sequence $\mathbf{a}$ is regular and thus by Proposition 7.2.7 $H_i(\mathbf{a}; M) \cong \operatorname{Tor}_i^A(A/(\mathbf{a}), M)$. In this case, the result is trivially true, since the right-hand side is an $A/(\mathbf{a})$-module.

To reduce the general case to this special one, we consider the homomorphism $f \colon \mathbb{Z}[x_1, \ldots, x_n] \to A$ that sends x_i to a_i. This makes M into a module over $\mathbb{Z}[x_1, \ldots, x_n]$, with $x_i m = a_i m$ for all $m \in M$. The map f induces a map of complexes $K_\bullet(\mathbf{x}) \to K_\bullet(\mathbf{a})$, where the terms of both complexes can be regarded as modules over $\mathbb{Z}[x_1, \ldots, x_n]$. The crucial point is that after tensoring with M, this map becomes an isomorphism, since both complexes have, in place i, a direct sum of $\binom{n}{i}$ copies of M. It follows that—as modules over $\mathbb{Z}[x_1, \ldots, x_n]$—$H_i(\mathbf{x}; M) \cong H_i(\mathbf{a}; M)$. If we regard the right-hand side as a module over A, multiplication by x_i becomes multiplication by a_i, hence the conclusion follows. $\square$

This already hints at a relation between I and $K_\bullet(\mathbf{a}; M)$. To generalize a little, we consider the case where $\mathbf{a}$ is weakly M-regular and contained in I (without necessarily generating it). In this case, we obtain a strengthening of Theorem 7.2.2.

Theorem 7.3.7. *Let M be an A-module, $\mathbf{a} = a_1, \ldots, a_r$ a weakly M-regular sequence. Assume that $(a_1, \ldots, a_r) \subset I = (b_1, \ldots, b_s)$. Then $H_i(\mathbf{b}; M) = 0$ for $i > s - r$.*

Proof. First, assume that $a \in I$ is not a zero divisor on M. By Proposition 7.3.6, multiplication by a is 0 on $H_i(\mathbf{b}; M)$ for all i. It follows that the long exact sequence of Proposition 7.1.14 splits into short exact sequences of the form

$$(7.3.1) \qquad 0 \longrightarrow H_i(\mathbf{b}; M) \longrightarrow H_i\left(\mathbf{b}; \tfrac{M}{aM}\right) \longrightarrow H_{i-1}(\mathbf{b}; M) \longrightarrow 0 .$$

We can now prove the result by induction on r, the case $r = 0$ being trivial. For the inductive step, let $r > 0$, and notice that $a_2, \ldots, a_r$ is weakly regular for $M' = M/a_1 M$. By induction, we know that $H_i(\mathbf{b}; M') = 0$ for $i > s-r+1$, and by the above exact sequence we obtain that $H_{i-1}(\mathbf{b}; M) = 0$ for the same range, which is the desired conclusion. $\square$

We are now ready to derive our first result on depth sensitivity.

Theorem 7.3.8. *Let A be a Noetherian ring, M a finitely generated A-module. Let $\mathbf{b} = b_1, \ldots, b_s$ be a sequence of elements of A, generating $I = (b_1, \ldots, b_s)$, and assume that $IM \neq M$. Then*

$$\mathrm{depth}_I(M) = \min \left\{ i \mid H^i(\mathbf{b}; M) \neq 0 \right\}.$$

If moreover we assume that $\mathbf{b}$ is M-regular, the result applies and yields $\mathrm{depth}_I(M) = s$. We record this special case, which is interesting in itself.

Corollary 7.3.9. *Let A be a Noetherian ring, M a finitely generated A-module. If $\mathbf{b} = b_1, \ldots, b_s$ is M-regular, then the ideal $I = (b_1, \ldots, b_s)$ does not contain any longer M-regular sequences.*

Proof of Theorem 7.3.8. By Koszul duality (Proposition 7.1.15),

$$\min \left\{ i \mid H^i(\mathbf{b}; M) \neq 0 \right\} = n - \max \left\{ i \mid H_i(\mathbf{b}; M) \neq 0 \right\},$$

so we will prove the theorem using Koszul homology.

Assuming that $\mathrm{depth}_I(M) < \infty$, we first prove the result by induction on $r = \mathrm{depth}_I(M)$. If $r = 0$, by the above discussion, $I \subset P$, where P is a prime of the form $(0 : m)$ for some $m \in M$. By Example 7.1.6(c), an element of $H_s(\mathbf{b}; M)$ is exactly an element $m \in M$ such that $mb_i = 0$ for all i. We conclude that $H_s(\mathbf{b}; M) \neq 0$, which is the base case.

For the inductive step, assume that $r > 0$, and let $a_1, \ldots, a_r \subset I$ be M-regular. First, by Theorem 7.3.7, we have $H_i(\mathbf{b}; M) = 0$ for $i > s - r$. Therefore, it is enough to prove that $H_{s-r}(\mathbf{b}; M) \neq 0$.

Letting $M' = M/a_1 M$, we have $\mathrm{depth}_I(M') \leq r - 1$. (Why?) If we let $\mathrm{depth}_I(M') = r - k$, we get $H_{s-r+k}(\mathbf{b}, M') \neq 0$. Using the exact sequence (7.3.1), we deduce that $H_{s-r+k-1}(\mathbf{b}; M) \neq 0$. This implies that we must have $k = 1$, and $H_{s-r}(\mathbf{b}; M) \neq 0$ as desired. $\qquad \square$

As a byproduct of the above proof, we obtain the following result.

Proposition 7.3.10. *Let A be a Noetherian ring, M a finitely generated A-module, $I \subset A$ an ideal such that $IM \neq M$. If $a \in I$ is not a zero divisor on M, then*

$$\mathrm{depth}_I M = 1 + \mathrm{depth}_I \frac{M}{aM}.$$

Corollary 7.3.11. *Let A be a Noetherian ring, M a finitely generated A-module, $I \subset A$ an ideal such that $IM \neq M$. If $d = \mathrm{depth}_I M$, every M-regular sequence in I can be extended to an M-regular sequence of length d.*

Proof. This follows immediately by induction from Proposition 7.3.10. $\quad \square$

We note one more consequence of Theorem 7.3.8. In the same setting, assume that $\mathbf{b}$ is a regular sequence. Then by Proposition 7.2.7, we have isomorphisms $\mathrm{Ext}^i(A/I, M) \cong H^i(\mathbf{b}; M)$. It follows that $\mathrm{depth}_I(M)$ can also be characterized as the smallest integer i such that $\mathrm{Ext}^i(A/I, M) \neq 0$. Notice that this characterization does not mention $\mathbf{b}$ at all, but we still need to require that I is generated by a regular sequence. As it turns out, this fact holds in greater generality, a fact known as the depth sensitivity of Ext.

Theorem 7.3.12 (Rees). *Let A be a Noetherian ring, M a finitely generated A-module, $I \subset A$ an ideal such that $IM \neq M$. Then*

$$\mathrm{depth}_I(M) = \min \left\{ i \mid \mathrm{Ext}^i_A(A/I, M) \neq 0 \right\}.$$

Proof. We use induction on $d = \mathrm{depth}_I M$. When $d = 0$, we have seen that I is contained in a prime P associated to M, so that $P = \mathrm{Ann}\, m$ for some

$m \in M$. Hence, we can define a map $f \colon A/I \to M$ by $f(1) = m$, showing that $\operatorname{Hom}(A/I, M) \neq 0$.

For the inductive step, assume that $d > 0$ and choose $a \in I$ that is not a zero divisor on M. Then we have the exact sequence

$$0 \longrightarrow M \xrightarrow{\ \cdot a\ } M \longrightarrow \tfrac{M}{aM} \longrightarrow 0,$$

whence a long exact sequence of $\operatorname{Ext}^i(A/I, -)$ groups. In this sequence, all maps $\operatorname{Ext}^i(A/I, M) \to \operatorname{Ext}^i(A/I, M)$ are given by multiplication by $a \in I$, so they are 0. If we denote by

$$e(M) := \min \left\{ i \mid \operatorname{Ext}^i_A(A/I, M) \neq 0 \right\}$$

the right-hand side, then it follows that $e(M) = 1 + e(M/aM)$. On the other hand, $\operatorname{depth}_I M = 1 + \operatorname{depth}_I M/aM$ by Proposition 7.3.10, so we are done by induction. $\qquad\square$

By the long exact sequence for the Ext groups, we immediately obtain that the depth of modules is well behaved in exact sequences.

Corollary 7.3.13. *Let A be a Noetherian ring, and*

$$0 \longrightarrow M_1 \longrightarrow M_2 \longrightarrow M_3 \longrightarrow 0$$

an exact sequence of finitely generated A-modules. Then for all ideals $I \subset A$ such that $IM_i \neq M_i$ for $i = 1, 2, 3$ (for instance, this holds if $I \subset \mathcal{J}(A)$) we have the inequalities

$$\operatorname{depth}_I M_2 \geq \min \left\{ \operatorname{depth}_I M_1, \operatorname{depth}_I M_3 \right\},$$
$$\operatorname{depth}_I M_1 \geq \min \left\{ \operatorname{depth}_I M_2, \operatorname{depth}_I M_3 + 1 \right\},$$
$$\operatorname{depth}_I M_3 \geq \min \left\{ \operatorname{depth}_I M_2, \operatorname{depth}_I M_1 - 1 \right\}.$$

Remark 7.3.14. Some authors *define* the depth of M with respect to I as the minimum integer i such that $\operatorname{Ext}^i(A/I, M) \neq 0$. This may not agree with our definition in the case where Rees's theorem does not hold.

7.4. The Auslander–Buchsbaum formula

One of the key results about depth is the Auslander–Buchsbaum formula [**AB57**], which links it to the projective dimension. We are going to prove it as a consequence of Rees's theorem. For a fancy proof using the Ischebeck spectral sequence, see Exercise 22 in Chapter 9.

Theorem 7.4.1 (Auslander–Buchsbaum formula). *Let A be a local Noetherian ring, M a finitely generated A-module with $\operatorname{pd} M < \infty$. Then*

$$\operatorname{pd} M + \operatorname{depth} M = \operatorname{depth} A.$$

Proof. Denote by $\mathcal{M}$ the maximal ideal, $k = A/\mathcal{M}$. Recall that by Kaplansky's theorem (Theorem 5.1.2), every projective A-module is free. We prove the result by induction on $p = \operatorname{pd} M$.

When $p = 0$, M is free, say $M = A^n$. In this case, a sequence is A-regular if and only if it is M-regular, hence $\operatorname{depth} M = \operatorname{depth} A$.

When $p = 1$, we take a minimal free resolution of M of the form

$$0 \longrightarrow A^n \xrightarrow{\ f\ } A^m \longrightarrow M \longrightarrow 0.$$

By Proposition 5.6.5, f is expressed by some $n \times m$ matrix with coefficients in $\mathcal{M}$.

In the associated long exact sequence of $\operatorname{Ext}(k, -)$ groups, consider the map $f_i \colon \operatorname{Ext}^i(k, A^n) \to \operatorname{Ext}^i(k, A^m)$. If we identify $\operatorname{Ext}^i(k, A^n)$ with $\operatorname{Ext}^i(k, A)^n$—and do the same for $\operatorname{Ext}^i(k, A^m)$—we can express f_i via an $n \times m$ matrix, which in fact is the same matrix as f. Since this matrix has coefficients in $\mathcal{M}$, the map f_i is 0. This allows us to split the long exact sequence into short exact sequences of the form

$$0 \longrightarrow \operatorname{Ext}^i(k, A)^n \longrightarrow \operatorname{Ext}^i(k, M) \longrightarrow \operatorname{Ext}^{i+1}(k, A)^m \longrightarrow 0 .$$

Using the characterization of depth given by Rees's theorem (Theorem 7.3.12), we conclude that $\operatorname{depth} A = \operatorname{depth} M + 1$.

For the inductive step, assume that $p > 1$, and consider an exact sequence

$$0 \longrightarrow M' \longrightarrow A^m \longrightarrow M \longrightarrow 0 ,$$

where $\operatorname{pd} M' = p - 1$. By induction, we know that $\operatorname{pd} M' + \operatorname{depth} M' = \operatorname{depth} A$, so we only neet to prove that $\operatorname{depth} M' = \operatorname{depth} M + 1$. But this follows again from Rees's theorem and the long exact sequence for Ext. $\square$

Remark 7.4.2. The above quick proof is from [**Mat86**, Theorem 19.1]. For another proof that uses induction on $\operatorname{depth} M$ instead, see [**dJea20**, Tag 090U].

Example 7.4.3. Finite projective dimension is required for the Auslander–Buchsbaum formula. To see this, let k be a field, $A = k[[x, y]]/(x^2, xy)$. Then $\operatorname{depth} A = 0$, since every noncostant power series is a zero divisor. Alternatively, $\operatorname{Hom}(k, A) \neq 0$, which implies that $\operatorname{depth} A = 0$ by Rees's theorem, Theorem 7.3.12. On the other hand, let $B = A/(x) \cong k[[y]]$. Then $\operatorname{depth} B = 1$, which means that $\operatorname{pd}_A B = \infty$ by the Auslander–Buchsbaum formula.

To better understand the formula, it is convenient to draw a link between depth and height (see Section A.7). With the techniques of Proposition 7.3.5, it is very easy to prove a lower bound for the height.

Proposition 7.4.4. *Let A be a Noetherian ring, $I \subset A$ an ideal containing a regular sequence $a_1, \ldots, a_n$. Then $\operatorname{ht} I \geq n$.*

Proof. Use induction on n, the case $n = 0$ being trivial. By hypothesis, a_1 is not a divisor of zero, hence it does not belong to any minimal prime of A. In the ring $A/(a_1)$, the sequence $a_2, \ldots, a_n$ is regular. It follows that $\operatorname{ht} I/(a_1) \geq n - 1$. A chain of primes ending at a minimal prime of $I/(a_1)$ can be lifted to a similar chain in A, and then it can be enlarged by adding a minimal prime, so $\operatorname{ht} I \geq n$. □

Combining this with Krull's principal ideal theorem (Theorem A.7.6), we obtain:

Corollary 7.4.5. *Let A be a Noetherian ring, $I \subset A$ an ideal generated by a regular sequence $a_1, \ldots, a_n$. Then $\operatorname{ht} I = \operatorname{depth}_I A = n$.*

In the special case where A is local we obtain:

Corollary 7.4.6. *Let $A, \mathcal{M}$ be a local Noetherian ring. Then $\operatorname{depth} A \leq \dim A$.*

Using this, we obtain a useful consequence of the Auslander–Buchsbaum formula.

Corollary 7.4.7. *Let A be a local Noetherian ring of finite global dimension. Then $\operatorname{gl.dim} A \leq \dim A$.*

This is clarified in Theorem 8.2.1.

7.5. Multiplicities revisited

In this section, we use the Koszul complex to revisit the concept of multiplicity. Let A be a semilocal Noetherian ring of dimension d, with Jacobson radical $\mathcal{M} = \mathcal{J}(A)$. Let Q be an ideal of definition for A, that is, assume that $\mathcal{M}^k \subset Q \subset \mathcal{M}$ for some $k > 0$, and let M be a finitely generated A-module. By Theorem A.8.3, the Hilbert function $\chi_M^Q(n) = \ell(M/Q^n M)$ agrees (for large n) with a rational polynomial of degree $\dim M := \dim A/\operatorname{Ann}(M)$.

In fact, we can write

$$\chi_M^Q(n) = \frac{e}{d!}n^d + a_{d-1}n^{d-1} + \cdots + a_0$$

for some rational numbers a_i and some $e \in \mathbb{N}$. The number $e = e(Q, M)$ is called the *multiplicity* of M at Q. When A is local and $Q = \mathcal{M}$, we simplify the notation to $e(M) = e(\mathcal{M}, M)$.

Our aim in this section is to prove the following result from [**AB58**].

Theorem 7.5.1 (Auslander–Buchsbaum). *Let A be a semilocal ring of dimension d, $\mathbf{a} = a_1, \ldots, a_d$ a system of parameters and $Q = (a_1, \ldots, a_d)$. Then for all finitely generated A-modules M we have*

$$e(Q, M) = \chi(H_\bullet(\mathbf{a}; M)) := \sum_{i=0}^{d} (-1)^d \ell\left(H_i(\mathbf{a}; M)\right),$$

where $H_i(\mathbf{a}; M)$ is the i-th Koszul homology.

We first remark that the lengths appearing in the formula are finite. In fact, by Proposition 7.3.6, $H_i(\mathbf{a}; M)$ is a finitely generated module over A/Q, and $\ell(A/Q) < \infty$ since Q is an ideal of definition. It follows that $\ell(H_i(\mathbf{a}; M))$ is finite as well, so at least the right-hand side is well defined. We will give a proof of the theorem based on a spectral sequence argument from [**dJea20**, Tag 0*AZU*] (already hinted at in the original paper).

Proof. For brevity, denote $K_n = K_n(\mathbf{a}; M)$. We define a decreasing filtration on the complex $K_\bullet$ by

$$F_p K_n = Q^{p-n} K_n$$

for $p \geq n$, and $F_p K_n = K_n$ otherwise. We claim that this makes $K_\bullet$ into a filtered complex. The reason is that the differential on a generator of K_n is a linear combination of generators of K_{n-1}, with coefficients the elements $a_1, \ldots, a_d \in Q$. It follows that

$$d_n F_p K_n \subset Q \cdot Q^{p-n} K_{n-1} = F_p K_{n-1}.$$

By definition, the filtration is canonical, and it is also exhaustive by Krull's intersection theorem, Theorem A.6.4.

We can then use Theorem 4.2.2 to obtain a spectral sequence $E^r_{p,q}$ with

$$E^0_{p,q} = \frac{F_p K_{p+q}}{F_{p+1} K_{p+q}} = \frac{Q^{-q} K_{p+q}}{Q^{-q+1} K_{p+q}}$$

for $q < 0$, and 0 otherwise. By the same theorem,

$$E^1_{p,q} = H_{p+q}\left(\frac{F_p K_\bullet}{F_{p+1} K_\bullet}\right).$$

Notice that this is 0 for $p < n = p+q$, that is $q > 0$. Since $K_\bullet$ is concentrated in degrees $0, \ldots, d$, this is also 0 unless $p + q$ is in this range. For $q \leq 0$ and $0 \leq p + q \leq d$, the module

$$E^0_{p,q} = K_{p+q}(\mathbf{a}) \otimes \frac{Q^{-q} M}{Q^{-q+1} M}$$

has finite length, and a fortiori $E^1_{p,q}$ has finite length. We claim that $E^1_{p,q}$ is also 0 for $p \gg 0$, so that the E^1 sheet has only finitely many nonzero entries, each having finite length. It will immediately follow that the same is true for all further sheets.

To prove the claim, it is simpler to rephrase it as $\ell(E^1) < \infty$, where as usual $E^1 = \bigoplus_{p,q} E^1_{p,q}$. The reason is that we can see E^1 as a graded module over the graded ring $\mathrm{Gr}_Q A$. The ring $\mathrm{Gr}_Q A$ is Noetherian, and as $\mathrm{Gr}_Q A$-modules we have

$$E^0 = \mathrm{Gr}_Q K(\mathbf{a}) \otimes_{\mathrm{Gr}_Q A} \mathrm{Gr}_Q(M),$$

since $\mathrm{Gr}_Q K(\mathbf{a})$ is free over $\mathrm{Gr}_Q A$. In particular, E^0 is finitely generated over $\mathrm{Gr}_Q A$, and the same is true for E^1. But E^1 is annihilated by Q, due to Proposition 7.3.6, so it is in fact a finitely generated module over A/Q, which implies that it has finite length.

Now, since length is additive on exact sequences, it is a simple exercise to check that if $\ell(E^r) < \infty$ for some r, then the alternating sum of lengths is conserved in the next sheet, that is,

$$\sum_{p,q}(-1)^{p+q}\ell(E^r_{p,q}) = \sum_{p,q}(-1)^{p+q}\ell(E^{r+1}_{p,q}).$$

The theorem will follow by comparing such alternating sums for E^1 and E^∞. Notice that Theorem 4.2.2 tells us that E^r abuts $E^\infty = \mathrm{Gr}\, H_\bullet(K)$, so each module $E^r_{p,q}$ eventually stabilizes for $r \gg 0$. Since there are only finitely many nonzero such modules, the sequence converges, justifying the comparison of the alternating sums.

It remains to compute such alternating sums. For E^∞, we get exactly

$$(7.5.1) \qquad \sum_n (-1)^n \ell(H_n(K_\bullet)) = \chi(H_n(\mathbf{a}; M)),$$

where $n = p + q$. To compute the same alternating sum for E^1, choose T such that $E^1_{p,q} = 0$ for $p \geq T$. So the sum is restricted to a finite region $R \subset \mathbb{Z}^2$ given by $q \leq 0$, $p \leq T$, $0 \leq p + q \leq d$. We change variables, letting $n = p + q$, $q' = -q = p - n$, so that $0 \leq n \leq d$ and $0 \leq q' \leq T - n$. The alternating sum becomes

$$(7.5.2)$$

$$\sum_{n=0}^{d} \sum_{q'=0}^{T-n} (-1)^n \ell\left(H_n\left(\frac{Q^{q'}(K_\bullet(\mathbf{a}) \otimes_A M)}{Q^{q'+1}(K_\bullet(\mathbf{a}) \otimes_A M)} \right) \right)$$

$$= \sum_{n=0}^{d} \sum_{q'=0}^{T-n} (-1)^n \binom{d}{n} \ell\left(\frac{Q^{q'} M}{Q^{q'+1} M} \right)$$

$$= \sum_{n=0}^{d} (-1)^n \binom{d}{n} \ell\left(\frac{M}{Q^{T+1-n} M} \right) = \sum_{n=0}^{d} (-1)^n \binom{d}{n} \chi_Q^M(T + 1 - n).$$

For a polynomial f and $r \in \mathbb{N}$, we have the equality

$$\sum_{n=0}^{r}(-1)^n \binom{r}{n} f(s-n) = (\Delta^r f)(s),$$

where Δ is the difference operator $(\Delta f)(s) = f(s) - f(s-1)$. This is easily proved by induction on r. In particular, when $r = d = \deg f$, we obtain

$$\sum_{n=0}^{r}(-1)^n \binom{d}{n} f(s-n) = d! \cdot f_d,$$

where f_d is the leading coefficient of f. Applying this to Hilbert polynomial χ_Q^M, we recognize in (7.5.2) the expression for $e(Q, M)$. Equating this to (7.5.1) gives the result. $\qquad\square$

Remark 7.5.2. Assume further that Q is generated by a regular sequence. We will see in the next chapter that this always holds if the ring A is *regular* and Q is the maximal ideal. In this case, by Proposition 7.2.7 we have an isomorphism $H_i(\mathbf{a}; M) \cong \mathrm{Tor}_i(A/Q, M)$, hence we can rewrite the multiplicity formula as

$$e(Q, M) := \sum_{i \geq 0}(-1)^i \ell\left(\mathrm{Tor}_i(A/Q, M)\right).$$

For a regular local ring A and two prime ideals P, Q, Serre defined the intersection numbers

$$\chi(A/P, A/Q) := \sum_{i \geq 0}(-1)^i \ell\left(\mathrm{Tor}_i(A/P, A/Q)\right),$$

under the assumption that $\ell(A/P \otimes_A A/Q) < \infty$. Even more generally, he worked with finitely generated A-modules M, N under the same assumption, defining

$$\chi(M, N) := \sum_{i \geq 0}(-1)^i \ell\left(\mathrm{Tor}_i(M, N)\right).$$

His aim was to develop a concept of intersection multiplicity suitable to develop intersection theory on regular varieties (and schemes). In order to make the theory work, though, Serre needed a few properties of intersection numbers, that have come to be known as Serre's multiplicity conjectures.

Conjecture (Serre). *Let A be a regular local ring, M, N finitely generated A-modules with $\ell(M \otimes_A N) < \infty$. Then*

    (i) *the inequality $\dim M + \dim N \leq \dim A$ holds;*

    (ii) *the intersection numbers $\chi(M, N) \geq 0$;*

    (iii) *if $\dim M + \dim N < \dim A$, then $\chi(M, N) = 0$; and*

    (iv) *if $\dim A/P + \dim A/Q = \dim A$, then $\chi(M, N) > 0$.*

The conjectures are present in [**Ser00**], and were proved by Serre under some additional conditions—for instance they hold if char $A =$ char k, where $k = A/\mathcal{M}$ is the residue field. The dimension inequality (i) was proved by Serre himself in the general case. The vanishing condition (iii) was proved in [**Rob85**] and [**GS87**] independently. The nonnegativity (ii) was proved in [**Gab95**] (unpublished, but see [**Hoc97**] for a presentation of the result). The strict positivity (iv) remains open in general.

7.6. Macaulay resultants

In this section, we will define the resultant of n homogeneous polynomials in n indeterminates. Using the Koszul complex, we are also going to give an algorithm to compute it. The Macaulay resultant is a fundamental tool in computational algebra and elimination theory, that allows us to generalize some of the techniques of [**Fer20**, Section 4.1]. Our presentation follows [**Jou91**] and [**AA97**].

Let us fix a field k, and consider r homogeneous polynomials $f_1, \ldots, f_r \in k[x_1, \ldots, x_n]$, having degrees $d_1, \ldots, d_r$. We would like to define some polynomial expression $R = \mathrm{Res}(f_1, \ldots, f_r)$ in the coefficients of $f_1, \ldots, f_r$ with the property that $f_1, \ldots, f_r$ have a common nontrivial zero in $\overline{k}$ if and only if R is 0. In the first volume, we were able to do this for $r = 2$, with an explicit construction as the determinant of a Sylvester matrix. In that situation, we did not even require f_1 and f_2 to be homogeneous.

Let us rephrase a bit what we are aiming to do. Our first requirement is that the resultant should be obtained by an explicit polynomial formula in the coefficients of $f_1, \ldots, f_r$. Because of this, it will be easier to first work with generic polynomials, and then specialize the formula. Given a multi-index $\alpha = (\alpha_1, \ldots, \alpha_n) \in \mathbb{N}^n$, we write $x^\alpha := x_1^{\alpha_1} \cdots x_n^{\alpha_n}$ and $\deg \alpha = \alpha_1 + \cdots + \alpha_n$. A homogeneous polynomial f_i of degree d_i then has the form

$$(7.6.1) \qquad\qquad f_i = \sum_{\deg \alpha = d_i} u_{i,\alpha} x^\alpha$$

for some coefficients $u_{i,\alpha}$. To work in the universal case, we will consider f_i defined by (7.6.1) over the polynomial ring $\mathbb{Z}[u_{i,\alpha}]$, where the $u_{i,\alpha}$ are indeterminates, one for each $i = 1, \ldots, n$ and α a multi-index with $\deg \alpha = d_i$. Our requirement that the resultant R is a polynomial expression in the coefficient of the f_i can be rephrased as saying that we are looking for a particular element of the ring $A_0 = \mathbb{Z}[u_{i,\alpha}]$. We also denote $A := A_0[x_1, \ldots, x_n]$, which we regard as a graded ring with $\deg x_i = 1$, so that $f_1, \ldots, f_r \in A$. The notation is consistent, since A_0 is the degree 0 part of A.

The result that will enable us to develop the theory of multivariate resultants is the following.

Theorem 7.6.1 (Macaulay). *Let $d_1, \ldots, d_n \in \mathbb{N}$ and consider for each i and each multi-index α with $\deg \alpha = d_i$ an indeterminate $u_{i,\alpha}$. Let R be a UFD, define rings $A_0 = R[u_{i,\alpha}]$ and $A = A_0[x_1, \ldots, x_n]$ and consider the elements $f_i \in A$ defined by (7.6.1). Define the ideal $I = (f_1, \ldots, f_n) \subset A$ and the subset*

$$T_0 = \{a \in A_0 \mid x^\alpha a \in I \text{ for all } \alpha \text{ with } \deg \alpha = s, \text{ for some } s \in \mathbb{N}\}$$

of the ring A_0. Then T_0 is a principal ideal of A_0.

Notice that in the theorem, the number r of polynomials agrees with the number n of variables. To explain where the ideal T_0 comes from; it is convenient to introduce some terminology.

Definition 7.6.2. Let A_0 be a ring, $A = A_0[x_1, \ldots, x_n]$ and consider an ideal $I \subset A$. An element $f \in A$ is a called a *inertia form*, with respect to I if $f x^\alpha \in I$ for all multi-indices α with $\deg \alpha = s$, for some $s \in \mathbb{N}$.

Clearly, an element of I is an inertia form with respect to I itself—such forms will be called *trivial*. The set T_0 defined above is then the set of inertia forms of degree 0.

Remark 7.6.3. For a fixed $s \in \mathbb{N}$, the elements that satisfy $f x^\alpha \in I$ for all α with $\deg \alpha = s$ form an ideal

$$I_s := \bigcap_{\deg \alpha = s} (I : x^\alpha) = (I : (x^\alpha)_{\deg \alpha = s}).$$

The inertia forms are the increasing union of such ideals I_s for $s \in \mathbb{N}$, hence they are an ideal as well.

Example 7.6.4. Let k be a field of characteristic 0, and consider polynomials $f_1, \ldots, f_n \in A = A_0[x_1, \ldots, x_n]$, with $\deg f_i = d_i$. The Jacobian determinant

$$J = J(f_1, \ldots, f_n) := \det \left(\frac{\partial f_i}{\partial x_j} \right)$$

is an inertia form for $(f_1, \ldots, f_n)$. To see this, we will use the Euler identity

$$d_j f_j = x_1 \frac{\partial f_j}{\partial x_1} + \cdots + x_n \frac{\partial f_j}{\partial x_n}.$$

Multiplying the first column of the matrix by x_1 and summing suitable multiples of the other columns, we obtain

$$x_1 J = \det \begin{pmatrix} d_1 f_1 & \frac{\partial f_1}{\partial x_2} & \cdots & \frac{\partial f_1}{\partial x_n} \\ \vdots & \vdots & & \vdots \\ d_n f_n & \frac{\partial f_n}{\partial x_2} & \cdots & \frac{\partial f_n}{\partial x_n} \end{pmatrix}$$

so that $x_1 J \in (f_1, \ldots, f_n)$. Similarly, $x_i J \in (f_1, \ldots, f_n)$ for all $i = 1, \ldots, n$.

The content of Theorem 7.6.1 is that for the ring of generic polynomials $A_0 = R[u_{i,\alpha}]$, the ideal of inertia forms of degree 0 is principal. To see why this is the case, we will introduce some other rings, and we will work in slightly more generality than required by the theorem.

Let us fix an integral domain R, fix degrees $d_i \in \mathbb{Z}_{>0}$, with corresponding indeterminates $u_{i,\alpha}$ with $\deg \alpha = d_i$, and let $A_0 = R[u_{i,\alpha}]$. We take the generic homogeneous polynomials f_i given by (7.6.1), living inside $A = A_0[x_1, \ldots, x_n]$. We also denote $I = (f_1, \ldots, f_r) \subset A$, and $B = A/I$. *This setup will be fixed for the rest of the section.*

Let us now fix an index $i = 1, \ldots, n$, and let v_j be the coefficient of $x_i^{d_j}$ in f_j, for $j = 1, \ldots, r$. We let A' be the ring generated over R by the coefficients of $f_1, \ldots, f_r$ *other than* $v_1, \ldots, v_r$. This way, we have $A_0 = A'[v_1, \ldots, v_r]$ by construction. The main technical result that we need is the following.

Lemma 7.6.5. *Keep notation as above. Then we have an isomorphism*

$$B_{x_i} \cong A'[x_1, \ldots, x_n, x_i^{-1}].$$

Proof. There is a natural map

$$A' \hookrightarrow A_0 \hookrightarrow A = A_0[x_1, \ldots, x_n] \to B = \frac{A}{(f_1, \ldots, f_r)}$$

which extends to a map

$$h \colon A'[x_1, \ldots, x_n, x_i^{-1}] \to B_{x_i}$$

by inverting x_i. We explicitly construct an inverse map. Let $g_j = f_j - v_j x_i^{d_j}$. Then the homomorphism

$$g \colon \quad A_0 = A'[v_1, \ldots, v_r] \longrightarrow A'[x_1, \ldots, x_n, x_i^{-1}]$$

$$v_j \longmapsto \frac{-g_j}{x_i^{d_j}}$$

is well defined and induces a map

$$\ell \colon A = A_0[x_1, \ldots, x_n] \to A'[x_1, \ldots, x_n, x_i^{-1}]$$

such that, by construction, $\ell(f_j) = 0$ for all $j = 1, \ldots, r$ (check this!). This induces a map $\ell \colon B_{x_i} \to A'[x_1, \ldots, x_n, x_i^{-1}]$, and it is a simple verification (do it!) that h and ℓ are mutual inverses. $\square$

Corollary 7.6.6. *The element $f \in A$ is an inertia form with respect to I if and only if $x_i^s f \in I$ for some $i = 1, \ldots, n$ and some $s \in \mathbb{N}$.*

Proof. As above, let $B = A/I$. Consider the commutative diagram of localizations

$$
\begin{array}{ccc}
B & \longrightarrow & B_{x_i} \\
\downarrow & & \downarrow \\
B_{x_j} & \longrightarrow & B_{x_i x_j}
\end{array}
$$

for some pair $i \neq j$. By Lemma 7.6.5, x_i is not a zero divisor in B_{x_j} (and conversely), hence the rightward and bottom maps in the square are injective. The condition $x_i^s f \in I$ means that $\overline{f}$ goes to 0 in B_{x_i}, and a fortiori in $B_{x_i x_j}$. By injectivity, $\overline{f}$ also goes to 0 in B_{x_j}, so that $x_j^t f \in I$ for some $t \in \mathbb{N}$. Since this holds for all $j = 1, \ldots, r$, it follows that $x^\alpha f \in I$ for all multi-indices α of degree high enough. The converse is immediate. $\square$

Corollary 7.6.7. *Let $T \subset A$ be the ideal of inertia forms. Then T is a prime ideal. It follows that $T_0 = T \cap A_0$ is a prime ideal of A_0.*

Proof. Define the ideal $S \subset B$ by

$$
S = \{ b \in B \mid \mathcal{M}^s b = 0 \text{ for some } s \in \mathbb{N} \},
$$

where $\mathcal{M} = (x_1, \ldots, x_n)$. Then $A/T \cong B/S$, so it is enough to show that B/S is an integral domain. By Corollary 7.6.6, S is the kernel of the localization map $B \to B_{x_i}$, hence we have an injection $B/S \hookrightarrow B_{x_i}$. By Lemma 7.6.5, B_{x_i} is an integral domain, and we are done. $\square$

While our treatment until this point has been completely elementary, for the next result we will make use of the notion of depth. For an elementary proof see [**Jou91**, Section 4.7].

Theorem 7.6.8 (Hurwitz). *Let $A_0 = R[u_{i,\alpha}]$, $f_1, \ldots, f_r \in A = A_0[x_1, \ldots, x_n]$ given by (7.6.1), and let $I = (f_1, \ldots, f_r)$. If $r < n$, then every inertia form with respect to I is trivial.*

Proof. Let $\mathcal{M}$ be the ideal of A generated by the indeterminates $x_1, \ldots, x_n$. Since $x_1, \ldots, x_n$ is a maximal regular sequence contained in $\mathcal{M}$, we have $\operatorname{depth}_{\mathcal{M}} A = n$ by Corollary 7.3.9.

We claim that $f_1, \ldots, f_r$ is a regular sequence. Letting, as usual, $B = A/I$, we then have $\operatorname{depth}_{\mathcal{M}} B = \operatorname{depth}_{\mathcal{M}} A - r = n - r > 0$ by Proposition 7.3.10. In particular, there exists $m \in \mathcal{M}$ which is regular on B. If f is an inertia form, the image $\overline{f} \in B$ must satisfy $fm^k = 0$ for some k, which means that $\overline{f} = 0$ and so f is trivial.

We can prove the claim by induction on r, the case $r = 0$ being empty. For $r > 0$, we denote $A_0' = R[u_{i,\alpha} \mid i < r]$, $A' = A_0'[x_1, \ldots, x_n]$, $I' = (f_1, \ldots, f_{r-1})$, $\mathcal{M}' = \mathcal{M} \cap A'$ and $B' = A'/I'$. We must check that f_r is not a zero divisor in $A/(f_1, \ldots, f_{r-1}) \cong B'[u_{r,\alpha}]$. For this, we look at f_r as a polynomial over B' in the indeterminates $u_{r,\alpha}$. The ideal generated by its coefficients is $(\mathcal{M}')^{d_r}$, as can be seen immediately by (7.6.1). If f_r was a zero divisor, then so would be every element of $\mathcal{M}'$ (prove this!). But by induction, $\mathrm{depth}_{\mathcal{M}'} B' = n - r + 1 > 0$, which means that there is $a' \in \mathcal{M}'$ which is not a zero divisor on B'. $\square$

Corollary 7.6.9. *Let* $A_0 = R[u_{i,\alpha}]$, $f_1, \ldots, f_r \in A = A_0[x_1, \ldots, x_n]$, *and let* $I = (f_1, \ldots, f_r)$. *If* $r = n$ *and* f *is a nontrivial inertia form with respect to* $I = (f_1, \ldots, f_n)$, *then* f *must involve all coefficients of all the polynomials* f_i.

Proof. Assume that f does not depend on some coefficient $u_{i,\alpha}$, say $i = n$. Let $g_n = f_n - u_{n,\alpha} x^\alpha$. Start from a relation

$$x_n^N f = h_1 f_1 + \cdots + h_n f_n$$

and make the substitution $u_{n,\alpha} \mapsto -g_n/x^\alpha$. This gives a relation in the ring $A_{x_1 \cdots x_n}$ of the form

$$x_n^N f = \widehat{h_1} f_1 + \cdots + \widehat{h_{n-1}} f_{n-1},$$

where $\widehat{h_i}$ denotes h_i after substitution, since f_n is mapped to 0. After clearing denominators, this becomes

$$(x_1 \ldots x_n)^t x_n^N f = l_1 f_1 + \cdots + l_{n-1} f_{n-1},$$

for some polynomials $l_1, \ldots, l_{n-1}$. Since $x_1, \ldots, x_n$ are not zero divisors in the ring B_{x_n} by Corollary 7.6.6, f is an inertia form for $f_1, \ldots, f_{n-1}$, hence it is trivial by Hurwitz's theorem, Theorem 7.6.8. $\square$

Remember that $A_0 = R[u_{i,\alpha}]$, where R is a UFD and the $u_{i,\alpha}$ are indeterminates. We fix any particular indeterminate $u = u_{i,\alpha}$, and let A' be the ring generated over R by all indeterminates $u_{i,\alpha}$ *other than* u, so that $A_0 = A'[u]$. Notice that A' is itself a UFD. Having fixed this notation, we will denote by $d_u a$, for $a \in A_0$ the degree of a as a polynomial in the variable u. With these preliminaries, we can prove our main result.

Proof of Theorem 7.6.1. We can assume that $T_0 \neq 0$, otherwise the result is trivial. Let s be the minimum degree of a nonzero element of T_0, and take $a \in T_0$ with $d_u a = s$. Assume that a is reducible, say $a = a_1 \cdots a_k$. By Corollary 7.6.7, the ideal T_0 is prime, so one of the a_i belongs to T_0—say $a_1 \in T_0$. Then $d_u a_1 \leq d_u a = s$, so $d_u a_1 = s$ by minimality. Up to replacing a with a_1, we will thus assume that a is irreducible. We claim that with these choices a generates T_0.

Take any $b \in T_0$. We apply the Euclidean division in $k[u]$, where $k = \mathcal{F}(A')$ is the fraction field of A'. We can thus write

$$\lambda b = aq + r,$$

where $\lambda \in A'$ and $r = 0$ or $d_u r < s$. Writing $r = \lambda b - aq$, we see that $r \in T_0$, and by minimality we must have $r = 0$. Since a is irreducible and A' is a UFD, a must divide either λ or b.

By Corollary 7.6.9, a does contain the u variable. Since $\lambda \in A'$ does not, it follows that a divides b, as desired. □

Notice that we did not prove that the ideal $T_0 \neq 0$. To prove that the ideal is not trivial, we explicitly exhibit some elements of it, that is, some nontrivial inertia forms of degree 0.

Example 7.6.10. Assume that $r = n$. Let $\delta = d_1 + \cdots + d_n - n$ and choose any integer $t \geq \delta + 1$. Let E_t be the set of multi-indices α with $\deg \alpha = t$, and fix $\alpha, \beta \in E_t$.

By assumption $\deg \alpha \geq \delta + 1$, hence $\alpha_i \geq d_i$ for some i. Let $i = i(\alpha)$ be the minimal such i. Then the polynomial $\frac{x^\alpha}{x_i^{d_i}} f_i$ is homogeneous of degree t, and we let $a_{\alpha\beta}$ be the coefficient of x^β in such a polynomial. Finally, define

$$D(t) = \det \left(a_{\alpha\beta}\right)_{\alpha,\beta \in E_t}.$$

This is a gigantic determinant, and is an element of A_0. Notice that any reordering of E_t will change both the order of rows and columns, so the determinant does not depend on the order that we put on E_t. We claim that $D(t)$ is a nontrivial inertia form, thereby proving that $T_0 \neq 0$ in the case that $r = n$.

To see this, multiply the column which corresponds to $\alpha = (t, 0, \ldots, 0)$ by x_1^t, and sum to it every other column β multiplied by x^β (thus obtaining a matrix with entries in $\mathbb{Z}[x_\alpha]$). The first column is $(f_{i(\beta)} x^\beta / x_{i(\beta)}^{d_{i(\beta)}})_{\beta \in E_t}$. By expanding this determinant, we find

$$x_1^t D(t) \in (f_1, \ldots, f_n),$$

showing that $D(t)$ is an inertia form.

Moreover, if we specialize each f_i to $x_i^{d_i}$, we find that the matrix defining $D(t)$ specializes to the identity, hence $D(t)$ specializes to 1. In particular, $D(t) \neq 0$, and since $D(t)$ has degree 0, it is a nontrivial inertia form.

By Theorem 7.6.1 and the above example, we conclude what follows. We start from the ring of generic coefficients $A_0 = \mathbb{Z}[u_{i,\alpha} \mid \deg \alpha = d_i]$ for fixed degrees $d_1, \ldots, d_n$, and we let $A = A_0[x_1, \ldots, x_n]$. We take the generic polynomials

$$f_i = \sum_\alpha u_{i,\alpha} x^\alpha.$$

The ideal $T_0 \subset A_0$ of inertia forms of degree 0 with respect to $f_1, \ldots, f_n$ is nonzero and principal. We denote by $R \in A_0$ a generator of T_0. Notice that R is uniquely determined up to sign, since ± 1 are the only invertible elements of A_0.

If C is any ring and $g_1, \ldots, g_n \in C[x_1, \ldots, x_n]$ are homogeneous of degrees $d_1, \ldots, d_n$, then there is a unique ring homomorphism $A \to C$ that sends 1 to 1 and f_i to g_i, obtained by specializing the coefficients $u_{i,\alpha}$. We will denote the image of R via this homomorphism by $R(g_1, \ldots, g_n)$. The next result gives a way to choose a preferred generator of T_0.

Proposition 7.6.11. *The polynomial R satisfies $R(x_1^{d_1}, \ldots, x_n^{d_n}) = \pm 1$.*

Proof. From Example 7.6.10, we see that the determinant $D(t)$ specializes to 1 when we choose $f_i = x_i^{d_i}$. For $t \gg 0$ (in fact, $t > \delta$), we have $D(t) \in T_0$, hence $R(x_1^{d_1}, \ldots, x_n^{d_n})$ divides 1. $\qquad\square$

Definition 7.6.12. The generator $R = \mathrm{Res}(f_1, \ldots, f_n)$ of T_0 that satisfies $R(x_1^{d_1}, \ldots, x_n^{d_n}) = 1$ is called the generic *Macaulay resultant* of $f_1, \ldots, f_n$. Given a ring C and polynomials $g_1, \ldots, g_n \in C[x_1, \ldots, x_n]$ homogeneous of degrees $d_1, \ldots, d_n$, the element $R(g_1, \ldots, g_n)$ is called the Macaulay resultant of $g_1, \ldots, g_n$, denoted $\mathrm{Res}(g_1, \ldots, g_n)$.

Remark 7.6.13. By specialization, it is clear that $\mathrm{Res}(g_1, \ldots, g_n)$ is an inertia form for $g_1, \ldots, g_n$ in any ring.

We will end the section with a geometric application. The following result is known as the main theorem of elimination theory.

Theorem 7.6.14. *Let k be an algebraically closed field, and take homogeneous polynomials $g_1, \ldots, g_n \in k[x_1, \ldots, x_n]$. Then $g_1, \ldots, g_n$ have a common nontrivial zero, that is, different from $(0, \ldots, 0)$—if and only if $\mathrm{Res}(g_1, \ldots, g_n) = 0$.*

Geometrically, the homogeneous polynomials $g_1, \ldots, g_n$ define a projective variety $V = V(g_1, \ldots, g_n) \subset \mathbb{P}_k^{n-1}$. The theorem says that $V \neq \emptyset$ if and only if $\mathrm{Res}(g_1, \ldots, g_n) = 0$.

Proof. Let $R = \mathrm{Res}(g_1, \ldots, g_n)$; notice that $R \in k$ is a scalar. Since R is an inertia form, if $R \neq 0$ we have the inclusion

$$(7.6.2) \qquad\qquad (x_1, \ldots, x_n)^N \subset (g_1, \ldots, g_n),$$

for large N, which implies that

$$V(g_1, \ldots, g_n) \subset V(x_1, \ldots, x_n) = \emptyset.$$

Conversely, if $V(g_1, \ldots, g_n)$ is empty, then we have (7.6.2) by the projective Nullstellensatz (Theorem A.9.5). In other words, 1 is an inertia form with respect to $g_1, \ldots, g_n$, hence a multiple of R, which implies that $R \neq 0$. $\qquad\square$

Corollary 7.6.15. *Fix degrees $d_1, \ldots, d_n \in \mathbb{N}$ and let $\mathbb{A}$ be the affine space of homogeneous polynomials $(g_1, \ldots, g_n)$ where $\deg g_i = d_i$ over the algebraically closed field k. Consider the Zariski closed set*

$$Z = \{(g_1, \ldots, g_n, a_1, \ldots, a_n) \mid g_i(a_1, \ldots, a_n) = 0\} \subset \mathbb{A} \times \mathbb{P}^{n-1},$$

and let $\pi_1 \colon Z \to \mathbb{A}$ be the projection on the first factor. Then $\pi_1(Z)$ is Zariski closed.

Proof. The set $\pi_1(Z)$ is described by the vanishing of the resultant. $\quad\square$

7.7. Computing the resultant

We still do not have a way to compute the resultant. Following [**Cha93**], we will use the Koszul complex to give an explicit algorithm. The way the Koszul complex appears is as follows. We keep the notation of the previous section, working over the generic ring $A_0 = \mathbb{Z}[u_{i,\alpha}]$, with $A = A_0[x_1, \ldots, x_n]$.

We have remarked in the proof of Theorem 7.6.8 that the generic polynomials $f_1, \ldots, f_n$ are a regular sequence, which we shorten by $\mathbf{f}$. By Theorem 7.2.2, the Koszul homology $H_i(\mathbf{f}) = 0$, except for $i = 0$, where we have $H_0(\mathbf{f}) = A/(f_1, \ldots, f_n) = B$. Notice that B is graded since the f_i are homogeneous, and the inertia forms are the increasing union of the annihilators $\bigcup_{k=0}^{\infty} \operatorname{Ann}_A B_k$. In the ring A_0 we then have

$$(\operatorname{Res}(f_1, \ldots, f_n)) = \operatorname{Ann}_{A_0} B_N$$

for any N large enough, by Theorem 7.6.1.

We are now going to introduce an invariant that gives the same result when computed on a complex (of a certain form) and on its homology. Applying this invariant to the Koszul complex $K_\bullet(\mathbf{f})$ will then allow us to compute the resultant.

Definition 7.7.1. Let A be a Noetherian UFD. A *divisor* over A is formal linear combination of prime ideals of A of height 1, with integer coefficients.

Given $a \in A$, factor it into primes as $a = p_1^{e_1} \cdots p_r^{e_r}$. The *divisor* associated to a is the formal sum

$$\operatorname{div} a := \sum_{i=1}^{r} e_i p_i,$$

where $e_i p_i$ just denotes the formal sum of e_i copies of p_i. Notice that this is a sum over primes of height 1 by Theorem A.7.6.

Let M be a finitely generated A-module which is torsion (that is, for every $m \in M$, we have $\operatorname{Ann}_A m \neq 0$). The *divisor* associated to M is the formal sum

$$\operatorname{div} M := \sum_{\substack{P \in \operatorname{Ass}(M) \\ \operatorname{ht} P = 1}} \ell(M_P) P.$$

Remark 7.7.2. When $A = R(V)$ is the coordinate ring of an affine variety V (or the projective coordinate ring of a projective variety V), a divisor over A can be seen as a formal linear combination of irreducible subvarieties of codimension 1.

Remark 7.7.3. The divisor function has the following additivity property: if

$$0 \longrightarrow M_1 \longrightarrow M_2 \longrightarrow M_3 \longrightarrow 0$$

is a short exact sequence of finitely generated torsion A-modules, then

(7.7.1) $\operatorname{div} M_2 = \operatorname{div} M_1 + \operatorname{div} M_3.$

In general, $\operatorname{Ass} M_2 \neq \operatorname{Ass} M_1 \cup \operatorname{Ass} M_3$, so this is not completely obvious. But notice that in the definition of $\operatorname{div} M_i$, we only consider associated primes of height 1. Since A is a domain, these are minimal elements in the set of associated primes of M_i. By Theorem A.3.4, the minimal primes in $\operatorname{Supp} M_i$ and $\operatorname{Ass} M_i$ are the same, and we know that $\operatorname{Supp} M_2 = \operatorname{Supp} M_1 \cup \operatorname{Supp} M_3$ by Theorem A.3.5. Then (7.7.1) follows at once from the same additivity property for the length.

Let us compute the divisor of B_N, regarded as an A_0-module, for $N \gg 0$. A prime appearing in the above sum has the form $P = \operatorname{Ann} f$ for some f in B_N. The resultant ideal certainly has this form by Proposition 7.6.6. Moreover, it is prime by Proposition 7.6.7, and it is generated by the element $R = \operatorname{Res}(f_1, \ldots, f_n)$. By the principal ideal Theorem A.7.6, we also have $\operatorname{ht} P = 1$, so P certainly appears in the sum.

In fact, we can be more precise and see that $\operatorname{div} B_N = (R)$. This follows at once from the next result.

Proposition 7.7.4. *Let A be a unique factorization domain (UFD), $I \subset A$ a nonzero ideal. Let x be the gcd of the elements of I, and factor $x = p_1^{e_1} \cdots p_r^{e_r}$. Then*

$$\operatorname{div} A/I = \operatorname{div} x = \sum_{i=1}^{r} e_i p_i.$$

Notice that the gcd of even infinitely many elements is well defined, since if $a \in I$ is any nonzero element, then the gcd is one of the (finitely many) divisors of a.

Proof. First, notice that a prime of height 1 must be principal by factorization, and conversely a principal prime has height 1 by Theorem A.7.6.

Hence we can only consider primes of the form $P = (p)$ for some prime element p. For $a \in A$, denote $\bar{a} \in A/I$ its class.

If p_i is any prime appearing in the factorization of x, then there is some $a \in I$ that factors as $a = p_i^{e_i} b$, where p_i does not appear in b. It follows that $(p_i) = \operatorname{Ann} \overline{p_i^{e_1-1}b}$, so in fact p_i is associated to A/I. Conversely, if $(p) = \operatorname{Ann} \bar{a}$ for some $a \in A$, then $ap \in I$—in particular, it is a multiple of x. Clearly $a \notin I$, hence p must be one of the p_i.

As for the multiplicity, we have

$$\ell\left(\frac{A}{I}\right)_{(p_i)} = \ell\left(\frac{A_{(p_i)}}{(p_i^{e_i})}\right) = e_i. \qquad \square$$

So far, we have verified that $\operatorname{div} B_N = (R)$, when B_N is regarded as an A_0-module. This will allow us to compute the resultant, thanks to the next result.

Theorem 7.7.5 (Demazure, Chardin [**Cha93**]). *Let $C_\bullet$ be a finite complex of A-modules concentrated in degrees 0 to n, where A is a Noetherian UFD. Assume that we have a decomposition of the form $C_i = E_{i+1} \oplus E_i$, where $E_0 = E_{n+1} = 0$, and that the differential $d_i \colon C_i \to C_{i-1}$ admits the decomposition*

$$d_i = \begin{pmatrix} a_i & \phi_i \\ b_i & c_i \end{pmatrix}$$

with respect to this decomposition, where $\phi_i \colon E_i \to E_i$ is injective. Then

(i) *the homology $H_i(C_\bullet)$ is a torsion A-module for all i; and*

(ii) *we have the equality*

$$\sum_{i=0}^{n} (-1)^n \operatorname{div} H_i(C_\bullet) = \sum_{i=0}^{n} (-1)^n \operatorname{div}(\det \phi_{i+1}).$$

Lemma 7.7.6. *Let A be a UFD, M a finitely generated free A-module, and let $\phi \colon M \to M$ be injective. Then*

(7.7.2) $$\operatorname{div}(\operatorname{coker} \phi) = \operatorname{div}(\det \phi).$$

Proof. Denote $N = \phi(M)$. Let $p \in A$ be a prime element, $P = (p)$. Notice that $M_P = N_P$, unless p divides $\det \phi$. So the set of primes appearing on the two sides of (7.7.2) are the same.

To compute the multiplicities, let $\{e_1, \ldots, e_n\}$ be a basis of M. Fix a prime $P = (p)$ where p divides $\det \phi$. The ring A_P is a principal ideal domain, hence by the structure theorem for finitely generated modules over a PID, up to automorphisms of M_P, we can diagonalize ϕ on M_P. It follows that in M_P we have $\phi(e_i) = p^{d_i} e_i$ for some $d_i \in \mathbb{N}$, and $\det \phi = p^{\sum d_i}$.

Then $\ell(M_P/N_P) = \sum_{i=1}^{n} d_i$, so the multiplicities in the two sums are the same. $\square$

Proof of Theorem 7.7.5. Let k be the fraction field of A. Notice d_i induces an automorphisms of $E_i \otimes_A k$. From this it is easy to see that the homology of $C_\bullet \otimes_A k$ vanishes. It follows that the homology of $C_\bullet$ is torsion.

Define a complex $D_\bullet$ by letting $D_i = C_i$, but with differentials given by the matrices $\begin{pmatrix} 0 & I \\ 0 & 0 \end{pmatrix}$ with respect to the decomposition $C_i = E_{i+1} \oplus E_i$.

For $i < n$, let $f_i \colon C_i \to C_i$ be given by $f_i = \begin{pmatrix} \phi_{i+1} & 0 \\ c_{i+1} & I \end{pmatrix}$, and let f_n be the identity. We claim that the f_i define a morphism of complexes $f \colon D_\bullet \to C_\bullet$, which can be directly verified by the equalities

$$\begin{pmatrix} a_i & \phi_i \\ b_i & c_i \end{pmatrix}\begin{pmatrix} \phi_{i+1} & 0 \\ c_{i+1} & I \end{pmatrix} = \begin{pmatrix} 0 & \phi_i \\ 0 & c_i \end{pmatrix} = \begin{pmatrix} \phi_i & 0 \\ c_i & I \end{pmatrix}\begin{pmatrix} 0 & I \\ 0 & 0 \end{pmatrix}.$$

The first equality uses the fact that $C_\bullet$ is a complex, hence $d_i d_{i+1} = 0$.

Noticing that f_i is injective and $\operatorname{coker} f_i = \operatorname{coker} \phi_{i+1}$, we obtain the exact sequence of complexes

$$0 \longrightarrow D_\bullet \longrightarrow C_\bullet \longrightarrow E_\bullet \longrightarrow 0 \,,$$

where $E_i = \operatorname{coker} \phi_{i+1}$. Let $e_i \colon E_i \to E_{i-1}$ be the differential. Then by Remark 7.7.3 we have the equalities

$$\operatorname{div} E_i = \operatorname{div} \operatorname{im} e_i + \operatorname{div} \ker e_i = \operatorname{div} \operatorname{im} e_i + \operatorname{div} \operatorname{im} e_{i+1} + \operatorname{div} H_i(E_\bullet).$$

A simple diagram chase shows that $H_i(E_\bullet) = H_i(C_\bullet)$, hence putting all together and taking the alternate sum over i we obtain

$$\sum_{i=0}^{n} (-1)^n \operatorname{div} H_i(C_\bullet) = \sum_{i=0}^{n} (-1)^n \operatorname{div} \operatorname{coker} \phi_{i+1},$$

and the conclusion follows from Lemma 7.7.6. $\square$

We can now use the above result to compute the resultant. Fix n integers $d_1, \ldots, d_n \in \mathbb{N}$. We set as usual $A_0 = \mathbb{Z}[u_{i,\alpha}]$ and let $A = A_0[x_1, \ldots, x_n]$ and $f_i = \sum_{\deg \alpha = d_i} u_{i,\alpha} x^\alpha$. All computations take place in the ring A of generic coefficients.

We look at the Koszul complex $K(\mathbf{f})$. We choose a degree N large enough so that $(\operatorname{Res}(f_1, \ldots, f_n))$ is the annihilator of the component of degree N of the ring $B = A/(f_1, \ldots, f_n)$, seen as a module over A_0. In fact, it is enough to choose any $N > \sum_{i=1}^{n}(d_i - 1)$. We also put a gradation on $K_r(\mathbf{f})$ by declaring $\deg e_{i_1} \wedge \cdots \wedge e_{i_r} = d_{i_1} + \cdots + d_{i_n}$. This way, the differentials of the

Koszul complex are graded, of degree 0. This allows us to split the complex into graded components, and we look at the component of degree N

$$0 \longrightarrow K_n(\mathbf{f})_N \longrightarrow \cdots \longrightarrow K_0(\mathbf{f})_N \longrightarrow 0.$$

To proceed further, we need to understand the ranks of the graded components of the Koszul complex. We will do this by finding a suitable generating function.

Lemma 7.7.7. *Let A_0 be a ring, $f_1, \ldots, f_n$ a regular sequence of homogeneous polynomials in $A_0[x_1, \ldots, x_n]$ and $K_{\bullet} = K_{\bullet}(f_1, \ldots, f_n)$ the associated Koszul complex. Denote $d_i = \deg f_i$, and $r_{p,t}$ the rank (over A) of the degree t part of K_p. Then*

(i) *the ranks satisfy*

$$\sum_{p,t} r_{p,t} X^t Y^p = \prod_{i=1}^{n} \frac{1 + X^{d_i} Y}{1 - X};$$

(ii) *letting $s_t = \sum_{p=0}^{n} (-1)^p r_{p,t}$, we have*

$$\sum_{t} s_t X^t = \prod_{i=1}^{n} \frac{1 - X^{d_i}}{1 - X}; \text{ and}$$

(iii) *the coefficient s_t vanishes for $t > \sum (d_i - 1)$.*

Proof. To prove (i), we need to understand how to obtain a monomial of degree t in K_p. By construction, K_p has a basis of elements of the form $e_{i_1} \wedge \cdots \wedge e_{i_p}$ for some $i_1 < \cdots < i_p$, having degree $d_{i_1} + \cdots + d_{i_p}$. Each such element can be multiplied by a monomial of degree k to obtain an element of K_p of degree $k + d_{i_1} + \cdots + d_{i_p}$. Thus $r_{p,t}$ counts all the ways we can choose $i_1 < \cdots < i_p$ with $d_{i_1} + \cdots + d_{i_p} \leq t$.

In the product

$$\prod_{i=1}^{n} (1 + X^{d_i} Y)$$

a monomial of degree p in Y is obtained by choosing exactly p of the terms involving the variables, and $n - p$ ones. Each such choice corresponds to a set of indices $i_1 < \cdots < i_p$, and the X degree of such monomial is precisely $d_{i_1} + \cdots + d_{i_p}$. By multiplying by

$$\frac{1}{1 - X} = 1 + X + X^2 + \cdots,$$

we count each such monomial in all degrees greater or equal than $d_{i_1} + \cdots + d_{i_p}$, proving (i).

We obtain (ii) by the substitution $Y = -1$. Finally, (iii) follows since

$$\frac{1 - X^{d_i}}{1 - X} = 1 + \cdots + X^{d_i - 1}$$

is a polynomial of degree $d_i - 1$. $\square$

Corollary 7.7.8. *Let A_0 be a Noetherian UFD, $\mathbf{f} = f_1, \ldots, f_n$ a regular sequence of homogeneous polynomials in $A = A_0[x_1, \ldots, x_n]$, of degrees $d_i = \deg f_i$. Then*

$$\mathrm{div}\left(\frac{A}{(f_1, \ldots, f_n)}\right)_N = \sum_{i=0}^{n} (-1)^n \, \mathrm{div}\det \phi_{i+1},$$

for $N > \sum(d_i - 1)$, where ϕ_i are suitable maps in a decomposition such as in Theorem 7.7.5.

Here $-_N$ denotes the degree N component of a module.

Proof. By Theorem 7.2.2, the complex $K_\bullet(\mathbf{f})$ is acyclic except in degree 0, where we have $H_0(\mathbf{f}) = A/(f_1, \ldots, f_n)$ by Example 7.1.6(c). By our choice of grading, each map in the Koszul complex is homogeneous of degree 0, so we can split the complex into many complexes, one for each degree. For such a complex in degree N, we will check that the hypothesis of Theorem 7.7.5 are satisfied.

Fix $N > \sum(d_i - 1)$ and denote by $K_\bullet'$ the degree N component of $K_\bullet(\mathbf{f})$. Notice that every module K_p' in the Koszul complex is free, with a certain basis $\mathcal{B}_i$. By Lemma 7.7.7,

$$(7.7.3) \qquad\qquad \sum_{p=0}^{n} (-1)^p \, \mathrm{rk}\, K_p' = 0.$$

We will define recursively a decomposition $\mathcal{B}_i = \mathcal{B}_i' \cup \mathcal{B}_i''$ in two sets, in such a way that the map on the spans

$$(7.7.4) \qquad\qquad \langle \mathcal{B}_{i+1}'' \rangle_{A_0} \to \langle \mathcal{B}_i' \rangle_{A_0}$$

is injective and the two sides have the same rank. We will then declare either side to be E_{i+1} and obtain the required decomposition.

Start from $\mathcal{B}_n' = \emptyset$ and $\mathcal{B}_n'' = \mathcal{B}_n$. Assuming that the split has been chosen up to $i + 1$, notice that the map induced on $\langle \mathcal{B}_{i+1}'' \rangle_{A_0}$ is injective by exactness. It follows that one of the minors of the restricted map is invertible over k, the fraction field of B. We can then choose $\mathcal{B}_i'$ so that (7.7.4) is an isomorphism over k, and $\mathcal{B}_i''$ its complement. At the end we find $\mathcal{B}_0'' = \emptyset$ by (7.7.3). $\square$

Combining Proposition 7.7.4 and Corollary 7.7.8, we see that

$$(\mathrm{Res}(f_1,\ldots,f_n)) = \mathrm{div}\left(\frac{A}{(f_1,\ldots,f_n)}\right)_N = \sum_{i=0}^{n}(-1)^i \, \mathrm{div}\det\phi_{i+1},$$

where ϕ_i is a certain injective endomorphism in the decomposition $K_i(\mathbf{f})_N = E_i \oplus E_{i+1}$ that we assume in Theorem 7.7.5. This decomposition is implicit, but we provided an explicit construction in the proof of Corollary 7.7.8. If we follow this construction, we arrive at the following algorithm to compute $\mathrm{Res}(f_1,\ldots,f_n)$.

(1) Fix an integer $N \geq \sum_{i=1}^{n}(d_i - 1) + 1$.

(2) Compute the matrix M_i for the degree N component in the i-th differential of the Koszul complex for $f_1,\ldots,f_n$.

(3) Start a loop from $i = n - 1$.

(4) If M_i does not have maximal rank, $\mathrm{Res}(f_1,\ldots,f_n) = 0$ and we are done.

(5) Otherwise, let M_i' be a square submatrix of M_i having maximal rank, and let $D_i = \det M_i' \neq 0$.

(6) Remove from M_{i-1} the columns corresponding to the image of M_i'.

(7) Decrease i by 1 and go back to (4), unless $i = 0$.

(8) Let $X = D_{n-1}D_{n-3}\cdots D_x$, $Y = D_{n-2}D_{n-4}\cdots D_y$, where x and y are either 0 or 1 depending on the parity of n, and compute $\mathrm{Res}(f_1,\ldots,f_n) = \pm X/Y$.

Remark 7.7.9. The final division in step (8) is a priori in the fraction field of A, but actually we know that it takes place in A_0. Moreover, one can compute it iteratively during the loop, by starting from $R_{n-1} = 1$, and computing at each step $R_i = D_i/R_{i-1}$—then R_0 is the resultant, up to sign.

Remark 7.7.10. We described the algorithm over the ring A of generic coefficients, but it is clear by specialization that it will work over any integral domain (the integrality condition is only needed to perform the division at the end), provided we are able to compute the matrices for the Koszul complex.

7.8. Exercises

1. Let $A, \mathcal{M}$ be a local Noetherian ring, $\mathbf{a}$ a regular sequence in A, $I = (\mathbf{a})$ the generated ideal. Prove that $K_\bullet(\mathbf{a})$ is a *minimal* free resolution of A/I.

2. Prove the graded version of Hilbert's syzygy theorem as follows. Let k be a field, $A = k[x_1,\ldots,x_n]$ and take a finitely generated graded A-module M. Take a free resolution $F_\bullet \to M$, where each F_i is recursively chosen with the smallest possible rank. Show that rk $F_i = \dim_k \operatorname{Tor}_i^A(M, k)$, and use the Koszul complex for k to show that $F_i = 0$ for $i > n$.

In Exercises 3–9 we will prove a result of Vasconcelos [**Vas67**] that completely characterizes ideals generated by regular sequences in local Noetherian rings. To do so, we will first flesh out an interesting result from [**AB57**] on the annihilator of modules having finite free resolutions.

3. Let A be a Noetherian ring, M a finitely generated A-module that admits a finite free resolution. Show that M_P is free over A_P for all associated primes P of A.

4. Let A, M be as in Exercise 3, and let $I = \operatorname{Ann} M$. Show that for all associated primes P of A either $M_P = 0$ or $I_P = 0$.

5 ([**AB57**]). Let A, M be as in Exercise 3. Show that if $\operatorname{Ann} M \neq 0$, then $\operatorname{Ann} M$ contains an element which is not a zero divisor in A.

6. Let A be a Noetherian ring, $I \subset A$ an ideal generated by a regular sequence. Show that I/I^2 is free over A/I, and $\operatorname{pd}_A I < \infty$.

7. Let $I \subset A$ be an ideal such that I/I^2 is free of rank r over A/I, and let $a \in I$ map to one of the basis elements. Denote $\overline{A} = A/(a)$, $\overline{I} = I/(a)$. Show that $\overline{I}/\overline{I}^2$ is free or rank $r - 1$ over $\overline{A}/\overline{I}$.

8. Let A be a Noetherian ring, $I \subset A$ an ideal such that I/I^2 is free over A/I. Assume that I does not consist entirely of zero divisors. Show that one can choose $a \in I$ which is not a zero divisor and such that it maps to a basis element in I/I^2.

9 ([**Vas67**]). Let A be a local Noetherian ring, $I \subset A$ an ideal. Show that I is generated by a regular sequence if and only if I/I^2 is free over A/I and $\operatorname{pd}_A I < \infty$. (Use induction on the rank of I/I^2.)

10. Let A be a local Noetherian ring, $I \subset A$ an ideal and M a finitely generated A-module. Assuming that $\operatorname{pd}_A A/I < \infty$ and $\operatorname{pd}_{A/I} M < \infty$, show that

$$\operatorname{pd}_A M = \operatorname{pd}_A A/I + \operatorname{pd}_{A/I} M.$$

11. Let M be a finitely generated A-module. Given $m \in M$, define the complex $K'(m)$ as

$$0 \longrightarrow A \longrightarrow M \longrightarrow \Lambda^2 M \longrightarrow \cdots,$$

where as usual we identify $A = \Lambda^0 M$, $M = \Lambda^1 M$. The map $\Lambda^i M \to \Lambda^{i+1} M$ is given by $a \mapsto m \wedge a$. Show that $K'(m)$ is indeed a complex. When $M = A^n$ and $m = a_1, \ldots, a_n$, show that K' is isomorphic to the dual of the Koszul complex $K(a_1, \ldots, a_n)$.

12. Let k be a field with char $k = 0$ and V a finite-dimensional vector space. Show that $\operatorname{Sym} V = \bigoplus_{n=0}^{\infty} \operatorname{Sym}^n V$ has the structure of a commutative k-algebra, and in fact there is an isomorphism $\operatorname{Sym} V^{\vee} \cong k[x_1, \ldots, x_n]$ if $x_1, \ldots, x_n$ are linear coordinates on V. Using this, show that the Koszul resolution for k gives rise to the exact sequence

$$\cdots \longrightarrow \Lambda^2 V \otimes \operatorname{Sym} V \longrightarrow V \otimes \operatorname{Sym} V \longrightarrow \operatorname{Sym} V \longrightarrow k \longrightarrow 0.$$

13. Let V be as in the previous exercise, and consider an element $T \in \operatorname{GL}(V)$ with eigenvalues $t_1, \ldots, t_n$. Denote by $e_k = \sum_{i_1 < \cdots < i_k} x_{i_1} \cdots x_{i_k}$ the k-th elementary symmetric polynomial and $s_k = \sum_{i_1 \leq \cdots \leq i_k} x_{i_1} \cdots x_{i_k}$ the k-th complete symmetric polynomial, which we regard as elements of $\mathbb{Z}[[x_1, x_2, \ldots]]$. Show that T acts on $\Lambda^k V$ with trace $e_k(\mathbf{t})$ and on $\operatorname{Sym}^k V$ with trace $s_k(\mathbf{t})$. Using the previous exercise, derive the identity of power series

$$\left(\sum_{i=0}^{\infty} s_i \right) \left(\sum_{j=0}^{n} (-1)^j e_j \right) = 1.$$

Taking n arbitrarily large, conclude that

$$\sum_{i=0}^{d} (-1)^i e_i s_{d-i} = 0$$

for all $d \geq 1$. These are called the *Newton identities* for the symmetric polynomials. This example if from [**JJ85**], where many other similar identities are derived from Koszul resoutions.

14. Let $f_1, \ldots, f_n \in A[x_1, \ldots, x_n]$ be linear polynomials, say

$$f_i = \sum_{i=1}^{n} a_{ij} x_j.$$

Prove that $R(f_1, \ldots, f_n) = \det(a_{ij})$.

15. Let f_1, f_2 be two homogeneous polynomials of degrees d_1, d_2 in the ring $A[x_1, x_2]$, where A is a UFD. Prove that $R(f_1, f_2)$, the Macaulay resultant, is the same as the standard resultant, defined as the determinant of the Sylvester matrix (see Section 4.1 in [**Fer20**]), of the polynomials obtained by dehomogeneization of f_1, f_2 (that is, by setting $x_2 = 1$).

16. Prove the multiplicative property of resultant:

$$R(f_1', f_2, \ldots, f_n) \cdot R(f_1'', f_2, \ldots, f_n) = R(f_1' f_1'', f_2, \ldots, f_n).$$

17. Let $f_1, \ldots, f_n$ be homogeneous polynomials of degrees $d_i = \deg f_i$, and let $B = \prod_{i=1}^{n} d_i$. Prove that $R(f_1, \ldots, f_n)$ is homogeneous of degree B/d_i in the coefficients of d_i. (You can prove this for the generic polynomials. Reduce to the case where f_i is a product of linear forms and then use Exercises 14 and 16).

18. Use the algorithm from Section 7.7 to verify that $R(f_1, \ldots, f_n)$ is a homogeneous polynomial in the coefficients of the f_i, of total degree $\sum_{i=1}^n B/d_i$, where $B = \prod_{i=1}^n d_i$. This is consistent with Exercise 17.

19. Let $A \to B$ be a flat local homomorphism of Noetherian rings, and denote by $\mathcal{M}_A$ the maximal ideal of A. Show that $\operatorname{depth} B = \operatorname{depth} A + \operatorname{depth} B/\mathcal{M}_A B$.

Exercises 20–24, verify the Serre conjectures for the case of local regular rings of equal characteristic (that is, $\operatorname{char} A = \operatorname{char} k$, with $k = A/\mathcal{M}$). We follow [**Ser00**, Section V.B.2].

20. Let $A, \mathcal{M}$ be a local Noetherian ring with residue field $k = A/\mathcal{M}$ be a field. Given two A-modules M, N, define the *completed tensor product*

$$M \widehat{\otimes} N := \varprojlim_{p,q} M_p \otimes_k N_q,$$

where $M_p = M/\mathcal{M}^p M$, $N_q = N/\mathcal{M}^q N$. Show that $M \widehat{\otimes} N$ is a module over $A \widehat{\otimes} A$, which is finitely generated if M and N are.

21. Let A be the ring $k[[x_1, \ldots, x_n]]$, where k is a field. Show that $A \widehat{\otimes} A \cong k[[x_1, \ldots, x_n, y_n, \ldots y_n]]$.

22. Let $A = k[[x_1, \ldots, x_n]]$ and let M, N be finitely generated A-modules. Show that

$$\dim_{A \widehat{\otimes} A} M \widehat{\otimes} N = \dim_A M + \dim_A N.$$

(Show that $\operatorname{Gr}_\mathcal{M} M \otimes \operatorname{Gr}_\mathcal{M} N \to \operatorname{Gr}_{\mathcal{M}'} M \widehat{\otimes} N$ is an isomorphism, where $\mathcal{M}'$ is the maximal ideal of $A \widehat{\otimes} A$.)

23. Let $A = k[[x_1, \ldots, x_n]]$ and $B = A \widehat{\otimes} A = k[[x_1, \ldots, x_n, y_n, \ldots y_n]]$. Regard M as a B-module using the x variables and N as a B-module using the y variables, while A is regarded as a quotient of B modulo $(x_1 - y_1, \ldots, x_n - y_n)$. Show that we have

$$\operatorname{Tor}_i^A(M, N) \cong \operatorname{Tor}_i^B(M \widehat{\otimes} N, A).$$

24. Let $A, \mathcal{M}$ be a regular local ring of equal characteristic (that is, $\operatorname{char} A = \operatorname{char} k$, with $k = A/\mathcal{M}$). Let M, N be two finitely generated A-modules such that $M \otimes_A N$ has finite length. Show that $\operatorname{Tor}_i^A(M, N)$ has finite length for all $i \geq 0$, and furthermore that the Serre conjectures from Section 7.5 hold. (Use Cohen's theorem, Theorem A.8.9, to reduce to the case where $A = k[[x_1, \ldots, x_n]]$, then use the previous exercise and an explicit Koszul resolution for A.)

Chapter 8

Regularity

In this chapter, we undertake the study of regular rings. These are the algebraic counterpart of smooth varieties. In [**Fer20**, Section 10.1] we had already introduced this concept, but our conclusions were limited by the tools developed in the first volume. A deeper study of the properties of regular rings requires the homological techniques that we have developed in the earlier chapters. In fact, some of the results of this chapter—such as Serre's characterization of local regular rings and the Auslander–Buchsbaum theorem—were considered the definitive proof that homological methods had become an indispensable instrument in the toolbox of a commutative algebraist.

In Section 8.1, we define regular local rings, and review their basic properties—in particular, we prove that they are integral domains. These results were already presented in the first volume, but we review them here.

Section 8.2 is devoted to Serre's celebrated theorem, characterizing local regular rings as those local Noetherian rings having finite global dimension; see Theorem 8.2.1. This paves the way to the study of regular rings by homological techniques. An easy consequence of this result is the fact that regularity is preserved by localization. This is a fact that is extremely difficult to prove by elementary means (but see [**Nag62**, Corollary 28.3]). As a consequence, we can define regular nonlocal rings: these will be the rings whose localizations are regular local rings.

In Section 8.3, we reap other useful consequences of Serre's theorem. In particular, we investigate how the notion of regular ring behaves under various operations: quotients, polynomial extensions and power series extensions.

Section 8.4 is dedicated to proving the Auslander–Buchsbaum theorem (not to be confused with the eponymous formula, Theorem 7.4.1). This states that a regular local ring is a UFD, and is yet another consequence of Serre's theorem.

In Section 8.5 we use the notion of regular ring to prove a criterion by Serre that characterizes normal rings, that is, rings whose localizations are integrally closed. Serre's criterion states that a Noetherian ring A is normal if and only if it satisfies two axioms that are phrased in terms of regularity and depth. These are easier to state when $A = R(V)$ is the coordinate ring of an affine variety. In this case, Serre's criterion states that A is normal if and only if the singular locus of V has codimension at least two, and moreover every regular function defined on $V \setminus W$, where W has codimension 2, can be extended to the whole of V.

Finally, in Section 8.6, we prove a theorem of Kunz that characterizes regular ring in prime characteristic in terms of the Frobenius endomorphism.

8.1. Regular local rings

Let A be a Noetherian local ring with maximal ideal $\mathcal{M}$. Recall from Section A.8.5 that A has finite dimension. Moreover, if $d = \dim A$, then d is also the smallest integer such that there exist elements $(a_1, \ldots, a_d)$ that generate an ideal of definition, that is, an ideal Q such that $\mathcal{M}^r \subset Q \subset \mathcal{M}$ for some r.

Definition 8.1.1. Let $A, \mathcal{M}$ be a Noetherian local ring. The minimum number of generators of $\mathcal{M}$ is called the *embedding dimension* of A, denoted embdim A.

By Nakayama's lemma, we have the equality embdim $A = \dim_k \mathcal{M}/\mathcal{M}^2$, where the latter is the dimension as a k-vector space, $k = A/\mathcal{M}$. Since $\mathcal{M}$ is itself an ideal of definition, we have $\dim A \leq$ embdim A.

Remark 8.1.2. Assume that $A = R(V)_P$ is the local ring at a point p of the affine variety V over the field k. In this case, we have $k = A/\mathcal{M}$, and the Zariski tangent space of V at the point p can be identified with $(\mathcal{M}/\mathcal{M})^\vee$. In particular, embdim A is the dimension of the tangent space of V at p.

Definition 8.1.3. Let $A, \mathcal{M}$ be a Noetherian local ring. We say that A is *regular* if $\dim A =$ embdim A. Otherwise, we say that A is *nonregular*, or *singular*. If A is regular of dimension d, and $\mathcal{M} = (a_1, \ldots, a_d)$, we say that $a_1, \ldots, a_d$ is a *regular system of parameters*.

Remark 8.1.4. Notice that by definition a regular ring is assumed to be Noetherian.

Example 8.1.5.

(a) Let A be a local ring of dimension 0. Then A is regular if and only if it is a field.

(b) More generally, let k be a field, $A = k[[x_1, \ldots, x_d]]$. Then A is local with maximal ideal $(x_1, \ldots, x_d)$. Since $\dim A = d$, A is regular of dimension d.

(c) Let $A, \mathcal{M}$ be a regular local ring of dimension 1. Then $\mathcal{M}$ has a single generator a. We are going to prove in Theorem 8.4.1 that A is a UFD, and in particular it is integrally closed. This means that A is a Dedekind ring (Section A.5), and since it is local it is in fact a DVR. Conversely, every DVR has dimension 1, and its maximal ideal is principal, so it is regular.

(d) Let A be the local ring in the origin of the curve described by the polynomial $f(x, y) = 0$ in the affine plane $\mathbb{A}^2$. Then A is regular if and only if at least one of the partial derivatives $f_x(0), f_y(0)$ is not zero. For instance, the curve $y^2 = x^3 + x$ is not singular in 0 (or anywhere else), while $y^2 = x^3$ is singular (a cusp) and $y^2 = x^3 + x^2$ is singular (a node). We depict these singularities below.

(e) For a nongeometric example, take $A = \mathbb{Z}[x]$ and the maximal ideal $\mathcal{M} = (p, x)$ for a prime p. Then $A_{\mathcal{M}}$ has dimension 2, and its maximal ideal is generated by two elements, so $A_{\mathcal{M}}$ is a regular local ring.

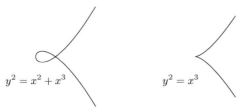

$$y^2 = x^2 + x^3 \qquad\qquad y^2 = x^3$$

Let $A, \mathcal{M}$ be a regular local ring, and let $a_1, \ldots, a_d$ be a minimal set of generators of $\mathcal{M}$. We will prove that $\mathbf{a}$ is in fact a regular sequence.

For this, it is convenient to introduce the associated graded ring

$$\mathrm{Gr}_{\mathcal{M}}(A) := \bigoplus_{r=0}^{\infty} \frac{\mathcal{M}^r}{\mathcal{M}^{r+1}}.$$

The crucial property is that this is essentially a polynomial ring.

Proposition 8.1.6. *Let $A, \mathcal{M}$ be a regular local ring with residue field $k = A/\mathcal{M}$, $a_1, \ldots, a_d$ a regular system of parameters. Then the map*

$$\phi\colon \; k[x_1, \ldots, x_d] \longrightarrow \mathrm{Gr}_{\mathcal{M}}(A)$$

$$x_i \longmapsto \overline{a_i}$$

is an isomorphism.

Notice that each graded piece $\mathcal{M}^r/\mathcal{M}^{r+1}$ is a module over $A/\mathcal{M} = k$. The map ϕ is defined by sending the indeterminate x_i to the class $\overline{a_i} \in \mathcal{M}/\mathcal{M}^2$, hence it respects degree.

Proof. By construction, ϕ is surjective and graded of degree 0. Notice that

$$\dim k[x_1, \ldots, x_d] = d = \dim \mathrm{Gr}_{\mathcal{M}}(A)$$

by Theorem A.8.6.

Assume that ϕ is not injective, say $I = \ker \phi \neq 0$. Then I is a graded ideal, so we can take a nonzero homogeneous element $f \in I$, say $\deg f = r$. The component of $k[x_1, \ldots, x_d]$ of degree n has dimension $\binom{d+n-1}{d-1}$. Since every multiple of f is sent to 0 by ϕ, we can bound the dimension of $\mathrm{Gr}_{\mathcal{M}}(A)_n$ by

$$\dim_k \mathrm{Gr}_{\mathcal{M}}(A)_n \leq \binom{d+n-1}{d-1} - \binom{d+n-r-1}{d-1}$$

for $n \geq r$. The right-hand side is a polynomial of degree at most $d-1$ in n, whence $\dim \mathrm{Gr}_{\mathcal{M}}(A) \leq d-1$, which is a contradiction. $\square$

Proposition 8.1.7. *Let A be a regular local ring. Then A is an integral domain.*

Proof. Take nonzero elements $a, b \in A$. By Krull's intersection theorem (Theorem A.6.4), $\bigcap_k \mathcal{M}^k = 0$, hence there exist integers r, s such that $a \in \mathcal{M}^r \setminus \mathcal{M}^{r+1}$ and $b \in \mathcal{M}^s \setminus \mathcal{M}^{s+1}$. Then we have nonzero classes $\overline{a} \in \mathrm{Gr}_{\mathcal{M}}(A)_r$ and $\overline{b} \in \mathrm{Gr}_{\mathcal{M}}(B)_s$. By Proposition 8.1.6, $\mathrm{Gr}_{\mathcal{M}}(A)$ is a domain, hence $\overline{a}\overline{b} \neq 0$, which implies that $ab \neq 0$. $\square$

Remark 8.1.8. Propositions 8.1.6 and 8.1.7 were already proved in [**Fer20**] as Proposition 10.1.8 and Theorem 10.1.9, respectively. We have proved them again here, since they are crucial for the rest of the chapter.

Corollary 8.1.9. *Let A be a regular local ring, $a_1, \ldots, a_d$ a regular system of parameters. Then $a_1, \ldots, a_d$ is a regular sequence.*

Proof. By Proposition 8.1.7, a_1 is not a divisor of 0. Since $A/(a_1)$ is again regular (why?), of dimension $d-1$, the rest of the proof goes through by induction. $\square$

8.2. Regularity and global dimension

This section is devoted to the proof of the celebrated result by Serre [**Ser56b**], that characterizes regular local rings by their global dimension (Section 5.5), and some of its consequences.

Theorem 8.2.1 (Serre). *Let A be a local Noetherian ring of dimension d. Then A is regular if and only if $\mathrm{gl.\,dim}\, A < \infty$, in which case $\mathrm{gl.\,dim}\, A = d$.*

The proof of Serre's theorem is based on the following technique to give a lower bound on the dimensions of the Tor vector spaces.

Proposition 8.2.2. *Let $A, \mathcal{M}$ be a local Noetherian ring, and assume that $\mathcal{M} = (a_1, \ldots, a_s)$, where s is chosen to be minimal. Let $k = A/\mathcal{M}$. Then*

$$\dim_k \operatorname{Tor}_i(k, k) \geq \binom{s}{i}$$

for $0 \leq i \leq s$.

Proof. Let $F_\bullet \to k$ be a minimal resolution as in Definition 5.6.4, and let $K_\bullet = K_\bullet(\mathbf{a})$ be the Koszul complex associated to the sequence $\mathbf{a} = a_1, \ldots, a_s$. By the dual of Proposition 3.2.4, we get a morphism of complexes $K_\bullet \to F_\bullet$. We claim that the homomorphisms $K_i \to F_i$ are injective. Since both F_i and K_i are finitely generated free A-modules, it will then follow that K_i is a direct summand of F_i (the inclusion $K_i \hookrightarrow F_i$ splits, since we can define a left inverse $F_i \to K_i$). In particular,

$$\binom{s}{i} = \operatorname{rk} K_i \leq \operatorname{rk} F_i = \dim_k \operatorname{Tor}_i(k, k),$$

where the last equality is Corollary 5.6.6.

We now turn to the claim on injectivity. Since both F_i and K_i are free, the claim is equivalent to saying that $K_i \otimes_A k \to F_i \otimes_A k$ is injective.

First, notice that the differential of the Koszul complex satisfies $dK_i \subset \mathcal{M}K_{i-1}$. Moreover, since s is chosen to be minimal, the classes $\overline{a_1}, \ldots, \overline{a_s}$ form a basis of $\mathcal{M}/\mathcal{M}^2$ as k-vector spaces. From this, one can verify using the formula for the Koszul differential that the induced map

$$k \otimes_A K_i \to \frac{\mathcal{M}}{\mathcal{M}^2} \otimes_A K_{i-1}$$

is injective (do it!). Then one can conclude that $K_i \otimes_A k \to F_i \otimes_A k$ is injective by induction, using the commutative square

$$\begin{array}{ccc}
k \otimes_A K_i & \longrightarrow & \frac{\mathcal{M}}{\mathcal{M}^2} \otimes_A K_{i-1} \\
\downarrow & & \downarrow \\
k \otimes_A F_i & \longrightarrow & \frac{\mathcal{M}}{\mathcal{M}^2} \otimes_A F_{i-1}.
\end{array}$$

Explicitly, assume that $K_{i-1} \otimes_A k \to F_{i-1} \otimes_A k$ is injective. Then, as argued above, $K_{i-i} \to F_{i-1}$ is injective as well, so the top and right maps in the square are injective. It follows that the left map is injective as well. To start the induction, we use the fact that K_0 has rank 1, hence $k \otimes_A K_0 \to k$ is an isomorphism. $\square$

In fact, in the singular case, a much stronger bound by Tate [**Tat57**, Theorem 8] holds.

Theorem (Tate). *Let $A, \mathcal{M}$ be a singular local Noetherian ring of dimension d, and assume that $\mathcal{M} = (a_1, \ldots, a_s)$, where s is chosen to be minimal. Let $k = A/\mathcal{M}$. Then*

$$\dim_k \operatorname{Tor}_i(k, k) \geq \binom{s}{i} + \binom{s}{i-2} + \cdots$$

for $i \in \mathbb{N}$.

Proof of Theorem 8.2.1. As usual, denote $\mathcal{M}$ the maximal ideal and $k = A/\mathcal{M}$. Assume that A is regular, and let $a_1, \ldots, a_d$ be a regular system of parameters. This is also a regular sequence by Corollary 8.1.9. By Proposition 5.5.2, $\operatorname{gl.\,dim} A = \operatorname{pd}_A k$. The Koszul complex associated to the regular sequence $a_1, \ldots, a_d$ is a free resolution of k of length d by Theorem 7.2.2. It follows that $\operatorname{gl.\,dim} A \leq d$. For the reverse inequality, by Propositions 7.2.7 and 7.1.15 we can compute

$$\operatorname{Ext}^d(k, A) = H^d(\mathbf{a}) = H_0(\mathbf{a}) = A/\mathcal{M} = k \neq 0,$$

hence $\operatorname{pd}_A k \geq d$. Alternatively, one can compute $\operatorname{pd}_A k = d$ by an immediate application of the Auslander–Buchsbaum formula, Theorem 7.4.1.

Conversely, assume that $\operatorname{gl.\,dim} A$ is finite. Let s be the minimal number of generators of $\mathcal{M}$. By Proposition 8.2.2 we have $\operatorname{Tor}^s(k, k) \neq 0$, which implies that $\operatorname{gl.\,dim} A \geq s$. (Why?) By Corollary 7.4.7 (an application of the Auslander–Buchsbaum formula), we get

$$s \leq \operatorname{gl.\,dim} A \leq \dim A = d,$$

whence $s = d$ and A is regular. $\qquad\square$

Remark 8.2.3. The proof that we presented is the original one by Serre. There is an alternative proof by Nagata simplified by Grothendieck. We present it in Exercises 2–4 (see also Exercise 21).

Remark 8.2.4. Let $A, \mathcal{M}$ be a regular local ring of dimension d. Then Proposition 8.2.2 applies with $s = d$. At the same time, a regular system of parameters $a_1, \ldots, a_d$ is a regular sequence by Corollary 8.1.9, hence the Koszul complex $K(\mathbf{a})$ is a resolution of $k = A/\mathcal{M}$. It follows that the Koszul complex is itself a *minimal* free resolution of k, and in particular

$$\dim_k \operatorname{Tor}_i^A(k, k) = \binom{d}{i}$$

for $i \leq d$.

Corollary 8.2.5. *Let A be a regular local ring, $P \subset A$ a prime. Then the localization A_P is again regular.*

Proof. This follows at once by Serre's theorem and the relation between the global dimension of a ring and its localizations, given in Theorem 5.5.4. □

Remark 8.2.6. The above result was one of the early successes of the homological techniques. Proving directly that the localization of a regular ring is again regular is extremely nontrivial to do by other means.

As a consequence of Corollary 8.2.5, we can define the notion of regularity for rings that are not necessarily local.

Definition 8.2.7. Let A be a Noetherian ring. We say that A is *regular* if for every prime ideal $P \subset A$, the localization A_P is a regular local ring.

The definition itself is easy, but in order to make sure that it does not change the meaning of regular local ring we need Corollary 8.2.5. Alternatively, one can give the same definition using only maximal ideals—in this case the definition clearly works for regular local rings, but proving that the two definitions are in fact equivalent is again an application of Corollary 8.2.5.

Remark 8.2.8. Let V be an affine variety with coordinate ring $A = R(V)$, and let $W \subset V$ be an irreducible subvariety, corresponding to the prime ideal $P \subset A$. Then Corollary 8.2.5 says that V is regular at all points of W if and only if the local ring A_P is regular. That is, we can check regularity in one fell swoop by looking at the ring A_P: V is regular everywhere along W if and only if it is regular at the generic point of W.

Example 8.2.9.

(a) Let $A = R(V)$ be the coordinate ring of the affine variety V over the algebraically closed field k. By the Nullstellensatz, the points of V correspond to the maximal ideal of A. Then A is regular if and only if V is nonsingular at all of its points.

(b) Let A be a Noetherian ring of dimension 1. By Example 8.1.5(c), A is regular if and only if all its localizations are DVR, in other words, if and only if A is a Dedekind ring (see Section A.5).

8.3. Change of rings

Thanks to Corollary 8.2.5, we know how regularity behaves under localization. We now want to understand how it behaves under other operations, such as quotients and polynomial extensions. Our first result deals with flat morphisms.

Proposition 8.3.1. *Let $f \colon A \to B$ be a faithfully flat morphism of Noetherian rings. If B is regular, then A is regular.*

Proof. We first reduce to the local case. By Corollary 8.2.5 we only need to check that $A_{\mathcal{M}}$ ir regular for all maximal ideals $\mathcal{M} \subset A$. Choose one such ideal $\mathcal{M}$; by Corollary 6.3.5 $\mathcal{M} \cdot B \neq B$, so we can choose a maximal $\mathcal{M}' \subset B$ such that $f^{-1}(\mathcal{M}') = \mathcal{M}$. By Corollary 6.2.7, the morphism $A_{\mathcal{M}} \to B_{\mathcal{M}'}$ is flat (hence faithfully flat by Corollary 6.3.7). It follows that we can assume from the start that A and B are local and f is a local morphism.

Now let M be a finitely generated A-module. Then M is flat over A if and only if $M \otimes_A B$ is flat over B. By Proposition 6.1.6, the same holds for projective modules. So if M is a finitely generated A-module, we can take a resolution by finitely generated free modules

$$\cdots \longrightarrow F_2 \longrightarrow F_1 \longrightarrow F_0 \longrightarrow M \longrightarrow 0,$$

which is a priori infinite. Tensoring it with B we obtain a resolution of $M \otimes_A B$. Since B is regular, by Serre's theorem, Theorem 8.2.1, there is a k independent from M such that $\ker(F_k \otimes_A B \to F_{k-1} \otimes_A B)$ is projective. But this is the same as $\ker(F_k \to F_{k-1}) \otimes_A B$, which implies that $\ker(F_k \to F_{k-1})$ is a projective A-module. In other words, gl. dim $A \leq k$, which means that A is regular by another application of Serre's theorem. $\qquad \square$

A typical example of application of this is completion. Let $A, \mathcal{M}$ be a local Noetherian ring, $\widehat{A}$ its completion with respect to the $\mathcal{M}$-adic topology. Then $A \to \widehat{A}$ is a flat morphism of local rings by Theorem 6.2.9, so the above Proposition applies. In fact, in this case we can be more precise by a direct argument.

Proposition 8.3.2. *Let $A, \mathcal{M}$ be a local Noetherian ring, $\widehat{A}$ its completion. Then A is regular if and only if $\widehat{A}$ is regular.*

Proof. The rings A and $\widehat{A}$ a both Noetherian local rings of the same dimension by Theorem A.7.3. Moreover, by Theorem A.6.3 there is an isomorphism $\mathcal{M}/\mathcal{M}^2 \cong \widehat{\mathcal{M}}/\widehat{\mathcal{M}}^2$, which means that A and $\widehat{A}$ have the same embedding dimension. $\qquad \square$

We now turn to polynomial extensions.

Proposition 8.3.3. *Let A be a Noetherian ring. Then A is regular if and only if $A[x]$ is regular.*

Proof. Assume that A is regular. Let $P \subset A[x]$ be a prime ideal, and $Q = A \cap P$. We need to prove that $A[x]_P$ is regular, but this ring is a further localization of $A_Q[x]$. By Corollary 8.2.5, we only need to prove that $A_Q[x]$ is regular, hence we can assume from the start that A is a regular local ring with a maximal ideal $\mathcal{M}$ and that $P \cap A = \mathcal{M}$.

Letting $k = A/\mathcal{M}$, notice that the image of P in $k[x]$ is generated by a single irreducible element $\overline{f}$. Lift this to a monic polynomial $f \in A[x]$. Then

$$\dim A[x]_P = \operatorname{ht} P = 1 + \operatorname{ht} \mathcal{M} = 1 + \dim A.$$

Since $\mathcal{M}$ is generated by $\dim A$ elements, P is generated by $1 + \dim A$ elements, $A[x]_P$ is regular.

For the converse, assume that $A[x]$ is regular, and let $\mathcal{M} \subset A$ be a maximal ideal. By Serre's theorem (Theorem 8.2.1), we need to prove that gl. $\dim A_{\mathcal{M}} < \infty$. By Theorem 5.5.7, this is equivalent to gl. $\dim A_{\mathcal{M}}[x] < \infty$. By Theorem 5.5.4, this is the supremum of the global dimension of all its localization at maximal ideals. All such localizations are in fact localizations of $A[x]$, so they are regular, hence have finite global dimension. These global dimensions are bounded above by $\dim A_{\mathcal{M}}[x] = \dim A_{\mathcal{M}} + 1$, which is finite, hence gl. $\dim A_{\mathcal{M}}[x]$ obeys the same bound. $\qquad \square$

For quotients, the things are a little more complex. Of course, taking a quotient cannot always preserve regularity: there are singular varieties inside regular ones and vice versa. Still, something can be said with a few assumptions.

Proposition 8.3.4. *Let A be a Noetherian ring, $a \in \mathcal{J}(A)$ a regular element. If $A/(a)$ is regular, then so is A.*

Proof. By Corollary 8.2.5, it is enough to prove that the localization $A_{\mathcal{M}}$ is a regular local ring for every maximal ideal $\mathcal{M} \subset A$. Thus, we can assume from the start that A is local with maximal ideal $\mathcal{M}$. Denote $A' = A/(a)$, $\mathcal{M}' = \mathcal{M}/(a)$, and let $d = \dim A'$.

Notice that $\dim A = d + 1$. For one thing, a maximal chain of primes in A' can be lifted to a chain of primes containing a in A. Since a is not a zero divisor, the smallest prime in this chain is not minimal by Theorem A.3.6. Hence $\dim A \geq d + 1$. At the same time, if $\mathcal{M}'$ is generated by classes $\overline{a_1}, \ldots, \overline{a_d}$, then $a, a_1, \ldots, a_d$ generate $\mathcal{M}$, which implies that $\dim A = d + 1$ and A is regular. $\qquad \square$

There is a partial converse to this, that requires one more assumption.

Proposition 8.3.5. *Let A be a regular ring, $a \in A$. Assume that a is not a zero divisor and that $a \in \mathcal{M} \setminus \mathcal{M}^2$ for every maximal ideal $\mathcal{M} \subset A$. Then $A/(a)$ is regular.*

Proof. As in the previous proof, we can reduce to the case where A is local with maximal ideal $\mathcal{M}$. Denote $A' = A/(a)$, $\mathcal{M}' = \mathcal{M}/(a)$, and let $d = \dim A$.

Then $\dim A' \geq d - 1$ (actually, there is equality). This can be seen in various ways: for instance, by Proposition 7.3.10 we have $\operatorname{depth} \mathcal{M}' = \operatorname{depth} \mathcal{M} - 1 = d - 1$, so that by Proposition 7.4.4 we get

$$\dim A' = \operatorname{ht} \mathcal{M}' \geq \operatorname{depth} \mathcal{M}' = d - 1.$$

It will then be enough to prove that $\mathcal{M}'$ is generated by $d - 1$ elements, which will imply that A' is regular and in fact $\dim A' = d - 1$.

Since $a \in \mathcal{M} \setminus \mathcal{M}^2$, the class $\bar{a} \in \mathcal{M}/\mathcal{M}^2$ is nonzero. By completing it to a basis of this vector space and then lifting to $\mathcal{M}$, we can obtain a set of generators $a = a_1, \ldots, a_d$ of $\mathcal{M}$. It follows that $\mathcal{M}'$ is generated by the residue classes of $a_2, \ldots, a_d$, as desired. $\qquad\square$

This result is especially useful if A is local, otherwise the hypothesis are a little difficult to satisfy. But there is an interesting case where we can ensure this.

Proposition 8.3.6. *Let A be a ring. Then the maximal ideals of $A[[x]]$ are all of the form $(\mathcal{M}, x)$ for some maximal ideal $\mathcal{M}$ of A.*

Proof. It is an elementary fact that an element r of a ring R belongs to the Jacobson radical $\mathcal{J}(R)$ if and only if $1 - rs$ is invertible for all $s \in R$ (if you have not seen this, prove it!). This condition is clearly satisfied by x in $A[[x]]$, so $x \in \mathcal{M}'$ for all maximal ideals $\mathcal{M}'$ of $A[[x]]$.

Fix one such ideal and let $\mathcal{M} = \mathcal{M}' \cap A$. First, we claim that $\mathcal{M}$ is itself maximal in A. To see this, we have an injection

$$\frac{A}{\mathcal{M}} \hookrightarrow \frac{A[[x]]}{\mathcal{M}'},$$

where the right-hand side is a field. Given $a \in A \setminus \mathcal{M}$, we find an inverse in this field. In other words, there is $b \in A[[x]]$ such that $ab = 1 + f$, with $f \in \mathcal{M}'$. Writing $b = b_0 + b_1 x + \cdots$, we find that $ab_0 - 1 = f - xg$ for some $g \in A[[x]]$. Since $x \in \mathcal{M}'$, we find that $ab_0 - 1 \in \mathcal{M}$, which means that a is invertible modulo $\mathcal{M}$.

At this point we know that $\mathcal{M}' \supset (\mathcal{M}, x)$. But then we must have equality, since

$$\frac{A[[x]]}{(\mathcal{M}, x)} \simeq \frac{A}{\mathcal{M}},$$

which is already a field. $\qquad\square$

Corollary 8.3.7. *Let A be a Noetherian ring. Then A is regular if and only if $A[[x]]$ is regular.*

8.4. Regularity and factorization

Another fundamental consequence of Serre's theorem is unique factorization in regular local rings, proved by Auslander and Buchsbaum [**AB59**], improving over a result of Nagata [**Nag58**].

Theorem 8.4.1 (Auslander–Buchsbaum). *Let A be a regular local ring. Then A is a UFD.*

The account that we give follows a simplified proof by Kaplansky [**Kap70**]. We start with a cancellation lemma.

Lemma 8.4.2. *Let A be a ring, M an A-module such that $M \oplus A^n \cong A^{n+1}$. Then $M \cong A$.*

Proof. By hypothesis, M is projective. For every prime ideal $P \subset A$, the A_P-module M_P is free by Kaplansky's theorem (Theorem 5.1.2), and clearly it has rank 1. It follows that

$$(\Lambda^i M)_P \cong \Lambda^i M_P \cong 0$$

for $i \geq 2$, hence $\Lambda^i M = 0$ for $i \geq 2$.

Taking the $(n+1)$-th exterior power we get

$$A \cong \Lambda^{n+1} A^{n+1} \cong \bigoplus_{i=0}^{n+1}(\Lambda^i M \otimes \Lambda^{n+1-i} A^n) \cong$$
$$\cong \Lambda^1 M \otimes \Lambda^n A^n \cong M \otimes A \cong M.$$

$\square$

Corollary 8.4.3. *Let $I \subset A$ be a stably free ideal. Then $I \cong A$ is principal.*

Proof. By hypothesis $I \oplus A^m \cong A^n$ for some $m, n \in \mathbb{N}$. By the lemma, we only need to check that $n = m + 1$, which follows if I_P has rank 1 as an A_P-module, for any prime P. By Kaplansky's theorem (Theorem 5.1.2), I_P is free, and it cannot be free of rank bigger than 1, since any two elements $a, b \in I_P$ are linearly dependent over A_P. $\square$

Another element in the proof of the Auslander–Buchsbaum theorem is the following criterion for unique factorization.

Lemma 8.4.4. *Let A be a Noetherian domain. Then the following are equivalent:*

(i) *A is a UFD; and*

(ii) *every prime ideal $P \subset A$ of height 1 is principal.*

Proof. If A is a UFD, every prime ideal P contains a prime element p by factorization. Since $(p) \subset P$ and $\operatorname{ht} P = 1$, we must have $P = (p)$.

Conversely, assume (ii). Since A is Noetherian, it is enough to show that every irreducible element is prime. Take an irreducible element $a \in A$ and a minimal prime $P \supset (a)$. By the principal ideal theorem (Theorem A.7.6), ht $P = 1$, hence P is principal, say $P = (p)$. Then p divides a, and since a is irreducible, $(p) = (a)$, so a is prime. □

Proof of Theorem 8.4.1. Let $d = \dim A$, which is finite by Theorem A.8.5. We prove the result by induction on d, the case $d = 0$ being trivial (A is a field).

Assume $d \geq 1$ and let $P \subset A$ be a prime ideal of height 1. By Lemma 8.4.4, it will be enough to show that P is principal. Let $a_1, \dots, a_d$ be a regular system of parameters, and denote $a = a_1$. Since $A/(a)$ is again regular—in particular a domain by Proposition 8.1.7—the ideal (a) is prime. If $a \in P$, then $(a) = P$ is principal, so we can assume that $a \notin P$.

We now work in the localization $B = A_a$. The ideal $P' = P \cdot B$ is again a prime ideal of B of height 1. Our aim is to prove that P' is principal, which implies that P is principal (check it!), and we are done.

Our first task is to prove that P' is projective, which by Proposition 5.1.15 is equivalent to saying that it is locally free. Choose any prime ideal Q' of B, say QB, where Q is a prime ideal of A. Then $a \notin Q$, hence $B_{Q'} \cong A_Q$. By Corollary 8.2.5, this is a regular local ring, and $\dim B_{Q'} < d$. By the inductive hypothesis, we can then assume that $B_{Q'}$ is a UFD. The ideal $P'B_{Q'}$ is either trivial or a principal ideal by Lemma 8.4.4, in particular it is always free. Hence P' is locally free as claimed.

Second, by Serre's theorem 8.2.1, P admits a finite resolution by finitely generated free A-modules. Hence the same is true for P', and by Proposition 5.7.5, P' is stably free. By Corollary 8.4.3, P' is principal, and we are done. □

8.5. Normal rings

Using the notion of regular ring, Serre also provided a general criterion to assess whether a ring is normal.

Definition 8.5.1. Let A be a ring. We say that A is *normal* if A_P is integrally closed for all prime ideals $P \subset A$.

Recall that a ring A is integrally closed if it is integrally closed in its total ring of fractions $S^{-1}A$, where S is the multiplicative set of elements that are not divisors of 0. Serre's criterion is formulated in terms of the following two conditions.

Definition 8.5.2. Let A be a Noetherian ring, $k \in \mathbb{N}$.

(1) We say that A satisfies axiom R_k if A_P is a regular local ring for all prime ideals $P \subset A$ of height at most k, and

(2) We say that A satisfies axiom S_k if $\operatorname{depth} A_P \geq \min\{k, \operatorname{ht} P\}$ for all prime ideals $P \subset A$.

Condition S_k can be reformulated by saying that $\operatorname{depth} A_P \geq \operatorname{ht} P$ for all primes $P \subset A$ such that $\operatorname{depth} A_P < k$.

Remark 8.5.3. By Proposition 7.4.4, we have $\operatorname{ht} A_P \geq \operatorname{depth} A_P$ in any case. Hence if S_k holds, we have $\operatorname{depth} A_P = \operatorname{ht} P$, provided $\operatorname{depth} A_P < k$.

Remark 8.5.4. Assume that $A = R(V)$ is the coordinate ring of the affine variety V, which for simplicity we assume irreducible. Condition R_k can be rephrased by saying that A is regular in codimension k. This means that if $W \subset V$ is the singular locus, then W has codimension at least k in V. This characterization follows immediately from Remark 8.2.8.

To start gaining familiarity with the R_k and S_k conditions, we prove a simple characterization of reduced rings.

Proposition 8.5.5. *Let A be a Noetherian ring. Then A is reduced if and only if it satisfies R_0 and S_1.*

Proof. This follows from primary decomposition. In particular, if A is a Noetherian ring and $\{P_1, \ldots, P_n\}$ is the set $\operatorname{Ass}(0)$ of primes associated to 0, $\mathcal{N}(A)$ is the nilradical and $D(A)$ is the set of divisors of 0, then by Theorem A.3.6, $\mathcal{N}(A) = \bigcap P_i$, while $D(A) = \bigcup P_i$. Moreover, for any multiplicative set $S \subset A$, we have $\mathcal{N}(S^{-1}A) = S^{-1}\mathcal{N}(A)$ (prove this!).

Assume that A is reduced. A prime P of height 0 is a minimal prime. Hence the local ring A_P has precisely one prime ideal. By the above remarks,

$$PA_P = \mathcal{N}(A_P) = \mathcal{N}(A)_P = 0,$$

so A_P is a field and A satisfies R_0. Moreover, all primes associated to 0 in A are minimal. In fact if $P_1, \ldots, P_n$ are the associated primes, we have the irredundant primary decomposition $0 = \mathcal{N}(A) = \bigcap P_i$, and since the decomposition is irredundant all P_i are minimal. Assume that P is a prime with $\operatorname{depth} A_P = 0$. This means that

$$P \subset D(A) = \bigcup_{i=1}^{n} P_i.$$

By prime avoidance, $P \subset P_i$ for some i. Since P_i is minimal, we have $P = P_i$ and $\operatorname{ht} P = 0$, proving S_1.

For the converse, assume that A satisfies R_0 and S_1. Consider the set of Ass(0) of primes associated to 0. By S_1, we know that an associated prime is in fact minimal. If $N = \mathcal{N}(A)$, then the image of N in A_P is 0 for all minimal primes P, since A_P is a field by R_0. Let S be the complement in A of the set of zero divisors. Then S is a multiplicative set—in fact, the complement of the union of minimal primes. It follows that the image of N in $S^{-1}A$ is 0. Since the map $A \to S^{-1}A$ is injective (it is the inclusion of A into its total ring of fractions), we get that $N = 0$, as desired.   □

Before moving to Serre's criterion, we are going to make a few remarks on the relation between integrally closed and normal rings.

Remark 8.5.6. Let A be an integrally closed ring. Then it is immediate to check that $S^{-1}A$ is integrally closed for all multiplicative sets S. In particular, an integrally closed ring is normal. The converse is true assuming that A is a domain (Exercise 6), and in fact when A is reduced; this will follow immediately from the proof of Serre's criterion.

The case of a reduced ring can be understood from the integral case, by the following results.

Lemma 8.5.7. *Let A be a reduced Noetherian ring, $P_1, \ldots, P_n$ its minimal primes. Then we have an isomorphism*

$$(8.5.1) \qquad\qquad \mathcal{F}(A) \cong \mathcal{F}(A/P_1) \times \cdots \times \mathcal{F}(A/P_n),$$

where $\mathcal{F}(-)$ denotes the total ring of fractions.

Notice that the right-hand side is a product of fields, which will simplify working with $\mathcal{F}(A)$.

Proof. This follows from the Chinese remainder theorem, once we check that the ideals $P_i\mathcal{F}(A)$ are pairwise coprime ideals whose intersection is 0. Notice that $\mathcal{F}(A) = S^{-1}A$, where S is the complement in A of $\bigcup P_i$, hence the ideals $P_i\mathcal{F}(A)$ are maximal, and in particular pairwise coprime. That the intersection is 0 follows from primary decomposition.   □

Corollary 8.5.8. *Let A be an integrally closed, reduced ring. Then A is a finite product of integrally closed domains.*

Proof. Consider the decomposition (8.5.1), and let e_i be the element of this product of fields having 1 in the i-th position. Notice that $e_i^2 = e_i$ is integral over A, hence $e_i \in A$. From this, we deduce that A is the direct product of the rings $Ae_i \subset \mathcal{F}(A/P_i)$, and these rings are integrally closed domains.   □

We can now state *Serre's criterion.*

Theorem 8.5.9 (Serre). *Let A be a Noetherian ring. Then A is normal if and only if it satisfies R_1 and S_2.*

Proof. Assume that A is normal—in particular, A is reduced, so it satisfies R_0 and S_1 by Proposition 8.5.5. Let $P \subset A$ be a prime of height 1, so $\dim A_P = 1$. Then A_P is an integrally closed Noetherian ring of dimension 1. If we knew that A_P is a domain, then A_P would be a Dedekind ring, which we have checked in Example 8.2.9(b) to be regular. By Corollary 8.5.8, it follows that A_P is a product of Dedekind rings, and since $\dim A_P = 1$, it is a Dedekind ring itself, hence regular. This means that A satisfies R_1.

To prove that it satisfies S_2, let $P \subset A$ be a prime with depth $P = 1$. In order to show that ht $P \leq 1$, we can localize at P, hence we can assume from the start that A is local with maximal ideal P. In particular A is integrally closed in its total ring of fractions $K = \mathcal{F}(A)$.

Choose $a \in P$ that is not a zero divisor. Then $P/(a)$ is a minimal prime of the ring $A/(a)$ (see the proof of Proposition 8.5.5), in particular we can write $P = \text{Ann}(\overline{b})$ for some $\overline{b} \in A/(a)$, which we lift to $b \in A$. Since a is not a zero divisor, we can consider $c = b/a \in K$. By construction, this satisfies $cP \subset A$.

If $cP \subset P$, then P is a finitely generated, faithful module over $A[c]$, hence c is integral over A by Theorem A.4.2. Since A is integrally closed in K, $c \in A$, which means that $\overline{b} = 0$—a contradiction. Since P is maximal, we must then have $cP = A$, or $P = c^{-1}A$. In particular, P is principal, so ht $P \leq 1$ by Krull's principal ideal theorem, Theorem A.7.6.

For the converse, assume that A satisfies R_1 and S_2, and let $K = \mathcal{F}(A)$ be its total ring of fractions. By the remarks before the theorem, we only need to prove that A is integrally closed in K. Let $a/b \in K$ be integral over A—in particular, b is not a zero divisor. By S_2, if P is an associated prime of the ideal (b), then ht $P = 1$. By R_1, it follows that A_P is a regular local ring of dimension 1, hence a DVR. In particular, it is integrally closed. Denoting by $\iota_P \colon A \to A_P$ the localization map, it follows that $\iota_P(a)/\iota_P(b) \in A_P$, or in other words $\iota_P(a) \in \iota_P(b)A_P$.

This is enough to get the conclusion, by primary decomposition. Let

$$(b) = \bigcap_{i=1}^{n} Q_i$$

be a primary decomposition of (b), where Q_i is P_i-primary. The point is that $\iota_{P_i}^{-1}(Q_i A_{P_i}) = Q_i$, since Q_i is P_i-primary. (Why?) It follows that $a \in \bigcap Q_i = (b)$, so in fact $a/b \in A$, as desired. $\square$

Corollary 8.5.10. *A regular ring is normal.*

Proof. A regular ring satisfies R_k, hence S_k, for all k.   □

Remark 8.5.11. Of course, the latter result follows immediately from the Auslander–Buchsbaum theorem (Theorem 8.4.1), but it is worth noting anyway, since the proof of Serre's criterion is much more elementary, being essentially based on primary decomposition.

Corollary 8.5.12. *Let A be a normal ring of dimension 1. Then A is regular.*

Remark 8.5.13. Let $A = R(V)$ be the ring of coordinates of the affine variety V. By the Auslander–Buchsbaum theorem, Theorem 8.4.1 (resp., Serre's criterion, Theorem 8.5.9), the local rings of A are UFD (resp., integrally closed). For this reason, one singles out varieties having this properties, which are called *factorial*, resp., *normal*. Factorial or normal varieties are considered to be a mild, controlled form of singularity. Of course, since these notions are local, they extend readily to arbitrary quasiprojective varieties.

Remark 8.5.14. In dimension 1, the notion of regular, factorial and normal variety agree with each other by Corollary 8.5.12. Given an affine curve C, one can always find another curve $\widehat{C}$ with a morphism $\widehat{C} \to C$ such that the associated map of rings $R(C) \to R(\widehat{C})$ is the integral closure of $R(C)$. It follows that $\widehat{C}$ is normal, hence regular—it is called the *normalization* of C. It is not difficult to see that this procedure generalizes to quasiprojective varieties by glueing the local normalizations. It follows that every curve admits a birational map from a smooth curve. This construction is called *resolution of singularities* for curves. Resolution of singularities exists also in higher dimensions in characteristic 0, but it is a much deeper result due to Hironaka ([**Hir64a**] and [**Hir64b**]; see also [**Kol07**]).

Example 8.5.15.

(a) Let k be a field, and let $A = k[x, y]/(f)$, where $f(x, y) = x^2 - y^3$ (a cusp). The only singular point is the origin, so let $B = A_{(x,y)}$. Then B is not normal, since $z = x/y$ satisfies $z^2 = y$, hence is integral over B, but $z \notin B$. (Why?)

(b) Take now $A = [k, y, z]/(f)$, where $f(x, y, z) = x^2 - y^2 - z^2$. This describes a cone singular in the origin, so let $B = A_{(x,y,z)}$. Now B is normal, since it satisfies R_1 and S_2 (check it!). On the other hand, we have the equality $y^2 = (x - z)(x + z)$, and it is easy to see that these are distinct factorizations of the same element, so B is not a UFD.

8.6. Regularity in characteristic p

This section is devoted the proof of Kunz theorem from [**Kun69**], which characterizes regular local rings in prime characteristic. A ring A has characteristic p, a prime number, if $p \cdot 1 = 0$ in A. In this case, the map

$$\phi\colon\ A \longrightarrow A$$

$$a \longmapsto a^p$$

is a homomorphism, called the *Frobenius map*.

Theorem 8.6.1 (Kunz). *Let $A, \mathcal{M}$ be a local Noetherian ring, char $A = p$, a prime number. Then A is regular if and only if the Frobenius morphism is flat.*

In this section, we only prove half the theorem: if ϕ is flat, then A is regular. The converse implication will be a simple application of miracle flatness, and we postpone it to Section 9.3. Our presentation follows [**dJea20**, Tag $0EBU$].

Before the proof of the theorem, we estalish some notation and make some preliminary remarks. We denote by ϕ the Frobenius map, $\mathcal{M}$ the maximal ideal, $k = A/\mathcal{M}$ the residue field. We also denote by A_ϕ the ring A itself, regarded as an A-module via ϕ.

We will need some remarks on tensor products with A_ϕ. First, since A_ϕ is flat, for every ideal $I \subset A$ there is an isomorphism $I \otimes A_\phi \cong IA_\phi = (\phi(I))$, where the last equality comes from the way we defined the structure of A-module on A_ϕ. If $I \subset J$ are ideals, by the exact sequence

$$0 \longrightarrow I \longrightarrow J \longrightarrow \frac{J}{I} \longrightarrow 0$$

and the flatness of A_ϕ, we get an isomorphism

$$\frac{J}{I} \otimes A_\phi \cong \frac{(\phi(J))}{(\phi(I))},$$

in particular, $A/I \otimes A_\phi \cong A/(\phi(I))$.

Let $a_1, \ldots, a_r \in A$, $I = (a_1, \ldots, a_r)$ the ideal they generate. For the purpose of this section, we will say that $a_1, \ldots, a_n$ are *independent* if I/I^2 is free of rank r as a A/I-module. Explicitly, this means that if $b_1, \ldots, b_r \in A$ satisfy

$$a_1 b_1 + \cdots + a_r b_r \in I^2,$$

then $b_1, \ldots, b_r \in I$. Notice that if this holds, then we can change $b_1, \ldots, b_r$ by an element of I and ensure that

$$(8.6.1) \qquad\qquad a_1 b_1 + \cdots + a_r b_r = 0.$$

Hence $a_1, \ldots, a_r$ are independent if, whenever (8.6.1) holds, then $b_1, \ldots, b_r \in I$. We need two simple lemmas regarding this property.

Lemma 8.6.2. *Assume that the elements $a_1, \ldots, a_{r-1}, a_r a_r'$ are independent. Then $a_1, \ldots, a_{r-1}, a_r$ are independent.*

Proof. Assume we have a dependence relation $\sum_{i=1}^r a_i b_i = 0$. Multiplying it by a_r', we get a dependence relation between the elements $a_1, \ldots, a_{r-1}, a_r a_r'$, in particular $b_r \in (a_1, \ldots, a_{r-1}, a_r a_r')$, say

$$b_r = c_1 a_1 + \cdots + c_{r-1} a_{r-1} + c a_r a_r'.$$

Substituting back in the original relation, we get

$$\sum_{i=1}^{r-1} (b_i + c_i a_r) a_i + c a_r^2 a_r' = 0,$$

which, again by independence, implies that $(b_i + c_i a_r) \in (a_1, \ldots, a_{r-1}, a_r a_r')$ for $i = 1, \ldots, r-1$. Thus, $b_i \in (a_1, \ldots, a_r)$ for $i < r$, and we are done. $\square$

Lemma 8.6.3. *Assume that $a_1, \ldots, a_{r-1}, a_r a_r'$ are independent and that $A/(a_1, \ldots, a_{r-1}, a_r a_r')$ has finite length. Then*

$$\ell \left(\frac{A}{(a_1, \ldots, a_{r-1}, a_r a_r')} \right) =$$
$$\ell \left(\frac{A}{(a_1, \ldots, a_{r-1}, a_r)} \right) + \ell \left(\frac{A}{(a_1, \ldots, a_{r-1}, a_r')} \right).$$

Proof. Consider the ideals $I = (a_1, \ldots, a_r)$, $I' = (a_1, \ldots, a_{r-1}, a_r')$ and $I'' = (a_1, \ldots, a_{r-1}, a_r a_r')$. It is enough to establish that there is an exact sequence

$$0 \longrightarrow \frac{A}{I} \stackrel{\cdot a_r'}{\longrightarrow} \frac{A}{I''} \longrightarrow \frac{A}{I'} \longrightarrow 0.$$

That the right-hand map is surjective is clear, since $I' \subset I''$. It is also clear that its kernel is generated by multiples of a_r'. Hence, all that remains is to check injectivity on the left.

Say $\bar{b}$ is in the kernel, so $b a_r' \in I''$, say

$$b a_r' = a_1 b_1 + \cdots + a_{r-1} b_{r-1} + a_r a_r' b_r.$$

Multiplying by a_r and rearranging, we get

$$a_1 b_1 a_r + \cdots + a_{r-1} b_{r-1} a_r + a_r a_r' (a_r b_r - b) = 0.$$

By independence, $a_r b_r - b \in (a_1, \ldots, a_{r-1}, a_r a_r')$, so that $b \in I$, as we wanted to prove. $\square$

Proof of Theorem 8.6.1, first half. Assume that ϕ is flat, and write $\mathcal{M}$ $= (a_1, \ldots, a_r)$, with r minimal. Letting $I = (\phi(\mathcal{M}))$, the result will follow from a Hilbert polynomial computation, for which we will need to compute $\ell(A/I)$.

The classes $\overline{a_1}, \ldots, \overline{a_r}$ form a basis of $\mathcal{M}/\mathcal{M}^2$ as k-vector spaces. By the above remarks on the flatness of A_ϕ, I/I^2 is free over A/I with basis $\overline{a_1^p}, \ldots, \overline{a_r^p}$. In our terminology, this means that $\overline{a_1^p}, \ldots, \overline{a_r^p}$ are independent, and by Lemma 8.6.2, $\overline{a_1^{e_1}}, \ldots, \overline{a_r^{e_r}}$ are independent for all $e_i \leq p$. Using induction and Lemma 8.6.3, we find that

$$\ell\left(\frac{A}{(a_1^{e_1}, \ldots, a_r^{e_r})}\right) = e_1 \cdots e_r,$$

for $e_i \leq p$; in particular $\ell(A/I) = p^r$.

For $n \gg 0$ we have the inclusions

$$\mathcal{M}^{pn+pr} \subset (\phi(\mathcal{M}^n)) \subset \mathcal{M}^{pn}.$$

We can use these to get bounds on the Hilbert polynomial $\chi = \chi_A^{\mathcal{M}}$. For this we need to compute the length of the quotient $A/\phi(\mathcal{M}^n) \cong A/\mathcal{M}^n \otimes A_\phi$. By Lemma 6.4.13, we get

$$\ell\left(\frac{A}{\phi(\mathcal{M}^n)}\right) = \ell\left(\frac{A}{\mathcal{M}^n}\right) \cdot \ell\left(\frac{A}{I}\right) = p^r \chi(n).$$

This gives us the inequalities

$$\chi(pn) \leq p^n \chi(n) \leq \chi(pn + pr).$$

Since χ is a polynomial of degree $\dim A$, this can only happen if $\dim A = r$, which means that A is regular. $\qquad\qquad\qquad\qquad\qquad\qquad\qquad\qquad\qquad\qquad\qquad\square$

8.7. Exercises

1. Let A be a regular local ring of dimension d, $I \subset A$ an ideal. If $\operatorname{depth} A/I = d - 1$, show that I is principal.

Exercises 2–4, present an alternative proof of one implication in Serre's theorem 8.2.1, originally proposed by Nagata and then simplified by Grothendieck.

2. Let $A, \mathcal{M}$ be a local Noetherian ring, and assume that all elements of $\mathcal{M} \setminus \mathcal{M}^2$ are zero divisors. Show that $\mathcal{M} \in \operatorname{Ass}(A)$.

3. Let $A, \mathcal{M}$ be a local Noetherian ring, $a \in \mathcal{M} \setminus \mathcal{M}^2$. Show that $\mathcal{M}/(a)$ is isomorphic to a direct summand of $\mathcal{M}/a\mathcal{M}$.

4. Let $A, \mathcal{M}$ be a local Noetherian ring with gl. dim $A < \infty$. Show that A is regular. (If there exists a regular element $a \in \mathcal{M} \setminus \mathcal{M}^2$, proceed by induction on dim A. Else, by Exercise 2, we have $\mathcal{M} = \mathrm{Ann}(b)$ for some $b \in A$. Derive an exact sequence involving $\mathrm{Tor}(k, -)$, and find a contradiction with Proposition 8.2.2.)

5. Assuming that the ring A has finite dimension, give a proof of Proposition 8.3.3 that A is regular if and only if $A[x]$ is regular based on Serre's theorem, Theorem 8.2.1.

6. Let A be a normal integral domain. Prove directly that A is integrally closed.

7. Let $A \to B$ be a flat local homomorphism of Noetherian rings. Show that if B satisfies condition R_k or S_k, then A also does. Conclude that the integral closure of an integral domain A is never flat over A, unless A is already integrally closed. For a different approach, see Exercise 5 in Chapter 6.

8. Let $A, \mathcal{M}$ be a regular local ring, $a_1, \dots, a_d$ a regular system of parameters. Prove that the ideal $(a_{i_1}, \dots, a_{i_k})$ is prime for all subset of k distinct indices $i_1, \dots, i_k \in \{1, \dots, d\}$.

9. Let A be a Noetherian ring such that every finitely generated A-module admits a finite free resolution. Prove that A is a UFD. (If P is a prime of height 1, prove that P is stably free.)

10. Let A be a reduced Noetherian ring, $P \subset A$ a minimal prime. Show that A_P is regular.

11. Let V be an affine variety over a field. Show that the locus of points $x \in V$ such that V is regular at x is open.

12. Let $f \colon A \to B$ be a local homomorphism of Noetherian rings, with A regular. Let $a_1, \dots, a_d \in A$ be a regular system of parameters, and assume that $f(a_1), \dots, f(a_d)$ is a regular sequence in B. Show that f is flat. (Keep in mind the local criterion of flatness)

13. Let k be a field, A a k-algebra, and $\mathbf{a} = a_1, \dots, a_d \in A$. Show that $\mathbf{a}$ is a regular sequence if and only if the map $k[x_1, \dots, x_d] \to A$ that sends x_i to a_i is flat. (Mimic the previous exercise.)

14. Let $f_1, \dots, f_r$ be a regular sequence of homogeneous polynomials in $A := k[x_1, \dots, x_t]$. Prove that $\mathrm{pd}_A(f_1, \dots, f_r) \leq r$.

Exercises 15–18, follow [**Mat86**, Chapter 20] to show that unique factorization is carried over to power series rings *in the regular case*.

15. Let A be a ring that satisfies the ascending chain condition on principal ideals and such that $(a) \cap (b)$ is principal for all $a, b \in A$. Prove that A is a UFD.

16. Let A be a UFD, $I \subset A$ a projective ideal. Prove that I is principal.

17. Let A be a regular UFD, $B = A[[x]]$. Given $f, g \in B$, set $(f) \cap (g) = x^r I$ for some ideal $I \subset B$. Prove that I is projective and $I \cap xB = xI$.

18. Let A be a regular UFD. Prove that $A[[x]]$ is a (regular) UFD as well.

19. Let $f \colon A \to B$ be a flat local homomorphism of Noetherian rings, having maximal ideals $\mathcal{M}_A$ and $\mathcal{M}_B$ respectively. Prove that if A and $B/\mathcal{M}_A B$ are regular, then B is regular.

20. Let $f \colon A \to B$ be a homomorphism of Noetherian rings such that A is regular and the fibers $k_P \otimes_A B$ are all regular, where $k_P = A_P/PA_P$ is the residue field of A at P. Prove that B is regular.

21. Use the theorem of Vasconcelos from Exercise 9 in Chapter 7 to prove one implication in Serre's theorem (Theorem 8.2.1)—namely that if $\operatorname{gl.dim} A < \infty$, then A is regular.

Chapter 9

Mild Singularities

In Section 8.4, we have seen that every local regular ring is a UFD, so in particular integrally closed. It follows that we can see rings that are locally UFDs or normal as a relaxation of regular rings, a kind of singularity that is somehow well behaved. In this chapter, we extend this line of reasoning by defining other classes of rings (in fact, a whole hierarchy of rings) that generalize regular rings in various aspects. By studying the properties of these more general rings, we sometimes get interesting insights in the regular case. For instance, we will be able to prove that primary ideals in local regular rings are irreducible, or that the chains of prime ideals in local regular rings have some nice structure. All the conditions that we study are local (with a slight exception for Gorenstein rings), so in most of the chapter we focus on local Noetherian rings.

We start with Cohen–Macaulay rings in Section 9.1. These are characterized (in the local case) by having depth equal to the dimension. In fact, Cohen–Macaulay rings have many equivalent characterizations, and for this reason they appear all over commutative algebra. One of the most useful ways to think of local Cohen–Macaulay rings is that they are the rings where systems of parameters and (maximal) regular sequences are one and the same thing, as we prove in Theorem 9.1.13. In the section we develop the basic properties of Cohen–Macaulay rings (in fact, since it is no more difficult, we work throughout the section with Cohen–Macaulay *modules*). In particular, we prove that they are stable under polynomial or power series extensions, localizations or quotients by regular sequences. In Section 9.2 we prove that chains of prime ideals in local Cohen–Macaulay rings are well behaved: in particular, if $P \subset Q$ are primes, then all maximal chains from P to Q have the same length. Moreover, all associated primes are minimal, and the length of all maximal chains in the ring are the same. These are nontrivial results already for regular rings, and in fact are one of the reasons why Cohen–Macaulay rings were introduced in the first place.

We then have the short Section 9.3, dedicated to the proof of the so-called miracle flatness theorem. This gives a very simple dimensional condition to ensure that a morphism from a local regular ring to a local Cohen–Macaulay ring is flat.

In Section 9.4 we start to investigate local Cohen–Macaulay rings from another point of view. Let $A, \mathcal{M}$ be a local Noetherian ring. The theme of this section is to understand the decomposition of an $\mathcal{M}$-primary ideal as an intersection of irreducible ideals. It turns out that the number of irreducible factors does not depend on the decomposition. Moreover, Northcott's theorem 9.4.9 ensures that if A is Cohen–Macaulay, the number of irreducible factors is the same for all ideals generated by a system of parameters. As a corollary, such ideals are always irreducible in a regular local ring. Conversely, a local Noetherian ring having this property is Cohen–Macaulay by the Northcott–Rees theorem, Theorem 9.4.12. We shall see in Theorem 9.7.12 that this property actually characterizes Gorenstein rings. A fundamental tool to prove these results is the socle of a module M, which is the largest submodule of M having the structure of a vector space over $k = A/\mathcal{M}$. Section 9.5 further develops some simple properties of this construct.

In Section 9.6, we shift to the study of an inequality between depth and injective dimension, with the aim of characterizing the cases of equality. One of the central results is Theorem 9.6.5, which ensures that if M is a finitely generated module over the local ring A, then $\mathrm{id}_A M = \mathrm{depth}\, A$ (unless $\mathrm{id}_A M = \infty$). In the special case where $M = A$, we obtain that a local Noetherian ring A has finite injective dimension if and only if $\dim A = \mathrm{depth}\, A = \mathrm{id}_A A$. These rings are called Gorenstein, and are the object of Section 9.7. As we mentioned, they are also characterized by the irreducibility of the ideals generated by systems of parameters.

The last class that we introduce, in Section 9.8, is that of complete intersection rings. These do not have such simple characterizations, but are convenient to compute with. A local ring A that can be written as the quotient B/I, where B is regular, is called a complete intersection if I can be generated by $\dim B - \dim A$ elements. In the geometric setting, this corresponds to asking that a variety is locally defined by a number of equations equal to its codimension in the ambient space. Most of the section is devoted to proving that this property is intrinsic, and does not depend on how we represent A as a quotient. Theorem 9.8.8 recaps all the implications between the properties of local rings that we have studied.

9.1. The Cohen–Macaulay property

We have observed in Corollary 7.4.6 that a local Noetherian ring A always satisfies $\operatorname{depth} A \leq \dim A$, with equality if A is regular (Corollary 8.1.9). It is immediate to extend this result to modules.

Proposition 9.1.1. *Let A be a local Noetherian ring, M a finitely generated A-module, $I = \operatorname{Ann}(M)$. Then $\operatorname{depth} M \leq \operatorname{depth} A/I$.*

Proof. It is then enough to show that if $a_1, \ldots, a_d$ is an M-regular sequence, then the classes $\overline{a_1}, \ldots, \overline{a_d}$ are a regular sequence in A/I. By induction, it is enough to do the case $d = 1$.

In other words, we need to show that a_1 is a divisor of 0 in M, assuming that $\overline{a_1}$ is a divisor of 0 in A/I. This is clear: assume that $\overline{a_1}\overline{b} = 0$ in A/I, where $b \in A \setminus I$. Take any $m \in M$ such that $bm \neq 0$; then $a_1 bm = 0$, showing that a_1 is a divisor of 0 in M. $\square$

Recalling that $\dim M = \dim A/\operatorname{Ann}(M)$ by definition and applying Corollary 7.4.6, we get the following result.

Corollary 9.1.2. *Let A be a local Noetherian ring, M a finitely generated A-module. Then $\operatorname{depth} M \leq \dim M$.*

Rings and modules that achieve equality are rather special.

Definition 9.1.3. Let A be a local Noetherian ring, M a finitely generated A-module. We say that M is Cohen–Macaulay (CM) if $\operatorname{depth} M = \dim M$. In particular, A is a local Cohen–Macaulay ring if $\operatorname{depth} A = \dim A$.

We are mostly interested in Cohen–Macaulay *rings*. Since every regular local ring is Cohen–Macaulay, we will see this condition as a relaxation of the regular condition, a mild form of singularity. But since the proofs are essentially the same, in this section we will mostly state our result for Cohen–Macaulay modules, with no additional effort. We postpone examples after Theorem 9.1.13, when we have more criteria to check whether a given ring is CM or not.

There are many characterizations of Cohen–Macaulay rings and modules, and in this section we shall see a few. To start with, we notice that this property is stable under localization, as it happens for regularity (Corollary 8.2.5).

Theorem 9.1.4. *Let A be a local Noetherian ring, M a Cohen–Macaulay A-module, $P \subset A$ a prime in the support of M. Then M_P is Cohen–Macaulay as an A_P-module.*

Notice that the requirement that $P \in \operatorname{Supp}(M)$ is only needed to avoid the trivial case that $M_P = 0$. To prove the result, we need a couple of lemmas on depth.

Lemma 9.1.5. *Let A be a Noetherian ring, $I \subset A$ an ideal, M a finitely generated A-module. Then there is a prime ideal $P \supset I$ such that*

$$\operatorname{depth}_I M = \operatorname{depth}_P M.$$

Proof. Choose a maximal M-regular sequence $a_1, \ldots, a_k \in I$, where $k = \operatorname{depth}_I M$. Letting $J = (a_1, \ldots, a_k)$, by definition I is contained in the set of divisors of 0 of the module M/JM. By Theorem A.3.6, this is the union of the associated primes $P \in \operatorname{Ass}(M/JM)$. By prime avoidance, I is contained in one such prime P. By construction, $\operatorname{depth}_P M = k$, unless $PM = M$.

The condition $PM = M$ implies that there exists $a \in P$ such that $1 + a \in \operatorname{Ann}(M)$, as can be seen by the classical proof of Nakayama's lemma. This is impossible since $P \in \operatorname{Ass}(M/JM)$. □

Lemma 9.1.6. *Let $A, \mathcal{M}$ be a local Noetherian ring, $I \subset A$ an ideal, M a finitely generated A-module. If $\operatorname{depth}_I M < \operatorname{depth}_{\mathcal{M}} M$, then there is a prime ideal $P \supset I$ such that $\operatorname{depth}_P M = \operatorname{depth}_I M + 1$.*

Proof. As in the previous proof, let $k = \operatorname{depth}_I M$, choose a maximal M-regular sequence $a_1, \ldots, a_k \in I$, and let $J = (a_1, \ldots, a_k)$. By the hypothesis, we find $a_{k+1} \in \mathcal{M}$ that is regular for M/JM. Clearly, $\operatorname{depth}_{I'} M \geq k + 1$, where $I' = (I, a_{k+1})$.

If $\operatorname{depth}_{I'} M \geq k + 2$, we could enlarge $a_1, \ldots, a_{k+1}$ to a bigger M-regular sequence $a_1, \ldots, a_{k+2}$ contained in I', by Corollary 7.3.11. Write

$$a_{k+2} = b a_{k+1} + c,$$

with $c \in I$. Then, since a_{k+2} is regular on $M/(a_1, \ldots, a_{k+1})M$, the same is true for c, so $a_1, \ldots, a_{k+1}, c$ is M-regular. By Corollary 7.2.5, the sequence $a_1, \ldots, a_k, c, a_{k+1}$ is *also* M-regular, which contradicts the fact that $\operatorname{depth}_I M = k$.

We conclude that $\operatorname{depth}_{I'} M = k + 1$, and then we can pass to a prime ideal containing it by Lemma 9.1.5. □

Proof of Theorem 9.1.4. First, notice that for any prime P we have $\operatorname{depth}_P M \leq \operatorname{depth}_{PA_P} M_P$ by Remark 7.3.3. In fact, we have equality, since if $a_1/b_1, \ldots, a_k/b_k$ is a sequence in A_P that is regular for M_P, then $a_1, \ldots, a_k$ is a sequence in A that is regular for M.

Now, let P be a prime such that M_P is not Cohen–Macaulay, and choose P maximal with respect to this condition. By the above remark, $\operatorname{depth}_P M < \dim A_P / \operatorname{Ann} M_P$.

By Lemma 9.1.6, we find $Q \supset P$ with $\mathrm{depth}_Q\, M = \mathrm{depth}_P\, M + 1$. Since $\dim A_Q / \mathrm{Ann}\, M_Q \geq \dim A_P / \mathrm{Ann}\, M_P + 1$, the module M_Q is not Cohen–Macaulay either, which contradicts the maximality of P. $\square$

It is helpful to explicitly note the form of this result when we specialize to the case $M = A$.

Corollary 9.1.7. *Let A be a local Cohen–Macaulay ring, $P \subset A$ a prime. Then A_P is Cohen–Macaulay, that is, $\mathrm{depth}_P\, A = \mathrm{ht}\, P$.*

In fact, the same property follows at once for *any* proper ideal.

Corollary 9.1.8. *Let A be a local Cohen–Macaulay ring, $I \subset A$ an ideal. Then $\mathrm{depth}_I\, A = \mathrm{ht}\, I$.*

Proof. By Lemma 9.1.5, $\mathrm{depth}\, I$ is the minimum of $\mathrm{depth}\, P$ for all primes $P \supset I$—the same is true by definition for the height. Hence the conclusion follows from Corollary 9.1.7. $\square$

As we saw for Noetherian rings, this leads to the following definition for the nonlocal case.

Definition 9.1.9. Let A be a Noetherian ring, M a finitely generated A-module. We say that M is Cohen–Macaulay (CM) if $M_{\mathcal{M}}$ is a Cohen–Macaulay module over $A_{\mathcal{M}}$ for every maximal ideal $\mathcal{M} \subset A$.

Remark 9.1.10. By Proposition 9.1.4, this is the same as asking that M_P is Cohen–Macaulay over A_P for all prime ideals $P \subset A$.

Remark 9.1.11. Some authors (for instance [**Nag62**]) require in addition that $\mathrm{ht}\, \mathcal{M} = \dim A$ for all maximal ideals $\mathcal{M} \subset A$. This has the advantage of making CM rings equidimensional; see Remark 9.2.11. On the other hand, with this convention it would not be always true that a regular ring is Cohen–Macaulay, something that we want to ensure.

Remark 9.1.12. By Corollary 9.1.7, A is Cohen–Macaulay if and only if it satisfies the condition S_k for all $k \in \mathbb{N}$.

The next result, that generalizes Corollary 8.1.9, is one of the most useful characterizations of CM modules.

Theorem 9.1.13. *Let A be a local Noetherian ring, M a finitely generated A-module, $I = \mathrm{Ann}\, M$. The following are equivalent:*

   (i) *M is Cohen–Macaulay;*

   (ii) *every system of parameters in A/I is an M-regular sequence;*

   (iii) *every maximal M-regular sequence in A/I is a system of parameters;*

(iv) *there is a system of parameters in A/I that is an M-regular sequence; and*

(v) *for every system of parameters $a_1, \ldots, a_d$ of A/I, M/I is not an associated prime of $M/(a_1, \ldots, a_{d-1})M$.*

Since this theorem is so important, we explicitly translate it in the special case $M = A$.

Corollary 9.1.14. *Let A be a local Noetherian ring. The following are equivalent:*

(i) *A is Cohen–Macaulay;*

(ii) *every system of parameters in A is a regular sequence;*

(iii) *every maximal regular sequence in A is a system of parameters;*

(iv) *there is a system of parameters in A that is a regular sequence; and*

(v) *for every system of parameters $a_1, \ldots, a_d$ of A, M is not an associated prime of $(a_1, \ldots, a_{d-1})$.*

To prove Theorem 9.1.13, we will need to compare different systems of parameters in a local ring. We start by discussing how to construct systems of parameters explicitly.

Remark 9.1.15. There is an inductive way to construct a system of parameters for a local Noetherian ring A, M of dimension d. If $d = 0$, just take the empty sequence. Otherwise, let S_0 be the union of the minimal primes of A. By prime avoidance, S_0 is properly contained in M. Choose any $a_1 \in M \setminus S_0$. Then by construction, $\dim A/(a_1) \leq d - 1$.

In fact, the dimension can decrease at most by 1 by taking a quotient by a single element. Otherwise, if $\dim A/(a_1) < d - 1$, we could lift a system of parameters for $A/(a_1)$ and add a_1 and obtain an ideal of definition for A that is generated by less than d elements. Hence, $\dim A/(a_1) = d - 1$.

If $d > 1$, we can continue this process: the union S_1 of all minimal primes of (a_1) does not cover M, so we can choose $a_2 \in M \setminus S_1$, so that $\dim A/(a_1, a_2) = d - 2$, and so on. After d steps, we obtain that $\dim A/(a_1, \ldots, a_d) = 0$, so that $a_1, \ldots, a_d$ is a system of parameters.

Lemma 9.1.16. *Let A, M be a local Noetherian ring, $\mathbf{a}$ and $\mathbf{b}$ two systems of parameters. Then there exists a sequence of systems of parameters, say $\mathbf{s}_1, \ldots, \mathbf{s_k}$ with $\mathbf{s}_1 = \mathbf{a}$ and $\mathbf{s_k} = \mathbf{b}$ such that every two consecutive systems of parameters in the sequence differ by just one element.*

Proof. Let $d = \dim A$—we prove the result by induction on d. If $d = 1$, the result is clear. Write $\mathbf{a} = a_1, \ldots, a_d$ and $\mathbf{b} = b_1, \ldots, b_d$ and consider the ideals $I = (a_2, \ldots, a_d)$ and $J = (b_2, \ldots, b_d)$. By prime avoidance, the union of the minimal primes of I and J is not the whole $\mathcal{M}$, so there exists $c \in \mathcal{M}$ that does not lie in any of these primes. As in the previous remark, it follows that $c, a_2, \ldots, a_d$ and $c, b_2, \ldots, b_d$ are still systems of parameters. We conclude by induction, working in $A/(c)$.   □

Proof of Theorem 9.1.13. By working in A/I, we can assume that $I = 0$; this will simplify the notation.

(i)$\Longrightarrow$(iii) Assume that M is Cohen–Macaulay of dimension d, and let $a_1, \ldots, a_d$ be a maximal M-regular sequence in A. By Theorem A.7.4, we have $\dim A/(a_1, \ldots, a_d) = 0$. It follows that $a_1, \ldots, a_d$ is a system of parameters.

(iii)$\Longrightarrow$(iv) Obvious.

(iv)$\Longrightarrow$(ii) There is a system of parameters $a_1, \ldots, a_d$ that is an M-regular sequence. We must prove that the same holds for every system of parameters, and by virtue of Lemma 9.1.16 it is enough to prove the result for a system of parameters that differs only in one element. Since regular sequences in a local ring are permutable (Corollary 7.2.5) and systems of parameters are unordered, we can assume that the target system of parameters is $a_1, \ldots, a_{d-1}, b$. By working in $A/(a_1, \ldots, a_{d-1})$, we can assume that $d = 1$.

Hence let $B, \mathcal{M}$ be a local ring of dimension 1, $a, b \in B$ two elements such that $\sqrt{(a)} = \sqrt{(b)} = \mathcal{M}$, and assume that a is not a divisor of 0 in M. Then $a^n = bc$ for some $c \in A$, and since a is not a zero divisor in M, neither b is one.

(ii)$\Longrightarrow$(v) Let $a_1, \ldots, a_d$ be a system of parameters. Then $\mathcal{M}$ cannot be an associated prime of $(a_1, \ldots, a_{d-1})$, otherwise by Theorem A.3.6 a_d could not be regular on $M/(a_1, \ldots, a_{d-1})M$.

(v)$\Longrightarrow$(iv) Since $\mathcal{M}$ is not an associated prime of $M/(a_1, \ldots, a_{d-1})M$ for a system of parameters $a_1, \ldots, a_d$, a fortiori $\mathcal{M}$ is not an associated prime of $M/(a_1, \ldots, a_k)M$ for any $k \leq d - 1$ such that $a_1, \ldots, a_k$ are part of a system of parameters. We then construct a system of parameters by the process of Remark 9.1.15. At each step, we can choose that a_k not be in the minimal primes of $(a_1, \ldots, a_{k-1})$, nor in the associated primes of $M/(a_1, \ldots, a_{k-1})M$. By Theorem A.3.6, it follows that $a_1, \ldots, a_d$ is M-regular.

(iv)$\Longrightarrow$(i) Let $d = \dim M = \dim A$, and let $\mathbf{a} = a_1, \ldots, a_d$ be a system of parameters that is M-regular. Then $\operatorname{depth} M \geq d$, which proves that M is Cohen–Macaulay.   □

The conditions of the theorem allow us to extend the definition of a Cohen–Macaulay module even without assuming finite generation.

Definition 9.1.17. Let $A, \mathcal{M}$ be a local Noetherian ring, M an A-module, $I = \operatorname{Ann} M$. We say that M is a *big Cohen–Macaulay* module if there exists a system of parameters for A/I that is M-regular. If *every* system of parameters if M-regular, we say that M is *balanced*.

Using Theorem 9.1.13, we can now give some examples (and nonexamples) of Cohen–Macaulay rings, other than the usual ones.

Example 9.1.18.

(a) Any local Noetherian ring of dimension 0 is Cohen–Macaulay.

(b) Let $A, \mathcal{M}$ be a local Noetherian ring of dimension 1. Then A is Cohen–Macaulay if and only if $\mathcal{M}$ is not an associated prime of 0. In particular, if A is reduced it is Cohen–Macaulay.

(c) Let A be a normal Noetherian ring of dimension 2. Then A is Cohen–Macaulay by Serre's criterion, Theorem 8.5.9.

(d) As an example of a 1-dimensional, non-CM ring, let k be a field and take $A = k[x, y]/(x^2, xy)$. Let $A_{\mathcal{M}}$ be its localization at the maximal ideal $\mathcal{M} = (x, y)$ of the point 0. Then $x \in A_{\mathcal{M}}$ is nonzero, but $fx = 0$ for either $f = x$ or $f = y$. It follows immediately that every noninvertible element of $A_{\mathcal{M}}$ is a divisor of zero.

(e) Take $A = k[x, y]/(y^2 - x^3)$ and let $A_{\mathcal{M}}$ be its localization at 0. We know that $A_{\mathcal{M}}$ is not integrally closed, but it is CM, since it is reduced.

(f) Now take $A = k[x^4, x^3 y, xy^3, y^4]$ and let $A_{\mathcal{M}}$ be its localization at 0. Then $A_{\mathcal{M}}$ is not Cohen–Macaulay. The reason is that x^4, y^4 is a system of parameters, but it is not a regular sequence. In fact, y^4 is a divisor of 0 modulo x^4, as the following computation shows:
$$(x^3 y)^2 y^4 = x^6 y^6 = x^4 (xy^3)^2.$$

Notice that A is an integral domain, and that in fact
$$A \cong \frac{k[a, b, c, d]}{(ad - bc, a^2 c - b^3, bd^2 - c^3, ac^2 - b^2 d)},$$
which is the coordinate ring of the variety cut out by these equations.

In the rest of the section, we are going to verify that the Cohen–Macaulay property is stable under the usual operations.

Proposition 9.1.19. *Let $A, \mathcal{M}$ be a local Noetherian ring, M a finitely generated A-module. Denote by $\widehat{A}$, respectively $\widehat{M}$, the completion of A,* *respectively M, with respect to the $\mathcal{M}$-adic topology. Then M is Cohen–Macaulay if and only if $\widehat{M}$ is.*

Proof. The crux of the proof is that completion preserves both systems of parameters and regular sequences. To see this, we first remark that by Krull's intersection theorem (Theorem A.6.4), the map $A \to \widehat{A}$ is injective, so we can regard A as a subring of $\widehat{A}$. Similarly, M is a submodule of $\widehat{M}$. Moreover, by Theorem A.6.2, if $J \subset A$ is an ideal, we have

$$\left(\widehat{\frac{M}{JM}} \right) \cong \frac{\widehat{M}}{\widehat{J}} = \frac{\widehat{M}}{J\widehat{M}},$$

so again we can regard M/JM as a submodule of $\widehat{M}/J\widehat{M}$.

Let $I = \operatorname{Ann} M$, $d = \dim A/I$, and take $\mathbf{a} = a_1, \dots, a_d \in A/I$. First, notice is that $\mathbf{a}$ is a system of parameters in A/I if and only if it is a system of parameters in $\widehat{A}/I\widehat{A} \cong \widehat{A/I}$. This is because dimension is preserved in completions.

Moreover, the completion map is flat by Theorem 6.2.9, hence faithfully flat by Corollary 6.3.7. By Proposition 7.2.8, if $\mathbf{a}$ is M-regular, then it is $\widehat{M}$-regular. The converse is also true because of the initial remark: if $a \in A$ is not a zero divisor on $\widehat{M}/J\widehat{M}$ for some ideal $J \subset A$, then, a fortiori, it is not a zero divisor on M/JM. Hence $\mathbf{a}$ is M-regular if and only if it is $\widehat{M}$-regular. The conclusion now follows from Theorem 9.1.13. $\square$

Proposition 9.1.20. *Let A be a local Noetherian ring, M a finitely generated A-module. If $a \in A$ is not a zero divisor on M, then M is Cohen–Macaulay if and only if M/aM is (either over A or over $A/(a)$).*

Proof. Combine Theorem A.7.4 with Proposition 7.3.10. $\square$

By induction, we immediately generalize this.

Corollary 9.1.21. *Let A be a local Noetherian ring, M a finitely generated A-module. If $\mathbf{a}$ is an M-regular sequence, then M is Cohen–Macaulay if and only if $M/(\mathbf{a})M$ is (either over A or over $A/(\mathbf{a})$).*

One direction generalizes to the nonlocal case.

Corollary 9.1.22. *Let A be a Noetherian ring, M a Cohen–Macaulay A-module and $\mathbf{a}$ an M-regular sequence. Then $M/(\mathbf{a})M$ is Cohen–Macaulay as a module over A or over $A/(\mathbf{a})$.*

Proof. Let $\mathcal{M} \subset A$ be a maximal ideal. By definition, $M_{\mathcal{M}}$ is CM over $A_{\mathcal{M}}$, so the same is true for $M_{\mathcal{M}}/(\mathbf{a})M_{\mathcal{M}}$, and we are done by commuting quotient and localization. $\square$

The obvious impediment to globalize the other implication is that we obtain the CM condition only over maximal ideals of A that contain $\mathbf{a}$. As special cases, we notice the behavior for polynomial and power series rings.

Proposition 9.1.23. *Let A be a Noetherian ring. Then A is Cohen–Macaulay if and only if $A[x]$ is.*

Proof. One direction follows at once from Corollary 9.1.22, so we only need to prove that if A is Cohen–Macaulay, then $A[x]$ is. Let $\mathcal{M}' \subset A[x]$ be a maximal ideal and $\mathcal{M} = \mathcal{M}' \cap A$. As in the beginning of the proof of Proposition 8.3.3, we reduce to the case where A is local with maximal ideal $\mathcal{M}$.

Letting $k = A/\mathcal{M}$, notice that the image of $\mathcal{M}'$ in $k[x]$ is generated by a single irreducible element $\overline{f}$. Lift this to a monic polynomial $f \in A[x]$. Then $\mathcal{M}' = (\mathcal{M}, f)$. Take a system of parameters $a_1, \ldots, a_d \in A$, which is a regular sequence by Theorem 9.1.13. Since the inclusion $A \to A[x]$ and localization are flat maps, $a_1, \ldots, a_d$ is also a regular sequence in $A[x]_{\mathcal{M}'}$, by Proposition 7.2.8. Clearly, f is not a divisor of 0 modulo $(a_1, \ldots, a_d)$, so $(a_1, \ldots, a_d, f)$ is regular and

$$\operatorname{depth} A[x]_{\mathcal{M}'} \geq d + 1 = \dim A[x]_{\mathcal{M}'}.$$

$\square$

Proposition 9.1.24. *Let A be a Noetherian ring. Then A is Cohen–Macaulay if and only if $A[[x]]$ is.*

Proof. One direction follows at once from Corollary 9.1.22, so we only need to prove that if A is Cohen–Macaulay, then $A[[x]]$ is. Once we have done a reduction to the local case, as in Proposition 9.1.23, it is enough to notice that if A is local, then $A[[x]]$ is local as well, and then apply Proposition 9.1.20. $\square$

9.2. Catenary rings

When working with chains of prime ideals, we often have to be careful of the way we phrase inductive arguments. The reason is that for a ring A of finite Krull dimension d, it is not necessarily the case that any maximal chain from a mimimal to a maximal prime will have length d. A priori, there could be chains having different lengths, the longest one having d steps.

This is most clearly seen in the geometric setting. If $A = R(V)$ is the coordinate ring of the affine variety V, a prime in A corresponds to an irreducible subvariety of V. If, say V is the union of components of different dimensions—for instance, a line meeting a plane—there will be maximal chains of different lengths. But even if we assume that V is equidimensional, which means that $\dim A/P$ is the same for all minimal primes P, it is not a priori clear that all maximal chains should have the same length.

Definition 9.2.1. Let A be a ring. We say that A is *catenary* if for all pairs of prime ideals $P \subset Q \subset A$, any chain of ideals

$$P = P_0 \subset P_1 \subset \cdots \subset P_n = Q$$

can be enlarged to a maximal chain from P to Q, and the finite length of such a maximal chain only depends on P and Q. If the chain cannot be enlarged to a longer chain by inserting other primes, we will say that it is *saturated*.

A ring A is called *universally catenary* if $A[x_1, \ldots, x_n]$ is catenary for all $n \in \mathbb{N}$.

Remark 9.2.2. Let A be a catenary ring. Then any localization or any quotient of A is still catenary, by the correspondence between prime ideals of A and its localizations and quotients.

One of the main results of this section is that a Cohen–Macaulay ring is catenary, hence universally catenary by Proposition 9.1.23. In particular, regular rings are universally catenary. Since the catenary property is preserved by quotients, any finitely generated k-algebra, where k is a field is catenary. For more properties of catenary rings, see Exercises 15–19.

Remark 9.2.3. The definition of catenary ring ensures that there cannot be two maximal paths of different lengths from the prime P to $Q \supset P$. But this still does not ensure that all maximal chains, globally, have the same lengths, as the example of a plane and a disjoint line shows. This prompts us to give another definition.

Definition 9.2.4. Let A be a Noetherian ring, $I \subset A$ an ideal. We say that I is *unmixed* if for every prime $P \in \mathrm{Ass}(I)$, we have $\mathrm{ht}\, P = \mathrm{ht}\, I$.

Remark 9.2.5. If I is unmixed, every associated prime of I is in fact minimal over I.

Remark 9.2.6. Consider the geometric case $A = R(V)$, so that the ideal I defines a subvariety $W \subset V$. Then saying that I is unmixed amounts to saying that all irreducible components of W have the same codimension in V.

The other important point of this section is that in a Cohen–Macaulay ring an ideal generated by number of elements equal to its height is unmixed. This, together with the catenary property, puts rigid constraints to the lattice structure of prime ideals in a Cohen–Macaulay ring. In fact, the notion of Cohen–Macaulay ring arose first in the proof by Cohen that a regular ring is catenary [**Coh46**].

We now turn to the proofs of these important results. The starting point is the following basic property of ideals in a Cohen–Macaulay ring.

Proposition 9.2.7. *Let A be a local Cohen–Macaulay ring, $I \subset A$ an ideal. Then*

$$\operatorname{ht} I + \dim A/I = \dim A.$$

Proof. By definition of dimension and height, it is enough to prove the equality for a minimal prime P of I. Let $d = \dim A$ and $r = \operatorname{ht} P$. By Corollary 9.1.7, we find a regular sequence $\mathbf{a} = a_1, \ldots, a_r$ contained in P.

By Proposition 7.3.11, we can enlarge $\mathbf{a}$ to a maximal regular sequence, which is then a system of parameters by Theorem 9.1.13. In particular, $A/(\mathbf{a})$ is a Cohen–Macaulay ring of dimension $d - r$. It follows that $P/(\mathbf{a})$ is a minimal prime of the ring $A/(\mathbf{a})$. In particular,

$$d - r \leq \operatorname{depth} \frac{A}{P} \leq \dim \frac{A}{P} \leq d - r,$$

hence we must have equality. $\square$

Theorem 9.2.8. *Let A be a Cohen–Macaulay ring. Then A is universally catenary.*

Proof. By Proposition 9.1.23, it is enough to prove that A is catenary. Further, when we consider chains from P to Q, we can localize at Q, hence we can assume that A is local with maximal ideal Q. In this case, the longest chain from P to Q will have length $\dim A/P = \operatorname{ht} Q - \operatorname{ht} P$ by Proposition 9.2.7.

Now consider any chain from P to Q. We can enlarge it to be maximal in a finite number of steps since $\dim A$ is finite. We need to prove that every such maximal chain has length $\operatorname{ht} Q - \operatorname{ht} P$. As in the proof of Proposition 9.2.7, we can take quotient by a regular sequence of length $\operatorname{ht} P$, while still working in a CM ring. It follows that we can reduce to the case where P is a minimal prime. Thus, we need to prove that if A is local, Q is maximal and P is minimal, every saturated chain from P to Q has length $\dim A$.

Consider such a saturated chain

$$P = P_0 \subset P_1 \subset \cdots \subset P_n = Q.$$

The first claim is that $\operatorname{ht} P_1 = 1$. If this was not true, we would have $\dim A_{P_1} \geq 2$, but PA_{P_1} is a minimal prime of that ring that satisfies $\dim A_{P_1}/PA_{P_1} = 1$, which contradicts Proposition 9.2.7. In a similar way, we prove inductively that $\operatorname{ht} P_k = k$ for $k = 0, \ldots, n$. In particular, $n = \operatorname{ht} Q$, which is what we needed to prove. $\square$

Corollary 9.2.9. *Let k be a field. Then every finitely generated k-algebra is catenary.*

This shows that the rings that appear as coordinate rings of algebraic varieties are catenary. So are the finitely generated algebras over $\mathbb{Z}$ or over a Dedekind ring. In fact, it is not easy to find a Noetherian ring that is not catenary. [**Nag62**, Appendix $A1$, Example 2] gives an example of a Noetherian ring that is catenary, but not universally catenary.

We can reap another useful consequence of Proposition 9.2.7.

Theorem 9.2.10 (Unmixedness theorem). *Let A be a CM ring, $I = (a_1, \ldots, a_r)$ an ideal, and assume that $\operatorname{ht} I = r$. Then I is unmixed, that is, $\operatorname{ht} P = r$ for all $P \in \operatorname{Ass}(I)$. In particular, every associated prime of I is minimal.*

Proof. Let P be an associated prime of I. By localizing at P, we can assume that A is local with maximal ideal P. Since $\operatorname{ht} I = r$, we can extend $a_1, \ldots, a_r$ to become a system of parameters for A, which is then a regular sequence by Theorem 9.1.13. It follows that A/I is again Cohen–Macaulay, and we can reduce to the case where $I = 0$. In other words, we need to prove that every associated prime of 0 in a CM ring is minimal. This is clear: if $P \in \operatorname{Ass}(0)$, then P only contains divisors of 0 by Theorem A.3.6, so $\operatorname{ht} P = \operatorname{depth}_P A = 0$ by Corollary 9.1.7. $\square$

Remark 9.2.11. Let us interpret the results of this section from a geometric point of view. Let $A = R(V)$ be the coordinate ring of the affine variety V over a field k. By Corollary 9.2.9, A is catenary, so if $W_1 \subset W_2 \subset V$ are irreducible subvarieties, we can always find an increasing chain of subvarieties from W_1 to W_2 whose dimension grows 1 by 1.

If moreover A is Cohen–Macaulay, then V does not have embedded components. This means that if we take a primary decomposition of 0, then all primes in that decomposition correspond maximal irreducible components of V.

What is not guaranteed is that V is equidimensional, that is, all maximal components have the same dimension. This is easy to see by taking, for instance, a plane and a disjoint line which is a regular, hence CM, variety. On the other hand, by Proposition 9.2.7, this is true *locally*. That is, if $p \in V$ is a point, every irreducible component passing through p has dimension equal to $\dim V$. For instance, this implies that a line *meeting* a point is not CM at the point of intersection. In fact, V can only have components of different dimension if it is disconnected.

This is the reason for the alternative definition from Remark 9.1.11. In fact, if we require in addition that every maximal ideal has the same height, then Proposition 9.2.7 generalizes immediately to non local Cohen–Macaulay rings.

Theorem 9.2.10 admits a converse.

Proposition 9.2.12. *Let A be a Noetherian ring. Assume that every ideal $I \subset A$ of height r generated by r elements is unmixed. Then A is Cohen–Macaulay.*

Proof. It is immediate to reduce to the local case. For this the proof is akin to the implication (v) $\implies$ (iv) in Theorem 9.1.13.

Let $d = \dim A$, $r = \operatorname{depth} A$, and take a maximal regular sequence $a_1, \ldots, a_r$. Letting $I = (a_1, \ldots, a_r)$, we have $\operatorname{ht} I = r$, either by Corollary 7.4.5 or by Corollary 9.1.8. By the hypothesis, I is unmixed. If $r < d$, this means that the maximal ideal $\mathcal{M}$ is not an associated prime of I, and by Theorem A.3.6 this implies that there is $a_{r+1} \in \mathcal{M}$ that is regular on A/I, which is a contradiction. $\square$

9.3. Miracle flatness

This section is devoted to the proof of the so-called miracle flatness theorem, and some of its consequences. We have seen in Section 6.2 that flatness has a geometric interpretation. Namely, let V, W be affine varieties, $A = R(V)$ and $B = R(W)$ their coordinate rings. A morphism $f \colon W \to V$ corresponds to a map of rings $f^* \colon A \to B$. The condition that f^* is flat ensures that the fibers of f vary in a controlled manner. In particular, in Theorem 6.3.12, we have seen that this guarantees that the fibers of f all have the same dimension. We will now see that, in the local case and under some regularity conditions, this property is also enough to guarantee flatness.

Theorem 9.3.1 (Miracle flatness). *Let $f \colon A \to B$ be a local morphism of local Noetherian rings, having maximal ideals $\mathcal{M}_A$ and $\mathcal{M}_B$ respectively. Assume that A is regular, B is Cohen–Macaulay, and*

$$\dim B = \dim A + \dim B/\mathcal{M}_A B.$$

Then f is flat (hence faithfully flat).

The parenthetical remark is Corollary 6.3.7.

Proof. From [dJea20, Tag 00R4]. We use induction on $d := \dim A$. If $d = 0$, then A is a field, in which case B is free over A, hence flat.

Now assume $d > 0$, hence $\dim B > 0$, and let $Q_1, \ldots, Q_r$ be the minimal primes of B. By Proposition 9.2.7, we have $\dim B/Q_i = \dim B > \dim B/\mathcal{M}_A B$ for all i, hence $\mathcal{M}_A B$ is not contained in any minimal prime of B.

Let $P_i = Q_i \cap A \subsetneq \mathcal{M}_A$. By prime avoidance, we can take an element $a \in \mathcal{M}_A \setminus \mathcal{M}_A^2$ such that $a \notin P_i$ for $i = 1, \ldots, r$. By Theorem 9.2.10, all associated primes of 0 in B are minimal, so a is not contained in any such ideal. By Theorem A.3.6, this means that a is not a divisor of 0 in B.

We conclude the following: $A/(a)$ is regular of dimension $d-1$ (Proposition 8.3.5), while B/aB is Cohen–Macaulay of dimension $\dim B - 1$ (Proposition 9.1.20). By induction, the induced map $A/(a) \to B/aB$ is flat. This implies that f is flat, by the strong form of the local criterion of flatness, Theorem 6.4.6.

To apply it, we have to check that $\operatorname{Tor}_1^A(B, A/(a)) = 0$, which is clear by, say, Example 3.4.12(a), and that for all ideals $J \subset A$, $J \otimes_A B$ is Hausdorff in the (a)-adic topology, which follows from Krull's intersection theorem, Theorem A.6.4. $\qquad\square$

As a simple consequence of miracle flatness, we can now finish the proof of Kunz's theorem.

Proof of Theorem 8.6.1, second half. Denote by ϕ the Frobenius map, $\mathcal{M}$ the maximal ideal. Assume that A is regular, and take a system of parameters $(a_1, \ldots, a_d)$. Then $a_1^p, \ldots, a_d^p$ is again a system of parameters, in particular $\dim A/\phi(\mathcal{M}) = 0$, so ϕ is flat by Theorem 9.3.1. $\qquad\square$

9.4. Irreducible decompositions

Let A be a Noetherian ring, $I \subset A$ an ideal. The theory of primary decomposition allows us to write I as a finite intersection of primary ideals, say $I = \bigcap_i Q_i$. If we choose the decomposition to be irredundant, then the radicals of these ideals, $P_i = \sqrt{Q_i}$ (which are prime) are uniquely determined by I, and in fact they are the primes associated to I. Finally, if P_i is a *minimal* prime of I, then Q_i is uniquely determined as well.

In fact, the proof of primary decomposition tells us a little more: we can decompose I as an intersection of *irreducible* ideals, and every irreducible ideal is primary. All of this is summarized in Theorem A.3.1. A natural question is to investigate the decomposition of I as an intersection of irreducible ideals, which we will call an *irreducible decomposition* of I. By primary decomposition, one is reduced to understanding the irreducible decomposition of a *primary* ideal. The aim of this section is to understand such a decomposition; in particular we will obtain a clear picture of the difference between primary and irreducible ideals. This topic appears here because the irreducible decomposition is especially well behaved in Cohen–Macaulay rings, and in fact the properties of irreducible decompositions can be used to characterize such rings.

So, let Q be a primary ideal, $P = \sqrt{Q}$. If

(9.4.1) $$Q = \bigcap_{i=1}^{n} E_i,$$

where the E_i are irreducible, then all E_i are P-primary, by primary decomposition. To understand the decomposition (9.4.1), we can then localize at P, so we can assume from the start that A is a local ring, $\mathcal{M}$ its maximal ideal and $\sqrt{Q} = \mathcal{M}$. In other words, we reduced to understanding irreducible decompositions for ideals of definitions in a local Noetherian ring.

Theorem 9.4.1 (Noether, Gröbner). *Let $A, \mathcal{M}$ be a local Noetherian ring, $Q \subset A$ an ideal of definition, and assume that Q admits the irredundant irreducible decomposition (9.4.1). Then n is determined by Q—that is, two such decompositions have the same number of factors—and in fact*

$$n = \ell\left(\frac{(Q : \mathcal{M})}{Q}\right).$$

The statement is due to the work Noether on primary decomposition [**Noe21**], while the precise formula for n is due to Gröbner [**Gro35**]. In fact, much as in the theory of primary decomposition, we can generalize this result to the module case, with no essential differences.

Definition 9.4.2. Let M be an A-module, $N \subset M$ a submodule. We say that N is *irreducible* (in M) if we cannot write $N = N_1 \cap N_2$ as a nontrivial intersection of submodules, where $N \subsetneq N_i$ for $i = 1, 2$. An *irreducible decomposition* of N in M is a way to write

$$N = \bigcap_{i=1}^{n} N_i,$$

where each N_i is an irreducible submodule. The decomposition is called *irredundant* if we cannot omit any of the N_i, that is,

$$N \subsetneq N_1 \cap \cdots \cap \widehat{N_i} \cap \cdots \cap N_n$$

for all $i = 1, \ldots, n$.

Theorem 9.4.3. *Let A be a local Noetherian ring, M a finitely generated A-module, and $N \subset M$ a submodule such that M/N has finite length. Assume that N admits the irredundant irreducible decomposition*

$$N = \bigcap_{i=1}^{n} N_i.$$

Then n depends only on N, and in fact

(9.4.2) $$n = \ell\left(\frac{(N : \mathcal{M})}{N}\right).$$

Here, of course, $(N \colon \mathcal{M}) = \{m \in M \mid \mathcal{M}m \subset N\}$. To prove Theorem 9.4.3, we might as well take the quotient by N, hence assume from the start that $N = 0$ and M has finite length.

Definition 9.4.4. Let $A, \mathcal{M}$ be a local ring, M an A-module of finite length. The *socle* of M, denoted $\operatorname{soc} M$ (or $\operatorname{soc}_A M$ to emphasize the ring A) is the A-module

$$\operatorname{soc} M = (0 \colon \mathcal{M}) = \operatorname{Ann}_M \mathcal{M} = \{m \in M \mid \mathcal{M}m = 0\}.$$

Remark 9.4.5. Let $k = A/\mathcal{M}$ be the residue field. By definition, $\operatorname{soc} M$ is the largest submodule of M that inherits the structure of a k-vector space. In particular, $\ell_A \operatorname{soc} M = \dim_k \operatorname{soc} M$.

Definition 9.4.6. The number n of factors in the irreducible decomposition (9.4.2) is called the *index of reducibility* of N in M, denoted $\operatorname{ir}_M N$. Confusingly, this is also sometimes called *the index of* ir*reducibility*.

Theorem 9.4.3 can then be summarized by the equation

$$\operatorname{ir}_M N = \ell \left(\operatorname{soc} \frac{M}{N} \right).$$

Lemma 9.4.7. *Let A be a local Noetherian ring, M an A-module of finite length, $M \neq 0$. Then $\operatorname{soc} M$ is a nonzero, essential submodule of M.*

Proof. Let $N \subset M$ be a nontrivial submodule. Then $N \cap \operatorname{soc} M = \operatorname{soc} N$, so it is enough to prove that $\operatorname{soc} M \neq 0$ whenever $M \neq 0$. Let $M_i = \mathcal{M}^i M$, so that we have the descending chain

$$M = M_0 \supset M_1 \supset \cdots .$$

By Nakayama's lemma, the submodules M_i are distinct unless they are 0. Since M has finite length, there is a first index k such that $M_k = 0$. Then $M_{k-1} \subset \operatorname{soc} M$, and $M_{k-1} \neq 0$. $\square$

It is convenient to prove the case $n = 1$ of Theorem 9.4.3 separately.

Lemma 9.4.8. *Let A be a local Noetherian ring, M a finitely generated A-module and $N \subset M$ a submodule such that M/N has finite length. Then N is irreducible if and only if*

$$\ell \left(\frac{(N \colon \mathcal{M})}{N} \right) = 1.$$

Proof. By taking the quotient, we can assume that $N = 0$. Assume that 0 is reducible, so $0 = N_1 \cap N_2$, where $N_i \neq 0$. By Lemma 9.4.7, $N_i \cap \operatorname{soc} M$ is a nonzero subspace of $\operatorname{soc} M$. If $\operatorname{soc} M$ is 1-dimensional, we must have $\operatorname{soc} M \subset N_i$ for $i = 1, 2$, which is a contradiction. Conversely, if $\dim_k \operatorname{soc} M \geq 2$, then we can choose two 1-dimensional subspaces $V_1, V_2 \subset \operatorname{soc} M$ such that $V_1 \cap V_2 = 0$. Then the V_i are also submodules, so 0 is reducible in M. $\qquad \square$

Proof of Theorem 9.4.3. As we have remarked, we can assume that $N = 0$. Assume that $0 = \bigcap_{i=1}^{n} N_i$, where each N_i is irreducible. By Lemma 9.4.8,

$$\ell\left(\frac{(N_i \colon \mathcal{M})}{N_i}\right) = 1$$

for all i. Since $\operatorname{soc} M = (0 \colon \mathcal{M}) \subset (N_i \colon \mathcal{M})$, we have an injection

$$\frac{\operatorname{soc} M}{\operatorname{soc} M \cap N_i} \hookrightarrow \frac{(N_i \colon \mathcal{M})}{N_i}.$$

Since the right-hand side has length 1, we conclude that either $\operatorname{soc} M \subset N_i$, or $\operatorname{soc} M \cap N_i$ has codimension 1 in $\operatorname{soc} M$. Since the intersection of all subspaces $\operatorname{soc} M \cap N_i$ is 0, we see that $n \geq \dim_k \operatorname{soc} M = \ell(\operatorname{soc} M)$.

Moreover, if the decomposition is irredundant, we must have equality. We prove this by induction on n, the case $n = 1$ being Lemma 9.4.8. Assume $n > 1$. Since the intersection of all N_i is 0, there is at least one of them that does not contain $\operatorname{soc} M$, say N_1. For $i = 2, \ldots, n$, let $N_i' = N_i \cap N_1$. Since N_1 does not contain $\operatorname{soc} M$,

$$\dim_k \operatorname{soc} N_1 = \dim_k(N_1 \cap \operatorname{soc} M) = \dim_k \operatorname{soc} M - 1$$

by the first part of the proof. In N_1 we have the irredundant decomposition

$$0 = \bigcap_{i=2}^{n} N_i',$$

so by induction we have $n - 1 = \dim_k \operatorname{soc} N_1 = \dim_k \operatorname{soc} M - 1$, and we are done. $\qquad \square$

Having proved that the index of irreducibility is well defined, we now see that it provides an invariant of Cohen–Macaulay rings. The following result is proved in [**Nor57**].

Theorem 9.4.9 (Northcott). *Let A be a local Cohen–Macaulay ring, Q_1, Q_2 two ideals generated by systems of parameters. Then $\operatorname{ir}_A Q_1 = \operatorname{ir}_A Q_2$.*

Proof. By Lemma 9.1.16, we can assume that the two systems of parameters differ in just one element, say $Q_1 = (a_1, \ldots, a_{d-1}, a)$ and $Q_2 = (a_1, \ldots, a_{d-1}, b)$. By Theorem 9.1.13, these are regular sequences. Looking at Gröbner's formula in Theorem 9.4.1, we can prove the result after taking the quotient by $(a_1, \ldots, a_{d-1})$. The quotient ring is again CM by Corollary 9.1.21.

We are thus left with the 1-dimensional case: A is a local Cohen–Macaulay ring with $\dim A = 1$, $a, b \in A$ are regular elements and we need to prove that

$$\ell\left(\frac{(a \colon \mathcal{M})}{(a)}\right) = \ell\left(\frac{(b \colon \mathcal{M})}{(b)}\right).$$

Since b is regular, the map $m_b \colon A \to (b)$ given by multiplication by b is an isomorphism. We claim that $m_b(a \colon \mathcal{M}) = (ab \colon \mathcal{M})$. One inclusion is obvious. For the other one, take $c \in (ab \colon \mathcal{M})$, so that $cx \in (ab)$ for all $x \in \mathcal{M}$. Taking $x = a$, we get $ca = aby$, and since a is not a zero divisor $c = by$, where $y \in (a \colon \mathcal{M})$, proving that $(ab \colon \mathcal{M}) \subset m_b(a \colon \mathcal{M})$.

Since m_b is an isomorphism sending (a) to (ab) and $(a \colon \mathcal{M})$ to $(ab \colon \mathcal{M})$, we get

$$\ell\left(\frac{(a \colon \mathcal{M})}{(a)}\right) = \ell\left(\frac{(ab \colon \mathcal{M})}{(ab)}\right).$$

The result now follows by symmetry. $\qquad\square$

As a corollary, we derive the following result from [**Gro51**].

Corollary 9.4.10 (Gröbner). *Let A be a regular local ring. Then every ideal generated by a system of parameters in A is irreducible.*

Proof. A regular local ring is CM, so Theorem 9.4.9 applies, and we only need to remark that the maximal ideal $\mathcal{M}$ is irreducible and generated by a system of parameters. $\qquad\square$

Remark 9.4.11. The reader should not confuse an ideal generated by a system of parameters with an ideal of definition. By definition, a system of parameters generates an ideal of definition. But a system of parameters is defined as a set of $\dim A$ elements that generate an ideal of definition, and of course not every ideal of definition is generated by $\dim A$ elements, for instance, this holds for the maximal ideal precisely when A is regular. An ideal generated by a system of parameters is sometimes called an *ideal of parameters*.

The theorem of Northcott has a kind of weak converse, due to Northcott and Rees [**NR54**].

Theorem 9.4.12 (Northcott–Rees). *Let A be a local Noetherian ring, and assume that every ideal of A generated by a system of parameters is irreducible. Then A is Cohen–Macaulay.*

First, we will need a simple but useful consequence of Lemma 9.4.8.

Lemma 9.4.13. *Let $A, \mathcal{M}$ be a local Noetherian ring, $Q' \subsetneq Q$ two ideals of definition. If Q' is irreducible, then $(Q' \colon \mathcal{M}) \subset Q$.*

Proof. In our situation, Q/Q' has finite length, hence we can find a series of composition

$$Q' = Q_0 \subset Q_1 \subset \cdots \subset Q_k = Q.$$

By Nakayama's lemma, $\mathcal{M}Q_1$ is properly contained in Q_1, and since Q_1/Q_0 is simple, $\mathcal{M}Q_1 \subset Q'$. In other words, $Q_1 \subset (Q' : \mathcal{M})$. By Lemma 9.4.8, $(Q' : \mathcal{M})/Q'$ is simple, hence we have $(Q' : \mathcal{M}) = Q_1 \subset Q$. □

Proof of Theorem 9.4.12. By point (v) in Theorem 9.1.13, it is enough to prove that $\mathcal{M}$ is not an associated prime of $(a_1, \ldots, a_{d-1})$ for any system of parameters $a_1, \ldots, a_d$. For every $k > 0$, the set $a_1, \ldots, a_{d-1}, a_d^k$ is still a system of parameters. (Why?) Let $Q_k = (a_1, \ldots, a_{d-1}, a_d^k)$ be the ideal it generates, and denote $I = (a_1, \ldots, a_{d-1})$.

By Krull's intersection theorem (Theorem A.6.4) applied to A/I, we find that $I = \bigcap_{k=1}^\infty Q_k$. Moreover, the inclusions $Q_{k+1} \subsetneq Q_k$ are strict, otherwise a_d is nilpotent modulo I, which contradicts the choice that $a_1, \ldots, a_d$ is a system of parameters. By hypothesis, all ideals Q_k are irreducible, so Lemma 9.4.13 guarantees that $(Q_{k+1} : \mathcal{M}) \subset Q_k$. It follows that

$$(I : \mathcal{M}) = \left(\bigcap_{k=2}^\infty Q_k : \mathcal{M} \right) = \bigcap_{k=2}^\infty (Q_k : \mathcal{M}) \subset \bigcap_{k=1}^\infty Q_k = I,$$

so $(I : \mathcal{M}) = I$.

If $\mathcal{M}$ is associated to I, there exists $a \in A$ such that $\mathcal{M} = (I : a)$. This is a contradiction, since $a \in (I : \mathcal{M})$ but $a \notin I$. □

9.5. The socle

In the previous section, we introduced the socle as a tool to prove results on irreducible decompositions of submodules. We will now see that this object has many interesting properties. Recall from Definition 9.4.4 that the socle of the A-module M is just the annihilator of $\mathcal{M}$ in M. Here, A is a local Noetherian ring and M is a nonzero A-module of finite length. We have already observed that $\operatorname{soc} M$ has the structure of vector space over $k = A/\mathcal{M}$, and as a consequence $\operatorname{soc} M$ is a nonzero, essential submodule of M (Lemma 9.4.7). We will now see other characterizations of it.

Proposition 9.5.1. *Let A be a local Noetherian ring, M an A-module of finite length. Then $\operatorname{soc} M$ is the sum of all simple submodules of M.*

Proof. Being a k-vector space, $\operatorname{soc} M$ is generated by 1-dimensional subspaces, which are clearly simple. Hence it is enough to check that every simple submodule of M is contained in $\operatorname{soc} M$. If $N \subset M$, then $\mathcal{M}N \subsetneq N$ by Nakayama's lemma, so if N is simple we must have $\mathcal{M}N = 0$—that is, $N \subset \operatorname{soc} M$. □

Proposition 9.5.2. *Let A be a local Noetherian ring, M an A-module of finite length. Then $\operatorname{soc} M$ is the smallest essential submodule of M.*

Proof. We have seen in Lemma 9.4.7 that $\operatorname{soc} M$ is indeed essential, so we must only check that it is contained in every essential submodule. If $N \subset M$ is essential and $S \subset M$ is simple, then $N \cap S \neq 0$, so $S \subset N$, and the conclusion follows from Proposition 9.5.1. $\square$

Remark 9.5.3. We only really uses the fact that $\ell(M) < \infty$ in the proof of Lemma 9.4.7. In fact, one could define the socle of an A-module M, not necessarily of finite length, as the sum of all simple submodules. The above proof would then show that this is the same as the intersection of all essential submodules. It is only in the finite length case, though, that one has a simple characterization such as $\operatorname{soc} M = \operatorname{Ann}_M \mathcal{M}$. In any case, we will only need the notion of a socle in the finite length case.

9.6. Modules of finite injective dimension

When we introduced Cohen–Macaulay rings, we started from the inequality between depth and dimension in local rings. We now turn to a relation between depth and injective dimension.

Let $A, \mathcal{M}$ be a local Noetherian ring, $k = A/\mathcal{M}$ the residue field and M a finitely generated A-module. Recall from Kaplansky's theorem (Theorem 5.4.4) that $\operatorname{id}_A M$ is the largest number n such that $\operatorname{Ext}_A^n(k, M) \neq 0$, or ∞ if no such n exists. On the other hand, by Rees's theorem (Theorem 7.3.12), $\operatorname{depth} M$ is the *smallest* number n such that $\operatorname{Ext}_A^n(k, M) \neq 0$. The following consequence is now immediate.

Proposition 9.6.1. *Let A be a local Noetherian ring, M a finitely generated A-module. Then $\operatorname{depth} M \leq \operatorname{id} M$.*

It is worth noting that in fact one has a stronger result.

Theorem ([FFGR75], [Rob76]). *Let A be a local Noetherian ring with residue field k, M a finitely generated A-module. Then $\operatorname{Ext}_A^i(k, M) \neq 0$ for $\operatorname{depth} M \leq i \leq \operatorname{id}_A M$.*

Of course, it is entirely possible that $\operatorname{id} M = \infty$, in which case Proposition 9.6.1 does not tell anything interesting. But in case it is finite, a natural way to study the relation between depth and injective dimension is by induction, using the change of ring theorems. As a starting point, we consider the case where M is injective.

Theorem 9.6.2. *Let $A, \mathcal{M}$ be a local Noetherian ring. There exists a finitely generated, injective A-module if and only if $\dim A = 0$.*

Proof. Assume that $\dim A > 0$ and that M is a finitely generated, injective A-module. By Theorem A.3.4 $\operatorname{Ass}(M) \neq \emptyset$, so there is a prime P that admits a monomorphism $A/P \hookrightarrow M$. In particular, there is a prime P such that $\operatorname{Hom}(A/P, M) \neq 0$, and we can even choose P to be strictly contained in $\mathcal{M}$. In fact, if $\mathcal{M}$ is the only associated prime, then any $P \subset \mathcal{M}$ will satisfy $\operatorname{Hom}(A/P, M) \neq 0$, via the composition

$$\frac{A}{P} \to \frac{A}{\mathcal{M}} \hookrightarrow M.$$

On the other hand, choose any $a \in \mathcal{M} \setminus P$, so that multiplication by a gives a monomorphism $A/P \to A/P$. Since M is injective, multiplication by a is surjective on $\operatorname{Hom}(A/P, M)$, that is, $\operatorname{Hom}(A/P, M) = a \operatorname{Hom}(A/P, M)$. By Nakayama's lemma, $\operatorname{Hom}(A/P, M) = 0$, which is a contradiction.

For the converse, assume that $\dim A = 0$. Let $k = A/\mathcal{M}$ and take $M = E(k)$, the injective hull of k. Since M is injective by definition, we only need to check that it is finitely generated. Since A is Artinian, there is a smallest $n \in \mathbb{N}$ such that $\mathcal{M}^n = 0$. We will prove the claim by induction on n. If $n = 1$, then $A = k$ is a field and the result is obvious.

For the inductive step, let $I = \mathcal{M}^{n-1}$ and $N = (0 \colon {}_M I)$, which is naturally a module over A/I. One checks easily (do it!) that N is injective over A/I—hence, it is the injective hull of the residue field of A/I. By induction, N is finitely generated. Since $\mathcal{M}M \subset N$ and A is Noetherian, $\mathcal{M}M$ is finitely generated as well. Let $\mathcal{M} = (a_1, \ldots, a_r)$ and consider the map

$$f \colon \quad M \longrightarrow a_1 M \oplus \cdots \oplus a_r M,$$

$$m \longmapsto (a_1 m, \ldots, a_r m).$$

By construction, $\ker f = (0 \colon {}_M \mathcal{M})$. This is a k-vector space containing k. Since M is an essential extension of k, we must have $k = \ker f$ (otherwise, there would be a complement to k). Since the image and the kernel of f are finitely generated, M is finitely generated as well. $\qquad\square$

From the proof, we extract a fact that is worth noting.

Corollary 9.6.3. *Let $A, \mathcal{M}$ be a local Artinian ring, $k = A/\mathcal{M}$. Then the injective hull $E(k)$ is finitely generated over A.*

Remark 9.6.4. Notice that by Matlis's theorem, Theorem 5.3.9, if there is a finitely generated injective module over A, that must be necessarily $E(k)$. This particular module does not come out of nowhere in the proof of Theorem 9.6.2: it is the only module that has a chance to work.

We now turn to higher dimensions. Let $A, \mathcal{M}$ be a local Cohen–Macaulay ring of dimension d. We can find a regular sequence $a_1, \ldots, a_d \subset \mathcal{M}$, so that $B = A/(a_1, \ldots, a_d)$ has dimension 0. As such, it admits a finitely generated injective B-module, namely $E_B(k)$, where $k = A/\mathcal{M} = B/\mathcal{M}B$ is the common residue field of A and B, and E_B denotes injective hull as B-modules. By repeated applications of the first change of rings theorem (Theorem 5.4.7), we find that $\mathrm{id}_A E_B(k) = d$. In particular, $E_B(k)$ is a finitely generated A-module of finite injective dimension. In fact, such a module can only exist under the assumption that A is Cohen–Macaulay. This result is known as *Bass's conjecture* [**Bas63b**], and was eventually proved by work of Peskine–Szpiro [**PS73**] and Roberts [**Rob87**], [**Rob89**].

Theorem (Peskine–Szpiro, Roberts). *Let A be a local Noetherian ring. Then A admits a finitely generated module of finite injective dimension if and only if A is Cohen–Macaulay.*

This theorem is outside the scope of this text. In any case, the surprise is that the injective dimension of modules, if finite, does not depend on the module at all. The reader should compare the next result to the Auslander–Buchsbaum formula, Theorem 7.4.1, which gives a similar result for the projective dimension.

Theorem 9.6.5. *Let $A, \mathcal{M}$ be a local Noetherian ring, M a nonzero finitely generated A-module. Assume that $\mathrm{id}\, M < \infty$. Then $\mathrm{id}\, M = \mathrm{depth}\, A$.*

Our proof follows [**Kap70**]. For a nice proof using the Ischebeck spectral sequences, see Exercise 22.

Lemma 9.6.6. *Let $A, \mathcal{M}$ be a local Noetherian ring, M, N finitely generated A-modules. Assume that $\mathrm{Ext}^1(N, M) = 0$ and $\mathrm{pd}\, N = 1$. Then $M = 0$.*

Proof. By Proposition 5.6.5, we can take a minimal free resolution

$$0 \longrightarrow F_1 \longrightarrow F_0 \longrightarrow N \longrightarrow 0 \,,$$

that will satisfy $F_1 \subset \mathcal{M}F_0$. Since $\mathrm{Ext}^1(N, M) = 0$, every morphism $F_1 \to M$ extends to F_0, and so has image contained in $\mathcal{M}M$. Given an element $m \in M$, we can take a morphism $F_1 \to M$ that sends a generator of F_1 to m—it follows that $M = \mathcal{M}M$, so $M = 0$ by Nakayama's lemma. $\square$

Proof of Theorem 9.6.5. Let $d = \mathrm{depth}\, A$. Let $\mathbf{a} = (a_1, \ldots, a_d)$ be a regular sequence contained in $\mathcal{M}$, so that the Koszul complex $K(\mathbf{a})$ is a free resolution of $S := A/(\mathbf{a})$ by Theorem 7.2.2, in fact a minimal free resolution by Proposition 5.6.5. We conclude that $\mathrm{pd}\, S = d$.

Now let $N = \ker K_{d-1}(\mathbf{a}) \to K_{d-2}(\mathbf{a})$. We have $\mathrm{pd}\, N = 1$ by construction. If we assume by contradiction that $\mathrm{id}\, M < d$, then $\mathrm{Ext}^1(N, M) = 0$ by dimension shifting (Proposition 3.3.2). By Lemma 9.6.6, we find $M = 0$, a contradiction, hence $\mathrm{id}\, M \geq d$.

For the converse inequality, notice that depth $S = 0$, so k is isomorphic to a submodule of S (either by noticing that $\mathcal{M}$ is an associated prime of S, or by Rees's theorem, Theorem 7.3.12). We can write this as an exact sequence

$$0 \longrightarrow k \longrightarrow S \longrightarrow T \longrightarrow 0 \,,$$

from which we derive a long exact sequence of $\mathrm{Ext}^i(-, M)$ groups. Assume by contradiction that $n = \mathrm{id}\, M > d$. Then $\mathrm{Ext}^n(S, M) = 0$ (since $\mathrm{pd}\, S = d$) and $\mathrm{Ext}^{n+1}(T, M) = 0$ (since $\mathrm{id}\, M = n$). From the long exact sequence, we find that $\mathrm{Ext}^n(k, M) = 0$, which contradicts Kaplansky's theorem (Theorem 5.4.4). $\qquad\square$

9.7. Gorenstein rings

We are going to specialize the results of the previous section in the case where $M = A$ itself. By Proposition 9.6.1, we have depth $A \le \mathrm{id}\, A$. If moreover the latter is finite, by Proposition 9.6.5, we have $\mathrm{id}\, A = \mathrm{depth}\, A \le \dim A$. At this point it is helpful to isolate the class of such rings.

Definition 9.7.1. Let A be Noetherian ring. We say that A is a *Gorenstein ring* if $\mathrm{id}_A A < \infty$.

By Serre's theorem (Theorem 8.2.1), a regular local ring has finite global dimension, in particular it is Gorenstein. Being Gorenstein is almost a local property; in fact:

Proposition 9.7.2. *Let A be a Noetherian ring. If A is Gorenstein, then A_P is Gorenstein for all prime ideals $P \subset A$. Conversely, if $A_{\mathcal{M}}$ is Gorenstein for all maximal ideals $\mathcal{M} \subset A$ and $\dim A < \infty$, then A is Gorenstein.*

Proof. Since A is Noetherian, a finitely generated A-module is k-finitely presented for all k. From Proposition 5.1.10 we then have $\mathrm{Ext}^k_{A_P}(M_P, N_P) = \mathrm{Ext}^k_A(M, N)_P$ for all k. By Kaplansky's theorem, Theorem 5.4.4, $\mathrm{id}\, A_P$ can be computed by looking at the Ext groups $\mathrm{Ext}^i(A_P/PA_P, A_P)$. Since both sides are finitely generated, the first claim follows.

For the second claim, assume that $A_{\mathcal{M}}$ is Gorenstein for all maximal ideals $\mathcal{M} \subset A$. By Proposition 9.6.5,

$$\mathrm{id}_{A_{\mathcal{M}}} A_{\mathcal{M}} \le \dim A_{\mathcal{M}} \le \dim A$$

is uniformly bounded for all maximal ideals $\mathcal{M}$. We can then reverse the above argument and conclude that $\mathrm{Ext}^k(A/I, A) = 0$ for $k > \dim A$ and all ideals I, so A is Gorenstein. $\qquad\square$

Corollary 9.7.3. *A regular ring of finite dimension is Gorenstein.*

The relation $\mathrm{id}\, A = \mathrm{depth}\, A \le \dim A$ can in fact be made more precise.

Theorem 9.7.4. *Let A be a local Gorenstein ring. Then A is Cohen–Macaulay, so*

$$\operatorname{id} A = \operatorname{depth} A = \dim A.$$

Notice that this would be an immediate consequence of Bass's conjecture, but it is a considerably easier result.

Proof. By induction on $d = \operatorname{depth} A = \operatorname{id} A$. If $d = 0$, then A is injective, hence $\dim A = 0$ by Theorem 9.6.2. If $d > 0$, we can take any regular element $a \in A$. By the third change of rings theorem (Theorem 5.4.10), $A/(a)$ is Gorenstein as well, hence

$$\operatorname{id} A/(a) = \operatorname{depth} A/(a) = \dim A/(a).$$

All these quantities are 1 less than in A, by Theorem 5.4.10, Proposition 7.3.10 and Theorem A.7.4 respectively, hence the conclusion for A follows. $\square$

Corollary 9.7.5. *Let A be a local Gorenstein ring with residue field k. Then $\operatorname{Ext}^i(k, A) \neq 0$ precisely when $i = \dim A$.*

Proof. The least i for which $\operatorname{Ext}^i(k, A) \neq 0$ is $\operatorname{depth} A$; the biggest such i is $\operatorname{id} A$. Now apply Theorem 9.7.4. $\square$

For Gorenstein rings, we have an analogue of Propositions 8.3.4 and 9.1.20.

Proposition 9.7.6. *Let A be a Noetherian ring, $a \in A$ a regular element. Then A is Gorenstein if and only if $A/(a)$ is Gorenstein.*

Proof. This is an immediate consequence of Theorem 5.4.9. $\square$

Because of this, many questions on Gorenstein rings reduce to the 0-dimensional case, after taking a quotient by a maximal regular sequence. It is thus helpful to have a characterization of such rings. For this, we need a lemma. We will not prove it right now, it will be an immediate consequence of (a special case of) Matlis duality in the next chapter (Corollary 10.1.5).

Lemma 9.7.7. *Let $A, \mathcal{M}$ a local Artinian ring, k its residue field. Then $\ell(A) = \ell(E(k))$.*

Proposition 9.7.8. *Let $A, \mathcal{M}$ be a local Artinian ring, $k = A/\mathcal{M}$ the residue field and $E(k)$ its injective hull. The following are equivalent:*

   (i) *A is Gorenstein;*

  (ii) *A is injective over itself;*

 (iii) *$A \cong E(k)$ as A-modules;*

(iv) *the socle* soc A *is* 1-*dimensional; and*

(v) 0 *is irreducible in* A.

Proof. If A is Gorenstein, it is injective by, say, Theorem 9.7.4. It follows that (i) is equivalent to (ii).

If A is injective, by Matlis's theorem, Theorem 5.3.9, it is the direct sum of copies of $E(k)$. By Lemma 9.7.7, we find that $A \cong E(k)$, hence (ii) is equivalent to (iii).

Since both k and soc $E(k)$ are essential in $E(k)$, it is clear that they meet, and since k is 1-dimensional we must have soc $E(k) = k$. Hence, (iii) implies (iv).

Conversely, assume that soc A is 1-dimensional, so soc $A \cong k$. By Lemma 9.4.7, A is an essential extension of k. Since $E(k)$ is injective, the inclusion $k \hookrightarrow E(k)$ can be extended to a morphism $A \to E(k)$, which is again injective by Lemma 2.7.2 (this is Remark 2.7.3). Using again Lemma 9.7.7, we find that $A \cong E(k)$, so (iv) implies (iii).

Finally, the equivalence between (iv) and (v) is Gröbner's theorem, Theorem 9.4.1. □

Using this criterion, we can give some examples of Gorenstein rings.

Example 9.7.9.

(a) Let k be a field, and consider
$$A = k[x, y, z]/(x^2, y^2, xz, yz, z^2 - xy).$$
As a k-vector space, A is a 5-dimensional, with basis $1, x, y, z, z^2$; in particular, it is a 0-dimensional ring. The ring A is local with maximal ideal $\mathcal{M} = (x, y, z)$, since $\mathcal{M}$ is the radical of $(x^2, y^2, xz, yz, z^2 - xy)$. The socle of A is generated by z^2, so A is Gorenstein by condition (iv) of the criterion.

(b) Now take $A = k[x, y]/(x^2, y^2, xy)$, which is again a local Artinian ring. This time, soc A is generated by x and y, so $\dim_k$ soc $A = 2$ and A is not Gorenstein. On the other hand, A is CM, like all local Artinian rings.

Using this characterization, we can immediately strengthen Corollary 9.7.5. We will need a convenient lemma about dimension shifting of Ext groups.

Lemma 9.7.10 (Rees). *Let A be a ring, M, N two A-modules. Let $a \in$ Ann M be regular and N-regular. Then*
$$\operatorname{Ext}_A^i(M, N) \cong \operatorname{Ext}_{\frac{A}{(a)}}^{i-1}\left(M, \frac{N}{aN}\right)$$
for all $i \geq 1$.

Proof. We use the base change spectral sequence from Theorem 4.5.5 with $B = A/(a)$. Notice that M has the structure of a B-module by the hypothesis that $a \in \operatorname{Ann} M$. To apply this sequence, we need to compute the groups $\operatorname{Ext}_A^p(A/(a), N)$.

Since a is regular, we have the exact sequence

$$0 \longrightarrow A \xrightarrow{\cdot a} A \longrightarrow \frac{A}{(a)} \longrightarrow 0.$$

This gives a long exact sequence of $\operatorname{Ext}(-, N)$ groups, from which we find that $\operatorname{Ext}_A^i(A/(a), N) = 0$ for $i \geq 2$. The initial segment is the exact sequence

$$0 \to \operatorname{Hom}\left(\tfrac{A}{(a)}, N\right) \to \operatorname{Hom}(A, N) \xrightarrow{\cdot a} \operatorname{Hom}(A, N) \to \operatorname{Ext}_A^1\left(\tfrac{A}{(a)}, N\right) \to 0.$$

Since a is N-regular, the first group is 0. We can also identify $\operatorname{Hom}(A, N)$ with N, leaving us with

$$0 \longrightarrow N \xrightarrow{\cdot a} N \longrightarrow \operatorname{Ext}_A^1\left(\tfrac{A}{(a)}, N\right) \longrightarrow 0.$$

In other words, we get the isomorphism $\operatorname{Ext}_A^1\left(\tfrac{A}{(a)}, N\right) \cong N/aN$.

The spectral sequence then degenerates to an isomorphism

$$\operatorname{Ext}_{\frac{A}{(a)}}^q\left(M, \frac{N}{aN}\right) \cong \operatorname{Ext}_A^{q+1}(M, N),$$

which is what we wanted to prove. $\qquad\square$

Proposition 9.7.11. *Let $A, \mathcal{M}$ be a local Noetherian ring of dimension d with residue field k. Then A is Gorenstein if and only if $\operatorname{Ext}_A^i(k, A) = 0$ for $i \neq d$ and $\operatorname{Ext}_A^d(k, A) \cong k$.*

Proof. We already know that if A is Gorenstein, then $\operatorname{Ext}_A^i(k, A) = 0$ for $i \neq d$ by Corollary 9.7.5. We prove the other implications by induction on d.

Start from the base case $d = 0$. Notice that a homomorphism $k \to A$ is defined by the image of 1, which must be an element annihilated by $\mathcal{M}$. Hence,

$$\operatorname{Hom}(k, A) \cong (0 :_A \mathcal{M}) = \operatorname{soc} A.$$

By Proposition 9.7.8, A is Gorenstein if and only if this is 1-dimensional.

For the inductive step, notice that under either hypothesis we can conclude that A is Cohen–Macaulay. We can then choose a regular element $a \in \mathcal{M}$. By Rees's lemma (Lemma 9.7.10), we obtain

$$\operatorname{Ext}_A^i(k, A) \cong \operatorname{Ext}_{\frac{A}{(a)}}^{i-1}(k, A/(a)).$$

Moreover, A is Gorenstein if and only if $A/(a)$ is, by Proposition 9.7.6. The conclusion then follows by induction. $\qquad\square$

As another application of this criterion, we are going to characterize local Gorenstein rings by the irreducibility of ideals generated by systems of parameters.

Theorem 9.7.12. *Let A be a local Noetherian ring. Then A is Gorenstein if and only if every ideal of A generated by a system of parameters is irreducible.*

Proof. Under either hypothesis, A is Cohen–Macaulay (Theorems 9.7.4 and 9.4.12). Let $a_1, \ldots, a_d$ be a system of parameters; by Theorem 9.1.13 it is also a regular sequence. Hence the quotient $B = A/(a_1, \ldots, a_d)$ is a ring of dimension 0 that is Gorenstein if and only if A is, by Theorem 9.7.6. The conclusion then follows from Proposition 9.7.8. $\qquad\square$

Another useful property of Gorenstein local rings is that modules of finite injective dimension and modules of finite projective dimension are one and the same thing.

Proposition 9.7.13. *Let A be a local Gorenstein ring, M a finitely generated A-module. Then $\operatorname{pd}_A M < \infty$ if and only if $\operatorname{id}_A M < \infty$.*

Proof. If M has finite projective dimension, it admits a finite resolution by finitely generated free modules. Since A is Gorenstein, such modules have finite injective dimension, and so has M. This can be seen by splitting the resolution into short exact sequences and applying the long exact sequence for Ext.

Conversely, let M be a finitely generated A module with $\operatorname{id} M < \infty$. We will prove that $\operatorname{pd} M < \infty$ by induction on $d = \dim A = \operatorname{id} M$ (by Theorem 9.6.5). If $d = 0$, M is injective, so by Matlis's theorem (Theorem 5.3.9) it is a direct sum of modules of the form $E(k)$, where k is the residue field of A. By Proposition 9.7.8, $E(k) \cong A$, so M is in fact free.

For the inductive step, take a regular element $a \in A$. We would like to apply the third change of ring theorems (Theorems 5.2.13 and 5.4.10), but we cannot guarantee that a is M-regular. To remedy this, take a short exact sequence

$$0 \longrightarrow K \longrightarrow F \longrightarrow M \longrightarrow 0 \,,$$

where F is finitely generated free. In particular, F has finite injective and projective dimension, so $\operatorname{id} M < \infty$ if and only if $\operatorname{id} K < \infty$ and $\operatorname{pd} M < \infty$ if and only if $\operatorname{pd} K < \infty$. What we have gained is that K is a submodule of

F, so a is K-regular. By the change of ring theorems

$$\operatorname{id}_{\frac{A}{(a)}} \frac{K}{aK} = \operatorname{id}_A K - 1 \quad \text{and} \quad \operatorname{pd}_{\frac{A}{(a)}} \frac{K}{aK} = \operatorname{pd}_A K.$$

By Proposition 9.7.6, the ring $A/(a)$ is Gorenstein, which by induction implies $\operatorname{pd}_{A/(a)} K/aK < \infty$, and by the above $\operatorname{pd}_A M < \infty$. □

Remark 9.7.14. This property characterizes Gorenstein rings. In fact, A itself has $\operatorname{pd}_A A = 0$. If A has the above property, $\operatorname{id}_A A$ is finite and so A is Gorenstein.

In fact, Foxby showed in [**Fox77**] that a single such module suffices.

Theorem 9.7.15 (Foxby). *Let A be a local Noetherian ring, M a finitely generated A-module. Assume that $\operatorname{pd}_A M < \infty$ and $\operatorname{id}_A M < \infty$. Then A is Gorenstein.*

The main ingredients of Foxby's theorem are the Ischebeck spectral sequences from Section 4.6, and in particular a useful numerical consequence. To state it, we need to introduce some notation. Let $A, \mathcal{M}$ be a local Noetherian ring, $k = A/\mathcal{M}$. For a finitely generated A-module M, denote

$$i_M(s) = \dim_k \operatorname{Ext}_s^A(k, M)$$

$$p_M(s) = \dim_k \operatorname{Tor}_s^A(k, M)$$

and consider the generating function for these numbers:

$$I_M(x) = \sum_{s=0}^{\infty} i_M(s)x^s$$

$$P_M(x) = \sum_{s=0}^{\infty} p_M(s)x^s.$$

Notice that I_M is a polynomial if M has finite injective dimension; and conversely, by Theorem 5.4.4. Similarly, P_M is a polynomial if M has finite projective dimension, and conversely; we do not need this last remark, which we leave as Exercise 21.

Proposition 9.7.16. *Let A be a local Noetherian ring, M a finitely generated A-module.*

(i) *If $\operatorname{pd}_A M < \infty$, then $I_M(x) = I_A(x)P_M(x^{-1})$.*

(ii) *If $\operatorname{id}_A M < \infty$, then $P_M(x) = I_A(x)I_M(x^{-1})$.*

Proof. The proof of the two results are entirely similar, so we just prove the first one. Expanding the notation, this amounts to the relation

$$\dim_k \operatorname{Ext}^n(k, M) = \sum_{p+q=n} \dim_k \operatorname{Ext}^p(k, A) \cdot \dim_k \operatorname{Tor}_{-q}(k, M).$$

This is just a consequence of the Ischebeck spectral sequence theorem (Theorem 4.6.2), applied with $M = k$, $N = M$.

First, the sequence degenerates at the page E^2. This is because all objects involved in the spectral sequence are k-vector spaces. To construct the sequence, we choose projective resolutions for M and N and build a double complex $C_{\bullet,\bullet}$ out of it. If we choose the minimal free resolutions, then all maps in the double complex $C_{\bullet,\bullet}$ have image in $\mathcal{M}C_{\bullet,\bullet}$, thanks to Proposition 5.6.5. It follows that all maps on E^2 are 0, so the sequence degenerates.

To prove the result we only need to compute the dimensions. Since $\mathrm{Ext}^p(k, A)$ is a k-vector space, say of dimension d, then

$$\mathrm{Tor}_q(\mathrm{Ext}^p(k, A), M) \cong \bigoplus_{i=1}^{d} \mathrm{Tor}_q(k, M),$$

so that taking dimensions we get

$$\dim_k \mathrm{Tor}_q(\mathrm{Ext}^p(k, A), M) = \dim_k \mathrm{Ext}^p(k, A) \cdot \dim_k \mathrm{Tor}_q(k, M),$$

and we get the desired relation. $\square$

Proof of Theorem 9.7.15. Under the assumption, we have both equations from Proposition 9.7.16. Let $d = \dim A$, so that $\deg I_M = d$ from Theorem 9.6.5 and $\deg P_M \leq d$ by the Auslander–Buchsbaum formula, Theorem 7.4.1.

Since $\deg I_M = d$, we can expand the first equality as

$$p_M(n) = i_A(n)i_M(0) + \cdots + i_A(n + d)i_M(d).$$

This is 0 for $n > \mathrm{pd}\, M$, and since all terms are nonnegative, they must individually vanish. In particular, $i_A(n + d)i_M(d) = 0$, and since $i_M(d) \neq 0$, this means that $i_A(n + d) = 0$ for $n > \mathrm{pd}\, M$, which means that I_A is a polynomial. $\square$

9.8. Complete intersections

We end the chapter by introducing yet another class of rings, complete intersections. They are especially convenient to give examples of Gorenstein, singular rings. We mostly introduce new terminology, and the bulk of our work will be showing that the definitions make sense. Our starting point is geometric.

Definition 9.8.1. Let V be an affine or projective variety of codimension d in the ambient space. We say that V is a *complete intersection* if the ideal $I(V)$ can be generated by d elements (homogeneous in the projective case). We say that V is a *set theoretic complete intersection* if it can be cut by d equations—in other word there is an ideal generated by d elements such that its zero locus is V.

We notice a few simple things about this definition.

Remark 9.8.2. Since V is the zero locus of $I(V)$, a complete intersection is a set theoretic complete intersection.

Remark 9.8.3. By Krull's principal ideal theorem, Theorem A.7.6, d is the minimum possible number of elements that can cut V.

Remark 9.8.4. Denote by $V(I)$ the zero locus of an ideal $I \subset k[x_1, \ldots, x_n]$ and by $I(V)$ the ideal of polynomials vanishing on the set V. If the ground field k is algebraically closed, for every ideal I we have $I(V(I)) = \sqrt{I}$ by the Nullstellensatz. Hence, the difference between a complete intersection and a set theoretic complete intersection is that the latter allows to generate I by d elements *up to radical*.

Remark 9.8.5. In the projective case, we asked that $I(V)$ is generated by d homogeneous elements, but this is actually redundant. In fact, on the ring $k[x_1, \ldots, x_n]$ if a homogeneous ideal is generated by d elements, it is generated by d *homogeneous* elements (Exercise 24).

It is quite natural to extend this definition to the context of general rings. Things are made slightly harder by the fact that the definition of complete intersection requires an ambient space.

Definition 9.8.6. Let A be a local ring, and assume that we are given a surjective homomorphism $f \colon B \to A$, where B is a regular local ring. We say that A is a *complete intersection* if $\ker f$ is generated by a regular sequence. If V is an algebraic variety, we say that V is a *local complete intersection* if the local rings of V at all points are complete intersections.

This definition is actually temporary: we will refine it in Definition 9.8.15.

Remark 9.8.7. This may look different from the geometric definition, but in fact it is equivalent to requiring that $\ker f$ is generated by $\dim B - \dim A$ elements. To wit, if $\mathbf{a} = a_1, \ldots, a_d$ is a regular sequence, the dimension of $B/(\mathbf{a})$ drops by d, thanks to Theorem A.7.4. Conversely, assume that $\dim B/(\mathbf{a}) = \dim B - d$. Then $\mathbf{a}$ is part of a system of parameters, hence it is a regular sequence by Theorem 9.1.13.

At this point, the definition seems to depend on the map $B \to A$, and we are not even sure what to do for rings that are not quotients of a regular ring. We will fix this in short order, but first we notice a few things. First, a regular local ring is a complete intersection by definition. Second, by Proposition 9.7.6, every complete intersection ring is Gorenstein. It follows that local complete intersections sit nicely between regular and Gorenstein rings. It is very easy to produce examples of complete intersections, and we get for free all the nice properties of Gorenstein rings—in particular, they are CM. Putting these observations together with Theorems 9.7.4 and 9.2.8 we get the following classic list of implications.

Theorem 9.8.8. *For a local Noetherian ring A, each one of these condition implies the following conditions:*

    (i) *A is regular,*

    (ii) *A is a complete intersection,*

    (iii) *A is Gorenstein,*

    (iv) *A is Cohen–Macaulay,*

    (v) *A is universally catenary, and*

    (vi) *A is catenary.*

In general, none of the implications is reversible. To see this, we discuss some examples and non examples of complete intersection rings. In all examples, we will fix a field k.

Example 9.8.9.

    (a) Let $A = k[x, y]/(y^2 - x^3 - x^2)$ and B its localization at the origin. This is the local ring of a node, which is singular, but is a complete intersection by definition.

    (b) Let $A = k[x, y, z]/I$, where $I = (x^2, y^2, xz, yz, z^2 - xy)$. We have seen in Example 9.7.9(a) that A is Gorenstein. Yet, A is not a complete intersection. This amounts to checking that I is not generated by 3 elements, which we leave to the reader.

    (c) We have seen in Example 9.7.9(b) that $k[x, y]/(x^2, y^2, xy)$ is Cohen–Macaulay, but not Gorenstein.

    (d) Let A be any finitely generated k-algebra. Then A is universally catenary, since k is, by Proposition 9.2.8. If we take, for instance, $A = k[x, y]/(x^2, xy)$, then A is not CM, as we have checked in Example 9.1.18(d).

    (e) Giving an example of a ring that is catenary but not universally catenary is highly nontrivial. Nagata gave an example in [**Nag62**, Appendix A, Example 2].

We should also give some examples that highlight the relation between the various geometric notions.

Example 9.8.10.

    (a) Let V be a collection of three noncollinear points in $\mathbb{A}^2$. Then V is regular, hence a local complete intersection. On the other hand, it is not a complete intersection. If it was, it would be the intersection of a line and a cubic, but the points are not collinear.

(b) For another simple example, one can take the disjoint union of two regular varieties of different dimension.

(c) The twisted cubic C is the image of the map $f \colon \mathbb{P}^1 \to \mathbb{P}^3$ given by $f([x:y]) = [x^3 : x^2y : xy^2 : y^3]$. One can verify (Exercise 26) that C is a set theoretic complete intersection, but it not a complete intersection, although its intersections with any affine chart are.

We now turn to making sure that the definition is actually well posed. The first step is proving that a local regular ring, which is tautologically a complete intersection when presented as a quotient of itself, is also a complete intersection when presented as a quotient of another ring.

Lemma 9.8.11. *Let A, B be regular local rings, and let $f \colon A \to B$ be a local surjective morphism. Then $\ker f$ is generated by a regular sequence.*

Proof. Write $B = A/P$, where $P = \ker f$, and denote by $\mathcal{M}_A, \mathcal{M}_B$ the maximal ideals. Let $k = A/\mathcal{M}_A = B/\mathcal{M}_B$. We have a surjective map of k-vector spaces

$$\frac{\mathcal{M}_A}{\mathcal{M}_A^2} \to \frac{\mathcal{M}_B}{\mathcal{M}_B^2}$$

whose kernel is $(P + \mathcal{M}_A^2)/\mathcal{M}_A^2$. So if we write $d_A = \dim A$, $d_B = \dim B$, we get

$$\dim_k \frac{P + \mathcal{M}_A^2}{\mathcal{M}_A^2} = d_A - d_B.$$

Let $d = d_A - d_B$ and choose elements $a_1, \ldots, a_d \in P$ that are linearly independent modulo $\mathcal{M}^2$. We can complete them to a basis of $\mathcal{M}/\mathcal{M}^2$, showing that $a_1, \ldots, a_d$ are part of a (regular) system of parameters for A. By Theorem 9.1.13, they are a regular sequence.

All that is left is to show that $P = (a_1, \ldots, a_d)$. To see this, let $Q = (a_1, \ldots, a_d) \subset P$. By Theorem A.7.4 and Proposition 8.3.5, A/Q is regular of dimension d_B. By Proposition 8.1.7, both P and Q are primes, and they have the same height by Proposition 9.2.7, so $P = Q$. $\square$

We now turn to the case where a complete intersection ring has two different presentations, one dominating the other.

Lemma 9.8.12. *Let $f \colon A \to B$ and $g \colon B \to C$ be surjective local morphisms of local rings, $h = g \circ f$. Assume that A and B are regular. Then $\ker g$ is generated by a regular sequence if and only if $\ker h$ is generated by a regular sequence.*

Proof. By Lemma 9.8.11, $\ker f$ is generated by a regular sequence, so if $\ker g$ is generated by a regular sequence, the same is true for $\ker h$.

Conversely, assume that $\ker h$ is generated by a regular sequence. If $\dim A = \dim B$, then f is an isomorphism (why?), and there is nothing to prove. Otherwise, take a regular element $a \in \ker f$. Letting $A' = A/(a)$, we can factor the above maps as a composition

$$A \to A' \to B \to C.$$

By Theorem A.7.4 and Proposition 8.3.5, A' is regular of dimension $\dim A - 1$. Moreover, a can be extended to a regular sequence generating $\ker h$, hence $A' \to C$ satisfies the hypothesis of the lemma. Since $\dim A' < \dim A$, we are done by induction. □

To deal with the general case, one may think that, starting with two unrelated surjections $A \to C$ and $B \to C$, with A and B regular, there could be a way to produce another regular ring that dominates A and B. A natural choice would be the fibered product

$$A \times_C B = \{(a,b) \mid f(a) = g(b)\},$$

but this is essentially never regular, except in trivial cases [**NTV18**]. Instead, we will pass to the completion and make use of Cohen's theorem, Theorem A.8.9.

Lemma 9.8.13. *Let A, B be local Noetherian rings, with B regular, and assume that there is a surjective map of rings $f\colon B \to A$ such that $\ker f$ is generated by a regular sequence. Denote by $\widehat{A}, \widehat{B}$ the completions of A, B with respect to their maximal ideals. Then there is an induced map $\widehat{f}\colon \widehat{B} \to \widehat{A}$, and $\ker \widehat{f}$ is generated by a regular sequence.*

Proof. Let $I = \ker f$, so that $A \cong B/I$, and let $a_1, \dots, a_d \subset I$ a regular sequence that generates I. By Krull's intersection theorem (Theorem A.6.4), we can see B as a subring of $\widehat{B}$. The map $B \to \widehat{B}$ is flat by Theorem 6.2.9, so $a_1, \dots, a_d$ is still a regular sequence in $\widehat{B}$ by Proposition 7.2.8. Then

$$\widehat{A} \cong \widehat{\left(\frac{B}{I}\right)} \cong \frac{\widehat{B}}{\widehat{I}},$$

and $\widehat{I} = I \cdot \widehat{B}$ is generated by $a_1, \dots, a_d$. □

We are now ready to see that our definition is well posed, at least in the complete case.

Theorem 9.8.14. *Let A be a complete local Noetherian ring. The following are equivalent:*

(i) *there is a local regular ring B and a surjection $f \colon B \to A$ such that $\ker f$ is generated by a regular sequence; and*

(ii) *for all local regular rings B and surjections $f \colon B \to A$, $\ker f$ is generated by a regular sequence.*

We will give a partial proof, covering the equicharacteristic case.

Proof. Let $f \colon B \to A$ be a surjection, where B is regular. By Cohen's theorem, there is a regular ring C and a surjection

$$g \colon C[[x_1, \ldots, x_t]] \to A,$$

where $t = \operatorname{embdim} A$. More precisely, let $\mathcal{M}$ be the maximal ideal of A and $k = A/\mathcal{M}$ be the common residue field of A and B. If A is equicharacteristic (that is, $\operatorname{char} A = \operatorname{char} k$), then we can take $C = k$. Else, we can take for C a complete ring which is the image of a DVR, and such that $C \cap \mathcal{M} = p \cdot C$, where $p = \operatorname{char} k$.

Using the completeness of B, we can lift g to a map

$$\widetilde{g} \colon C[[x_1, \ldots, x_t]] \to B.$$

First, we lift the restriction $g|_C$ to B. If A is equicharacteristic, then so is B, in which case $C = k$ and both A and B are k-algebras. Otherwise, C is the image of a DVR. We will not handle this case, as things get a little too technical.

Once we have a lift of $g|_C$, defining the rest of $\widetilde{g}$ is simple. Let $a_i = g(x_i)$ and choose any b_i such that $f(b_i) = a_i$. Then we can define $\widetilde{g}$ by letting $\widetilde{g}(x_i) = b_i$. This uniquely defines $\widetilde{g}$ on polynomials in the x_i, hence on power series, using the fact that B is complete and Hausdorff (the latter due to Theorem A.7.6).

The map $\widetilde{g}$ needs not be surjective, but we can make it surjective by adding finitely many variables $x_{t+1}, \ldots, x_s$ mapping to generators of $\mathcal{M}_B$, as we did above. We end up with the commutative diagram

$$\begin{array}{ccc} C[[x_1, \ldots, x_s]] & \longrightarrow & B \\ \downarrow & & \downarrow{\scriptstyle f} \\ C[[x_1, \ldots, x_t]] & \overset{g}{\longrightarrow} & A. \end{array}$$

Using Lemma 9.8.12 twice, we see that $\ker f$ is generated by a regular sequence if and only if $\ker g$ is. $\quad\square$

We now see how the definition of complete intersection ring is well defined. Start with a local Noetherian ring A with a surjection $f\colon B \to A$, where B is a regular local ring. By Lemma 9.8.13, this induces a surjective map between the completions $\widehat{f}\colon \widehat{B} \to \widehat{A}$, and $\ker \widehat{f}$ is generated by a regular sequence if $\ker f$ is. But then in the complete case, the definition of complete intersection does not depend on the particular choice of a surjection from a regular ring by Theorem 9.8.14.

We still have to consider the possibility that a local Noetherian ring A is not the quotient of a regular local ring at all. To remedy this, notice that by Cohen's theorem, every local *complete* Noetherian ring is the quotient of a regular one. We will thus modify our original definition.

Definition 9.8.15. Let A be a local ring. Let $\widehat{A}$ be its completion, and take a surjective homomorphism $f\colon B \to \widehat{A}$, where B is a regular local ring. This exists by virtue of Cohen's theorem. We say that A is a *complete intersection* if $\ker f$ is generated by a regular sequence.

A ring that is a complete intersection in the sense of Definition 9.8.6 is also a complete interesection according to this new definition, by Lemma 9.8.13. The fact that the definition is well posed follows from Theorem 9.8.14.

9.9. Exercises

1. Find a counterexample to the converse implication in Corollary 9.1.22— that is, a Noetherian ring A, an A-module M and an M-regular sequence $\mathbf{a}$ such that $M/(\mathbf{a})M$ is Cohen–Macaulay, but M is not.

2. Let A be a local Cohen–Macaulay ring, $Q \subset A$ an ideal generated by a system of parameters. Prove that $e(Q, A) = \ell(A/Q)$.

3. Conversely, assume that $A, \mathcal{M}$ is a local Noetherian ring, and there exists an ideal $Q \subset A$ generated by a system of parameters such that $e(Q, A) = \ell(A/Q)$. Prove that A is Cohen–Macaulay. (Letting $d = \dim A$, notice that the graded ring $\mathrm{Gr}_Q(A)$ is a quotient of $(A/Q)[x_1, \ldots, x_d]$, and compare the Hilbert polynomials.)

4. Let A be a local Cohen–Macaulay ring of dimension d. Prove that A is regular if and only if every Cohen–Macaulay A-module M of dimension d is free.

In Exercises 5–11, we are going to develop some of the theory of CM rings and modules in the graded case. We will assume that $A = \bigoplus_{i=0}^{\infty} A_i$ is a graded Noetherian ring, and that $A_0 = k$ is a field. We say that elements $a_1, \ldots, a_d$ are a system of parameters if they are homogeneous of positive degree and $A/(a_1, \ldots, a_d)$ has dimension 0.

5. Prove that A always admits a system of parameters.

6. Let $a_1, \ldots, a_d \in A$ be homogeneous of positive degree. Prove that they are a system of parameters if and only if $A/(a_1, \ldots, a_d)$ is a finite-dimensional vector space over k, or equivalently if $\mathcal{M}^t \subset (a_1, \ldots, a_d)$ for some t. Moreover, show that in this case $a_1, \ldots, a_d$ are algebraically independent over k.

7. Prove the homogeneous form of prime avoidance: let I be a homogeneous ideal of the graded ring A, and assume that every homogeneous element of I has positive degree and lies in $\bigcup_{i=1}^n P_i$ for some prime ideals $P_i \subset A$. Then I is contained in one of the P_i.

8. Let $A = k[x_1, \ldots, x_n]$ where the x_i are homogeneous elements of positive degree (not necessarily 1), and M a finitely generated graded A-module. Show that M is free over A if and only if $x_1, \ldots, x_n$ is M-regular.

9. Keeping the notation as in the introduction to Exercise 5, let $\mathcal{M} = \bigoplus_{i=1}^{\infty} A_i$ be the irrelevant ideal of A. Show that the following are equivalent:

(i) there exists a system of parameters that is a regular sequence,

(ii) every system of parameters is a regular sequence,

(iii) there exist a homogeneous system of parameters $a_1, \ldots, a_d$ such that A is free over $k[a_1, \ldots, a_d]$, and

(iv) for every homogeneous system of parameters $a_1, \ldots, a_d$, A is free over $k[a_1, \ldots, a_d]$.

In this case, we say that A is a graded Cohen–Macaulay ring.

10. Let A be as in the previous exercise. Show that A is a graded Cohen–Macaulay ring if and only if the local ring $A_{\mathcal{M}}$ is Cohen–Macaulay.

Conversely, start from a local Noetherian ring $A, \mathcal{M}$ and construct the associated graded ring $\mathrm{Gr}_{\mathcal{M}}(A) := \bigoplus_{i=0}^{\infty} \mathcal{M}^i/\mathcal{M}^{i+1}$. Show that A is Cohen–Macaulay if and only if $\mathrm{Gr}_{\mathcal{M}}(A)$ is.

11. Let A be a graded Noetherian ring, with A_0 a field. Show that A is graded Cohen–Macaulay is and only if it is Cohen–Macaulay in the usual sense. (Write A as a finitely generated free module over a polynomial ring. Alternatively, write A as a quotient of a polynomial ring, and use the Auslander–Buchsbaum formula.)

12. Let $A = k[x_1, \ldots, x_n]$ and G a finite group. Assume that G acts on A by k-algebra automorphisms that preserve the degree, and denote A^G the fixed ring. Finally, assume that $|G|$ does not divide char k. Show that A^G is a CM ring. (Show that A^G is a direct summand of A. This is a very special case of [**HE71**].)

13. Let $A \to B$ be a flat local homomorphism of Noetherian rings, and denote by $\mathcal{M}_A$ the maximal ideal of A. Show that B is CM if and only if both A and $B/\mathcal{M}_A B$ are. (See Exercise 19 in Chapter 7.)

14. Let $A \to B$ be a flat homomorphism of Noetherian rings. Assume that A is CM and for all prime ideals $P \subset A$ the fiber $k_P \otimes_A B$ is CM, where $k_P = A_P/PA_P$ is the residue field at P. Show that B is CM as well. Derive from this another proof of Proposition 9.1.23.

Exercises 15–19, develop in more details the properties of catenary rings. Many of these are from [**Mat86**, Chapter 31].

15. We start by understanding the simplest violation to being catenary. Let A be a Noetherian ring and $P \subset A$ a prime. Prove that there can be at most a finite number of primes $Q \supset P$ such that $\operatorname{ht} Q/P = 1$ in A/P, but $\operatorname{ht} Q > \operatorname{ht} P + 1$. This result is due to Ratliff [**Rat72**] and McAdam. (Let $n = \operatorname{ht} P$ and write P as a minimal prime of an ideal generated by n elements. Use primary decomposition for I, together with the following observation: if there exists an infinite family of primes Q_i with $\operatorname{ht} Q_i/P = 1$, then $\bigcap_i Q_i = P$.)

16. Let A be a Noetherian ring, $P \subset Q$ prime ideals such that $\operatorname{ht} P = h$, $\operatorname{ht} Q/P = d$, where $d > 1$. Prove that there exist infinitely many intermediate primes P', $P \subset P' \subset Q$ such that $\operatorname{ht} P' = h + 1$ and $\operatorname{ht} Q/P' = d - 1$. (Try to guarantee that $\operatorname{ht} P'/P = 1$ first.)

17. Prove Ratliff's criterion [**Rat72**]: the Noetherian local domain A is catenary if and only if $\operatorname{ht} P + \dim A/P = \dim A$ for all prime ideals $P \subset A$.

18. Let A be a Noetherian ring of dimension $d \leq 2$. Prove that A is catenary.

19. Let A be a Noetherian UFD of dimension $d \leq 3$. Prove that A is catenary. (From [**Mur20**]: use Ratliff's criterion from Exercise 17. Given a nonzero prime P, find $Q \subset P$ of height 1, and notice that Q is principal. Then apply Exercise 18 to A/Q.)

20. Use Exercise 9 in Chapter 7 to give an alternative proof of Lemma 9.8.13.

21. Let A be a local Noetherian ring, M a finitely generated A-module. Prove that
$$\operatorname{pd}_A M = \sup \left\{ s \mid \operatorname{Tor}_s^A(k, M) \neq 0 \right\}.$$

22. Use Foxby's polynomial formulas (Proposition 9.7.16) to give another proof of the Auslander–Buchsbaum formula (Theorem 7.4.1) and of Theorem 9.6.5. After doing this exercise, it should be apparent why the injective dimension has the simple relation $\operatorname{id} M = \operatorname{depth} A$, while the projective dimension needs the additional term $\operatorname{pd} M = \operatorname{depth} A - \operatorname{depth} M$.

23. By the previous exercise, you have a proof of Theorems 7.4.1 and 9.6.5 that does not depend on the change of ring theorems. Use this to give an alternative proof of the third change of ring theorems (Theorems 5.2.13 and 5.4.10).

24. Let $A = k[x_1, \ldots, x_n]$ be a polynomial ring over a field. If a homogeneous ideal $I \subset A$ is generated by d elements, then it can be generated by d *homogeneous* elements.

25. Let A be a Noetherian ring, M a CM A-module. Prove that $M[x]$ is a CM $A[x]$-module and $M[[x]]$ is a CM $A[[x]]$-module.

26. Prove the claims in Example 9.8.10(c). (For the first one, consider the minors of the matrix $\begin{pmatrix} u & v & w \\ v & w & z \end{pmatrix}$. If C was a complete intersection, it would be the intersection of a plane and a cubic surface.)

Chapter 10

Local Cohomology and Duality

In this chapter, we review various forms of duality that connect many of the cohomology theories that we have set up. The inspiration for this developments comes from geometry, in the form of Poincaré duality and Serre duality.

For a compact, orientable, real manifold M of dimension d, Poincaré duality gives an isomorphism $H^i(M, \mathbb{Z}) \cong H_{d-i}(M, \mathbb{Z})$. Serre duality works in the context of proper, Cohen–Macaulay schemes over a field k, but for simplicity we will state it for a regular projective variety X over k of dimension d. In this case, one has a sheaf ω_X of regular, algebraic d-forms, constructed in the same way as De Rham differential forms in the differentiable setting. We say that a sheaf S on X is coherent if there exists a covering of X such that on an open set U of the covering one has a finite presentation

$$\mathcal{O}_U^p \longrightarrow \mathcal{O}_U^q \longrightarrow S|_U \longrightarrow 0.$$

Serre duality then gives an isomorphism $\text{Ext}^i(S, \omega_X) \cong H^{d-i}(S)^\vee$ for all such sheaves S and all $i = 0, \ldots, d$. In particular, for $i = 0$, we see that H^d is represented by ω_X.

When $k = \mathbb{C}$ and $S = \Omega_X^p$, the sheaf of algebraic p-forms, these two results related by the GAGA Theorem from [**Ser56a**], which guarantees that the cohomology $H^i(S)$ is the same, whether we regard X as a topological space with the Zariski topology or with the Euclidean one; a resolution of Ω_X^p similar to (3.3.3), which allows to equate $H^q(\Omega_X^p)$ to the Dolbeault cohomology $H^{p,q}(X)$; finally, the Hodge decomposition

$H^k(X, \mathbb{C}) = \bigoplus_{p+q=k} H^{p,q}(X)$, valid for a compact complex manifold equipped with a Kähler metric—in particular, a projective one. A good reference for all of this is [**Voi02**] or [**GH94**].

This chapter is devoted to the local analogues of these results. We start with the simplest form of duality, Matlis duality, which is more algebraic in nature. Matlis duality extends the usual duality for finite-dimensional vector spaces to finitely generated modules over a local Noetherian ring. The central object is the functor $-^\vee = \mathrm{Hom}_A(-, E(k))$, where A is a local Noetherian ring, k its residue field and $E(k)$ the injective hull. In Section 10.1 we introduce the language of Matlis duality, and prove it in the case of local rings of dimension 0. The general case is treated in Section 10.2, where we see that the natural setting for this type of duality is that of a complete local Noetherian ring. As a particular case, we obtain a duality between Ext and Tor groups, giving isomorphisms $\mathrm{Tor}_i^A(M, N)^\vee \cong \mathrm{Ext}_A^i(M, N^\vee)$ for two A-modules M, N.

A different form of duality between Ext and Tor groups is briefly presented in Section 10.3, which is independent of the rest of the chapter. Here we define, for some A-modules M, a dualizing module ω_M. This is related to M by the relation $\mathrm{Tor}_i^A(\omega_M, N) \cong \mathrm{Ext}_A^{d-i}(M, N)$ for all A-modules N, where $d = \mathrm{pd}\, M$.

The rest of the chapter is devoted to setting up the most sophisticated of the dualities that we consider, Grothendieck local duality. This is the local analogue of Serre duality, which we mentioned at the beginning, and even stating the result requires some preliminaries. In Section 10.4 we introduce the concept of the canonical module ω_A of a local Noetherian ring A of dimension d. This is defined by the property that $\mathrm{Ext}_A^i(k, \omega_A) = 0$ for $i \neq d$ and $\mathrm{Ext}_A^d(k, \omega_A) \cong k$. In this section, we prove that ω_A is in fact unique up to isomorphism, and we show that it exists if and only if A is Cohen–Macaulay and is the quotient of a Gorenstein ring. We also see that local Gorenstein rings are characterized by the property that $\omega_A \cong A$.

The next ingredient that we need is local cohomology, which is the subject of Section 10.5. Given a ring A, an ideal I, and an A-module M, we have cohomology groups $H_I^i(M)$, which are defined as usual as derived functors. It turns that these groups can also be defined as a direct limit $H_I^i(M) \cong \varinjlim_{n \in \mathbb{N}} \mathrm{Ext}^i(A/I^n, M)$ of Ext groups. Section 10.6 studies some further properties of local cohomology—in particular the fact that in the local Noetherian case, the cohomology $H_{\mathcal{M}}^i(M)$ with respect to the maximal ideal $\mathcal{M}$ can only be nonzero when $\mathrm{depth}\, A \leq i \leq \dim A$. This also allows us to characterize CM and Gorenstein rings in terms of local cohomology.

Section 10.7 finally proves the promised duality. This takes the form $H_{\mathcal{M}}^i(M) \cong \mathrm{Ext}_A^{d-i}(M, \omega_A)^\vee$ for a local CM ring $A, \mathcal{M}$ of dimension d, having a canonical module ω_A, where $-^\vee$ denotes Matlis duality. As a consequence, we prove that the groups $H_\mathcal{M}^i(M)$ are always Artinian in the local case, and a nonvanishing result for local cohomology.

10.1. Duality for Artinian rings

We start with Matlis duality, which is a generalization of the duality on finite-dimensional vector spaces over a field. Given a field k, denote by Vect_k^* the category of finite-dimensional vector spaces over k. In this case, there is a functor $D\colon \mathrm{Vect}_k^* \to (\mathrm{Vect}_k^*)^{op}$ which is an equivalence of categories, and in fact such that D^2 is isomorphic to the identity. Of course, we can be explicit: $D(V) = V^\vee = \mathrm{Hom}_k(V, k)$ for a vector space V of finite dimension.

We now turn to a local Noetherian ring $A, \mathcal{M}$ and the category Mod_A^* of finitely generated A-modules, which is the same as the category Noe_A of Noetherian A-modules. Since we are looking for a contravariant functor, we expect that increasing chains will be taken to decreasing ones, so we take as target the category Art_A of Artinian A-modules. We thus look for an antiequivalence $D\colon \mathrm{Noe}_A \to \mathrm{Art}_A^{op}$ such that D^2 is isomorphic to the identity (here, we are abusing notation, by also writing $D\colon \mathrm{Art}_A^{op} \to \mathrm{Noe}_A$ for the inverse functor).

Let M be any Noetherian A-module. Then we must have

$$D(M) \cong \mathrm{Hom}(A, D(M)) \cong \mathrm{Hom}(D^2(M), D(A)) \cong \mathrm{Hom}(M, D(A)),$$

so if D exists at all, $D = \mathrm{Hom}(-, E)$, where $E = D(A)$. This justifies the abuse of notation: the functor $\mathrm{Hom}(-, E)$ is defined on all of Mod_A, and we are considering its restrictions on Noe_A and Art_A respectively.

What else can we say about E? For one thing, it satisfies

$$\mathrm{End}(E) = \mathrm{Hom}(D(A), D(A)) \cong D(D(A)) \cong A.$$

Next, notice that if D is an antiequivalence, it preserves limits and colimits by Proposition 1.5.4, so it is exact. In particular, it transforms quotients into subobjects. Since every nontrivial quotient of A has $k = A/\mathcal{M}$ as a quotient, it follows that $D(k)$ is essential in $E = D(A)$. If we assume that A is Artinian, D also preserves the length, which immediately implies that $D(k) \cong k$. We will assume this also in the general case.

Next, we remark that $D = \mathrm{Hom}(-, E)$ is an exact functor, at least on Noe_A. In other words, E satisfies the defining property of injective modules, restricted to monomorphisms of *finitely generated* A-modules. By Baer's criterion (Theorem 2.6.12), E is in fact injective. Since k is essential in E, it follows that the only choice for E is $E(k)$, the injective hull of k. We can summarize all of this as follows.

Proposition 10.1.1. *Let A be a local Noetherian ring with residue field k. Assume that there exists an equivalence $D\colon \mathrm{Noe}_A \to \mathrm{Art}_A^{op}$ with inverse $D'\colon \mathrm{Art}_A \to \mathrm{Noe}_A^{op}$ such that $D(k) \cong k$. Then $D = D' = \mathrm{Hom}(-, E)$ (suitably restricted), where $E = E(k)$ is the injective hull of k. Finally, $\mathrm{End}(E) \cong A$.*

Let us turn this logic around, and try to define a functor by this formula. Recall from Section 5.3 that $E_A(M)$ denotes the injective hull of the A-module M.

Definition 10.1.2. Let $A, \mathcal{M}$ be a local Noetherian ring with residue field k, $E = E(k)$ the injective hull of k. For a module M, we define the *Matlis dual* of M as $M^\vee = \mathrm{Hom}_A(M, E)$. We also call E the *Matlis module* of A.

This defines a contravariant functor on A-modules. For a map $f\colon M \to N$ of A-modules, we will denote by $f^\vee\colon N^\vee \to M^\vee$ the map induced on the dual modules. By definition, the Matlis module is injective, hence the dual functor is exact on Mod_A.

Our first remark concerns the relation between the Matlis duals of different rings.

Lemma 10.1.3. *Let $f\colon A \to B$ be a surjective homomorphism of rings, M a B-module. Then $\mathrm{Hom}_A(B, E_A(M)) \cong E_B(M)$.*

Proof. Since f is surjective, we can identify $\mathrm{Hom}_A(B, M)$ and $\mathrm{Hom}_B(B, M)$. Clearly, $M \cong \mathrm{Hom}_B(B, M)$ is a submodule of $\mathrm{Hom}_A(B, E_A(M))$. Hence, we only need to check that the B-module $\mathrm{Hom}_A(B, E_A(M))$ is injective, and M is essential inside it. By the adjunction in Example 1.5.2(d) we have

$$\mathrm{Hom}_B(N, \mathrm{Hom}_A(B, E_A(M))) \cong \mathrm{Hom}_A(N, E_A(M))$$

for all B-modules N. Since $\mathrm{Hom}_A(-, E_A(M))$ is exact, so is the functor $\mathrm{Hom}_B(-, \mathrm{Hom}_A(B, E_A(M)))$, whence $\mathrm{Hom}_A(B, E_A(M))$ is injective over B.

Let $N \subset \mathrm{Hom}_A(B, E_A(M))$ be a nonzero submodule, and take a nonzero $n \in N$, say $n\colon B \to E_A(M)$. Since M is essential in $E_A(M)$, the A-submodule generated by $n(1)$ meets M, and since f is surjective this means that $n(b) \in M$ for some $b \in B$ such that $m = n(b) \neq 0$. Up to rescaling n we can assume that $b = 1$. For all $b' \in B$, $n(b') = b'm$, hence n lies in $\mathrm{Hom}_B(B, M) \cong M$, showing that M is indeed essential. $\square$

In the rest of the section, we will focus on the special case where A is Artinian. In this case, $\mathrm{Noe}_A = \mathrm{Art}_A$ is the category of A-modules of finite length. Length will turn out to be a crucial invariant for the proof of Matlis duality in this case, much like dimension in the case of vector spaces.

Proposition 10.1.4. *Let A be a local Artinian ring with residue field k. Then $k^\vee \cong k$.*

Proof. By Lemma 10.1.3, $k^\vee = \operatorname{Hom}_A(k, E_A(k)) \cong E_k(k) = k$. $\square$

Corollary 10.1.5. *Let A be a local Artinian ring, $M \in \operatorname{Noe}_A$. Then $\ell(M) = \ell(M^\vee)$. In particular, $\ell(A) = \ell(E_A(k))$.*

Proof. By Theorem A.3.7, M has a finite chain of composition in which all successive quotients are isomorphic to k. The conclusion then follows from Proposition 10.1.4 and the fact that $\operatorname{Hom}(-, E_A(k))$ is exact. $\square$

We are now ready to prove Matlis duality in the Artinian case.

Theorem 10.1.6. *Let A be a local Artinian ring, $E = A^\vee$ its Matlis module. Then the Matlis functor $D = \operatorname{Hom}(-, E)$ is an exact antiequivalence on the category Noe_A of finitely generated A-modules such that D^2 is isomorphic to the identity.*

Proof. The only thing to prove is that D^2 is isomorphic to the identity. There is a natural map $\iota_M \colon M \to M^{\vee\vee} = \operatorname{Hom}(M^\vee, E)$ that sends $m \in M$ to the evaluation on m,

$$e_m \colon \quad M^\vee \longrightarrow E$$

$$f \longmapsto f(m).$$

Since $\ell(M) = \ell(M^{\vee\vee})$ by Corollary 10.1.5, we just need to check that ι_M is injective.

Noticing that ι defines a natural transformation between the identity and the double dual, we will prove this by induction on $\ell(M)$. If $\ell(M) = 1$, then $M \cong k$, and this case is Proposition 10.1.4.

Else, we can take a submodule $0 \subsetneq N \subsetneq M$, and obtain the commutative diagram

$$
\begin{array}{ccccccccc}
0 & \longrightarrow & N & \longrightarrow & M & \longrightarrow & M/N & \longrightarrow & 0 \\
 & & \downarrow{\iota_N} & & \downarrow{\iota_M} & & \downarrow{\iota_{M/N}} & & \\
0 & \longrightarrow & N^{\vee\vee} & \longrightarrow & M^{\vee\vee} & \longrightarrow & (M/N)^{\vee\vee} & \longrightarrow & 0,
\end{array}
$$

and the conclusion follows by induction and the five lemma (Lemma 2.5.1). $\square$

Corollary 10.1.7. *Let A be a local Artinian ring, $E = A^\vee$ its Matlis module. Then $E^\vee \cong A$.*

There are two ways in which this result can be generalized to higher dimensions: Matlis duality and local duality. In view of the latter, we introduce some notation.

Definition 10.1.8. Let A be a local Artinian ring with residue field k. The Matlis dual $E_A(k)$ of A is also called the *canonical module* of A, and denoted ω_A.

Remark 10.1.9. Assume further that A is Gorenstein. Then $A \cong E_A(k) = \omega_A$ by Proposition 9.7.8 (that is now proved). In this case, the duality functor takes the simple form $D = \mathrm{Hom}(-, A)$.

Remark 10.1.10. Let M be a finitely generated A-module. By Proposition 9.5.2, $\mathrm{soc}\, M$ is the smallest essential submodule of M. By duality, $(\mathrm{soc}\, M)^\vee$ is a quotient of $M^\vee$, say $M^\vee/N$, and N is the largest submodule such that every other nonzero quotient $M^\vee/N'$ has the property that $M^\vee/(N+N') \neq 0$. By Nakayama's lemma, this is $M^\vee/\mathcal{M}M^\vee$, so

$$(\mathrm{soc}\, M)^\vee \cong \frac{M^\vee}{\mathcal{M}M^\vee}.$$

In particular, $(\mathrm{soc}\, A)^\vee \cong \omega_A/\mathcal{M}\omega_A$.

Example 10.1.11.

(a) Let k be a field, $A = k[[x,y]]/I$ a finite-dimensional k-algebra. If $\mathrm{soc}\, A$ is 1-dimensional, then A is Gorenstein, and then we can visualize Matlis duality quite explicitly, especially if I is generated by monomials. For instance, take $I = (x^2, y^3)$. Then there is a single nonzero monomial that is annihilated both by x and y, namely xy^2, and $\mathrm{soc}\, A$ is generated by it. If we draw the monomials that do not vanish in A, we get the symmetric picture (taken from [**Eis95**, Section 21.1])

$$xy^2$$

$$xy \qquad\qquad y^2$$

$$x \qquad\qquad y$$

$$1.$$

(b) Now take an example with $\dim_k \operatorname{soc} A > 1$. This certainly produces a non symmetric picture, for the obvious fact that there is a single monoidal of lowest degree, namely 1. Taking for instance $I = (x^3, xy^2, y^3)$, we get

$$x^2y$$

$$x^2 \qquad xy \qquad y^2$$

$$x \qquad\qquad y$$

$$1.$$

10.2. Matlis duality

We now come back to the more general case of a local Noetherian ring A. We will see that we cannot expect this to be an actual antiequivalence unless A is complete.

Proposition 10.2.1. *Let A be a local Noetherian ring, $E = A^\vee$ its Matlis module. Then $E^\vee \cong \widehat{A}$, the completion of A.*

Of course, here and in the sequel, we mean completion with respect to the $\mathcal{M}$-adic topology, where $\mathcal{M}$ is the maximal ideal.

Proof. Define an increasing chain of submodules $E_n = (0 : {}_E \mathcal{M}^n)$, so that $\bigcup E_n = E$ by Remark 5.3.10. Notice that

$$E_n \cong \operatorname{Hom}_A\left(\frac{A}{\mathcal{M}^n}, E\right) \cong E_{\frac{A}{\mathcal{M}^n}}(k),$$

by Lemma 10.1.3.

Take an element $f \in E^\vee = \operatorname{Hom}(E, E)$. Then f restricts to morphisms $f_n \colon E_n \to E_n$ that are compatible, hence it defines an element in $\varprojlim \operatorname{Hom}(E_n, E_n)$. Conversely, any compatible sequence of morphisms $f_n \colon E_n \to E_n$ glues to a morphism $f \colon E \to E$. In other words,

$$E^\vee = \operatorname{Hom}(E, E) \cong \varprojlim \operatorname{Hom}(E_n, E_n) \cong \varprojlim \frac{A}{\mathcal{M}^n} = \widehat{A},$$

where the third isomorphism is Corollary 10.1.7. $\qquad\square$

Guided by this, we are going to prove Matlis duality for complete local Noetherian rings. In fact, one immediately reduces to the complete case by the following result.

Lemma 10.2.2. *Let $A, \mathcal{M}$ be a local Noetherian ring, $\widehat{A}$ its completion, and k the common residue field. Then $E_A(k) \cong E_{\widehat{A}}(k)$.*

Proof. We know by Remark 5.3.10 that every element $m \in E_A(k)$ is annihilated by a power of $\mathcal{M}$, say $\mathcal{M}^t m = 0$. Given $b \in \widehat{A}$, we can find $a \in A$ such that $a - b \in \mathcal{M}^t \widehat{A}$. Then we can define $bm = am$. It is easy to see that this gives $E_A(k)$ the structure of a $\widehat{A}$-module. Since it is an essential extension of k as an A-module, a fortiori it is essential as a $\widehat{A}$-module. Remark 2.7.3 yields a monomorphism $E_A(k) \hookrightarrow E_{\widehat{A}}(k)$. We claim that, conversely, $E_{\widehat{A}}(k)$ is essential over k as A-modules. Granting this, the conclusion follows immediately.

To check the claim, take a nonzero $m \in E_{\widehat{A}}(k)$. Then $\widehat{A}m$ meets $k \setminus \{0\}$, say $bm \in k^*$ for some $b \in \widehat{A}$. Reasoning as in the beginning, we find $a \in A$ such that $am = bm$, and we are done. $\qquad \square$

An immediate consequence of this is that if M is an A-module, then $M^\vee = \mathrm{Hom}_A(M, E_A(k))$ has the structure of a $\widehat{A}$-module. So applying the dual functor once immediately lands in $\mathrm{Mod}_{\widehat{A}}$. Still, we can still develop some useful results in the general case.

Lemma 10.2.3. *Let $A, \mathcal{M}$ be a local Noetherian ring, M a nonzero A-module. Then $M^\vee \neq 0$.*

Proof. If M is finitely generated, this follows immediately from the vector space case. The module $M/\mathcal{M}M$ is nonzero by Nakayama's lemma, hence it admits a nonzero functional with value in $k = A/\mathcal{M}$. The composition

$$M \to \frac{M}{\mathcal{M}M} \to k \to E(k)$$

is nonzero.

In the general case, choose a finitely generated submodule $N \subset M$ and a nonzero homomorphism $f \colon N \to E(k)$. Then f extends to M since $E(k)$ is injective. $\qquad \square$

Since the functor $-^\vee \colon \mathrm{Mod}_A \to \mathrm{Mod}_A$ is exact, the lemma implies immediately that it is faithful.

Lemma 10.2.4. *Let A be a local Noetherian ring. Then its Matlis module E is Artinian over $\widehat{A}$.*

Proof. If $M \subset E$ is an $\widehat{A}$-submodule, we have a surjection $E^\vee \to M^\vee$. By Proposition 10.2.1, $E^\vee \cong \widehat{A}$, and the kernel of this surjection is the annihilator of M in $\widehat{A}$.

A descending sequence $\{M_n\}$ of submodules corresponds to an ascending sequence of such ideals, and since $\widehat{A}$ is Noetherian, this stabilizes. This means that the sequence $\{M_n^\vee\}$ stabilizes as well.

Now if $M_n \supset M_{n+1}$ are such that $M_n^\vee = M_{n+1}^\vee$, we must have $M_n = M_{n+1}$ since the dual functor is faithful. It follows that the chain $\{M_n\}$ stabilizes as well, and E is Artinian. $\qquad\square$

Lemma 10.2.5. *Let $A, \mathcal{M}$ be a local Noetherian ring with Matlis module E, M an Artinian A-module. Then there is a monomorphism $M \to E^n$ for some n.*

Proof. Let $V = \operatorname{soc} M$, which has the structure of a vector space over $k = A/\mathcal{M}$. Note that V is itself Artinian, hence finite-dimensional, say $\dim V = n$. By Lemma 9.4.7, M is an essential extension of V. Since E is injective, the inclusion of V into $E(V) = E^n$ extends to M, and the extension remains injective by Lemma 2.7.2. $\qquad\square$

We can now prove that the dual functor takes values in the expected categories.

Proposition 10.2.6. *Let $A, \mathcal{M}$ be a local Noetherian ring, M an A-module.*

(i) *If M is Noetherian, then $M^\vee$ is Artinian as an $\widehat{A}$-module.*

(ii) *If M is Artinian, then $M^\vee$ is Noetherian as an $\widehat{A}$-module.*

Proof. For the first item, take a surjection $A^n \to M$. This induces a monomorphism $M^\vee \hookrightarrow (A^n)^\vee = E^n$, where E is the Matlis module. Since E is Artinian over $\widehat{A}$ by Lemma 10.2.4, $M^\vee$ is Artinian as well.

The second item is similar, using the inclusion $M \hookrightarrow E^n$ of Lemma 10.2.5. There is a surjection $(E^n)^\vee \to M^\vee$, so it is enough to prove that $E^\vee$ is Noetherian over $\widehat{A}$, but this is Proposition 10.2.1. $\qquad\square$

We are now ready to prove Matlis duality in full generality.

Theorem 10.2.7. *Let A be a complete local Noetherian ring, E its Matlis module, and consider the functor $-^\vee \colon \operatorname{Mod}_A \to \operatorname{Mod}_A^{op}$ and its restrictions $F = -^\vee \colon \operatorname{Noe}_A \to \operatorname{Art}_A^{op}$ and $G = -^\vee \colon \operatorname{Art}_A \to \operatorname{Noe}_A^{op}$. Then F and G are exact antiequivalences of categories. Explicitly, this means that $M \cong M^{\vee\vee}$, naturally in M, if M is either Noetherian or Artinian.*

Proof. We have already proved most of the theorem, we only need the last claim. Notice that every module M admits a natural morphism $M \to M^{\vee\vee}$ and we only need to prove that this is an isomorphism when M is Noetherian or Artinian.

Assume that M is Noetherian and take a finite presentation

$$A^m \longrightarrow A^n \longrightarrow M \longrightarrow 0.$$

Take the double dual. Using the fact that $-^\vee$ is exact and Proposition 10.2.1, we get the diagram

$$
\begin{array}{ccccccc}
A^m & \longrightarrow & A^n & \longrightarrow & M & \longrightarrow & 0 \\
\downarrow & & \downarrow & & \downarrow & & \\
A^m & \longrightarrow & A^n & \longrightarrow & M^{\vee\vee} & \longrightarrow & 0,
\end{array}
$$

whence $M \to M^{\vee\vee}$ is an isomorphism by the five lemma (Lemma 2.5.1).

The case where M is Artinian is identical, but starts from the embedding $M \to E^n$ of Lemma 10.2.5. □

Corollary 10.2.8. *Let A be a complete local Noetherian ring. Then the dual functor gives an exact antiequivalence from the category of A-modules of finite length to itself.*

An important example of Matlis duality is given by the following relation between Tor and Ext groups.

Proposition 10.2.9. *Let A be a local Noetherian ring, M, N A-modules.*

(i) *There is a natural isomorphism $\mathrm{Tor}_i^A(M, N)^\vee \cong \mathrm{Ext}_A^i(M, N^\vee)$.*

(ii) *If moreover M is finitely generated, $\mathrm{Ext}_A^i(M, N)^\vee \cong \mathrm{Tor}_i^A(M, N^\vee)$.*

Proof. For the first item, we compute $\mathrm{Tor}_i^A(M, N)$ via a free resolution $F_\bullet \to M$. Then $\mathrm{Tor}_i^A(M, N) = H_i(F_\bullet \otimes_A N)$. Since the dual functor is exact, we obtain

$$
\mathrm{Tor}_i^A(M, N)^\vee \cong H_i((F_\bullet \otimes_A N)^\vee) = H_i(\mathrm{Hom}(F_\bullet \otimes_A N, E)) \cong
$$
$$
\cong H_i(\mathrm{Hom}(F_\bullet, \mathrm{Hom}(N, E))) = H_i(\mathrm{Hom}(F_\bullet, N^\vee)) = \mathrm{Ext}_A^i(M, N^\vee),
$$

where the equality in the middle is the adjunction from Example 1.5.2(d).

For the second item, we do the same, but this time we can assume that all modules in the free resolution $F_\bullet$ are finitely generated. This time, we need to compute the dual of $\mathrm{Ext}_A^i(M, N) = H_i(\mathrm{Hom}(F_\bullet, N))$. Using again the exactness of the dual functor, we obtain

$$
\mathrm{Ext}_A^i(M, N)^\vee \cong H_i(\mathrm{Hom}(\mathrm{Hom}(F_\bullet, N), E)) \cong
$$
$$
\cong H_i(F_\bullet \otimes_A \mathrm{Hom}(N, E)) = H_i(F_\bullet \otimes_A N^\vee) = \mathrm{Tor}_i^A(M, N^\vee).
$$

Here we have used the isomorphism

$$
\mathrm{Hom}_A(\mathrm{Hom}_A(F, N), E) \cong F \otimes_A \mathrm{Hom}_A(N, E),
$$

which is valid for a finitely generated free A-module F. This is because both expressions are additive functors in F, which are clearly isomorphic when $F = A$. □

10.3. Poincaré duality

There is another form of duality between the Ext and Tor functors, which, following [**Kra06**], we will call Poincaré duality. This does not require the local hypothesis, but only works for a suitable class of modules.

Definition 10.3.1. Let A be a Noetherian ring, M a finitely generated A-module of finite projective dimension. We say that M satisfies *Poincaré duality* in dimension $d \in \mathbb{N}$ if $\mathrm{Ext}_A^i(M, A) = 0$ for all $i \neq d$. In this case, we call $\omega_M := \mathrm{Ext}_A^d(M, A)$ the *dualizing module* of M.

The reason for the name is the following duality result, which is an immediate consequence of the Ischebeck spectral sequence.

Theorem 10.3.2. *Let M be an A-module that satisfies Poincaré duality in dimension d, with dualizing module ω_M. Then $\mathrm{Tor}_i^A(\omega_M, N) \cong \mathrm{Ext}_A^{d-i}(M, N)$ for all A-modules N and all $i \in \mathbb{N}$.*

Proof. Use the spectral sequence from Theorem 4.6.2, and notice that by the hypothesis it degenerates at E^2. $\qquad\square$

Of course, the result is not of much use unless we can produce A-modules that satisfy the condition. We will briefly cover a couple of results in this direction.

Proposition 10.3.3. *Let M be an A-module that satisfies Poincaré duality in dimension d. Then ω_M satisfies Poincaré duality in dimension d as well, and $\omega_{\omega_M} \cong M$.*

Proof. Clearly, ω_M is finitely generated. We can apply the other Ischebeck spectral sequence from Theorem 4.6.4, taking $N = A$, which takes the form $E_{d,q}^2 = \mathrm{Ext}_A^q(\omega_M, A)$ and $E_{p,q}^2 = 0$ for $p \neq d$. This degenerates at E^2, hence we get $\mathrm{Ext}_A^q(\omega_M, A) \cong \mathrm{Tor}_{q-d}(M, A)$, which is 0 unless $q = d$. Finally, $\mathrm{Ext}_A^d(\omega_M, A) \cong M \otimes A \cong M$. $\qquad\square$

Proposition 10.3.4. *Let $\mathbf{a} = a_1, \ldots, a_d$ be a regular sequence in the Noetherian ring A. Then $M = A/(\mathbf{a})$ satisfies Poincaré duality in dimension d, and $\omega_M \cong M$.*

Proof. This is an immediate consequence of Proposition 7.2.7. $\qquad\square$

Of course, in this case, we get Poincaré duality just by the duality of the Koszul resolution, Proposition 7.1.15.

10.4. Canonical modules

In this section we introduce canonical modules, which are central objects to set up duality theory.

Definition 10.4.1. Let A be a local Noetherian ring of dimension d, with residue field k. We say that the A-module M is a *canonical module* (or *dualizing module*) for A if $\mathrm{Ext}_A^i(k, M) = 0$ for all $i \neq d$, while $\mathrm{Ext}_A^d(k, M) \cong k$. We will see in a moment that the canonical module is unique up to isomorphism—we will denote it by ω_A.

Despite similar notation, the reader should not confuse the canonical module ω_A with the dualizing module ω_M used in Poincaré duality.

Example 10.4.2.

  (a) Let A be a Gorenstein ring. By Proposition 9.7.11, A is a canonical module for itself.

  (b) Let A be a local Artinian ring, so $\dim A = 0$. By Kaplansky's theorem (Theorem 5.4.4), a canonical module M for A is an injective module such that $\mathrm{Hom}(k, M) \cong k$. This is the same as saying that M is an injective hull for k. (Why?) It follows that $\omega_A \cong E(k)$, the Matlis module of A.

Remark 10.4.3. Let A be a ring admitting a canonical module ω_A. By Rees's theorem (Theorem 7.3.12), we have $\mathrm{depth}\, \omega_A = \dim A \geq \dim \omega_A$, which means that ω_A is a Cohen–Macaulay A-module of maximal dimension $\dim \omega_A = \dim A$. A Cohen–Macaulay A-module M such that $\dim M = \dim A$ is called a *maximal Cohen–Macaulay module*.

Remark 10.4.4. Assume that A admits a canonical module ω_A. By Kaplansky's theorem (Theorem 5.4.4), ω_A has finite injective dimension. By Theorem 9.6.5, we have $\mathrm{depth}\, A = \mathrm{id}\, \omega_A = \dim A$, hence A itself is a Cohen–Macaulay ring.

Remark 10.4.5. Let us rephrase the definition of a canonical module. Let A be a local Noetherian ring of dimension d. Combining Kaplansky's theorem (Theorem 5.4.4) and Rees's theorem (Theorem 7.3.12), the condition that $\mathrm{Ext}_A^i(k, \omega_A) = 0$ for $i \neq d$ is equivalent to

$$\mathrm{depth}\, \omega_A = \mathrm{id}\, \omega_A = d.$$

The additional condition that $\mathrm{Ext}_A^d(k, \omega_A) \cong k$ is a normalization condition. We will see in a moment that it is equivalent to $\mathrm{Hom}_A(\omega_A, \omega_A) \cong A$.

To better understand the properties of canonical modules we will, as usual, argue by induction with respect to a regular sequence. Let A be a local Noetherian ring, M a finitely generated A-module. Assume that depth $A > 0$ and depth $M > 0$, and take an element $a \in A$ that is both regular and M-regular—this exists by Remark 5.2.14. Then we have

$$\operatorname{depth} \frac{M}{aM} = \operatorname{depth} M - 1$$

$$\operatorname{id}_{A/(a)} \frac{M}{aM} = \operatorname{id}_A M - 1$$

$$\dim \frac{A}{(a)} = \dim A - 1$$

Moreover, we have an isomorphism

$$\operatorname{Ext}^{d-1}_{\frac{A}{(a)}} \left(k, \frac{M}{aM} \right) \cong \operatorname{Ext}^d_A(k, M),$$

by Rees's lemma (Lemma 9.7.10). Combining all of this shows that M/aM is canonical for $A/(a)$ if and only if M is canonical for A. By induction, we get:

Proposition 10.4.6. *Let A be a local Noetherian ring, M a finitely generated A-module. If $\mathbf{a}$ is a sequence that is both regular and M-regular, then $M/(\mathbf{a})M$ is canonical for $A/(\mathbf{a})$ if and only if M is canonical for A.*

In particular, assume that A admits the canonical module M. Let $d = \dim A$. Then depth $A = \operatorname{depth} M = d$, so we can take a sequence $\mathbf{a}$ of length d which is regular and M-regular. The quotient $M/(\mathbf{a})M$ is canonical for $A/(\mathbf{a})$, which has dimension 0. Combining with Example 10.4.2(b), we get:

Corollary 10.4.7. *Let A be a local Noetherian ring of dimension d, admitting the canonical module M. Denote by k the residue field. Let $\mathbf{a}$ be a maximal regular and M-regular sequence. Then*

$$\frac{M}{(\mathbf{a})M} \cong E_{\frac{A}{(\mathbf{a})}}(k).$$

The last piece that we need to prove uniqueness of the canonical module is a simple vanishing result.

Lemma 10.4.8. *Let A, M be a local Noetherian ring of dimension d admitting a canonical module M. Let N be a finitely generated A-module with depth $N = e$. Then $\operatorname{Ext}^i_A(N, M) = 0$ for $i > d - e$.*

Proof. By induction on e. If $e = 0$, this follows because M has injective dimension d. For the inductive step, we take an N-regular element $a \in M$ and look at the exact sequence of $\operatorname{Ext}(-, M)$ groups

$$\cdots \longrightarrow \operatorname{Ext}^i_A(N, M) \overset{\cdot a}{\longrightarrow} \operatorname{Ext}^i_A(N, M) \longrightarrow \operatorname{Ext}^{i+1}_A \left(\tfrac{N}{aN}, M \right) \longrightarrow \cdots.$$

Since depth $N/aN = e - 1$, the last displayed group vanishes for $i + 1 > d - e + 1$—that is, $i > d - e$. The vanishing of $\mathrm{Ext}_A^i(N, M)$ then follows from Nakayama's lemma. $\qquad\square$

Theorem 10.4.9. *Let A be a local Noetherian ring. The canonical module of A, if it exists at all, is unique up to isomorphism.*

Proof. Let M, N be two canonical modules. Choose a sequence $\mathbf{a} = a_1, \ldots, a_d$ of length $d = \dim A$ that is regular, M-regular and N-regular (see Remark 5.2.14). By Corollary 10.4.7, we get an isomorphism

$$\frac{M}{\mathbf{a}M} \cong E_{\frac{A}{(\mathbf{a})}}(k) \cong \frac{N}{\mathbf{a}N},$$

where $k = A/\mathcal{M}$.

For every initial segment $\mathbf{a}' = a_1, \ldots, a_i$ of $\mathbf{a}$, the quotients $M/(\mathbf{a}')M$ and $N/\mathbf{a}'N$ are canonical for $A/(\mathbf{a}')$. By Lemma 10.4.8,

$$\mathrm{Ext}_{\frac{A}{(\mathbf{a}')}}^1\left(\frac{M}{\mathbf{a}'M}, \frac{N}{\mathbf{a}'n}\right) = 0,$$

and this allows us to inductively lift the isomorphism $M/\mathbf{a}M \cong N/\mathbf{a}N$ to maps $M/\mathbf{a}'M \to N/\mathbf{a}'N$, until we get a lift $f \colon M \to N$.

We now notice that f is surjective by Nakayama's lemma. Letting $K = \ker f$, we have $(K + \mathbf{a}M)/\mathbf{a}M = 0$, which means $K \subset \mathbf{a}M$. Moreover, $(K :_M \mathbf{a}) = K$, since $\mathbf{a}$ is regular on N. Putting the two things together, we get $(\mathbf{a})K = K$, whence $K = 0$ by another application of Nakayama's lemma. $\qquad\square$

We now turn to the question of existence.

Proposition 10.4.10. *Let $f \colon A \to B$ be a local homomorphism of CM rings, and assume that B is finitely generated as an A-module. If A admits the canonical module ω_A, then B does as well, and $\omega_B \cong \mathrm{Ext}_A^t(B, \omega_A)$, where $t = \dim A - \dim B$.*

Proof. As a first step, we reduce to the case where $\dim A = \dim B$. For this, it is enough to find a regular sequence $\mathbf{a} = a_1, \ldots, a_t$ contained in $\ker f$. In fact, letting $A' = A/(\mathbf{a})$, we will get an induced morphism $A' \to B$, and A' has canonical module $\omega_{A'} = \omega_A/(\mathbf{a})\omega_A$ by Proposition 10.4.6. Finally, Rees's lemma (Lemma 9.7.10) gives us an isomorphism $\mathrm{Ext}_A^t(B, \omega_A) \cong \mathrm{Hom}_{A'}(B, \omega_{A'})$.

It remains to find such regular sequence, which amounts to proving that $\mathrm{depth}_{\ker f} A \geq t$. By Corollary 9.1.8 and Proposition 9.2.7, we have $\mathrm{depth}_{\ker f} A = \mathrm{ht}\,\ker f = t$, which is what we need.

We now assume that $\dim A = \dim B = d$. We can take a sequence **a** in A which is regular and such that the image in B is regular as well (Remark 5.2.14). Taking the quotient and considering Proposition 10.4.6 again, we reduce to the case where $d = 0$.

In this case, we know that both A and B have canonical modules, which are simply their Matlis modules. Moreover, the group $\mathrm{Hom}_A(B, \omega_A)$ has the structure of a B-module, and is injective. This follows from Lemma 2.6.16 because $\mathrm{Hom}_A(B, -)$ is a left adjoint by Example 1.5.2(g). Since B is finitely generated over A, $\mathrm{Hom}_A(B, \omega_A)$ is a finitely generated B-module, hence a finite sum of copies of ω_B by Matlis's theorem, Theorem 5.3.9. Finally, a simple count of dimensions over the residue field of B shows that $\omega_B \cong \mathrm{Hom}_A(B, \omega_A)$. $\qquad\square$

Corollary 10.4.11. *Let B be a local CM ring that is a quotient of a Gorenstein ring A. Then B has a canonical module $\omega_B = \mathrm{Ext}_A^t(B, A)$, where $t = \dim A - \dim B$.*

In particular, by Cohen's theorem (Theorem A.8.9), every local *complete* CM ring has a canonical module.

We will now see that the Corollary is sharp: a CM ring A has a canonical module if and only if it is the image of a Gorenstein ring. To see this, we are going to introduce a simple construction. Let M be an A-module, and consider the A-module $A \oplus M$. This can be endowed with a commutative multiplication by defining $m \cdot n = 0$ for $m, n \in M$ and otherwise using the multiplication by an element of A. In more explicit terms, we let

$$(a, m) \cdot (b, n) = (ab, an + bm)$$

for $(a, m), (b, n) \in A \oplus M$. This gives $A \oplus M$ the structure of a commutative ring, called the *extension* of A by M, which we denote $A \ltimes M$.

There is a natural projection $\pi \colon A \ltimes M \to A$ which is a ring homomorphism. In particular, $\pi^{-1}(I) = I \ltimes M$ is an ideal whenever I is. Moreover, $P \ltimes M$ is prime if P is a prime ideal. The next result summarize a few properties of this construction.

Lemma 10.4.12. *Let A be a Noetherian ring, M a finitely generated A-module.*

(i) *The ring $A \ltimes M$ is Noetherian;*

(ii) $\dim A \ltimes M = \dim A$;

(iii) *if A is local with maximal ideal $\mathcal{M}$, $A \ltimes M$ is local with maximal ideal $\mathcal{M} \ltimes M$; and*

(iv) *if A is local with maximal ideal $\mathcal{M}$, then*

$$\mathrm{soc}(A \ltimes M) = \{(a, m) \mid a \in \mathrm{soc}\, A \cap \mathrm{Ann}\, M; m \in \mathrm{soc}\, M\}.$$

Proof. The first claim is immediate. For (ii), it is enough to check that all prime ideals $Q \subset A \ltimes M$ have the form $Q = P \ltimes M$ for a prime ideal $P \subset A$. In fact, notice that the ideal $J = 0 \ltimes M$ satisfies $J^2 = 0$, hence $J \subset Q$. Let $P = \pi(Q)$. Since $Q \supset 0 \ltimes M$, it follows immediately that $Q \supset P \ltimes M$, and then we must have equality. Finally, P is prime since Q is.

For (iii) take $(a, m) \in A \ltimes M$ and assume that $a \notin \mathcal{M}$. Then a is invertible, and (a, m) admits the inverse $(a^{-1}, -a^{-2}m)$, so the ideal $\mathcal{M} \ltimes M$ is indeed the unique maximal ideal.

Finally, (iv) is an immediate verification once we know the explicit description of the maximal ideal. □

We can now prove a kind of converse to Corollary 10.4.11.

Theorem 10.4.13. *Let A be a local Noetherian ring admitting a canonical module. Then A is Cohen–Macaulay and is the quotient of a Gorenstein ring. In fact, $A \ltimes \omega_A$ is Gorenstein.*

Proof. We have already remarked that A is necessarily CM, hence we only need to prove that $B := A \ltimes \omega_A$ is Gorenstein. Let $d = \dim A = \dim B$ and take a regular and ω_A-regular sequence $\mathbf{a}$ of length d. Because of the way multiplication is defined, $\mathbf{a}$ is also regular for B. Moreover,

$$\frac{B}{\mathbf{a}B} \cong \frac{A}{\mathbf{a}A} \ltimes \frac{\omega_A}{\mathbf{a}\omega_A}.$$

From Proposition 10.4.6, we know that $\omega_A/(\mathbf{a})\omega_A$ is canonical for $A/(\mathbf{a})$. Moreover, B is Gorenstein if and only if $B/\mathbf{a}B$ is, by Proposition 9.7.6. We conclude that it is enough to prove the result when $d = 0$.

In this case, using Proposition 9.7.8, it is enough to prove that $\operatorname{soc} B$ is 1-dimensional. By definition, ω_A satisfies $\operatorname{soc} \omega_A \cong \operatorname{Hom}_A(k, \omega_A) \cong k$, where k is the residue field of A. Moreover, $\omega_A \cong E_A(k)$, which means that $\operatorname{Ann} \omega_A = 0$ by Matlis duality, Theorem 10.1.6. Using Lemma 10.4.12(iv), we see that $\operatorname{soc} B \cong k$, as desired. □

10.5. Local cohomology

The main technical tool that we will use in this chapter is the concept of local cohomology. Our presentation follows many sources, such as [**Hun07**], [**va05**], and [**Sch98**].

Definition 10.5.1. Let A be a ring, $I \subset A$ an ideal. For an A-module M, we denote by $\Gamma_I(M)$ the subset

$$\Gamma_I(M) = \{m \in M \mid I^n m = 0 \text{ for some } n \in \mathbb{N}\} = \bigcup_{n=0}^{\infty} (0 :_M I^n).$$

The set $\Gamma_I(M)$ is a submodule of M, and is called the submodule of *sections with support on I*.

Remark 10.5.2. To make sense of the terminology, take a prime $P \in \operatorname{Supp}(\Gamma_I(M))$, which means that $\Gamma_I(M)_P \neq 0$. This can only happen if $I \subset P$. Assume that $A = R(V)$ is the coordinate ring of the affine variety V over an algebraically closed field, so that points in V correspond to maximal ideals in A. Then the points in the support of $\Gamma_I(V)$ all lie in the subvariety of V defined by the ideal I.

It is immediate to check that if $f \colon M \to N$ is A-linear, it induces a map $\Gamma_I(M) \to \Gamma_I(N)$, which we denote by $\Gamma_I(f)$. This makes $\Gamma_I \colon \operatorname{Mod}_A \to \operatorname{Mod}_A$ into a functor, which is clearly additive.

Lemma 10.5.3. *The functor* $\Gamma_I \colon \operatorname{Mod}_A \to \operatorname{Mod}_A$ *is left exact.*

Proof. If $N \subset M$ are two A-modules, clearly $\Gamma_I(N) \subset \Gamma_I(M)$. Moreover, assume that $m \in \Gamma_I(M)$ maps to 0 in $\Gamma_I(M/N)$. Since $\Gamma_I(M/N) \subset M/N$, m goes to 0 in M/N, that is, $m \in N \cap \Gamma_I(M) = \Gamma_I(N)$. $\qquad\square$

This immediately prompts us to define the right derived functors.

Definition 10.5.4. Let $I \subset A$ be an ideal, M an A-module. The functor $H_I^i \colon \operatorname{Mod}_A \to \operatorname{Mod}_A$ is the i-th right derived functor of Γ_I. The module $H_I^i(M)$ is called the i-th *local cohomology module* of M with support in I.

Remark 10.5.5. Assume that $\sqrt{I}$ is finitely generated, so that $(\sqrt{I})^n \subset I$ for some n. Then $\Gamma_I = \Gamma_{\sqrt{I}}$, so the same holds for the derived functors H_I^i. It follows that if A is Noetherian, the cohomology functors H_I^i only depend on I up to radical.

Remark 10.5.6. Every element in $H_I^i(M)$ is annihilated by I^k for some k. This follows at once by writing $H_I^i(M)$ as the homology of a complex obtained by taking a resolution of M and applying Γ_I, and noticing that the same is true for Γ_I by definition.

We shall now compute some simple examples, and in doing so verify that Γ_I is not right exact, so its right derived functors are nontrivial.

Example 10.5.7.

(a) Let us start from a PID A and a prime ideal $(p) \subset A$. Here, things are simple, because an A-module is injective if and only if it is divisible. Given an A-module M, $\Gamma_{(p)}(M)$ is the submodule of elements that are p^n-torsion for some n. Choose $M = A$ itself and denote k the fraction field of A, so that we have the injective resolution

$$0 \longrightarrow A \longrightarrow k \longrightarrow k/A \longrightarrow 0 \; .$$

Since A and k have no torsion, if we apply $\Gamma_{(p)}$ to this resolution we compute $H^i_{(p)}(A) = 0$ for $i \neq 1$ and $H^1_{(p)}(A) = A_p/A$, where $A_p = A[1/p]$ is the localization at powers of p.

(b) Keeping A a PID, we now choose $M = A/(a)$ for some $a \in A$. Taking the resolution

$$0 \longrightarrow \frac{A}{(a)} \longrightarrow \frac{k}{(a)} \longrightarrow \frac{k}{A} \longrightarrow 0$$

and applying $\Gamma_{(p)}$, we find that computation is easy in two cases. If $a = p^e$ is a power of p, then $H^0_{(p)}(M) = M$ and $H^i_{(p)}(M) = 0$ for $i \geq 1$. If instead a is relatively prime to p, then $H^i_{(p)}(M) = 0$ for $i \neq 1$, while $H^1_{(p)}(M) = A_p/aA_p$.

Since A is a PID, every finitely generated A-module is the direct sum of cyclic A-modules that are either of the form A or $A/(p^e)$ or $A/(q^e)$ for some prime $q \neq p$. Hence, with the above examples we can compute $H^i_{(p)}(M)$ for all finitely generated A-modules M.

(c) Assume that I is nilpotent, so that $I^e = 0$ for some e. Then Γ_I is the identity functor, so all functors H^i_I for $i \geq 1$ vanish.

(d) Let A be a Noetherian ring, $I \subset A$ an ideal and E an injective A-module. Clearly, $H^i_I(E) = 0$ for $i \geq 1$, regardless of the Noetherian hypothesis. By Matlis's theorem (Theorem 5.3.9), E is a direct sum of modules of the form $E(A/P)$, where P is a prime ideal and $E(-)$ denotes the injective hull. Hence, we only need to compute $H^0_I(E(A/P)) = \Gamma_I(E(A/P))$.

By Remark 5.3.10, P is the only associated prime of $E(A/P)$, and every element of $E(A/P)$ is annihilated by a power of P. So if $I \subset P$, we get $H^0_I(E(A/P)) = E(A/P)$. If instead $I \not\subset P$, choose any $a \in I \setminus P$. Then a is not a zero divisor on $E(A/P)$, so $H^0_I(E(A/P)) = 0$.

We notice one consequence of the last example.

Proposition 10.5.8. *Let A be a Noetherian ring, E an injective A-module. Then $\Gamma_I(E)$ is injective for all ideals $I \subset A$.*

We now start to link local cohomology to other cohomology theories.

Remark 10.5.9. We can write

$$\Gamma_I(M) = \varinjlim_{n \in \mathbb{N}} (0 :_M I^n) \cong \varinjlim_{n \in \mathbb{N}} \mathrm{Hom}(A/I^n, M).$$

By Remark 3.5.1, direct limits commute with right derived functors, so we get an identification

$$H_I^i(M) \cong \varinjlim_{n \in \mathbb{N}} \operatorname{Ext}^i(A/I^n, M).$$

More generally, if I_n is any decreasing sequence of ideals that are cofinal with the powers I^n, we get

$$H_I^i(M) \cong \varinjlim_{n \in \mathbb{N}} \operatorname{Ext}^i(A/I_n, M).$$

We can use this remark to draw a link between local cohomology and Koszul cohomology. Let $I \subset A$ be a finitely generated ideal, say $I = (a_1, \ldots, a_k)$ and consider the sequence $\mathbf{a} = a_1, \ldots, a_k$. By Proposition 7.2.7, we have a natural map $\operatorname{Ext}_A^i(A/I, M) \to H^i(\mathbf{a}; M)$. This need not be an isomorphism unless $\mathbf{a}$ is regular, but it becomes one if we pass to the limit over powers of I.

Proposition 10.5.10. *Let A be a Noetherian ring, $\mathbf{a} = a_1, \ldots, a_k$. Denote $I = (a_1, \ldots, a_k)$, $\mathbf{a}^n = a_1^n, \ldots, a_k^n$ and $I_n = (\mathbf{a}^n) \subset I^n$. Then the induced map*

$$H_I^i(M) \cong \varinjlim_{n \in \mathbb{N}} \operatorname{Ext}_A^i(A/I_n, M) \to \varinjlim_{n \in \mathbb{N}} H^i(\mathbf{a}^n; M)$$

is an isomorphism.

Proof. Notice that the isomorphism on the left follows by Remark 10.5.9, given that I_n and I^n are cofinal systems of ideals. If we define

$$F^i(M) := \varinjlim_{n \in \mathbb{N}} H^i(\mathbf{a}^n; M),$$

then we have a natural transformation of functors $\phi_i \colon H_I^i \to F^i$. We thus get two families of functors $\operatorname{Mod}_A \to \operatorname{Mod}_A$, both satisfying a long exact sequence arising from a short exact sequence of A-modules. This is clear for local homology, and follows from Remark 3.5.1 for F^i. In the language introduced before Exercise 13 of Chapter 3, the families $\{H_I^i\}$ and $\{F^i\}$ are both δ-functors.

We want to prove that all ϕ_i are isomorphisms. Using the long exact sequences and dimension shifting, it will be enough to prove that ϕ_0 is an isomorphism and that both H_I^i and F^i vanish for $i \geq 1$ on injective modules.

We proceed to the first claim, which amounts to the map

$$\Gamma_I(M) \to \varinjlim_{n \in \mathbb{N}} H^0(\mathbf{a}^n; M)$$

being an isomorphism. By Koszul duality and Example 7.1.6(c), we have $H^0(\mathbf{a}^n; M) \cong (0 :_M I_n)$, so the isomorphism holds by the fact that $\{I_n\}$ and $\{I^n\}$ are cofinal families.

The second claim is obvious for H_I^i, which is a derived functor, hence we only need to prove that $F^i(M) = 0$ when M is injective. Here is where we use the Noetherian hypothesis. By Theorem 5.3.9, M is a direct sum of modules of the form $E(A/P)$, where P is a prime ideal, so it suffices to verify vanishing on such modules. We now distinguish two cases, according to whether $\mathbf{a} \subset P$ or not.

Denote $E = E(A/P)$, and assume first that $\mathbf{a} \subset P$. In this case, every a_i is a torsion element on E. Consider a map in the direct system that defines $F^i(E)$, say

$$f_{s,t} \colon K^i(\mathbf{a}^s; E) \to K^i(\mathbf{a}^t; E)$$

for some pair of integers $s \leq t$. The map $f_{s,t}$ is derived from the map $g_{t,s} \colon K_i(\mathbf{a}^t) \to K_i(\mathbf{a}^s)$, which we spell out explicitly. The reader can verify that, in order to make all diagrams commute, we should take $g_{t,s} = \Lambda^i h_{t,s}$, where

$$h_{t,s} \colon A^n \to A^n$$

is the map defined by the diagonal matrix with entries a_i^{t-s}. The upshot of all of this is that the matrix for $f_{s,t}$ has entries in I.

Now take an element of $F^i(E)$, which is represented by a map $\phi \colon K_i(\mathbf{a}^s) \to E$ for some $s \in \mathbb{N}$. Since E is P-torsion, the same holds for ϕ, hence $f_{s,t}(\phi) = 0$ for t large enough, and ϕ represents 0 in $F^i(E)$.

The other case is that $\mathbf{a} \not\subset P$, say $a_1 \notin P$. By divisibility, E is a module over A_P, and we have an isomorphism

$$K^i(\mathbf{a}^s, E) = \mathrm{Hom}_A(K_i(\mathbf{a}), E) \cong \mathrm{Hom}_{A_P}(K_i(\mathbf{a}) \otimes_A A_P, E).$$

But a_1 is invertible in A_P, which means that $K_i(\mathbf{a}) \otimes_A A_P$ is acyclic. Since E is injective, the same is true applying $\mathrm{Hom}(-, E)$, hence $F^i(E) = 0$ also in this case, and we are done. $\square$

We can simplify our conclusion by getting rid of the direct limit in the right-hand side. First, by Koszul duality (Proposition 7.1.15), $H^i(\mathbf{a}; M) \cong H_{n-i}(\mathbf{a}; M)$. The latter is the homology of the complex $K_\bullet(\mathbf{a}) \otimes M$. Since the tensor product commutes with direct limits, it is enough to study the complex $\varinjlim_n K_\bullet(\mathbf{a}^n)$. By Corollary 7.1.10,

$$K_\bullet(a_1, \ldots, a_n) \cong K_\bullet(a_1) \otimes \cdots \otimes K_\bullet(a_n),$$

so it will be enough to understand the behavior of $\varinjlim_n K_\bullet(a^n)$ for a single element $a \in A$.

Lemma 10.5.11. *Let A be a ring, $a \in A$. The limit $\varinjlim_n K_\bullet(a^n)$ is the complex*

$$0 \longrightarrow A \longrightarrow A_a \longrightarrow 0$$

given by the localization map at powers of a.

Proof. Let us picture two consecutive elements in the direct system. By Example 7.1.6(a), these are

$$
\begin{array}{ccccccccc}
0 & \longrightarrow & A & \xrightarrow{\ \cdot a^n\ } & A & \longrightarrow & 0 \\
& & \mathrm{id}_A \downarrow & & \downarrow \cdot a & & \\
0 & \longrightarrow & A & \xrightarrow{\ \cdot a^{n+1}\ } & A & \longrightarrow & 0.
\end{array}
$$

The direct limit is then

$$
0 \longrightarrow A \longrightarrow B \longrightarrow 0,
$$

where B is the direct limit of the diagram

$$(10.5.1) \qquad A \xrightarrow{\ \cdot a\ } A \xrightarrow{\ \cdot a\ } A \xrightarrow{\ \cdot a\ } \cdots .$$

It remains to identify this direct limit with the localization A_a. Consider the map

$$f_i \colon\ A \longrightarrow A_a$$

$$x \longmapsto \frac{x}{a^i}$$

and regard A_a as a cocone over the diagram (10.5.1), where the i-th copy of A is sent to A_a via f_i. Then the universal property of localization translates into the universal property of this cocone, making A_a into the direct limit of the diagram, as we wanted to prove. $\qquad\qquad\square$

To generalize this to more than one element, it is convenient to introduce some notation.

Definition 10.5.12. Let A be a ring, $\mathbf{a} = a_1, \ldots, a_n \in A$, and let M be an A-module. The *Čech complex* (or *stable Koszul complex*) associated to $\mathbf{a}$ and M, denoted $C_{\mathbf{a}}(M)$, is the complex

$$0 \to M \to \bigoplus_{i_0} M_{a_{i_0}} \to \bigoplus_{i_0 < i_1} M_{a_{i_0} a_{i_1}} \to \cdots \to M_{a_1 \ldots a_n} \to 0,$$

where in the k-th position we have the direct sum of all localizations of M at powers of a product of k distinct elements among the a_i. The maps in the complex are induced from maps

$$M_{a_{i_0} \cdots a_{i_k}} \to M_{a_{j_0} \cdots a_{j_{k+1}}},$$

which are the natural maps given by localization in the case that $\{i_0, \ldots, i_k\} \subset \{j_0, \ldots, j_{k+1}\}$, and 0 otherwise. If $\{j_0, \ldots, j_{k+1}\}$ is obtained by $\{i_0, \ldots, i_k\}$ through the addition of j_r, we add a sign $(-1)^r$ to the localization map; this makes $C_{\mathbf{a}(M)}$ into a complex.

Remark 10.5.13. It is immediate to verify that $C_{\mathbf{a}}(M) \cong C_{\mathbf{a}}(A) \otimes M$. With a little more effort, one can check, using the definition of the tensor product of complexes in Definition 7.1.7, that

$$C_{\mathbf{a}'} \cong C_{\mathbf{a}} \otimes C_{a_{k+1}}$$

for sequences $\mathbf{a} = a_1, \ldots, a_k$ and $\mathbf{a}' = a_1, \ldots, a_{k+1}$. By induction, it follows that

$$C_{\mathbf{a}} \cong C_{a_1} \otimes \cdots \otimes C_{a_k}.$$

Notice that this is the same kind of recursive structure that one has for the Koszul complex; see Corollary 7.1.10.

Proposition 10.5.14. Let A be a ring, $\mathbf{a} = a_1, \ldots, a_k \in A$. There is a natural isomorphism

$$\varinjlim_{n} K_{\bullet}(\mathbf{a}^n) \cong C_{\mathbf{a}}.$$

Proof. Using the decomposition as a tensor product in the above remark and Corollary 7.1.10, one reduces to the case of a single element. Here we make use again of the fact that tensor product commutes with direct limits. The case of a single element is Lemma 10.5.11. $\qquad \square$

Corollary 10.5.15. Let A be a Noetherian ring, $\mathbf{a} = a_1, \ldots, a_k$. Denote $I = (a_1, \ldots, a_k)$. Then there is a natural isomorphism

$$H_I^i(M) \cong H_i(C_{\mathbf{a}}(M)).$$

Proof. Combine Propositions 10.5.10 and 10.5.14, and use the fact that taking homology commutes with direct limits (either by direct verification or by Theorem 1.4.16). $\qquad \square$

10.6. Properties of local cohomology

We keep our study of local cohomology, starting with some simple results on change of rings.

Proposition 10.6.1. Let $f \colon A \to B$ be a morphism of Noetherian rings, M an A-module, N a B-module and $I \subset A$ an ideal.

   (i) *There is an isomorphism of A-modules $H_I^i(N) \cong H_{I \cdot B}^i(N)$.*
   (ii) *If f is flat, there is an isomorphism of B-modules $H_I^i(M) \otimes_A B \cong H_{I \cdot B}^i(M \otimes_A B)$.*

Proof. Both claims will follow from Proposition 10.5.10. Write $I = (\mathbf{a})$, where $\mathbf{a} = a_1, \ldots, a_n$. It will be enough to check the analogous claims for the Koszul cohomology $H^i(\mathbf{a}^n, M)$ for $n > 0$. Let $b_i = f(a_i)$ and denote $\mathbf{b} = b_1, \ldots, b_n$.

The first claim follows by the isomorphisms

$$K^\bullet(\mathbf{a}^n; N) = \operatorname{Hom}_A(K_\bullet(\mathbf{a}^n), N) \cong \operatorname{Hom}_B(K_\bullet(\mathbf{b}^n), N) = K^\bullet(\mathbf{b}^n; N).$$

The second claim is similar, using

$$K^\bullet(\mathbf{a}^n; M) \otimes_A B = \operatorname{Hom}_A(K_\bullet(\mathbf{a}^n), M) \otimes_A B \cong$$
$$\cong \operatorname{Hom}_B(K_\bullet(\mathbf{b}^n), M \otimes B) = K^\bullet(\mathbf{b}^n; M \otimes B).$$

Since B is A-flat, tensoring with B is exact, hence the conclusion follows. □

Since localization is flat by Proposition 6.2.3, we conclude:

Corollary 10.6.2. *If S is a multiplicative set in the Noetherian ring A, then $S^{-1}H_I^i(M) \cong H_{S^{-1}I}^i(S^{-1}M)$ for all ideals I and A-modules M.*

Similarly for completion:

Corollary 10.6.3. *Let A be a local Noetherian ring, M a finitely generated A-module. For all $i \in \mathbb{N}$ we have $H_{\mathcal{M}}^i(M) \cong H_{\widehat{\mathcal{M}}}^i(\widehat{M})$, where $\widehat{-}$ denotes completion with respect to the $\mathcal{M}$-adic topology.*

Proof. By Proposition 10.6.1,

$$H_{\widehat{\mathcal{M}}}^i(\widehat{M}) \cong H_{\mathcal{M}\widehat{A}}^i(M \otimes_A \widehat{A}) \cong H_{\mathcal{M}}^i(M) \otimes_A \widehat{A}.$$

To remove the outer tensor product, notice that

$$\varinjlim_n \operatorname{Ext}_A^i(A/\mathcal{M}^n, M) \otimes \widehat{A} \cong \varinjlim_n \operatorname{Ext}_A^i(A/\mathcal{M}^n, M)$$

since $\operatorname{Ext}_A^i(A/\mathcal{M}^n, M)$ is annihilated by $\mathcal{M}^n$, and then use Remark 10.5.9. □

Our next result is an analogue of the usual Mayer–Vietoris exact sequence in topology. Let $I \subset J$ be ideals of A. Then for each A-module M we have a natural inclusion $\iota_{J,I} \colon \Gamma_J(M) \hookrightarrow \Gamma_I(M)$, which in turn induces natural transformations $H_J^i \to H_I^i$ for all i.

Lemma 10.6.4. *Let A be a Noetherian ring, I, J ideals of A. Then the sequence*

$$0 \longrightarrow \Gamma_{I+J}(M) \longrightarrow \Gamma_I(M) \oplus \Gamma_J(M) \longrightarrow \Gamma_{I \cap J}(M) \longrightarrow 0$$

is left exact for all A-modules M, and exact when M is injective.

Here the first map is $(\iota_{I+J,I}, \iota_{I+J,J})$, while the second one is $\iota_{I,I \cap J} - \iota_{J,I \cap J}$.

Proof. The first claim is a simple diagram chasing. To prove exactness on the right when M is injective, we can assume by Theorem 5.3.9 that $M = E_A(P)$ for some prime ideal $P \subset A$, in which case the conclusion follows by Example 10.5.7(d). □

Given an injective resolution $M \to E_\bullet$, the above lemma yields a short exact sequence of complexes

$$0 \longrightarrow \Gamma_{I+J}(E_\bullet) \longrightarrow \Gamma_I(E_\bullet) \oplus \Gamma_J(E_\bullet) \longrightarrow \Gamma_{I \cap J}(E_\bullet) \longrightarrow 0 \ .$$

Taking the long exact sequence in homology immediately gives us the *Mayer–Vietoris* theorem.

Theorem 10.6.5. *Let A be a Noetherian ring, I, J ideals of A. Then for all A-modules M we have the long exact sequence*

$$0 \longrightarrow H^0_{I+J}(M) \longrightarrow H^0_I(M) \oplus H^0_J(M) \longrightarrow H^0_{I \cap J}(M)$$

$$\hookrightarrow H^1_{I+J}(M) \longrightarrow H^1_I(M) \oplus H^1_J(M) \longrightarrow H^1_{I \cap J}(M) \longrightarrow \cdots .$$

We next turn to some vanishing results. Vanishing of local cohomology with support in I is strictly related to the number of generators of I, up to radical.

Definition 10.6.6. Let $I \subset A$ be an ideal. The least number n such that there exist $a_1, \ldots, a_n \in A$ with $\sqrt{I} = \sqrt{(a_1, \ldots, a_n)}$ is called the *arithmetic rank* of I, denoted ar I. If no such n exists, we put ar $I = \infty$.

Remark 10.6.7. By Krull's principal ideal theorem (Theorem A.7.6), if A is Noetherian, then $\mathrm{ar}(I) \geq \mathrm{ht}(I)$ for all ideals I.

Proposition 10.6.8. *Let $I \subset A$ be an ideal with ar $I = n$, M an A-module. Then $H^i_I(M) = 0$ for $i > n$.*

Proof. Since $H^i_I(M) = H^i_{\sqrt{I}}(M)$, we can assume that I is generated by n elements, in which case this follows immediately from Proposition 10.5.10, since a Koszul complex for n elements has length n. $\qquad \square$

In itself, this would not be of much use, unless we are able to show that the arithmetic rank is reasonably well behaved.

Lemma 10.6.9. *Let A be a Noetherian ring, $I \subset A$ an ideal. For all integers $r \geq 1$, there exists elements $a_1, \ldots, a_r \in I$ such that, given a prime P with $\mathrm{ht}\, P \leq r - 1$, we have $P \supset I$ if and only if $P \supset (a_1, \ldots, a_r)$.*

Proof. By induction on r. For $r = 1$, consider all minimal primes P such that $I \not\subset P$. Since A is Noetherian, there are finitely many of them by primary decomposition, and by prime avoidance we find $a_1 \in I$ such that $a_1 \notin P$ for all such primes P, hence the claim holds.

Assume that we have chosen $a_1, \ldots a_{r-1} \in I$ that satisfy the claim. We proceed in a similar way, by considering the finitely many primes P of height $r - 1$ that are minimal over $(a_1, \ldots, a_{r-1})$ and such that $I \not\subset P$. For the same reasons as above, we can choose $a_r \in I$ that does not lie in any such prime. Then, we argue as above in the ring $A/(a_1, \ldots, a_{r-1})$ to see that $a_1, \ldots, a_r$ satisfy the claim.   □

Proposition 10.6.10. *Let A be a Noetherian ring of dimension d, $I \subset A$ an ideal. Then $\operatorname{ar} I \leq d + 1$. If moreover A is local, $\operatorname{ar} I \leq d$.*

Proof. Every prime in A has height at most d, so by Lemma 10.6.9 we find $a_1, \ldots, a_{d+1} \in I$ such that a prime P contains I if and only if it contains $(a_1, \ldots, a_{d+1})$. Since $\sqrt{I}$ is the intersection of all primes containing I, it follows that $\sqrt{I} = \sqrt{(a_1, \ldots, a_{d+1})}$.

In the local case, there is a single maximal ideal, hence the first d elements $a_1, \ldots, a_d$ already satisfy the property that a prime P contains I if and only if it contains $(a_1, \ldots, a_d)$. Hence, in this case, $\operatorname{ar} I \leq d$.   □

Corollary 10.6.11. *Let A be a Noetherian ring, $I \subset A$ an ideal and M an A-module. Then $H_I^i(M) = 0$ for $i > \dim A$.*

Proof. By Corollary 10.6.2, it is enough to consider the case where A is local. But in the local case, $\operatorname{ar} I \leq \dim A$, hence the result follows from Proposition 10.6.8.   □

If $J = \operatorname{Ann}(M) \neq 0$, M can also be seen as a module over A/J. Combining the above result with Proposition 10.6.1, we immediately obtain the following variant.

Corollary 10.6.12. *Let A be a Noetherian ring, $I \subset A$ an ideal and M a finitely generated A-module. Then $H_I^i(M) = 0$ for $i > \dim M$.*

We have also a result on depth sensitivity of local cohomology, akin to Theorem 7.3.8 for Koszul cohomology and Rees's theorem (Theorem 7.3.12) for the Ext functors.

Theorem 10.6.13. *Let A be a Noetherian ring, $I \subset A$ an ideal and M an A-module. Then*

$$\min \left\{ i \mid H_I^i(M) \neq 0 \right\} = \min \left\{ i \mid \operatorname{Ext}_A^i(A/I, M) \neq 0 \right\}.$$

Proof. Consider the functors $F, G \colon \operatorname{Mod}_A \to \operatorname{Mod}_A$ given by $F = \Gamma_I$ and $G = \operatorname{Hom}(A/I, -)$. Given an A-module N, a homomorphism $A/I \to N$ takes values in $\Gamma_I(N)$ by definition—in other words, $G \circ F = G$. This prompts to use the Grothendieck spectral sequence from Theorem 4.5.2. All we have to check is that $F(E)$ is G-acyclic when E is injective, which is Proposition 10.5.8. We thus obtain a spectral sequence

$$E_{p,q}^2 = \mathrm{Ext}^q\left(\frac{A}{I}, H_I^p(M)\right) \implies \mathrm{Ext}^{p+q}\left(\frac{A}{I}, M\right)$$

for all A-modules M. Now if $H_I^p(M) = 0$ for all $p < k$, then $\mathrm{Ext}^p(A/I, M) = 0$ for all such p as well. Morever, if p_0 is the minimum value for which $H_I^{p_0}(M) \neq 0$, then $\mathrm{Hom}(A/I, H_I^{p_0}(M)) \neq 0$ by Remark 10.5.6, and by the spectral sequence again we conclude that $\mathrm{Ext}^{p_0}(A/I, M) \neq 0$. $\square$

Together with Theorem 7.3.12, this immediately gives:

Corollary 10.6.14. *Let A be a Noetherian ring, $I \subset A$ an ideal and M a finitely generated A-module. Assume that $IM \neq M$. Then*

$$\mathrm{depth}_I(M) = \min\left\{i \mid H_I^i(M) \neq 0\right\}.$$

Corollary 10.6.15. *Let $A, \mathcal{M}$ be a local Noetherian ring, M a finitely generated A-module. Then M is Cohen–Macaulay if and only if we have $H_{\mathcal{M}}^i(M) = 0$ for all i except $i = \dim M$, while $H_{\mathcal{M}}^{\dim M}(M) \neq 0$.*

Proof. Combine Corollary 10.6.14 and Corollary 10.6.12. $\square$

We will see in Corollary 10.7.5 that $H_{\mathcal{M}}^{\dim M}(M) \neq 0$ anyway, even if M is not Cohen–Macaulay. A similar characterization exists for Gorenstein rings.

Proposition 10.6.16. *Let $A, \mathcal{M}$ be a local Noetherian ring of dimension d, $k = A/\mathcal{M}$. Then A is Gorenstein if and only if $H_{\mathcal{M}}^i(A) = 0$ for $i \neq d$ and $H_{\mathcal{M}}^d(A) \cong E(k)$, the injective hull of k.*

Proof. By Corollary 10.6.15, we can assume that A is CM, and the only nonvanishing local cohomology group is $H_{\mathcal{M}}^d(A)$. In particular, $\mathrm{Ext}_A^i(k, A) = 0$ for $i < d$. Recall from Proposition 9.7.11 that in this case A is Gorenstein if and only if $\mathrm{Ext}_A^d(k, A) \cong k$.

Now use the spectral sequence from the proof of Theorem 10.6.13, which gives us the isomorphism

$$\mathrm{Hom}(k, H_{\mathcal{M}}^d(A)) \cong \mathrm{Ext}^d(k, A).$$

If $H_{\mathcal{M}}^d(A) \cong E(k)$, the left-hand side is $\mathrm{Hom}(k, E(k)) \cong \mathrm{soc}\, E(k) \cong k$. So $H_{\mathcal{M}}^d(A) \cong E(k)$ implies that A is Gorenstein.

Conversely, assume that A is Gorenstein, and take an injective resolution $A \to I_\bullet$. By Proposition 3.3.2, $H_{\mathcal{M}}^d(A) \cong H_{\mathcal{M}}^0(I_d)$, which by Example 10.5.7(d) must be $E(k)^m$ for some $m \geq 1$. If $m > 1$, then $\dim_k \mathrm{Ext}^d(k, A) > 1$ as well, which is a contradiction. $\square$

10.7. Grothendieck local duality

We now have all the ingredients necessary to prove local duality, a fundamental result that links local cohomology to the Ext groups. This form of duality will involve the canonical module ω_A of a local Cohen–Macaulay ring A, and in particular will take a simple form when A is Gorenstein, in which case $\omega_A \cong A$.

It turns out that it is easier to prove the result Gorenstein rings *first*, and then generalize this to CM rings making use of Corollary 10.4.11 and Theorem 10.4.13, which allow us to express the canonical module of a CM ring in terms of Ext groups involving a Gorenstein ring that maps surjectively onto it.

Theorem 10.7.1 (Local duality for Gorenstein rings). *Let $A, \mathcal{M}$ be a local Gorenstein ring of dimension d, and denote by $-^\vee$ the Matlis dual functor. For a finitely generated A-module M and $0 \leq i \leq d$ there is an isomorphism, natural in M,*

$$H_{\mathcal{M}}^i(M) \cong \mathrm{Ext}_A^{d-i}(M, A)^\vee.$$

If moreover A is complete, $H_{\mathcal{M}}^i(M)^\vee \cong \mathrm{Ext}_A^{d-i}(M, A)$.

This allows us to compute local cohomology, which is a quite subtle invariant, and in general is not even finitely generated, in terms of the more familiar Ext groups.

Proof. The last statement follows from Matlis duality, hence we only need to prove the first one. Denote by E the Matlis dual of A, and choose a system of parameters $\mathbf{a} = a_1, \ldots, a_d$.

Consider the Čech complex $C_{\mathbf{a}}$. By Corollary 10.5.15 we can compute $H_{\mathcal{M}}^i(M)$ as the Čech cohomology $H_i(C_{\mathbf{a}}(M))$. Since A is Gorenstein, we have depth $A = d$ by Theorem 9.7.4, hence $H_i(C_{\mathbf{a}}) = 0$ for $i \neq d$, while $H_d(C_{\mathbf{a}}) \cong E$ by Proposition 10.6.16.

In other words, $C_{\mathbf{a}}$ is a finite left resolution of E. In addition, each module in $C_{\mathbf{a}}$ is a localization of A, hence flat by Proposition 6.2.3. We can then use this complex to compute the Tor functors of E, and in particular obtain

$$H_{\mathcal{M}}^i(M) \cong H_i(C_{\mathbf{a}}(M)) \cong \mathrm{Tor}_{d-i}^A(M, E) \cong \mathrm{Ext}_A^{d-i}(M, A)^\vee,$$

where the last isomorphism is Proposition 10.2.9. □

To extend the result to CM rings, we will just need a simple local cohomology computation.

Corollary 10.7.2. *Let $A, \mathcal{M}$ be a local Cohen–Macaulay ring of dimension d, having a canonical module ω_A. Then $H_{\mathcal{M}}^d(A) \cong \omega_A^\vee$.*

Proof. By Theorem 10.4.13, we have a surjective morphism $f\colon B \to A$, where B is a local Gorenstein ring. Denote $d_A = d$ and $d_B = \dim B$. Similarly, put $\mathcal{M}_A = \mathcal{M}$ and let $\mathcal{M}_B$ be the maximal ideal of B. Notice that $f(\mathcal{M}_B) = \mathcal{M}_A$ and A and B have a common residue field k.

By Corollary 10.4.11, we can compute $\omega_A \cong \operatorname{Ext}_B^{d_B - d_A}(A, B)$. Using Proposition 10.6.1 and local duality for B, we can then compute

$$H_{\mathcal{M}_A}^{d_A}(A) \cong H_{\mathcal{M}_B}^{d_A}(A) \cong \operatorname{Hom}_B(\operatorname{Ext}_B^{d_B - d_A}(A, B), E_B(k)) \cong$$
$$\cong \operatorname{Hom}_B(\omega_A, E_B(k)) \cong \operatorname{Hom}_A(\omega_A, E_A(k)) = \omega_A^\vee.$$

To justify the last step, notice that $E_A(k) \cong \operatorname{Hom}_B(A, E_B(k))$ by Lemma 10.1.3, hence for any A-module M we have

$$\operatorname{Hom}_A(M, E_A(k)) \cong \operatorname{Hom}_A(M, \operatorname{Hom}_B(A, E_B(k))) \cong$$
$$\cong \operatorname{Hom}_B(M, E_B(k))$$

by the adjunction in Example 1.5.2(g).   □

Theorem 10.7.3 (Local duality for CM rings). *Let $A, \mathcal{M}$ be a local Cohen–Macaulay ring of dimension d having a canonical module ω_A, and denote by $-^\vee$ Matlis duality. For a finitely generated A-module M and $0 \le i \le d$ there is an isomorphism, natural in M,*

$$H_{\mathcal{M}}^i(M) \cong \operatorname{Ext}_A^{d-i}(M, \omega_A)^\vee.$$

If moreover A is complete, $H_{\mathcal{M}}^i(M)^\vee \cong \operatorname{Ext}_A^{d-i}(M, \omega_A)$.

Proof. The proof is identical to Theorem 10.7.1, using Corollary 10.7.2 in place of Proposition 10.6.16.   □

We end this section—and the book—with a few simple consequences of local duality.

Corollary 10.7.4. *Let $A, \mathcal{M}$ be a local Noetherian ring, M a finitely generated A-module. Then $H_{\mathcal{M}}^i(M)$ is Artinian for all $i \in \mathbb{N}$.*

Proof. By Corollary 10.6.11, we need only consider $i \le d = \dim A$. If A is Gorenstein, we have $H_{\mathcal{M}}^i(M) \cong \operatorname{Ext}_A^{d-i}(M, A)^\vee$, which is Artinian by Matlis's duality theorem, Theorem 10.2.7.

In the general case, we can assume that A is complete by Corollary 10.6.3. In this case, A is a quotient of a complete regular ring $B, \mathcal{M}_B$ by Cohen's theorem (Theorem A.8.9), and then $H_{\mathcal{M}}^i(M) \cong H_{\mathcal{M}_B}^i(M)$.   □

Corollary 10.7.5. *Let $A, \mathcal{M}$ be a local Noetherian ring, M a finitely generated A-module of dimension d_M. Then $H_{\mathcal{M}}^{d_M}(M) \ne 0$.*

Proof. As in the previous corollary, we can assume that A is complete and, via Cohen's Theorem and Proposition 10.6.1, even that A is regular. In this case, we find $H^d_{\mathcal{M}}(M) \cong \mathrm{Ext}^{d_A - d_M}_A(M, A)^{\vee}$ by Theorem 10.7.1, where $d_A = \dim A$. By Lemma 10.2.3, it is enough to check that $\mathrm{Ext}^{d_A - d_M}_A(M, A) \neq 0$.

Denote $t = d_A - d_M$, and let $I = \mathrm{Ann}\, M$, so that $\mathrm{ht}\, I = t$ by Proposition 9.2.7. By Proposition 9.1.8, we have $\mathrm{depth}_I A = t$ as well, so that we can take a maximal regular sequence $\mathbf{a} = a_1, \ldots, a_t$ in I. The quotient $A/(\mathbf{a})$ is a complete intersection, hence Gorenstein, and M can be regarded as an $A/(\mathbf{a})$-module. Up to replacing A by $A/(\mathbf{a})$, we can then assume that $d_M = d_A$.

Under this assumption, it remains to check that $\mathrm{Hom}_A(M, A) \neq 0$. Let P be a minimal prime of A in the support of M, which exists since $\dim M = \dim A$. We are going to check that

$$(\mathrm{Hom}_A(M, A))_P = \mathrm{Hom}_{A_P}(M_P, A_P) \neq 0.$$

The ring A_P is still Gorenstein (why?) and has dimension 0. Moreover, $M_P \neq 0$ by our choice of P, hence $\mathrm{Hom}_{A_P}(M_P, A_P) \neq 0$ by Matlis duality. $\qquad\square$

10.8. Exercises

1. Give a direct proof of Corollary 10.6.2, using the definition of H^i_I as derived functors of Γ_I.

2. Let A be a local ring having a canonical module ω_A. Prove that the completion $\widehat{\omega_A}$ is a canonical module for $\widehat{A}$.

3. Let A be a local ring having a canonical module ω_A. Prove that for a prime ideal $P \subset A$, the localization $(\omega_A)_P$ is a canonical module for A_P.

4. Let $f : A \to B$ be a surjective local map of Noetherian rings, where A is Gorenstein and B is Cohen–Macaulay. Show that ω_B, the canonical module of B, is in fact a dualizing module for B, regarded as an A-module, in the sense of Section 10.3.

5. Let A be a local CM ring of dimension d, M a CM A-module of dimension r. Assume that A admits the canonical module ω_A. Prove that $\mathrm{Ext}^i(M, \omega_A) = 0$ for $i \neq d - r$, while $\omega_M := \mathrm{Ext}^{d-r}(M, \omega_A)$ is again a CM module of dimension r. Moreover, $\omega_{\omega_M} \cong M$. Link this to the results in Section 10.3.

6. Let $A, \mathcal{M}$ be a local Noetherian ring, $\mathbf{a} \subset \mathcal{M}$ a regular sequence. Use local cohomology to give an alternative proof that $A/\mathbf{a}$ is CM (resp., Gorenstein) if and only if A is CM (resp., Gorenstein).

7. Let $A, \mathcal{M}$ be a local CM ring of dimension d. Prove directly that $H^d_{\mathcal{M}}(A) \neq 0$ by exhibiting a nonzero element, using the identification with the cohomology of the Čech complex of a system of parameters.

8. Let A be a Noetherian ring, $I \subset A$ an ideal, $a \in A$. Show that for all finitely generated A-modules M there is a long exact sequence

$$0 \longrightarrow H^0_{(I,a)}(M) \longrightarrow H^0_I(M) \longrightarrow H^0_I(M_a) \longrightarrow H^1_{(I,a)}(M) \longrightarrow \cdots .$$

(Show that for E an injective A-module there is a split short exact sequence

$$0 \longrightarrow \Gamma_a(E) \longrightarrow E \longrightarrow E_a \longrightarrow 0 ,$$

then apply Γ_I.)

9. Let k be a field, $I \subset A := k[x, y, z, w]$ the ideal of two skew lines, say $I = (x, y) \cap (z, w) = (xz, xw, yz, yw)$. Show that I is generated up to radical by 3 elements. Use the Mayer-Vietoris sequence for I to compute that $H^3_I(A) \neq 0$. Conclude that $\operatorname{ar}(I) = 3$, so I cannot be generated up to radical by two equations, even though $\operatorname{ht} I = 2$.

10. Let $A, \mathcal{M}$ a local Noetherian ring, M, N two finitely generated A-modules. Show that there exists a spectral sequence

$$E^2_{p,q} = \operatorname{Ext}^p_A(M, H^q_{\mathcal{M}}(N)) \implies \operatorname{Ext}^{p+q}_A(M, N).$$

(Use the double complex $C_\bullet(\mathbf{a}) \otimes \operatorname{Hom}(F_\bullet, N))$, where $C_\bullet(\mathbf{a})$ is the Čech complex of the system of parameters $\mathbf{a}$ and $F_\bullet \to M$ is a minimal free resolution.)

11. Let $A, \mathcal{M}$ a local CM ring of dimension d with the canonical module ω_A, M, N two finitely generated A-modules. Combine the previous exercise with local duality to show that there exists a spectral sequence

$$E^2_{p,q} = \operatorname{Tor}^A_p(M, \operatorname{Ext}^{d-q}_A(N, \omega_A)) \implies \operatorname{Ext}^{p+q}_A(M, N).$$

12. Let $A \subset B$ be rings. Define the *conductor* ideal

$$\mathfrak{f}(B/A) := \operatorname{Ann}_A \frac{B}{A}.$$

Show that $\mathfrak{f}(B/A)$ is the largest set that is a common ideal of *both* A and B.

13. Let A be a local Gorenstein domain of dimension 1, and B its integral closure. Assume that B is finitely generated as an A-module. Show that B/A and $A/\mathfrak{f}(B/A)$ have finite length, and in fact

$$\ell\left(\frac{B}{A}\right) = \ell\left(\frac{A}{\mathfrak{f}(B/A)}\right).$$

(Consider the exact sequence

$$0 \longrightarrow A \longrightarrow B \longrightarrow B/A \longrightarrow 0$$

and apply $\mathrm{Hom}_A(-, A)$ to deduce $A/\mathfrak{f}(B/A) \cong \mathrm{Ext}^1_A(B/A, A)$.)

Exercises 14–17, prove a theorem of Murthy [**Mur64**] that links unique factorization to the Gorenstein property.

14. Let A be a regular local ring, M a CM A-module with $\mathrm{pd}_A M = h$. Prove that $\mathrm{Ext}^i_A(M, A) = 0$ for $i < h$ and $N := \mathrm{Ext}^h_A(M, A)$ is a CM A-module with $\mathrm{pd}_A N = h$.

15. Let A be a regular local ring, M a finitely generated A-module and P an associated prime of M. Prove that $\mathrm{ht}\, P \le \mathrm{pd}_A M$.

16. Let A be a regular local ring, $P \subset A$ a prime ideal, and assume that $B := A/P$ is a Cohen–Macaulay ring. Let $r = \mathrm{ht}\, P$. Prove that $M := \mathrm{Ext}^r_A(B, A)$ is either isomorphic to B or to an unmixed ideal $I \subset B$ of height 1. If moreover B is a UFD, then $M \cong B$ in any case. (By induction on $\dim B$, prove that M is torsion free and has rank 1, so it is isomorphic to an ideal $I \subset B$. If $J \subset A$ is the corresponding ideal of A, prove that $\mathrm{pd}_A A/J \le r + 1$, and use the previous exercise.)

17. Prove the following theorem of Murthy: let A be a local CM ring that is a quotient of a regular local ring. If A is a UFD, then A is Gorenstein.

18. Let A be a local CM ring having a canonical module ω_A. Prove that the functor $D := \mathrm{Hom}(-, \omega_A)$ is an exact functor from the category of maximal CM modules to itself, and satisfies $D^2 \cong \mathrm{id}$.

19. Use Matlis duality to give an alternative proof of one implication in Theorem 9.6.2—namely, if A is a local Noetherian ring that admits a finitely generated injective module, then A is Artinian. (Show that the Matlis dual of A is finitely generated.)

20. Let A be a local CM ring. Prove that M is a canonical module for A if and only if it is a maximal CM module satisfying $\mathrm{Hom}_A(M, M) \cong A$.

21. Let $B = A/I$, where A is a regular local ring, and set $d = \dim B$. Consider a minimal free resolution $F_\bullet \to B$ as A-modules. Show that B is Gorenstein if and only if $F_i = 0$ for $i > d$, while $F_d \cong B$.

22. Let B be as in the previous exercise, and assume that B is indeed Gorenstein. Show that the free resolution $F_\bullet$ is symmetric, in the sense that $\mathrm{rk}\, F_k = \mathrm{rk}\, F_{d-k}$ for all $k = 0, \ldots, d$.

Consider the ring A of Example 9.8.9(b), which is Gorenstein but not a complete intersection. It is presented as $A = B/I$, where B is local regular (namely B is the localization in 0 of $k[x, y, z]$). In the next two exercises, following [**Hun98**], we want to show that this can only happen if ht $I \geq 3$— in particular we could not produce such an example from $B = k[x, y]$.

23. Let B be a regular local ring and $I \subset B$ an ideal such that B/I is Gorenstein but not a complete intersection. Show that I cannot have height 0 or 1. (For the case of height 1, use the fact that A is unmixed and B a UFD.)

24. Let B be a regular local ring and $I \subset B$ an ideal such that B/I is Gorenstein but not a complete intersection. Show that I cannot have height 2. (Consider a minimal free resolution $F_\bullet \to A$, and notice that it is symmetric. Counting ranks, deduce that $F_1 \cong B^2$ and conclude that A is a complete intersection.)

Background Material

This appendix recalls some facts about rings and modules that are assumed in the body of the text. As with the remainder of this text, all rings throughout this appendix are assumed to be commutative. The appendix is meant just for convenience of reference, and for reasons of space we do not give proofs or discuss the motivation behind the definitions—in fact, we only survey the results that are explicitly used in the book. The reader that is not familiar with this material should consult either the first volume [**Fer20**], or another reference, such as [**AM69**] or [**AK13**].

A.1. Basics

Here we review some results that are used often and are so common that they are not explictly referenced. Prime avoidance appears as Proposition 1.1.20 in [**Fer20**] or 1.11 in [**AM69**].

Theorem A.1.1 (Prime avoidance). *Let* $P_1, \ldots, P_n \subset A$ *be prime ideals and* $I \subset A$ *be any ideal. If*

$$I \subset P_1 \cup \cdots \cup P_n,$$

then $I \subset P_k$ *for some* k.

We will also use the following fact, that is an immediate application of Zorn's lemma. This is Proposition 2.3.7 in [**Fer20**] (or see the discussion in Section 3.1 of [**Eis95**]).

Theorem A.1.2. *Let* A *be a ring,* $P \subset A$ *an ideal. Then* P *contains a minimal prime of* A.

For a ring A, we define the *Jacobson radical* $\mathcal{J}(A)$ as the intersection of all maximal ideals of A. If A has a single maximal ideal $\mathcal{M}$, we say that A is *local*. We often denote A by $A, \mathcal{M}$ for brevity. Notice that some authors assume that local rings are Noetherian, but we do not do so. The following is Theorem 1.3.19 in [**Fer20**] or Proposition 2.6 in [**AM69**].

Theorem A.1.3 (Nakayama's lemma). *Let M be a finitely generated A-module, $J = \mathcal{J}(A)$ the Jacobson radical. If $JM = M$, then $M = 0$.*

This is often applied in other forms. For instance, if $N \subset M$ is a submodule and $M = JM + N$, then $M = N$, as one can see by applying the result to M/N. In particular, if $S \subset M$ is a subset whose image generates M/JM, then S generates M (let N be the submodule generated by S).

A.2. Finiteness conditions

An A-module M is called Noetherian (resp., Artinian) if every infinite increasing (resp., decreasing) chain of submodules in M is eventually constant. Hilbert basis theorem (Theorem 2.3.3 in [**Fer20**] or 7.5 in [**AM69**]) states:

Theorem A.2.1 (Hilbert basis theorem). *Let A be a ring. Then A is Noetherian if and only if the polynomial ring $A[x]$ is Noetherian.*

A similar result holds for the power series ring $A[[x_n]]$ (Theorem 7.5.22 in [**Fer20**] or Corollary 10.27 in [**AM69**]):

Theorem A.2.2. *Let A be a ring. Then A is Noetherian if and only if the power series ring $A[[x]]$ is Noetherian.*

If M is an A-module that is both Artinian and Noetherian, every chain of distinct submodules of M is finite. The following is Theorem 2.5.6 in [**Fer20**] or Proposition 6.7 in [**AM69**].

Theorem A.2.3. *If M is Artinian and Noetherian, every maximal chain of submodules of M has the same length.*

We define the length of M, $\ell(M)$, as the length of any such maximal chain. By definition, if

$$0 \longrightarrow M_1 \longrightarrow M_2 \longrightarrow M_3 \longrightarrow 0$$

is a short exact sequence, we have $\ell(M_2) = \ell(M_1) + \ell(M_3)$.

A principal ideal domain (PID) is a ring such that all its ideals are principal; in particular it is Noetherian. By Theorem 2.1.5 of [**Fer20**] or 7.1 in [**Lan02**], we have:

Theorem A.2.4. *Let A be a PID, F a free A-module. Then any submodule of F is free.*

In the finitely generated case, this immediately implies Theorem 2.1.7 of [**Fer20**], or 7.3 and 7.7 in [**Lan02**].

Theorem A.2.5. *Every finitely generated module over a principal ideal domain A is isomorphic to*

$$A^r \oplus A/(a_1) \oplus \cdots \oplus A/(a_k)$$

for a suitable choice of $r \in \mathbb{N}$ and $a_j \in A$.

A.3. Primary decomposition

Let $I, J_1, \ldots, J_k$ be ideals of the ring A. If

(A.3.1) $$I = J_1 \cap \cdots \cap J_k,$$

we call (A.3.1) a *decomposition* of I. This decomposition is called *irredundant* if none of the ideals J_i can be omitted. An ideal I that does not admit nontrivial decompositions is called *irreducible*. An ideal $Q \subset A$ is called *primary* if for all $a, b \in A$, if $ab \in Q$, then $a \in Q$ or $b \in \sqrt{Q}$. In this case, the ideal $P = \sqrt{Q}$ is a prime ideal. Combining Proposition 3.2.10, Corollary 3.2.13 and Theorem 3.2.19 in [**Fer20**], or Theorem 4.5 and Lemma 7.11 and 7.12 in [**AM69**], we get the following result.

Theorem A.3.1. *Let A be a Noetherian ring. Then*

(i) *every irreducible ideal of A is primary;*

(ii) *every ideal $I \subset A$ admits a decomposition by irreducible (hence primary) ideals; and*

(iii) *if*

(A.3.2) $$I = Q_1 \cap \cdots \cap Q_k,$$

is a primary, irredundant decomposition if I, where the primes $P_i = \sqrt{Q_i}$ are all distinct, then the set $\{P_i\}$ is uniquely determined by I.

Definition A.3.2. Let M be an A-module. The support of M, denoted $\mathrm{Supp}(M)$, is the set of primes $P \subset A$ such that $M_P \neq 0$, or equivalently $\mathrm{Ann}(M) \subset P$. A prime ideal P is called *associated* to M if $P = \mathrm{Ann}(m)$ for some $m \in M$. We denote by $\mathrm{Ass}(M)$ the set of associated primes.

For an ideal I, by a prime associated to I we mean an element of $\mathrm{Ass}(A/I)$, where A/I (and not I) is regarded as an A-module. We can then complement the above result with Corollary 3.2.20 from [**Fer20**] or Proposition 7.17 from [**AM69**].

Theorem A.3.3. *Let A be a Noetherian ring, $I \subset A$ an ideal. Then the primes $P_i = \sqrt{Q_i}$ in an irredundant primary decomposition (A.3.2) are exactly the primes associated to I.*

The relation between the support and the set of associated primes is described by the following result from Lemma 3.3.4 and Corollaries 2.5.14 and 3.3.15 in [**Fer20**]. Alternatively, (17.3), (17.13), and (17.18) in [**AK13**].

Theorem A.3.4. *For every A-module M we have $\mathrm{Ass}(M) \subset \mathrm{Supp}(M)$. If moreover A is Noetherian, $\mathrm{Ass}(M) \neq \emptyset$. If in addition M is finitely generated, then the minimal elements of the sets $\mathrm{Ass}(M)$ and $\mathrm{Supp}(M)$ are the same.*

The minimal elements of either $\mathrm{Ass}(M)$ or $\mathrm{Supp}(M)$ are called the minimal primes of M. For an ideal I, these are the minimal elements in the set of primes containing I. In particular, the minimal primes of 0 in A are the primes that are minimal in the ring A. One of the advantages of this relation is that support behaves nicely in exact sequences, Proposition 3.3.3 in [**Fer20**], or (13.27) in [**AK13**].

Theorem A.3.5. *Let A be a Noetherian ring and*

$$0 \longrightarrow M_1 \longrightarrow M_2 \longrightarrow M_3 \longrightarrow 0$$

an exact sequence of A-modules. Then $\mathrm{Supp}(M_2) = \mathrm{Supp}(M_1) \cup \mathrm{Supp}(M_3)$.

Associated primes can be used to describe the divisors of 0 in M. By Corollaries 3.3.13 and 3.3.14 in [**Fer20**] or (17.15) in [**AK13**], we obtain:

Theorem A.3.6. *Let A be a Noetherian ring, M a finitely generated A-module. Denote by $Z(M)$ the set of zero divisors on M, that is*

$$Z(M) = \{a \in A \mid am = 0 \text{ for some } m \neq 0, m \in M\}.$$

Then,

$$Z(M) = \bigcup_{P \in \mathrm{Ass}(M)} P.$$

In particular, the set of zero divisors in A is

$$Z(A) = \bigcup_{P \in \mathrm{Ass}(0)} P.$$

The above result should be compared to the fact that in any ring A, the nilradical $\mathcal{N}(A)$ is the intersection of all primes of A.

Associated primes can also be used to filter a module. The following combines Lemma 2.5.15 and Theorem 2.5.16 in [**Fer20**]; else see (17.20) in [**AK13**].

Theorem A.3.7. *Let A be a Noetherian ring, M a finitely generated A-module. Then there exists a finite chain of submodules*

$$0 = M_0 \subset M_1 \subset \cdots \subset M_k = M$$

such that each quotient $M_{i+1}/M_i \cong A/P_i$ for some prime ideal $P_i \subset A$. Moreover, such primes are exactly the associated primes of M. In particular, M has finite length if and only if each P_i is maximal.

A.4. Integral extensions

Definition A.4.1. Let $A \subset B$ be rings. The element $b \in B$ is said to be *integral* over A if it satisfies a monic polynomial equation with coefficients in A. The subset $\overline{A} \subset B$ of elements that are integral over A is called the *integral closure* of A in B. If $\overline{A} = B$, we say that B is an *integral extension* of A. Conversely, if $\overline{A} = A$, we say that A is *integrally closed* in B. When A is an integral domain with fraction field k, we say that A is integrally closed to mean integrally closed in k.

Integral elements can be characterized in several ways (Theorem 5.1.6 in [**Fer20**] or Proposition 5.1 in [**AM69**]).

Theorem A.4.2. *Let $A \subset B$ be rings, $b \in B$. The following conditions are equivalent:*

(i) *b is integral over A;*

(ii) *the ring $A[b]$ is finitely generated as an A-module;*

(iii) *there exists a ring C containing b, with $A \subset C \subset B$ such that C is finitely generated as an A-module; and*

(iv) *there exists a faithful $A[b]$-module M which is finitely generated as an A-module.*

As a consequence, we have Corollary 5.1.9 of [**Fer20**] or 5.3 of [**AM69**]. This just stems from the observation that if $b, b' \in B$ are integral over A, then $A[b, b']$ is finitely generated as an A-module, and hence $b + b'$ and bb' are integral as well.

Corollary A.4.3. *Let $A \subset B$ be rings. The integral closure of A in B is a subring of B.*

Prime ideals in integral extensions are related by the Going Down Theorem 5.2.13 in [**Fer20**], or 5.16 in [**AM69**].

Theorem A.4.4 (Going down). *Let $A \subset B$ be an integral extension of integral domains, and assume that A is integrally closed.*

Let $P_1 \supsetneq \cdots \supsetneq P_n$ be a chain of primes of A and $Q_1 \supsetneq \cdots \supsetneq Q_m$ ($m \leq n$) a chain of primes of B such that $Q_i \cap A = P_i$ (we say that Q_i lies over P_i). Then we can extend this to a chain $Q_1 \supsetneq \cdots \supsetneq Q_n$ of primes of B such that Q_i lies over P_i for all $i = 1, \ldots, n$.

A.5. Dedekind rings

Definition A.5.1. A *Dedekind ring* is a Noetherian, integrally closed integral domain of dimension 1. In particular, the only prime ideals of A are 0 and the maximal ideals.

Dedekind rings are especially useuful because of prime factorization for ideals. The following appears in [**Fer20**] as Corollary 3.4.8, taking into account also Theorem 5.1.19. Alternatively, it is Corollary 9.4 in [**AM69**]

Theorem A.5.2. *Let A be a Dedekind ring. Then each ideal $I \subset A$ admits a factorization $I = P_1^{e_1} \cdots P_r^{e_r}$, where each P_i is a distinct prime, which is unique up to reordering of the P_i.*

It can be easily shown that principal ideal domains, in particular $\mathbb{Z}$, are Dedekind rings. We can then produce more such rings using the following result, which is a particular case of the Krull–Akizuki theorem, Corollary 5.4.5 in [**Fer20**], or Theorem 11.13 in [**Eis95**].

Theorem A.5.3. *Let A be a Dedekind ring, K its field of fractions. Take a finite extension $K \subset L$ of fields and let $B \subset L$ be the integral closure of A in L. Then B is itself a Dedekind ring.*

A local Dedekind ring is called a *discrete valuation ring* (DVR). Explicitly, this is a local Noetherian domain A whose only prime ideals are 0 and the maximal ideal $\mathcal{M}$. Every other ideal is then a power of $\mathcal{M}$. This immediately implies that $\mathcal{M}$ is principal. If $\mathcal{M} = (a)$, we say that a is a *uniformizer* of A. A ring is Dedekind if and only if all its localizations are DVRs.

If A is a Dedekind ring with fraction field K, we consider finitely generated A-submodules of K. Each such submodule has the form I/a for some ideal $I \subset A$ and $a \in A$—we call it a *fractional ideal*. Fractional ideals form a multiplicative monoid, of which principal fractional ideals (that is, those for which I is principal) are a submonoid. Their quotient is a group, see Corollary 3.4.13 in [**Fer20**] or 9.9 in [**AM69**].

Theorem A.5.4. *Let A be a Dedekind ring. The quotient of fractional ideals of A modulo the principal ones is a group, called the ideal class group, and denoted $G(A)$.*

More explicitly, this means that given an ideal $I \subset A$, there exists an ideal $J \subset A$ such that $I \cdot J$ is principal. We also have some results on finitely generated modules over Dedekind rings. Combining Proposition 3.5.3 and Corollary 3.5.10 in [**Fer20**] (else, see [**Sch03**, Corollary 2.10]), we get:

Theorem A.5.5. *Let A be a Dedekind ring. Then every ideal of A is projective. Moerover, if M is a finitely generated A-module, then $M \cong T \oplus P$, where T is a torsion module and P is projective.*

A.6. Topological rings

We recall a few basic facts about completion from Section 7.5 of [**Fer20**]. Fix a ring A and an ideal I. The I-adic topology on an A-module M is defined by letting $\{I^n \cdot M\}$ be a fundamental system of neighborhoods of $0 \in M$, and defining the neighborhoods of other points by translation. The *completion* of M with respect to this topology is the Abelian group $\widehat{M} := \varprojlim M/I^n M$, with the obvious transition maps. In particular, $\widehat{A}$ inherits by continuity a ring structure, and $\widehat{M}$ is then a module over $\widehat{A}$.

If $M \subset N$, the I-adic topology on M can be different from the subspace topology inherited from the I-adic topology on N. But there is the following fundamental result, Corollary 7.5.15 in [**Fer20**] or Proposition 10.9 in [**AM69**].

Theorem A.6.1 (Artin–Rees). *If A is Noetherian and N is finitely generated, then for every submodule $M \subset N$, the I-adic topology on M and the subspace topology inherited from the I-adic topology on N are the same.*

Combining this with the exactness of completion (for the induced topology on submodules and quotient modules), one immediately obtains the following result (Corollary 7.5.17 in [**Fer20**], Proposition 10.12 in [**AM69**]).

Theorem A.6.2. *Let A be a Noetherian ring, $I \subset A$ an ideal and*

$$0 \longrightarrow M_1 \longrightarrow M_2 \longrightarrow M_3 \longrightarrow 0$$

an exact sequence of finitely generated A-modules. If $\widehat{-}$ denotes completion with respect to the I-adic topology, then

$$0 \longrightarrow \widehat{M_1} \longrightarrow \widehat{M_2} \longrightarrow \widehat{M_3} \longrightarrow 0$$

is exact.

An immediate consequence is Corollary 7.5.18 in [**Fer20**] or Proposition 10.15 in [**AM69**].

Theorem A.6.3. *Let A be a Noetherian ring, $I \subset A$ an ideal, and denote by $\widehat{-}$ the completion in the I-adic topology. Then for all $n \in \mathbb{N}$, $\widehat{I^n} \cong I^n \cdot \widehat{A}$, and there is an isomorphism $I^n/I^{n+1} \cong \widehat{I^n}/\widehat{I^{n+1}}$.*

Let M be an A-module, $I \subset A$ an ideal, and endow M with the I-adic topology. Letting $\widehat{M}$ be the completion, the map $M \to \widehat{M}$ is injective if and only if M is Hausdorff. Explicitly, this amounts to saying that $\bigcap_{n\in\mathbb{N}} I^n M = 0$. As an example, this holds for all A-modules M if $I^n = 0$ for some n. A fundamental criterion on this topic if Krull's intersection theorem, Corollary 7.5.24 in [**Fer20**] or 10.19 in [**AM69**].

Theorem A.6.4. *Let A be a Noetherian ring, $I \subset \mathcal{J}(A)$ an ideal, M a finitely generated A-module. Then $\bigcap_{n\geq 0} I^n \cdot M = 0$.*

As a special case, if A is local with maximal ideal $\mathcal{M}$, then $\bigcap_{n\geq 0} \mathcal{M}^n M = 0$.

A.7. Dimension and height

Definition A.7.1. Let A be a ring. The (Krull) *dimension* of A, denoted $\dim A$, is the supremum of the lengths n of chains of the form

$$P_0 \subsetneq P_1 \subsetneq \cdots \subsetneq P_n$$

of prime ideals of A. The dimension of an A-module M is by definition $\dim A/\operatorname{Ann}(M)$—we denote it $\dim M$.

We recall a couple of fact about the Krull dimension that are used in the text, see Propositions 9.5.5, 9.5.6 and 9.5.3 in [**Fer20**]. Alternatively, Exercise 11.7, Corollary 11.19 in [**AM69**].

Theorem A.7.2. *Let A be a Noetherian ring. Then*

$$\dim A[x] = \dim A[[x]] = 1 + \dim A.$$

In particular, $\dim k[x_1, \ldots, x_n] = \dim k[[x_1, \ldots, x_n]] = n$ for all fields k.

Theorem A.7.3. *Let $A, \mathcal{M}$ be a local Noetherian ring, $\widehat{A}$ its $\mathcal{M}$-adic completion. Then $\dim A = \dim \widehat{A}$.*

The following is [**Fer20**, Proposition 9.5.2] or [**AM69**, Corollary 11.18].

Theorem A.7.4. *Let A be a local Noetherian ring, $a \in A$ not a zero divisor or a unit. Then $\dim A/(a) = \dim A - 1$.*

We will also need a result on relative dimension, Lemma 9.5.4 in [**Fer20**] or Theorem 10.10 in [**Eis95**].

Theorem A.7.5. *Let $f \colon A \to B$ be a morphism between Noetherian rings. For a prime $Q \subset B$, let $P = f^{-1}(Q)$. Then*

$$\dim B_Q \leq \dim A_P + \dim(B/PB)_Q.$$

More generally, let A be a Noetherian ring, $P \subset A$ a prime ideal. Recall from Section 9.4 of [Fer20] that the *height* of P is the length n of the longest chain of prime ideals

$$P_0 \subset \cdots \subset P_n = P$$

ending at P, or equivalently $\dim A_P$. This is always finite by Theorem A.8.5. In fact, there is a more precise result, known as the Krull Hauptidealsatz, or principal ideal theorem (Theorem 9.4.2 in [Fer20] or Corollary 11.16 in [AM69]) (the name Hauptidealsatz should refer to the case $n = 1$, but it is customary to apply it also to this more general version).

Theorem A.7.6 (Krull). *Let A be a Noetherian ring, $I \subset A$ an ideal generated by n elements. Then every minimal prime of I has height at most n.*

To simplify such statements, it is convention to denote the height of I, $\operatorname{ht} I$, as the minimum of the heights of the minimal primes P of I. The above statement can then be simplified to the claim that $\operatorname{ht} I \leq n$ for an ideal generated by n elements.

A.8. Local theory of rings

Definition A.8.1. Let A be a ring. We say that A is *semilocal* if it has finitely many maximal ideals $\mathcal{M}_1, \ldots, \mathcal{M}_n$. We denote by $\mathcal{M} = \bigcap \mathcal{M}_i$ its Jacobson radical. If $n = 1$, we say that A is *local*, in which case $\mathcal{M} = \mathcal{M}_1$ is maximal.

Definition A.8.2. Let $Q \subset A$ be an ideal. We say that Q is an *ideal of definition* if $\mathcal{M}^k \subset Q \subset \mathcal{M}$ for some $k \in \mathbb{N}$. If $Q = (a_1, \ldots, a_d)$ and d is the minimal number of generators of an ideal of definition, we say that $a_1, \ldots, a_d$ is a *system of parameters*.

Notice that if A is local, Q is an ideal of definition if and only if Q is $\mathcal{M}$-primary. More generally, for an A-module M, we can consider the least number $d = \delta M$ such that

$$\ell\left(\frac{M}{a_1 M + \cdots + a_d M}\right) < \infty.$$

Such number is called the *Chevalley dimension* of M.

Assume that A is Noetherian and let Q be an ideal of definition. Then the ring A/Q is Artinian. For a finitely generated A-module M, we define the *Hilbert function* of M at Q by

$$\chi_M^Q(n) := \ell(M/Q^n M).$$

The fundamental result about this function is the following (Corollary 9.2.6 in [Fer20] or Proposition 11.14 in [AM69]).

Theorem A.8.3. *The function* $\chi_M^Q(n)$ *agrees with a rational polynomial for large n. This is called the* Hilbert *polynomial of M at Q, and sometimes denoted by the same symbol. Moreover, the degree of* $\chi_M^Q(n)$ *does not depend on Q. If we let d be the degree, then the first coefficient of* χ_M^Q *has the form* $e/d!$, *where* $e \in \mathbb{Z}$.

The main result on dimension theory is the following (Theorem 9.3.1 in [**Fer20**] or 11.14 in [**AM69**]).

Theorem A.8.4. *Let A be a semilocal Noetherian ring, M a finitely generated A-module. Then the following numbers are equal:*

(i) *the Krull dimension* $\dim M$;

(ii) *the degree dM of the Hilbert polynomial* χ_M^Q, *for any ideal of definition Q; and*

(iii) *the Chevalley dimension* δM.

In particular, these numbers are all finite. As a special case, we have Corollary 9.3.2, [**Fer20**], or 11.11, [**AM69**]:

Theorem A.8.5. *Let A be a local Noetherian ring. Then A has finite Krull dimension.*

The following result (Theorem 9.6.3 in [**Fer20**] or Corollary 12.5 in [**Eis95**]) lets us compare the dimension for graded and local rings.

Theorem A.8.6. *Let k be a field, $A = k[a_1, \ldots, a_n]$ be a graded ring, where the elements $a_1, \ldots, a_n$ have degree 1. Let $M = (a_1, \ldots, a_n)$ and consider the local ring A_M. Then $\dim A = \dim A_M$.*

Let A be a semilocal Noetherian ring of dimension d, M a finitely generated A-module. Fix an ideal of definition Q. Then we can write the Hilbert polynomial

$$(\text{A.8.1}) \qquad \chi_M^Q(n) = \frac{e}{d!}n^d + a_{d-1}n^{d-1} + \cdots + a_0,$$

where e is an integer.

Definition A.8.7. The number e in (A.8.1) is called the *multiplicity* of M at Q, denoted $e(Q, M)$. When A is local, we denote $e(M) := e(\mathcal{M}, M)$.

One of the most important results on the theory of local rings is the following structure theorem, 10.6.6 in [**Fer20**] or 7.7 in [**Eis95**] (in a special case). To state it, we first need a definition.

Definition A.8.8. Let $A, \mathcal{M}$ be a local Noetherian ring, $k = A/\mathcal{M}$. We say that the field $K \subset A$ is a *coefficient field* if the induced map $K \to k$ is surjective. We say that the subring $C \subset A$ is a *coefficient ring* if $C \to k$ is surjective and moreover C is a complete Hausdorff local ring with maximal ideal $\mathcal{M} \cap C$ generated by $p = \operatorname{char} k$.

Theorem A.8.9 (Cohen). *Let $A, \mathcal{M}$ be a complete local Noetherian ring, $d = \dim A$. Then A admits a coefficient field K or a coefficient ring C that is the image of a DVR. In the first case,*

$$A \cong k[[x_1, \ldots, x_d]].$$

In the second case, assuming $p^2 \notin \mathcal{M}$, we can take for C a complete DVR, and

$$A \cong C[[x_1, \ldots, x_{d-1}]].$$

As a consequence, A is always a quotient of a regular local ring.

A.9. Algebraic varieties

Let k be a field. We say that a subset $V \subset k^n$ is an (affine) algebraic variety if it is the zero locus of a set of polynomials, which can then be chosen to be finite, since $A := k[x_1, \ldots, x_n]$ is Noetherian. Given polynomials $f_1, \ldots, f_k \in A$, we write $V(f_1, \ldots, f_k)$ for their common zero locus. When we we consider k^n as an algebraic variety itself, we denote it by $\mathbb{A}_k^n$, or simply $\mathbb{A}^n$, the affine n-space.

The affine varieties form the closed sets for a topology on k^n, called the *Zariski topology*. We always consider $\mathbb{A}^n$ and its subvarieties endowed with this topology. Given a variety $V \subset \mathbb{A}^n$, we denote by $I(V)$ the ideal of A consisting of polynomials vanishing identically on V, and by $R(V) := A/I(V)$ the coordinate ring of V. Notice that elements of $R(V)$ give rise to well-defined functions $V \to k$. Moreover, $I(V)$ is always a radical ideal, so $R(V)$ is reduced.

Similarly, let $\mathbb{P}_k^n$ or simpy $\mathbb{P}^n$ denote the projective space of dimension n over k—that is, the set of lines through the origin in k^{n+1}. Elements of $A = k[x_0, \ldots, x_n]$ do *not* define functions $\mathbb{P}^n \to k$. Still, *homogeneous* elements of A have a well-defined zero locus. As in the affine case, we denote $V(f_1, \ldots, f_k)$ the common zero locus of the homogeneous polynomials $f_1, \ldots, f_k$.

Such sets are called projective varieties, and again are the closed sets for the Zariski topology on $\mathbb{P}^n$. Given a variety $V \subset \mathbb{A}^n$, we denote by $I(V)$ the ideal of A generated by homogeneous polynomials vanishing identically on V, and $R(V) = A/I(V)$. Notice that $R(V)$ is a reduced, graded ring.

We also have converse constructions: given a finitely generated reduced k-algebra A, we can write $A = k[x_1, \ldots, x_n]/I$, where I is a radical ideal. We can say a little more in the case where k is algebraically closed. This is the famous Nullstellensatz by Hilbert, Theorem 8.2.1 in [**Fer20**] (Corollary 7.10 in [**AM69**] in a different form, also Exercise 7.14).

Theorem A.9.1 (Nullstellensatz). *Let k be an algebraically closed field. Given an ideal $J \subset k[x_1, \ldots, x_n]$ we have $I(V(J)) = \sqrt{J}$.*

This important result has many equivalent formulations, which we summarize. They appear as Theorems 8.2.2 and 8.2.3 in [**Fer20**].

Theorem A.9.2. *Let k be an algebraically closed field, $J \subset k[x_1, \ldots, x_n]$ an ideal such that $V(J) = 0$. Then J is the whole ring.*

Theorem A.9.3. *Let k be an algebraically closed field. Then the maximal ideals of $k[x_1, \ldots, x_n]$ are all of the form $I(p)$, where $p \subset \mathbb{A}^n$ is a point.*

Let A be a finitely generated reduced k-algebra, which we have seen we can write as $k[x_1, \ldots, x_n]/I$. Since I is radical, we have $I = I(V(I))$ by the Nullstellensatz, hence $A \cong R(V(I))$, the coordinate ring of an affine variety. The discussion in Section 8.4 in [**Fer20**], or 1.6 in Eisenbud, can be summarized as follows.

Theorem A.9.4. *Let k be an algebraically closed field. The correspondence $V \to R(V)$ gives a duality between affine varieties and finitely generated reduced k-algebras.*

There is also an equivalent version of the Nullstellensatz for projective varieties, of which we state just one form for simplicity.

Theorem A.9.5 (Projective nullstellensatz). *The maximal homogeneous ideals of $A = k[x_0, \ldots, x_n]$ are all of the form $I(x)$, where $x \in \mathbb{P}^n_k$ is a point, except for the irrelevant ideal A_+ consisting of graded polynomials with 0 constant term.*

Bibliography

[AA97]   Abdallah Al-Amrani, *An introduction to Tragheitsformen*, Available at `https://hal.archives-ouvertes.fr/hal-00912907/file/Abdallahpaper.pdf`, 1997.

[AB57]   Maurice Auslander and David A. Buchsbaum, *Homological dimension in local rings*, Trans. Amer. Math. Soc. **85** (1957), 390–405, DOI 10.2307/1992937. MR86822

[AB58]   Maurice Auslander and David A. Buchsbaum, *Codimension and multiplicity*, Ann. of Math. (2) **68** (1958), 625–657, DOI 10.2307/1970159. MR99978

[AB59]   Maurice Auslander and D. A. Buchsbaum, *Unique factorization in regular local rings*, Proc. Nat. Acad. Sci. U.S.A. **45** (1959), 733–734, DOI 10.1073/pnas.45.5.733. MR103906

[AF12]   Frank W. Anderson and Kent R. Fuller, *Rings and categories of modules*, 2nd ed., Graduate Texts in Mathematics, vol. 13, Springer-Verlag, New York, 1992, DOI 10.1007/978-1-4612-4418-9. MR1245487

[AH19]   Tigran Ananyan and Melvin Hochster, *Small subalgebras of polynomial rings and Stillman's conjecture*, J. Amer. Math. Soc. **33** (2020), no. 1, 291–309, DOI 10.1090/jams/932. MR4066476

[AHRT02]   Jiří Adámek, Horst Herrlich, Jiří Rosický, and Walter Tholen, *Injective hulls are not natural*, Algebra Universalis **48** (2002), no. 4, 379–388, DOI 10.1007/s000120200006. MR1967087

[AK13]   Allen Altman and Steven Kleiman, *A term of commutative algebra*, Worldwide Center of Mathematics, 2013.

[AM69]   M. F. Atiyah and I. G. Macdonald, *Introduction to commutative algebra*, Addison-Wesley Publishing Co., Reading, Mass.-London-Don Mills, Ont., 1969. MR0242802

[Ann]   Toni Annala, *Diagram chasing in Abelian categories*, Available at `http://www.math.ubc.ca/~tannala/AC.pdf`.

[Bae40]   Reinhold Baer, *Abelian groups that are direct summands of every containing abelian group*, Bull. Amer. Math. Soc. **46** (1940), 800–806, DOI 10.1090/S0002-9904-1940-07306-9. MR2886

[Bak19]   Matt Baker, *Complementary sets of natural numbers and Galois connections*, Available at `https://mattbaker.blog/2019/09/12/`.

[Ban20]   Taras Banakh, *Classical set theory: theory of sets and classes*, 2020, Available at `arXiv:2006.01613`.

[Bas59]   Hyman Bass, *GLOBAL DIMENSIONS OF RINGS*, ProQuest LLC, Ann Arbor, MI, 1959. Thesis (Ph.D.)–The University of Chicago. MR2611482

[Bas63a]   Hyman Bass, *Big projective modules are free*, Illinois J. Math. **7** (1963), 24–31. MR143789

[Bas63b]   Hyman Bass, *On the ubiquity of Gorenstein rings*, Math. Z. **82** (1963), 8–28, DOI 10.1007/BF01112819. MR153708

[BEBE01]   L. Bican, R. El Bashir, and E. Enochs, *All modules have flat covers*, Bull. London Math. Soc. **33** (2001), no. 4, 385–390, DOI 10.1017/S0024609301008104. MR1832549

[Ber07]   George M. Bergman, *On diagram-chasing in double complexes*, Available at https://sbseminar.files.wordpress.com/2007/11/diagramchasingbergman.pdf.

[Bet20]   Alexander Betts, *Answer on MathOverflow*, available at https://mathoverflow.net/questions/365962/.

[Bor94]   Francis Borceux, *Handbook of categorical algebra. 2*, Encyclopedia of Mathematics and its Applications, vol. 51, Cambridge University Press, Cambridge, 1994. Categories and structures. MR1313497

[BT95]   Raoul Bott and Loring W. Tu, *Differential forms in algebraic topology*, Graduate Texts in Mathematics, vol. 82, Springer-Verlag, New York-Berlin, 1982. MR658304

[Buc55]   D. A. Buchsbaum, *Exact categories and duality*, Trans. Amer. Math. Soc. **80** (1955), 1–34, DOI 10.2307/1993003. MR74407

[Cha60]   Stephen U. Chase, *Direct products of modules*, Trans. Amer. Math. Soc. **97** (1960), 457–473, DOI 10.2307/1993382. MR120260

[Cha93]   M. Chardin, *The resultant via a Koszul complex*, Computational algebraic geometry (Nice, 1992), Progr. Math., vol. 109, Birkhäuser Boston, Boston, MA, 1993, pp. 29–39, DOI 10.1007/978-1-4612-2752-6_3. MR1230856

[Cla15]   Pete L. Clark, *Commutative algebra*, available at http://math.uga.edu/~pete/integral.pdf, 2015.

[Coh46]   I. S. Cohen, *On the structure and ideal theory of complete local rings*, Trans. Amer. Math. Soc. **59** (1946), 54–106, DOI 10.2307/1990313. MR16094

[Coh08]   Paul J. Cohen, *Set theory and the continuum hypothesis*, W. A. Benjamin, Inc., New York-Amsterdam, 1966. MR0232676

[Con13]   Keith Conrad, *Stably free modules*, Lecture notes, available at https://kconrad.math.uconn.edu/blurbs/linmultialg/stablyfree.pdf.

[Con17]   Jesús Conde-Lago, *A short proof of smooth implies flat*, Comm. Algebra **45** (2017), no. 2, 774–775, DOI 10.1080/00927872.2016.1175461. MR3562538

[dJea20]   Aise Johan de Jong and et al., *The stack project*, 2020, available at https://stacks.math.columbia.edu/.

[EEE01]   Edgar E. Enochs and Overtoun M. G. Jenda, *The flat cover conjecture and its solution*, International Symposium on Ring Theory (Kyongju, 1999), Trends Math., Birkhäuser Boston, Boston, MA, 2001, pp. 117–122. MR1851196

[EH66]   B. Eckmann and P. J. Hilton, *Exact couples in an abelian category*, J. Algebra **3** (1966), 38–87, DOI 10.1016/0021-8693(66)90019-6. MR191937

[Eis95]   David Eisenbud, *Commutative algebra*, Graduate Texts in Mathematics, vol. 150, Springer-Verlag, New York, 1995. With a view toward algebraic geometry, DOI 10.1007/978-1-4612-5350-1. MR1322960

[EM45]   Samuel Eilenberg and Saunders MacLane, *General theory of natural equivalences*, Trans. Amer. Math. Soc. **58** (1945), 231–294, DOI 10.2307/1990284. MR13131

[ES53]   B. Eckmann and A. Schopf, *Über injektive Moduln* (German), Arch. Math. (Basel) **4** (1953), 75–78, DOI 10.1007/BF01899665. MR55978

[Fer20]   Andrea Ferretti, *Commutative algebra*, Graduate Studies in Mathematics, vol. 233, American Mathematical Society, Providence, RI, 2023.

[FFGR75]   Robert Fossum, Hans-Bjørn Foxby, Phillip Griffith, and Idun Reiten, *Minimal injective resolutions with applications to dualizing modules and Gorenstein modules*, Inst. Hautes Études Sci. Publ. Math. **45** (1975), 193–215. MR396529

[Fol94]   Gerald B. Folland, *A course in abstract harmonic analysis*, Studies in Advanced Mathematics, CRC Press, Boca Raton, FL, 1995. MR1397028

[Fox77]   Hans-Bjørn Foxby, *Isomorphisms between complexes with applications to the homological theory of modules*, Math. Scand. **40** (1977), no. 1, 5–19, DOI 10.7146/math.scand.a-11671. MR447269

[Fre64]   Peter J. Freyd, *Abelian categories*, Harper and Row, 1964.

[Fri08]   Greg Friedman, *Survey article: An elementary illustrated introduction to simplicial sets*, Rocky Mountain J. Math. **42** (2012), no. 2, 353–423, DOI 10.1216/RMJ-2012-42-2-353. MR2915498

[Gab62]   Pierre Gabriel, *Des catégories abéliennes* (French), Bull. Soc. Math. France **90** (1962), 323–448. MR232821

[Gab72]   Michael Randy Gabel, *STABLY FREE PROJECTIVES OVER COMMUTATIVE RINGS*, ProQuest LLC, Ann Arbor, MI, 1972. Thesis (Ph.D.)–Brandeis University. MR2621719

[Gab95]   Ofer Gabber, *Non-negativity of serre's intersection multiplicities*, Exposé à L'IHES, 1995.

[Ger07]   Anton Geraschenko, *The salamander lemma*, Available at https://sbseminar.wordpress.com/2007/11/13/anton-geraschenko-the-salamander-lemma/.

[GH94]   Phillip Griffiths and Joseph Harris, *Principles of algebraic geometry*, Reprint of the 1978 original, Wiley Classics Library, John Wiley & Sons, Inc., New York, 1994, DOI 10.1002/9781118032527. MR1288523

[GM03]   Sergei I. Gelfand and Yuri I. Manin, *Methods of homological algebra*, 2nd ed., Springer Monographs in Mathematics, Springer-Verlag, Berlin, 2003, DOI 10.1007/978-3-662-12492-5. MR1950475

[Gov65]   V. E. Govorov, *On flat modules* (Russian), Sibirsk. Mat. Ž. **6** (1965), 300–304. MR0174598

[Gra62]   John W. Gray, *Category-valued sheaves*, Bull. Amer. Math. Soc. **68** (1962), 451–453, DOI 10.1090/S0002-9904-1962-10769-1. MR142601

[Gro35]   Wolfgang Gröbner, *Über irreduzible Ideale in kommutativen Ringen* (German), Math. Ann. **110** (1935), no. 1, 197–222, DOI 10.1007/BF01448025. MR1512936

[Gro51]   W. Gröbner, *Ein Irreduzibilitätskriterium für Primärideale in kommutativen Ringen* (German), Monatsh. Math. **55** (1951), 138–145, DOI 10.1007/BF01486922. MR43074

[Gro57]   Alexander Grothendieck, *Sur quelques points d'algèbre homologique* (French), Tohoku Math. J. (2) **9** (1957), 119–221, DOI 10.2748/tmj/1178244839. MR102537

[Gro64]   Alexander Grothendeick, *Éléments de géométrie algébrique. IV. Étude locale des schémas et des morphismes de schémas. I.*, Publications Mathématiques, no. 20, Institute des Hautes Études Scientifiques, 1964.

[GS87]   H. Gillet and C. Soulé, *Intersection theory using Adams operations*, Invent. Math. **90** (1987), no. 2, 243–277, DOI 10.1007/BF01388705. MR910201

[Göo08]   Kurt Gödel, *The Consistency of the Continuum Hypothesis*, Annals of Mathematics Studies, No. 3, Princeton University Press, Princeton, N. J., 1940. MR0002514

[Hat02]   Allen Hatcher, *Algebraic topology*, Cambridge University Press, Cambridge, 2002. MR1867354

[Hat17]   Allen Hatcher, *Vector bundles & K-theory*, available at https://pi.math.cornell.edu/~hatcher/VBKT/VBpage.html, 2017.

[HE71]   M. Hochster and John A. Eagon, *Cohen-Macaulay rings, invariant theory, and the generic perfection of determinantal loci*, Amer. J. Math. **93** (1971), 1020–1058, DOI 10.2307/2373744. MR302643

[Hil90]   David Hilbert, *Ueber die Theorie der algebraischen Formen* (German), Math. Ann. **36** (1890), no. 4, 473–534, DOI 10.1007/BF01208503. MR1510634

[Hir64a]   Heisuke Hironaka, *Resolution of singularities of an algebraic variety over a field of characteristic zero. I, II*, Ann. of Math. (2) **79** (1964), 109–203; ibid. (2) **79** (1964), 205–326, DOI 10.2307/1970547. MR0199184

[Hir64b]   Heisuke Hironaka, *Resolution of singularities of an algebraic variety over a field of characteristic zero. I, II*, Ann. of Math. (2) **79** (1964), 109–203; ibid. (2) **79** (1964), 205–326, DOI 10.2307/1970547. MR0199184

[Hir97]   Morris W. Hirsch, *Differential topology*, Graduate Texts in Mathematics, vol. 33, Springer-Verlag, New York, 1994. Corrected reprint of the 1976 original. MR1336822

[HM98]   Joe Harris and Ian Morrison, *Moduli of curves*, Graduate Texts in Mathematics, vol. 187, Springer-Verlag, New York, 1998. MR1631825

[Hoc97]   Melvin Hochster, *Nonnegativity of intersection multiplicities in ramified regular local rings following Gabber/De Jong/Berthelot*, available at http://www.math.lsa.umich.edu/~hochster/mult.pdf.

[Hoc72]   M. Hochster, *Nonuniqueness of coefficient rings in a polynomial ring*, Proc. Amer. Math. Soc. **34** (1972), 81–82, DOI 10.2307/2037901. MR294325

[Hun98]   Craig Huneke, *Hyman Bass and ubiquity: Gorenstein rings*, Algebra, K-theory, groups, and education (New York, 1997), Contemp. Math., vol. 243, Amer. Math. Soc., Providence, RI, 1999, pp. 55–78, DOI 10.1090/conm/243/03686. MR1732040

[Hun07]   Craig Huneke, *Lectures on local cohomology*, Interactions between homotopy theory and algebra, Contemp. Math., vol. 436, Amer. Math. Soc., Providence, RI, 2007, pp. 51–99, DOI 10.1090/conm/436/08404. Appendix 1 by Amelia Taylor. MR2355770

[Isc69]   Friedrich Ischebeck, *Eine Dualität zwischen den Funktoren Ext und Tor* (German), J. Algebra **11** (1969), 510–531, DOI 10.1016/0021-8693(69)90090-8. MR237613

[Ive86]   Birger Iversen, *Cohomology of sheaves*, Universitext, Springer-Verlag, Berlin, 1986, DOI 10.1007/978-3-642-82783-9. MR842190

[JJ85]   T. Józefiak and J. Weyman, *Symmetric functions and Koszul complexes*, Adv. in Math. **56** (1985), no. 1, 1–8, DOI 10.1016/0001-8708(85)90081-7. MR782539

[Jou91]   J.-P. Jouanolou, *Le formalisme du résultant* (French), Adv. Math. **90** (1991), no. 2, 117–263, DOI 10.1016/0001-8708(91)90031-2. MR1142904

[Kap70]   Irving Kaplansky, *Commutative rings*, Revised edition, University of Chicago Press, Chicago, Ill.-London, 1974. MR0345945

[Kol07]   János Kollár, *Resolution of singularities – Seattle lecture*, 2007, Available at arXiv:0508332.

[Kra06]   U. Krähmer, *Poincaré duality in Hochschild (co)homology*, New techniques in Hopf algebras and graded ring theory, K. Vlaam. Acad. Belgie Wet. Kunsten (KVAB), Brussels, 2007, pp. 117–125. MR2395770

[Kun69]   Ernst Kunz, *Characterizations of regular local rings of characteristic p*, Amer. J. Math. **91** (1969), 772–784, DOI 10.2307/2373351. MR252389

[Lam78]   T. Y. Lam, *Serre's conjecture*, Lecture Notes in Mathematics, Vol. 635, Springer-Verlag, Berlin-New York, 1978. MR0485842

[Lam94]   J. Lambek, *Some Galois connections in elementary number theory*, J. Number Theory **47** (1994), no. 3, 371–377, DOI 10.1006/jnth.1994.1043. MR1278405

[Lam01]   T. Y. Lam, *A first course in noncommutative rings*, 2nd ed., Graduate Texts in Mathematics, vol. 131, Springer-Verlag, New York, 2001, DOI 10.1007/978-1-4419-8616-0. MR1838439

[Lam06] T. Y. Lam, *Serre's problem on projective modules*, Springer Monographs in Mathematics, Springer-Verlag, Berlin, 2006, DOI 10.1007/978-3-540-34575-6. MR2235330

[Lan02] Serge Lang, *Algebra*, 3rd ed., Graduate Texts in Mathematics, vol. 211, Springer-Verlag, New York, 2002, DOI 10.1007/978-1-4613-0041-0. MR1878556

[Laz69] Daniel Lazard, *Autour de la platitude* (French), Bull. Soc. Math. France **97** (1969), 81–128. MR254100

[Lei06] Tom Leinster, *A universal banach space*, Available at `https://www.maths.ed.ac.uk/~tl/glasgowpssl/banach.pdf`, 2006.

[LM54] J. Lambek and L. Moser, *Inverse and complementary sequences of natural numbers*, Amer. Math. Monthly **61** (1954), 454–458, DOI 10.2307/2308078. MR62777

[Mas52] W. S. Massey, *Exact couples in algebraic topology. I, II*, Ann. of Math. (2) **56** (1952), 363–396, DOI 10.2307/1969805. MR52770

[Mat58] Eben Matlis, *Injective modules over Noetherian rings*, Pacific J. Math. **8** (1958), 511–528. MR99360

[Mat86] Hideyuki Matsumura, *Commutative ring theory*, Cambridge Studies in Advanced Mathematics, vol. 8, Cambridge University Press, Cambridge, 1986. Translated from the Japanese by M. Reid. MR879273

[May99] Jon Peter May, *A primer on spectral sequences*, Available at `http://www.math.uchicago.edu/~may/MISC/SpecSeqPrimer.pdf`.

[Mil13] James S. Milne, *Lectures on etale cohomology*, Available at `https://www.jmilne.org/math/CourseNotes/LEC.pdf`, 2013.

[Miy67] Takehiko Miyata, *Note on direct summands of modules*, J. Math. Kyoto Univ. **7** (1967), 65–69, DOI 10.1215/kjm/1250524308. MR214585

[ML71] Saunders MacLane, *Categories for the working mathematician*, Graduate Texts in Mathematics, Vol. 5, Springer-Verlag, New York-Berlin, 1971. MR0354798

[Mur20] Takumi Murayama, *Answer on MathOverflow*, available at `https://mathoverflow.net/questions/349818/`.

[Mur64] M. Pavaman Murthy, *A note on factorial rings*, Arch. Math. (Basel) **15** (1964), 418–420, DOI 10.1007/BF01589225. MR173695

[Nag58] Masayoshi Nagata, *A general theory of algebraic geometry over Dedekind domains. II. Separably generated extensions and regular local rings*, Amer. J. Math. **80** (1958), 382–420, DOI 10.2307/2372791. MR94344

[Nag62] Masayoshi Nagata, *Local rings*, Interscience Tracts in Pure and Applied Mathematics, No. 13, Interscience Publishers (a division of John Wiley & Sons, Inc.), New York-London, 1962. MR0155856

[Nee01] Amnon Neeman, *Triangulated categories*, Annals of Mathematics Studies, vol. 148, Princeton University Press, Princeton, NJ, 2001, DOI 10.1515/9781400837212. MR1812507

[Nee02] Amnon Neeman, *A counterexample to a 1961 "theorem" in homological algebra*, With an appendix by P. Deligne, Invent. Math. **148** (2002), no. 2, 397–420, DOI 10.1007/s002220100197. MR1906154

[Noe21] Emmy Noether, *Idealtheorie in Ringbereichen* (German), Math. Ann. **83** (1921), no. 1-2, 24–66, DOI 10.1007/BF01464225. MR1511996

[Nor57] D. G. Northcott, *On irreducible ideals in local rings*, J. London Math. Soc. **32** (1957), 82–88, DOI 10.1112/jlms/s1-32.1.82. MR83477

[NR54] D. G. Northcott and D. Rees, *Reductions of ideals in local rings*, Proc. Cambridge Philos. Soc. **50** (1954), 145–158, DOI 10.1017/s0305004100029194. MR59889

[NTV18] Saeed Nasseh, Ryo Takahashi, and Keller VandeBogert, *On gorenstein fiber products and applications*, 2018, Available at `arXiv:1701.08689`.

[NvO82]   C. Năstăsescu and F. van Oystaeyen, *Graded ring theory*, North-Holland Mathe-
          matical Library, vol. 28, North-Holland Publishing Co., Amsterdam-New York, 1982.
          MR676974

[Pap59]   Zoltán Papp, *On algebraically closed modules*, Publ. Math. Debrecen **6** (1959), 311–
          327. MR121390

[Pol]   Jason Polak, *Wild spectral sequences ep. 4: Schanuel's lemma*, available at `http://`
          `blog.jpolak.org/?p=1207`.

[Pos18]   Leonid Positselski, *Answer on MathOverflow*, available at `https://mathoverflow.net/`
          `questions/299014/`.

[Pos17]   Leonid Positselski, *Contraadjusted modules, contramodules, and reduced cotorsion
          modules*, Mosc. Math. J. **17** (2017), no. 3, 385–455, DOI 10.17323/1609-4514-2017-
          17-3-385-455. MR3711003

[PS73]   Christian Peskine and Lucien Szpiro, *Dimension projective finie et cohomologie locale*,
          Publ. Math. IHES **42** (1973), 47–119.

[PS09]   Irena Peeva and Mike Stillman, *Open problems on syzygies and Hilbert functions*,
          J. Commut. Algebra **1** (2009), no. 1, 159–195, DOI 10.1216/JCA-2009-1-1-159.
          MR2462384

[Qui76]   Daniel Quillen, *Projective modules over polynomial rings*, Invent. Math. **36** (1976),
          167–171, DOI 10.1007/BF01390008. MR427303

[Rat72]   L. J. Ratliff Jr., *Catenary rings and the altitude formula*, Amer. J. Math. **94** (1972),
          458–466, DOI 10.2307/2374632. MR311659

[RG71]   Michel Raynaud and Laurent Gruson, *Critères de platitude et de projectivité. Tech-
          niques de "platification" d'un module* (French), Invent. Math. **13** (1971), 1–89, DOI
          10.1007/BF01390094. MR308104

[Ric18]   Jeremy Rickard, *Answer on MathOverflow*, available at `https://mathoverflow.net/`
          `questions/292566/`.

[Rob76]   Paul Roberts, *Two applications of dualizing complexes over local rings*, Ann. Sci. École
          Norm. Sup. (4) **9** (1976), no. 1, 103–106. MR399075

[Rob85]   Paul Roberts, *The vanishing of intersection multiplicities of perfect complexes*, Bull.
          Amer. Math. Soc. (N.S.) **13** (1985), no. 2, 127–130, DOI 10.1090/S0273-0979-1985-
          15394-7. MR799793

[Rob87]   Paul Roberts, *Le théorème d'intersection* (French, with English summary), C. R. Acad.
          Sci. Paris Sér. I Math. **304** (1987), no. 7, 177–180. MR880574

[Rob89]   Paul Roberts, *Intersection theorems*, Commutative Algebra (Melvin Hochster, Craig
          Huneke, and Judith D. Sally, eds.), Springer New York, 1989, pp. 417–436.

[Roo61]   Jan-Erik Roos, *Sur les foncteurs dérivés de* lim. *Applications* (French), C. R. Acad.
          Sci. Paris **252** (1961), 3702–3704. MR132091

[Roo06]   Jan-Erik Roos, *Derived functors of inverse limits revisited*, J. London Math. Soc. (2)
          73 (2006), no. 1, 65–83, DOI 10.1112/S0024610705022416. MR2197371

[Run07]   Volker Runde, *A taste of topology*, Universitext, Springer, New York, 2005. MR2155623

[Rus16]   Jeremy Russell, *Applications of the defect of a finitely presented functor*, J. Algebra
          465 (2016), 137–169, DOI 10.1016/j.jalgebra.2016.07.016. MR3537819

[Sch98]   Peter Schenzel, *On the use of local cohomology in algebra and geometry*, Six lectures
          on commutative algebra (Bellaterra, 1996), Progr. Math., vol. 166, Birkhäuser, Basel,
          1998, pp. 241–292. MR1648667

[Sch03]   René Schoof, *Number theory*, Lecture notes, available at `http://www.mat.uniroma2.`
          `it/~schoof/tn.html`, 2003.

[Ser55]   Jean-Pierre Serre, *Faisceaux algébriques cohérents* (French), Ann. of Math. (2) **61**
          (1955), 197–278, DOI 10.2307/1969915. MR68874

[Ser56a] Jean-Pierre Serre, *Géométrie algébrique et géométrie analytique*, Annales de l'Institut Fourier **6** (1956), 1–42.

[Ser56b] Jean-Pierre Serre, *Sur la dimension homologique des anneaux et des modules Noethériens*, Proceedings of the International Symposium on Algebraic Number Theory, Tokyo & Nikko, 1956, pp. 175–189.

[Ser00] Jean-Pierre Serre, *Local algebra*, Translated from the French by CheeWhye Chin and revised by the author, Springer Monographs in Mathematics, Springer-Verlag, Berlin, 2000, DOI 10.1007/978-3-662-04203-8. MR1771925

[She09] Sherry, *Answer on MathOverflow*, available at `https://mathoverflow.net/questions/616/`.

[She74] Saharon Shelah, *Infinite abelian groups, Whitehead problem and some constructions*, Israel J. Math. **18** (1974), 243–256, DOI 10.1007/BF02757281. MR357114

[Shu08] Mike Shulman, *Set theory for category theory*, 2008, Available at `arXiv:0810.1279`.

[Sri13] Shashi Mohan Srivastava, *A course on mathematical logic*, 2nd ed., Universitext, Springer, New York, 2013, DOI 10.1007/978-1-4614-5746-6. MR3013686

[Sus76] A. A. Suslin, *Projective modules over polynomial rings are free* (Russian), Dokl. Akad. Nauk SSSR **229** (1976), no. 5, 1063–1066. MR0469905

[Swa62] Richard G. Swan, *Vector bundles and projective modules*, Trans. Amer. Math. Soc. **105** (1962), 264–277, DOI 10.2307/1993627. MR143225

[Tat57] John Tate, *Homology of Noetherian rings and local rings*, Illinois J. Math. **1** (1957), 14–27. MR86072

[va05] various authors, *Local cohomology at Snowbird*, Lecture notes, available at `https://www.math.purdue.edu/~walther/snowbird/main_keyless.pdf`, 2005.

[Vas67] Wolmer V. Vasconcelos, *Ideals generated by R-sequences*, J. Algebra **6** (1967), 309–316, DOI 10.1016/0021-8693(67)90086-5. MR213345

[Vas69] Wolmer V. Vasconcelos, *On finitely generated flat modules*, Trans. Amer. Math. Soc. **138** (1969), 505–512, DOI 10.2307/1994928. MR238839

[Vas73] Wolmer V. Vasconcelos, *Finiteness in projective ideals*, J. Algebra **25** (1973), 269–278, DOI 10.1016/0021-8693(73)90045-8. MR314828

[Voi02] Claire Voisin, *Hodge theory and complex algebraic geometry. II*, Translated from the French by Leila Schneps, Cambridge Studies in Advanced Mathematics, vol. 77, Cambridge University Press, Cambridge, 2003, DOI 10.1017/CBO9780511615177. MR1997577

[VS74] L. N. Vaserštein and A. A. Suslin, *Serre's problem on projective modules over polynomial rings, and algebraic K-theory* (Russian), Funkcional. Anal. i Priložen. **8** (1974), no. 2, 65–66. MR0347802

[Wat75] William C. Waterhouse, *Basically bounded functors and flat sheaves*, Pacific J. Math. **57** (1975), no. 2, 597–610. MR396578

[Wei80] Claudia Weill, *It's my turn*, 1980, Available at `https://www.youtube.com/watch?v=etbcKWEKnvg`.

[Wei95] Charles A. Weibel, *An introduction to homological algebra*, Cambridge Studies in Advanced Mathematics, vol. 38, Cambridge University Press, Cambridge, 1994, DOI 10.1017/CBO9781139644136. MR1269324

[Wof16] Eric Wofsey, *Answer on Math Stack Exchange*, available at `https://math.stackexchange.com/questions/1766047/`.

[Yen11] Ihsen Yengui, *Stably free modules over* $\mathbf{R}[X]$ *of rank* $> \dim \mathbf{R}$ *are free*, Math. Comp. **80** (2011), no. 274, 1093–1098, DOI 10.1090/S0025-5718-2010-02427-5. MR2772113

Index of Notation

Ab	category of Abelian groups, page 3
$\mathrm{Add}(\mathcal{A}, \mathcal{B})$	category of additive functors between $\mathcal{A}$ and $\mathcal{B}$, page 57
$\mathbb{A}_k^n$	affine n-space over k, page 393
$A \ltimes M$	extension of A by M, page 365
$\mathrm{ar}\, I$	arithmetic rank of the ideal I, page 374
$\mathrm{Bord}(n)$	bordism category of dimension n, page 5
$C_{\mathbf{a}(M)}$	Čech complex associated to the sequence $\mathbf{a}$ and the module M, page 371
$C(f)$	mapping cone of f, page 110
$\chi_M^Q(n)$	Hilbert function of M at the ideal of definition Q, page 391
$\mathcal{C}^{op}$	opposite category of $\mathcal{C}$, page 5
$\mathrm{depth}_I M$	depth of the module M with respect to I, page 262
Diff	category of differentiable manifolds, page 3
$\mathrm{div}\, a$	divisor associated to the element a, page 279
$\mathrm{div}\, M$	divisor associated to the torsion module M, page 280
$E(A)$	injective hull of the object A, page 90
$\mathrm{embdim}\, A$	embedding dimension of the local ring A, page 290
$e(Q, M)$	multiplicity of M at the ideal of definition Q, page 392
$E_{p,q}^r \implies E_{p,q}^\infty$	spectral sequence $E_{p,q}^r$ converges to $E_{p,q}^\infty$, page 152
$\mathrm{Ext}^i(A, B)$	i-th Ext functor between A and B, page 118
$\mathcal{E}xt^i(S, T)$	sheaf Ext of S and T, page 141
$\mathfrak{f}(B/A)$	conductor ideal of A in B, page 380
$G(A)$	ideal class group of A, page 388
$\Gamma_I(M)$	sections of M with support on I, page 366
$\mathcal{G}^i(S)$	i-th term in the Godement resolution of S, page 123
$\mathrm{gl.\, dim}\, A$	global dimension of the ring A, page 200
$\mathrm{gl.\, wdim}\, A$	global weak dimension of the ring A, page 231
$\mathrm{gr.\, gl.\, dim}\, A$	graded global dimension of the ring A, page 201
Grp	category of groups, page 3

Index

For a complete list of titles in this series, visit the
AMS Bookstore at **www.ams.org/bookstore/gsmseries/**.